SEISMIC IMAGING AND INVERSION

Application of Linear Inverse Theory

Extracting information from seismic data requires knowledge of seismic wave propagation and reflection. The commonly used method involves solving linearly for a reflectivity at every point within the Earth. The resulting reflectivity, however, is not an intrinsic Earth property, and cannot easily be extended to nonlinear processes which might provide a deeper understanding and a more accurate image of the subsurface.

In this book, the authors follow an alternative approach which invokes inverse scattering theory. By developing the theory of seismic imaging from basic principles, they relate the different models of seismic propagation, reflection, and imaging – thus providing links to reflectivity-based imaging on the one hand, and to nonlinear seismic inversion on the other. Full, three-dimensional algorithms are incorporated for scalar, acoustic, and elastic wave equations.

The comprehensive and physically complete linear imaging foundation developed in this volume presents new results at the leading edge of seismic processing for target location and identification. The book serves as a fundamental guide to seismic imaging principles and algorithms, and their foundation in inverse scattering theory, for today's seismic processing practitioners and researchers. It is a valuable resource for geoscientists wishing to understand the basic principles of seismic imaging, for scientific programmers with an interest in imaging algorithms, and for theoretical physicists and applied mathematicians seeking a deeper understanding of the subject. It will also be of interest to researchers in other related disciplines such as remote sensing, non-destructive evaluation and medical imaging.

ROBERT H. STOLT is currently a Geoscience Fellow at ConocoPhillips. He is an Honorary Member of the Society of Exploration Geophysicists (SEG) and of the Geophysical Society of Tulsa (GST). He obtained a PhD in theoretical physics at the University of Colorado in 1970, and joined Conoco in 1971. He spent 1979–80 at Stanford University as Consulting Professor and Acting Director of the Stanford Exploration Project. In 1980 he received the Reginald Fessenden Award for original contributions to geophysics, and in 1998 the DuPont Lavoisier Medal for technical achievement. From 1979 to 1985 he was SEG Associate Editor for seismic imaging and inversion, was SEG editor from 1985 to 1987, and SEG Publications Committee Chairman from 1987 to 1989. In 1994 he served as Technical Program Chairman of the Sixty-Fourth Annual SEG Meeting in Los Angeles. Dr Stolt has authored numerous scientific publications, including an earlier text on seismic migration.

ARTHUR B. WEGLEIN holds the Hugh Roy and Lillie Cranz Cullen Distinguished University Professorship in Physics at the University of Houston, with a joint professorship in the Department of Physics and the Department of Earth and Atmospheric Sciences. He is the founder and Director of the Mission-Oriented Seismic Research Program, which began in 2001 and is a consortium supported by the world's major oil and service companies, as well as various US government programs. Before joining the University of Houston, he worked at Arco's Research Laboratory in Plano, Texas, and at Schlumberger Cambridge Research Laboratory in the UK. Professor Weglein served as the SEG Distinguished Lecturer in 2003 and was awarded the SEG's Reginald Fessenden Award in 2010. In 2008, he received the Distinguished Townsend Harris Medal from the City College of the City University of New York in recognition of his contributions to exploration seismology.

SEISMIC IMAGING AND INVERSION

Application of Linear Inverse Theory

ROBERT H. STOLT AND ARTHUR B. WEGLEIN

CAMBRIDGE
UNIVERSITY PRESS

CAMBRIDGE
UNIVERSITY PRESS

University Printing House, Cambridge CB2 8BS, United Kingdom

One Liberty Plaza, 20th Floor, New York, NY 10006, USA

477 Williamstown Road, Port Melbourne, VIC 3207, Australia

4843/24, 2nd Floor, Ansari Road, Daryaganj, Delhi - 110002, India

79 Anson Road, #06-04/06, Singapore 079906

Cambridge University Press is part of the University of Cambridge.

It furthers the University's mission by disseminating knowledge in the pursuit of education, learning and research at the highest international levels of excellence.

www.cambridge.org
Information on this title: www.cambridge.org/9781108446662

First published 2012
First paperback edition 2017

A catalogue record for this publication is available from the British Library

Library of Congress Cataloging in Publication data
Stolt, Robert H.
Seismic imaging and inversion : application of linear inverse theory / Robert H. Stolt and, Arthur B. Weglein.
p. cm.
Includes bibliographical references and index.
ISBN 978-1-107-01490-9 (hardback)
1. Seismic reflection method. 2. Scattering (Mathematics) 3. Linear operators – Generalized inverses. I. Weglein, Arthur B. II. Title.
TN269.84.S76 2012
622′.1592015157246–dc23
2011041739

ISBN 978-1-107-01490-9 Hardback
ISBN 978-1-108-44666-2 Paperback

Contents

Preface and acknowledgments

In exploration seismology, a man-made energy source (on or near the Earth's surface for land exploration, or within the ocean water column in the case of marine off-shore exploration) generates a wave that propagates down into the Earth. When the downward propagating wave encounters a rapid change in Earth properties, a portion of the wave is reflected upward and another part is transmitted below the rapid variation and continues propagating down and deeper into the subsurface. The spatial locations of rapid variations in Earth properties are typically called "reflectors". The reflected wavefields that ultimately return to the Earth's surface, or, in the marine case, arrive near the air–water boundary, are recorded by large numbers of wavefield sensors. The collection of these recorded wavefields constitutes seismic reflection data. The objective of exploration seismology is to use the recorded data to make inferences about the subsurface that are relevant to determining the location and quantity of hydrocarbons. Among types of subsurface information that are useful in hydrocarbon prediction are (1) the spatial locations of any rapid variations in Earth properties, loosely called "locating reflectors", and (2) the sign and size of changes in specific Earth properties at those locations. The former of these two goals is called "imaging" or "migration", and the latter is typically called "inversion".

In this book, the first of a two-volume set, we present both the basic concepts behind, and the relationship between, wave-equation migration and inversion. In Volume I, relationships and algorithms between recorded data and the seismic image are constrained to be linear, and the relationship between the seismic image and the changes in Earth mechanical properties across the image is assumed to be linear, as well. For seismic imaging, this linearity implies that (1) the velocity model of the Earth is known between the measurement surface(s) where the energy sources and the recording sensors reside and the seismic image at depth, and (2) that an adequate wave theoretical model is available to back-propagate the waves accurately through that velocity model. For seismic inversion and target identification, the linearity implies (1) that all subsurface mechanical properties that influence the amplitude and phase of a wave propagating between the measurement surface(s) and the seismic image are known, and (2) that at the migrated image both the changes in Earth properties and the reflection angles being considered are small. The second of these two objectives requires a high-end form of amplitude-preserving back-propagation with a generalization and extension of the seismic imaging condition, and this has resulted in an evolution and merging of the originally distinct processes of migration and inversion into migration–inversion. The linearity also assumes that multiply reflected events (called multiples) have either been removed or can be ignored. The linearity assumptions in Volume I are behind all current leading-edge seismic processing algorithms employed in the petroleum industry, and are broadly assumed by research programs within industry research laboratories and within most leading academic consortia supported by the petroleum industry. To move beyond the constraints imposed by linearity is the province of Volume II.

Exploration geophysics has evolved over the years, with contributions from several disciplines. Before digital computers, seismic processing was analog and mechanical. Early digital processing was single-channel, amenable to signal-processing concepts imported from electrical engineering. As computational power improved, multichannel processing was introduced, and the field began to attract the attention of physicists and applied mathematicians. Wave-equation processing methods were introduced, and seismic imaging began to be recognized as a physical inverse scattering problem.

We believe that the best way to glean information from seismic data is to bring to bear the tools of inverse scattering theory. Linear inverse scattering theory, although relatively simple, is physically sound (within the limits of its approximation). That linear theory provided, for the first time, the framework to put imaging and inversion on a single firm and consistent multidimensional footing, which naturally led to the concept and methods of migration-inversion, and the mechanism for the necessary merging of two fields, migration and AVO (amplitude variation with offset). Each of these fields had previously acted (and sometimes still behave) as though the other was unnecessary or perhaps did not even exist. Filling that unfortunate and harmful chasm between two useful methodologies was a first and important conceptual and practical contribution brought to exploration seismology by inverse scattering theory.

Inverse scattering theory, as applied to seismic data, has been around for a while, and is well represented in the seismic literature. However, this is the first textbook that provides this comprehensive level of depth and completeness to the linear theory, as well as the broader nonlinear concepts and algorithms.

In Volume I, the introductory chapters present a number of different imaging techniques, algorithmically very different but sharing (and deriving from) a common set of imaging principles. For a deeper understanding of seismic imaging, we choose not to solve for specular reflectivity. Rather, we introduce the concept of scattering potential and invoke the Born approximation. We also introduce the reflectivity function as a derived quantity, defined by its relation to the scattering potential. Thus defined, the reflectivity function has a value at each subsurface point, whether or not a specular reflector exists at that point. Imaging operators invert the Born equation for either the scattering potential or the reflectivity function. At points where specular reflectors exist, the reflectivity function reduces to the specular reflectivity. Formulation of the imaging problem in terms of the scattering potential gives up nothing in terms of algorithms or capability, rather gains physical respectability while giving access to the toolbox of scattering theory. Treating a scattering potential as the generator or source of the reflection data allows imaging of smooth surfaces as well as, for example, diffractors and pinchouts, automatically and without a-priori information on which type of wave generator is at play, and without the need for interpreter intervention or two separate imaging methods/models.

Our approach is largely three-dimensional, viewing two-dimensional seismic acquisition and processing as a restriction of the general case. Primary focus is on the elastic wave equation, though with slight adjustments the formulas are also applicable to the acoustic and scalar wave equations. In addition to seismic migration, other imaging processes such as data continuation and residual migration are included in this volume. Among other new contributions, the volume provides the first linear inverse for the elastic isotropic wave equation in a three-dimensional Earth. No attempt, however, has been made to incorporate all modern migration algorithms – there are just too many. Most attention has been paid to *f–k* algorithms, on the one hand, and asymptotic imaging on the other. Other subjects not dealt with include anisotropic and lossy media, ray tracing algorithms, and determination of background velocities. Throughout the text the emphasis is on understanding how the mathematical physics speaks to the relationship between the recorded seismic wavefield, the source that generated the wave, and the properties of the Earth that the wave has experienced in its history, and staying focused on how that understanding can be used for the purposes of exploration seismology.

We would like to thank ConocoPhillips for support, for encouragement, and for permission to publish this textbook. We are deeply grateful to Sven Treitel for his continuous encouragement of this project, and for his positive interest and reinforcement as this book was being written. We would like to thank the M-OSRP sponsors, and NSF CMG award DMS-0327778 and DOE BES award DE-FG02-05ER15697 for their strong and constant support of this writing project. We thank

1

Introduction – modeling, migration, imaging, and inversion

1.1 Seismic data contains information

The fundamental assumption of this book is that seismic data contain decipherable information about Earth properties. We like to think of the Earth's subsurface as composed of regions or layers within which the properties change slowly, separated by boundaries across which properties vary rapidly, perhaps even discontinuously. The regions between boundaries propagate and sometimes distort seismic waves, while each boundary reflects some of the seismic energy. Thus, seismic data tend to contain sequences of discrete "events", each associable with a reflecting boundary, separated by relatively quiet intervals. Buried in the data is information on the location and composition of boundaries, and also on the properties of the regions between them.

Raw seismic data may be difficult to interpret, and seldom reveals the true location or amplitude of reflectors. Events may be obscured by noise and interference. Most locations in the subsurface are illuminated by many sources and receivers, from different directions and distances. Imposed on the data may be a source wavelet long enough to jumble events together and make them difficult to identify individually. Pairs of strong reflectors may produce multiple reflections that mask primary reflections from deeper reflectors. Seismic processing renders the data more interpretable by compressing the source wavelet, identifying and removing multiple reflections, aligning and compositing images of individual events, filtering noise, "migrating" events from apparent to actual locations, and "inverting" for Earth properties.

1.2 Models for propagation and reflection

Seismic migration is based on a simplified picture of seismic reflection data which assumes primary reflections only. Multiple reflections, if they cannot be ignored a priori, are assumed to have been removed from the data. In this picture, a seismic wave travels from a source at or near the Earth's surface to one or more reflectors in the Earth. At each reflector, a reflected wave is generated which travels to receivers placed at or near the Earth's surface.

To migrate seismic data requires physical descriptions of the processes of wave propagation and of reflection. Without a propagation model, one can only wonder how seismic energy got from source to reflector to receiver, and unravelling the process is impossible. Without a reflection model, one cannot interpret reflected energy or relate it to properties of the reflector. The two models provide a descriptive framework from which one can deduce where reflectors are, and an interpretive framework from which one can deduce reflector properties.

In the real Earth, propagation and reflection are intertwined, and treating them as separable entities is based as much on perception as on reality. Surprising as it may seem, this simplified picture usually works out fairly well.

Ideally, descriptions of seismic wave propagation and reflection stem from a common underlying physical model. In practice, however, it may be difficult to unify the two, and separate models are often employed for the two processes. This leads to a propagation model based on a simplified "background" velocity structure, engineered so as to not produce reflections, and a reflection model based on complex, highly localized changes in seismic properties.

This happens in part because of the bandlimited nature of seismic data. The data directly illuminate changes in seismic properties within a range of wavelengths. These changes give rise to what we perceive as reflections. Slower changes do not produce perceptible reflected energy, but do largely determine how seismic waves propagate to and from reflections. The physics of propagation and reflection partially decouple because they inhabit regions of different wavelength. Changes more rapid than the illuminated wavelengths also occur, but they do not directly produce perceptible reflections; rather, they can and do affect the effective seismic properties. They may in fact affect propagation and reflection in different ways, creating an apparent disparity between the reflection and propagation models.

Other factors may also favor separate models for propagation and reflection. The purpose of a seismic model is not to perfectly describe the contents of the real Earth, which is more complex than any model one is likely to devise. The model is meant to allow one to describe, analyze, and interpret seismic processes well enough to derive the information required. To do worse is to fail; to do better is to waste effort. It follows that the physical model that best describes propagation may be different than the one that best describes reflections.

Even so, the most satisfying approach would be to work from a single physical model that describes both reflections and propagation. The inverse scattering approach to seismic processing conforms to this ideal. For the most part, however, the theory and practice of seismic imaging retains two separate models. Seeing value in both approaches, this text is divided into two volumes, the first approaching the inverse seismic problem from the simpler point of view, the second following a more comprehensive approach.

Even within Volume I, there are multiple choices for propagation and reflection models. The best choice depends upon circumstance – upon the geology, upon the data, and upon the processing objectives. If one is interested primarily in the location of reflectors, or if data quality does not support amplitude recovery, then it makes little sense to impose all the trappings of a true-amplitude model on the processing. Generally, we support the principle of minimum complexity: one should use the simplest model that works for the circumstances at hand. While the perfectionist within may cry out for the most realistic model possible, that personality may need to be sedated. It is just as much in error to carry around an unnecessarily realistic Earth model as it is to insist on meaningless precision in calculations. Penurious instincts may need to be suppressed as well: dedicating inadequate resources guarantees failure before one begins.

1.3 Going forward to go back

Modeling, migration, imaging, and inversion of seismic data are dealt with in this Volume. Generally, we assume that other seismic processes, where necessary, have already been performed. We are mostly concerned with the inverse seismic problem, in which Earth properties are inferred from seismic data. However, to go backward successfully, one must be able to go forward, predicting the data for a given set of Earth properties. *Modeling*, as used here, refers to processes for simulating seismic data. The starting point for the simulation may be a map of Earth properties, or, less fundamentally, a map of reflectivity images or an image function.

1.4 Seismic and non-seismic imaging

Migration and inversion have to do with the inverse problem. *Imaging*, narrowly defined, might be considered synonymous with migration. From a broader perspective, virtually all seismic data contains images. If one defines an imaging process as an operation that forms, modifies, or manipulates seismic images, then imaging encompasses modeling, migration, and a host of other processes.

Seismic imaging has relatives in other fields, including radar, sonar, remote sensing, and various forms of medical imaging. All these fields share a common wave-theoretical basis, and have to some extent shared technology and algorithms. However, in some respects the seismic problem is, if not unique, nearly so. Most other imaging problems are very much in the far field, with targets many hundreds or thousands of wavelengths from sources and receivers. In contrast, a 30 hertz seismic wave traveling at 10 000 feet per second (fps) covers only three wavelengths per 1000 feet (300 m). We can usually get away with treating seismic propagation as far field, but not always. Propagation velocities in other disciplines are, with a few exceptions, much less variable than seismic velocities. Seismic data sets tend to be relatively large, and not amenable to processing in real time. These and other differences limit the possibilities for cross-disciplinary technology transfer.

1.5 Motivation for migration

While unmigrated seismic data contain reflector images, they are geometrically distorted; certainly vertically, since images appear in time rather than depth, and generally laterally as well (Claerbout, 1971, 1976; Berkhout, 1982; Stolt & Benson, 1986). *Migration* refers to a process which builds, moves, or "migrates" reflector images as close as possible to the geometric location of the actual reflectors. Figures 1.1 to 1.3 illustrate the process. Figure 1.1 shows a highly simplified two-dimensional dome, with prospective oil reservoirs in the cap and on the flanks.

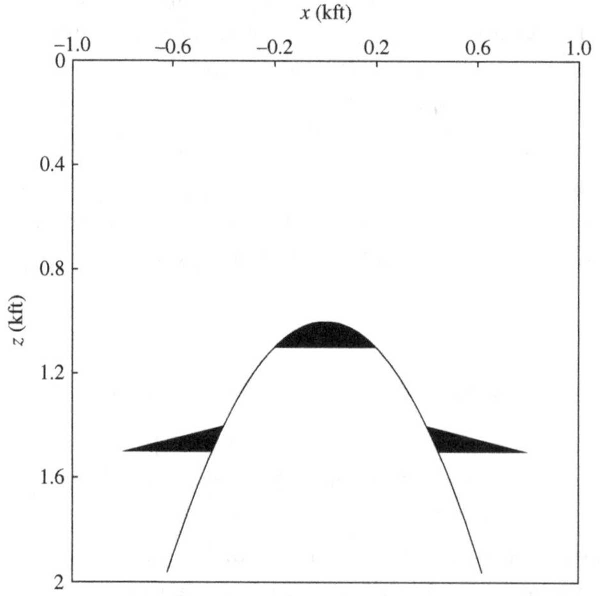

Figure 1.1 Two-dimensional dome model. Seismic velocity is a constant 10 000 fps. The model includes three "reservoirs", one in the dome cap and two on the flanks. 1 kft = 1000 feet.

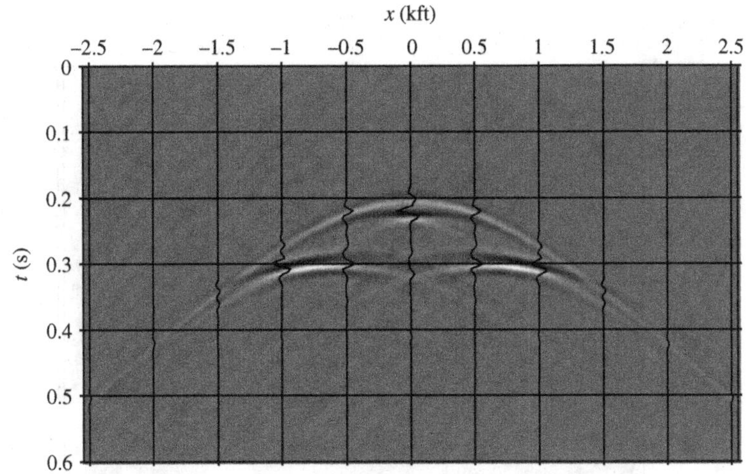

Figure 1.2 Synthesized zero-offset data from dome model. Source–receiver spacing is 20 feet. Time increments are 0.004 s. 1 kft = 1000 feet.

With the dome as initial model, and assuming a constant velocity of 10 000 fps throughout, Figure 1.2 shows the corresponding seismogram for a reflection experiment, in which bandlimited impulsive sources are spaced along the surface at 20 ft intervals, and a response from each source is recorded by a single receiver at the same location as the source. Visible on the seismogram is a dome-like structure much broader than the actual dome. There is some indication of a flat reflector just beneath the top of the dome, though its apparent size and shape do not correlate well with the bottom of the actual reservoir. Reflections from the two reservoirs flanking the dome have transformed into minidomes inside the central dome, leaving one to wonder what and where they really are. An explorer looking at this data would want to know several things: (1) can any potential oil reservoirs be identified; (2) can they be accurately located, and (3) can their size be accurately estimated? Meeting any of these objectives directly from these data would be difficult.

A migration of this data set is shown in Figure 1.3. Like the data it came from, the image is bandlimited, but the size and location of the dome and the reservoirs is accurately recovered. In more realistic situations, where structures are more complex, velocities are variable, and the data are undersampled, the importance of obtaining an accurate image can only increase. In areas of extremely complex geology, reflections may not even be visible in unmigrated data, in which case the only hope for an interpretable image is through a sophisticated imaging algorithm.

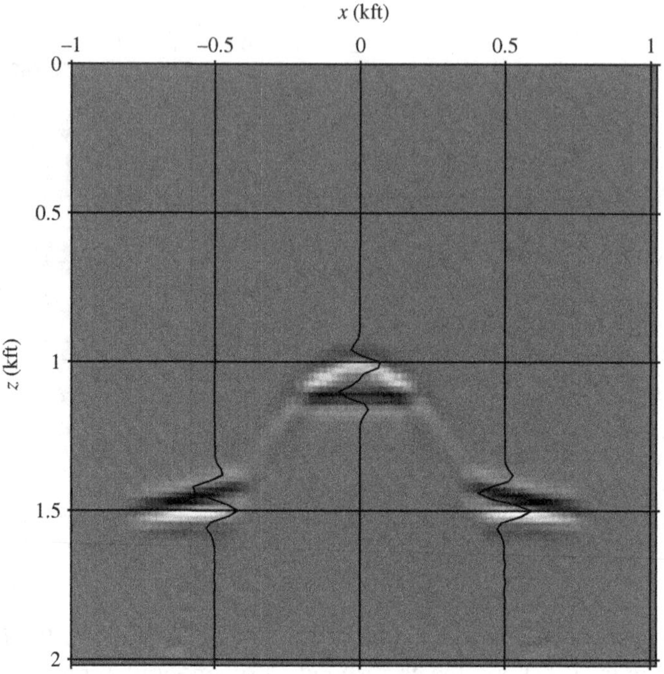

Figure 1.3 Migration of the dome model data. The dome and reservoirs have assumed their geometrically correct positions.

1.6 Time and depth migration

Time migration and depth migration are two types of migration which differ in their ability to image the objective (Schultz & Sherwood, 1980; Stolt & Benson, 1986). Depth migration strives for fidelity both laterally and vertically, whereas time migration leaves the vertical direction in traveltime units. Depth migration requires a detailed model of propagation velocities within the Earth, while time migration needs only an average, or rms (root-mean-squared) velocity structure. Given the right velocity field, depth migration can produce superior images, but time migration is less sensitive to velocity error.

1.7 Migration velocity

Even though time migration may be less sensitive to local variations in seismic velocity, neither form of migration can be expected to yield a good image using bad velocities. To illustrate sensitivity to velocity, the next two figures show migrations of the data in Figure 1.2. Figure 1.4 results from migrating with a velocity that is

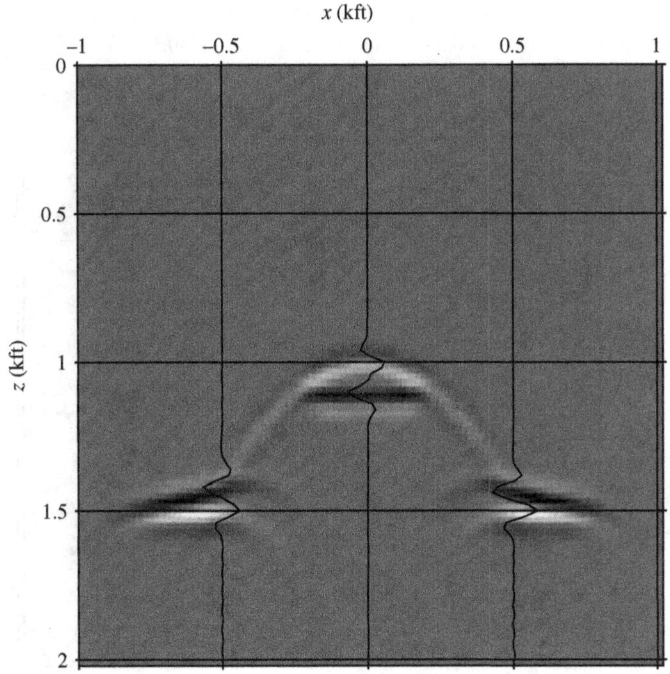

Figure 1.4 Undermigration of the dome model data. Migration velocity is 9500 fps. Neither the dome nor the reservoirs are correctly imaged.

5 percent too low. The dome is too wide, the reservoir terminations are misplaced, and downward-frowning edge diffractions are still very much in evidence.

Figure 1.5 shows a migration with a velocity that is 5 percent too high. The dome is too narrow, the reservoir terminations misplaced, not to mention obscured by upward-smiling diffractions.

These simple examples make it clear that even a small error in migration velocity will result in an inaccurate image. The news, however, is not all bad. Comparing the three migrated images, the correctly migrated one displays a crispness not enjoyed by the other two. It is often possible to spot an undermigrated (overmigrated) image by the presence of downward-pointing (upward-pointing) residual diffractions, and to make velocity adjustments accordingly. Of course, in real life, things are seldom that simple. Determination of correct migration velocities under differing circumstances is given little attention in this text, not because it is unimportant, but because it is a vast, difficult subject, worthy of a volume in itself (see e.g. Liu & Bleistein, 1995; Yilmaz, 2001; Fomel, 2003; Sava *et al.*, 2005).

The migration velocity issue is sometimes viewed as paradoxical, in that on the one hand one cannot perfectly migrate a data set without the correct velocity, while

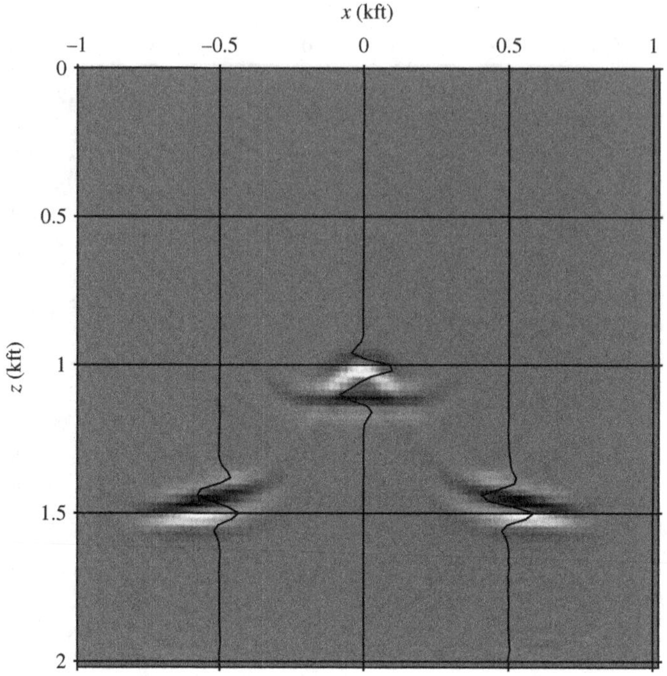

Figure 1.5 Overmigration of the dome model data. Migration velocity is 10 500 fps. Neither the dome nor the reservoirs are correctly imaged.

on the other hand, if the velocity structure were perfectly known, one would already have a structural image of the subsurface. A better view is to acknowledge that the velocity information is imbedded in the data. Extracting this information is a difficult, nonlinear process which is best done concurrently with the imaging. Once this interleaved process is complete, one should have both an accurate velocity map and an accurate seismic image.

1.8 Full-wave and asymptotic migration

Migration methods can be distinguished by whether they employ a full wave equation or an asymptotic or far-field approximation to it. Asymptotic methods assume that wavefronts can be treated as locally planar, with a well-defined direction and local wavenumber. The distinction between asymptotic and full-wave-equation methods is more equivocal than one might think, in that many ostensibly full-wave-equation algorithms have allowed some asymptotic assumptions to slip in, noticed or unnoticed, along the way. Nevertheless, the distinction is worth making. For an individual primary wavefront, the asymptotic approximation is for the most part pretty well satisfied. A complete asymptotic description of a wave in complex media, however, may be difficult or cumbersome. Multiple reflections, largely untreated in Volume I of this book, are more likely to violate the asymptotic assumption. On the plus side, asymptotic algorithms are simple and powerful, and are capable of extracting useful information even under conditions where the assumptions behind them are not completely met. Dealing with missing or limited data may be simplified with asymptotic approximations. It is easier to think (and interpret) in terms of localized events than of extended wavefields, even if a wavefield description is more complete and accurate. The asymptotic approximation certainly has a place in seismic imaging, hence is introduced without hesitation in Volume I.

1.9 Seismic migration and inversion

Seismic migration is an inverse process, and one would not be wrong to refer to seismic migration as seismic inversion. Typically, however, the term *inversion* is reserved for either (1) a process in which a reflector image (hopefully migrated) is used to predict quantitative changes in physical properties; or (2) a process which combines migration with property prediction. Sometimes the term *migration–inversion* is used to mean either the latter process or a migration algorithm in which the imaged reflection amplitudes are physically meaningful. Use of this term suggests that ordinary migration is indifferent to amplitude, which is not exactly the case. All migration algorithms produce amplitude as well as phase, and the amplitude is often interpretable, provided one understands what the algorithm has done. However, one must be aware that many migration algorithms distort amplitudes by making approximations or by using physically inadequate descriptions of wave propagation or reflection.

1.10 True amplitude migration

In a similar vein, one also finds the terms *true-amplitude migration* and *amplitude-preserving migration* (see, e.g., Black *et al.*, 1993; Schleicher *et al.*, 2007). Again, these terms are sometimes used interchangeably. Purists will claim that there is no such thing as true amplitude, let alone true-amplitude migration. Those who use the term may be thinking of "true" not in the sense of being absolutely correct, but in the sense of faithfully preserving the amplitude information. Standards may depend on need and level of ambition. Ideally, one would like an amplitude twice as large to correspond to twice the reflectivity. Practically, one might have to settle for this condition to be satisfied within some range of depths, or dips, or incident angles.

Sometimes, requirements are more rigorous: the amplitudes of multiple reflections depend on the absolute magnitudes of the constituent primaries.

1.11 Linear and nonlinear processes

A useful way to look at modeling and migration is as forward and inverse mappings between a model space and a data space. For convenience, we often treat these mappings as linear operations, though in reality, they are not. Multiple reflections are obviously nonlinearly related to the Earth model. Multiples aside, the amplitudes of primary reflections are also nonlinear. Consider, for example, a scalar plane wave traveling at velocity c_0. An abrupt change in velocity to c_1 will produce a reflection of strength $R = (c_1 - c_0) / (c_1 + c_0)$, nonlinear in the velocity change except in the limit as c_1 approaches c_0. One can argue that the data amplitude is proportional to reflection strength. This leads one to devise linear forward and inverse mappings between data and reflectivity, leaving the nonlinearities to be dealt with by subsequent processing and interpretation.

Subsurface complexity, especially if not fully known a priori, can also introduce nonlinearities. Depth migration requires that the velocity structure between a reflector and the sources and receivers be accurately specified. If the velocities are imperfectly specified, the resulting images may be distorted, spurious, or missing. Recovery, where possible, is a nonlinear operation.

In Volume I, we generally treat the forward and inverse mappings as linear, in the knowledge that Volume II will address the full, nonlinear problem.

At heart, seismic migration is a simple concept. Imaging algorithms, though diverse, share this underlying simplicity. Seismic data, however, are complex, and computational resources are finite. Expectations continue to rise, and as imaging algorithms become more ambitious they also become more complicated.

2

Basic migration concepts

2.1 Migration as a map

Migration is a mapping operator. If seismic data $\mathcal{D}$ can be viewed as the result of an operator $\mathfrak{D}$ acting on a set $\mathcal{R}$ of Earth properties, then seismic migration can be viewed as the result of application to the data of an operator $\mathcal{M}$, the inverse of $\mathfrak{D}$. Symbolically, we have the forward or modeling operation

$$\mathcal{D} = \mathfrak{D}\mathcal{R}, \tag{2.1}$$

and the inverse or migration operation,

$$\mathcal{R} = \mathcal{M}\mathcal{D}, \tag{2.2}$$

with

$$\mathcal{M} = \mathfrak{D}^{-1}. \tag{2.3}$$

Imaging operations such as migration are usually considered to be linear, though more general inverse operations can be contemplated.

Both migration and modeling have often been defined somewhat differently. With this convention, modeling and migration are each other's inverses. There is no real distinction between migration and migration–inversion here, in that *any* migration operator is the inverse of *some* forward modeling operator.

2.1.1 The data

The most general surface-seismic reflection data set is five-dimensional: there are two source position coordinates x_s and y_s, two receiver position coordinates x_g and y_g (the g standing for geophone), and a time coordinate t. Depth coordinates z_g and z_s, if not zero, would be determined by the horizontal coordinates. For land acquisition, if more than one elastic component of the data is recorded, there will be a discrete index (j, say) as well. Thus, if $\tilde{\boldsymbol{x}}_s = (x_s, y_s)$ and $\tilde{\boldsymbol{x}}_g = (x_g, y_g)$, then

$$\mathcal{D} \longleftrightarrow D_j\left(\tilde{\boldsymbol{x}}_g \,|\, \tilde{\boldsymbol{x}}_s; t\right). \tag{2.4}$$

(Here and below, we depict vector quantities in bold face and place a 'tilde' (~) over two-component vectors. The vertical bar separates source (input) and receiver (output) coordinates. The time coordinate has been removed from the source and receiver coordinates because it is not uniquely associated with either source or receiver, rather represents the time interval between source generation and signal reception.) Most acquisition is actually offshore, with sources and receivers immersed in water, limiting both sources and receivers to a single (compressional wave) component. A complete data set requires, for every source a dense 2-D grid of receivers (see Figure 2.1b), and for every receiver a dense 2-D grid of sources (see Figure 2.1c). In practice, few data sets are this complete. Typically, the azimuthal angle (see Figure 2.1a) or angle of alignment between sources and receivers is either fixed or restricted, reducing the effective number of dimensions in the data by one. Two-dimensional surveys keep (or try to keep) sources and receivers in a single vertical plane, further reducing the number of data dimensions to three.

What we have described as a complete seismic data set represents physically a collection of many experiments, in each of which a single source is fired, and data are gathered at many locations on or near the Earth's surface. Within each single-source experiment, changes in the data from one receiver to another are governed

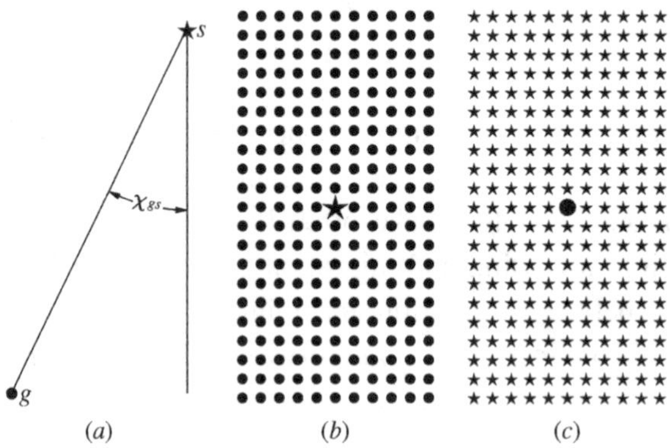

Figure 2.1 Source–receiver azimuth χ_{gs}. This is the angle made by a horizontal line drawn on the Earth's surface from source s to receiver g with respect to a reference direction, as depicted in (a). For the simpler acquisition geometries, the entire data set may have a single azimuth. In contrast, a complete 3-D data set would have (b) a 2-D field of receivers for every source, and (c) a 2-D field of sources for every receiver, with all possible azimuths represented.

by the seismic wave equation. Between these experiments, the source is moved and fired again, either after waiting a sufficiently long time that the response from the last experiment will not interfere, or, if more convenient to picture, shifting to an alternate universe identical to the last except for source location. Either way, as one moves the source, changes in the response at any fixed receiver are governed by the same wave equation that described changes with receiver location for a fixed source. Even though the physics of moving receivers is different than that of moving sources, the mathematics treats them identically, and we are justified in a symmetrical treatment of sources and receivers.

2.1.2 The image function

The exact set of physical properties in $\mathcal{R}$ may vary, depending upon objectives, approach, and level of ambition. In the simplest case, $\mathcal{R}$ is defined somewhat imprecisely as an image function. The image amplitude $\mathcal{R}(\boldsymbol{x})$ is related to some measure of the reflection strength at point $\boldsymbol{x}$.

Suppose for example a reflecting boundary exists at the surface $z = z_r$, as depicted on the left in Figure 2.2.

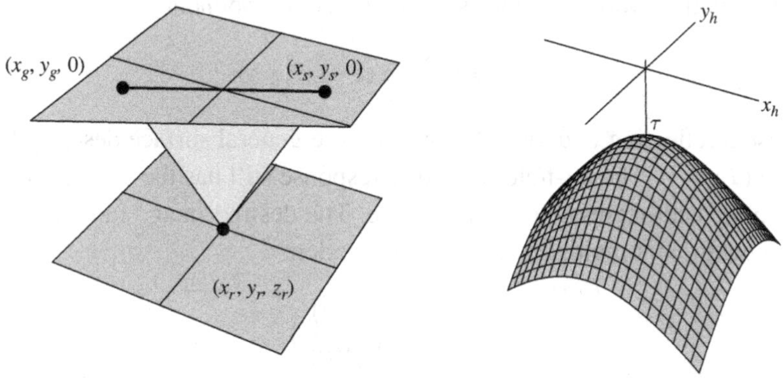

Figure 2.2 A family of rays and their traveltimes. Left: A ray from a surface ($z=0$) source at $\tilde{\boldsymbol{x}}_s = (x_s, y_s)$ strikes a horizontal reflector at $\boldsymbol{x}_r = (x_r, y_r, z_r)$ and returns to a surface receiver at $\tilde{\boldsymbol{x}}_g = (x_g, y_g)$. Since the surface is horizontal, the reflection point is at $(x_r, y_r) = \left(\left(x_g + x_s\right)/2, \left(y_g + y_s\right)/2\right)$, and the total distance traveled by the ray is $r = 2\left(x_h{}^2 + y_h{}^2 + z_r{}^2\right)^{1/2}$, with $x_h = \left(x_g - x_s\right)/2$ and $y_h = \left(y_g - y_s\right)/2$ the half-offset in x and y. Right: The corresponding traveltime surface as a function of the half-offset in x and y. The reflection response to a flat surface is proportional to a delta function centered on the depicted traveltime surface.

In the seismic far field, the impulse response to the reflecting boundary is proportional to the reflection coefficient of the boundary multiplied by a delta function denoting the two-way traveltime τ to and from the boundary:

$$D\left(\tilde{\boldsymbol{x}}_g | \tilde{\boldsymbol{x}}_s; t\right) \propto R\,\delta(t-\tau), \tag{2.5}$$

where, in the case of constant velocity c,

$$\tau = \frac{2}{c}\sqrt{{x_h}^2 + {y_h}^2 + {z_r}^2}, \tag{2.6}$$

with

$$x_h = \left(x_g - x_s\right) / \, 2, \tag{2.7}$$

and

$$y_h = \left(y_g - y_s\right) / \, 2, \tag{2.8}$$

as depicted on the right in Figure 2.2.

The data show an image of the reflecting boundary, located on the hypersurface $t = \tau$. Suppose we desire a migration operator which moves the reflector image to $z = z_r$, the spatial location of the boundary surface, without affecting phase. The desired image function in this case would be the reflection coefficient multiplied by a delta function whose support is on the reflecting boundary:

$$\mathcal{R}(\boldsymbol{x}) = R\delta\left(z - z_r\right). \tag{2.9}$$

Suppose a reflecting boundary follows a more general surface described by the function $b(\boldsymbol{x}) = 0$. The far-field reflection response still has the form (2.5), with a more complicated expression for traveltime. The desired image function is still a reflection coefficient multiplied by a delta function:

$$\mathcal{R}(\boldsymbol{x}) = R\delta\left(\frac{b(\boldsymbol{x})}{|\nabla b|}\right). \tag{2.10}$$

The support of the delta function lies along the reflector boundary $b(\boldsymbol{x}) = 0$. The gradient magnitude in the denominator of the delta function normalizes its argument as follows. If one defines a local coordinate system (x_1, x_2, x_3) at $\boldsymbol{x}$, oriented so that x_3 points in the direction of the gradient of b, and with $b(\boldsymbol{x}) = 0$ at $x_3 = 0$ (see Figure 2.3), then

$$\delta\left(\frac{b(\boldsymbol{x})}{|\nabla b|}\right) = \delta\left(x_3\right). \tag{2.11}$$

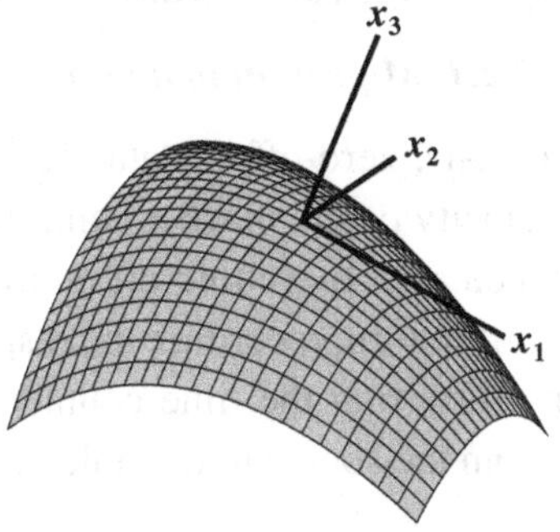

Figure 2.3 Illustration of a general curved surface defined by $b(\boldsymbol{x}) = 0$. At any point on the surface, one may define a local coordinate system (x_1, x_2, x_3) such that x_1 and x_2 are tangent and x_3 is normal to the surface.

More generally still, a reflecting region may not be a sharply defined boundary, and the image function may not be a delta function. Rather, $\mathcal{R}(\boldsymbol{x})$ is a general function of the spatial coordinates.

Considering the image function $\mathcal{R}(\boldsymbol{x})$ to be three-dimensional, there would appear to be at least enough degrees of freedom in the data to derive it. Extra degrees of freedom in the data can be used to reduce noise and to compensate for imperfections and incompleteness in the data. Bear in mind that more realistic imaging functions, as discussed in Chapter 8, may require some or all of this "extra" information in order to be completely reconstructed.

2.1.3 Mapping to points and lines

Which came first, the point, the line, or the plane? Put another way, which of the above is the most fundamental, basic, migrated image? We don't know, but it doesn't really matter – as it turns out (see Appendix D), any 2-D image can be constructed from either points or lines, and any 3-D image can be constructed from points, lines, or planes. Having assumed that migration is a linear operator $\mathcal{M}$, it follows that

$$\mathcal{R} = \mathcal{M}\left(\mathcal{D}_1 + \mathcal{D}_2\right) = \mathcal{M}\mathcal{D}_1 + \mathcal{M}\mathcal{D}_2 = \mathcal{R}_1 + \mathcal{R}_2. \tag{2.12}$$

That is, the migration of a sum of data sets is the sum of the migrations. This means that if we know how to migrate a point (or a line), we know how to migrate any image. The same migration operator that produces a point image will also (given the appropriate data set) produce a line image, or for that matter, any image.

2.2 Ray-theoretical migration concepts

2.2.1 Migration to a point

Consider the simple case of 2-D, zero-offset data. To make things simpler still, assume a single, constant velocity c, and place a point image (whatever that is) at location (x_r, z_r). We make a data set by setting off an impulsive source at each of a set of surface locations $(x, 0)$ in turn, recording the response at the same location for an adequate time interval, starting the time counter at the instant the source goes off. The traveltime τ_r from the source to the reflecting point and back again is a hyperbolic function of x:

$$\tau_r(x) = \frac{2}{c}\sqrt{(x - x_r)^2 + {z_r}^2}. \tag{2.13}$$

Thus, the data from this experiment should consist of an event following the hyperbolic traveltime curve equation (2.13). This is illustrated in Figure 2.4. On the left is a plot of the raypaths from points on the surface to the reflecting point. On the right is a cartoon of the data. The figure accurately depicts neither phase nor amplitude of the event, but that is not of concern to the current level of inquiry.

Migration can be described as an operation that turns the hyperbolic event on the right into the point on the left. An algorithm suggests itself: For every point in image space, sum along the corresponding hyperbola in data space. for the data in this example, this would build an event at (x_r, z_r). At other points, what would happen is less clear, though it is reasonable to expect the data would not add constructively. One might suspect that to avoid producing artifacts at points other than the reflecting point, some attention should be paid to phase, amplitude, and

Figure 2.4 A point reflector. (a), Zero-offset raypaths between the surface and the reflecting point. (b), Traveltime to the event (expressed in depth units) at a sequence of surface locations. It can be considered to be a depiction of the impulse response of the point reflector, simplified in that the actual response has an associated phase and an amplitude that depends upon source–receiver position.

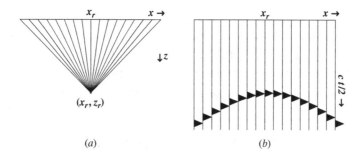

possible aliasing. Such refinements aside, we have a basic algorithm. Generalization to variable velocity, finite source–receiver separation, and three dimensions is straightforward, provided the traveltime between surface points and subsurface points can be computed: for every source, receiver, and possible reflection point, gather the data from the associated traveltime and sum into the image of the reflection point. Such a migration scheme is *ray-theoretical*, in that associating a traveltime between two points implies a raypath.

The hyperbolic-sum concept relies on destructive interference to prevent events from building where they are not supposed to. In this simplified analysis, where the event is represented as a spike, it may be less than obvious how the data can ever sum to zero. Real data, of course, are bandlimited, with no wavelengths below a certain threshold. Events should have equal parts of negative and positive amplitudes, and sums of incoherent data should be close to zero. Summation migration (often called *Kirchhoff migration*, because of a relation to the Kirchhoff diffraction integral) assumes that the bandwidth lower bound is high enough for cancellation to take place, hence is inherently a "high-frequency" approximation (Schneider, 1978; Bleistein *et al.*, 2001; Schleicher *et al.*, 2007).

2.2.2 Migration to a line

Consider an experiment to locate a seismic reflector in a medium of constant velocity c using a coincident source and receiver that can be moved along a line on the Earth's surface. After detonating several sources, the situation is depicted in Figure 2.5. On the left is the data collected from the experiment, with

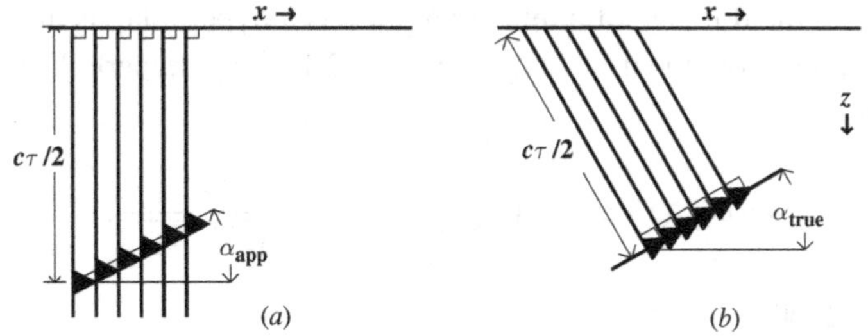

Figure 2.5 The response to a dipping reflector. The unmigrated data set on the left plots the response vertically at each source–receiver position, which displaces the response down dip from its true reflection point as seen on the plot on the right. The tangent of the apparent dip, α_{app}, equals the sine of the true dip, α_{true}. Migration of each trace is effected by moving the data up dip, keeping distance to the surface source–receiver point fixed, until the ray makes an angle α_{true} with the vertical.

source–receiver location plotted on the horizontal axis, and traveltime plotted on the vertical axis. On the right is the true reflector location, together with rays drawn from the source–receiver surface locations to the reflector. According to Snell's law (Jackson, 1962; Born and Wolf, 1980), the rays impact the reflector at right angles, so the points of impact are horizontally displaced up dip from the source–receiver positions. The unmigrated data on the left contains an image of the reflector, but it is displaced down dip and appears to be less steep than it really is. If Δx is the horizontal distance between adjacent source–receiver locations, and $\Delta r = c\Delta t/2$ the difference in length between adjacent raypaths, then from the left-hand plot we see that $\Delta r/\Delta x = \tan\alpha_{\text{app}}$, and from the right-hand plot that $\Delta r/\Delta x = \sin\alpha_{\text{true}}$. That is, the tangent of the apparent dip equals the sine of the true dip:

$$\tan\alpha_{\text{app}} = \sin\alpha_{\text{true}}. \tag{2.14}$$

This suggests a simple migration algorithm: first, identify the reflectors and measure the apparent dips α_{app}. For each source–receiver location, draw a circle centered on the source–receiver location, with radius equal to the distance to the reflector. The point on the circle tangent to a line dipping at angle $\alpha_{\text{true}} = \arcsin(\tan\,\alpha_{\text{app}})$ is the true reflector location.

This geometrical migration algorithm can be generalized to variable velocity, and is relatively simple to implement. In fact, its use pre-dates digital computers. One should acknowledge, however, some limitations. First, it only works where reflections can be identified and dip measured prior to migration. Second, the geometrical argument does not tell us what to do about amplitudes. However incomplete, the concept is useful for understanding migration as an attempt to move events up dip to their true locations.

What happens to this concept if the migrated image is a point instead of a line? Since the unmigrated image of a reflecting point is a hyperbola, its apparent dip is spatially variable. From the traveltime equation (2.13) for a hyperbola, the apparent slope tangent is

$$\tan\alpha_{\text{app}}(x) \equiv \frac{\partial}{\partial x}\left(\frac{c\tau_r}{2}\right) = \frac{x - x_r}{\sqrt{(x - x_r)^2 + {z_r}^2}}, \tag{2.15}$$

so, invoking (2.14),

$$\sin\alpha_{\text{true}}(x) = \frac{x - x_r}{\sqrt{(x - x_r)^2 + {z_r}^2}}, \tag{2.16}$$

or

$$\tan\alpha_{\text{true}}(x) = \frac{x - x_r}{z_r}. \tag{2.17}$$

With this formula, α_{true} is the angle from the horizontal made by a ray between points $(x, 0)$ and (x_r, z_r). Consequently, moving the event from $(x, c\tau_r/2)$ along a circle of radius $c\tau_r/2$ until the angle made by the ray with the vertical is α_{true}, one arrives at the point (x_r, z_r). The geometrical migration algorithm reduces the hyperbola to a point at (x_r, z_r), as illustrated in Figure 2.6.

How does the hyperbolic stacking concept, conceived for migrating to a point, work on lines? Figure 2.7 shows its application to the sloping reflector of Figure 2.5. In this figure, the vertical direction is in depth units, or $c/2$ times traveltime. Both the apparent and the actual reflector locations are depicted as dark lines, the

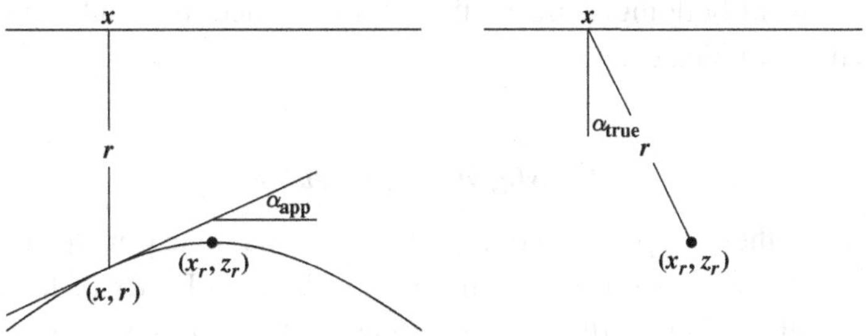

Figure 2.6 Migration of the hyperbolic diffraction pattern from a point reflector. From any point on the hyperbola, application of the formula $\tan \alpha_{\text{app}} = \sin \alpha_{\text{true}}$ maps the data onto the reflection point. On the left is the diffraction hyperbola, with apex at the reflector position (x_r, z_r). At any point (x, r) on the hyperbola, the apparent dip is the slope of the line tangent to the hyperbola at that point. On the right is the migrated image point at reflector position (x_r, z_r), which can be found by projecting a line from the surface position $(x, 0)$ a distance r in the direction α_{true}.

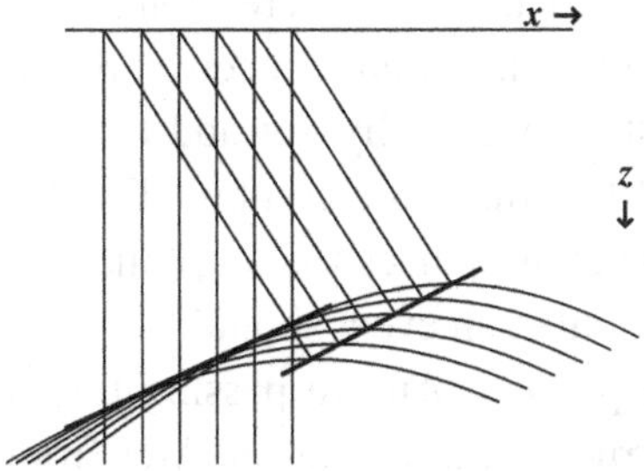

Figure 2.7 Migration of a reflecting line by hyperbolic summation. An image at the true reflector location (the rightward line) is formed by summing the data (the leftward line) along the hyperbolas shown. These hyperbolas are each tangent to some point on the apparent reflector line, allowing data to sum constructively to form the image. For hyperbolas corresponding to other points in the image plane, no point of tangency occurs, and no image can form.

leftward and rightward lines at the apparent and actual reflector locations, respectively. To form the image along the true reflector line, hyperbolas (equation (2.13)) are drawn with their apexes at points along the true reflector line. Each of these hyperbolas are tangent to some point on the apparent reflector line. As data is summed along any of these hyperbolas, data near the point of tangency will add constructively to form an image at the true reflector location. Thus, the hyperbolic stacking concept works for lines as well as points, which is good news for the principle of superposition.

The geometrical migration concept is closely related to the hyperbolic sum. Instead of performing the entire sum, one selects events a priori, finds the hyperbola tangent to each, and moves the data to the apex of the hyperbola. Avoiding the sum ought to avoid building artifacts, though in practice the results are no better than one's ability to pick events.

2.2.3 Migration to a curve

If the data are neither a hyperbola nor a straight line, the same principles still apply. Any curved line will, at every point, have a local slope. Either of the migration processes described above will carry that point up dip until the tangent of the old slope equals the sine of the new slope. In this process, hills (anticlines) become steeper and narrower, while valleys (synclines) become broader, as illustrated in Figure 2.8.

2.2.4 Migration by broadcast

Another way to view migration is as a process that broadcasts data along arcs. Looking at a trace in isolation, suppose there is an event at time t. The reflector that generated the event must lie somewhere along the circle of radius $ct/2$ centered on the surface location of the source–receiver. Since one doesn't know where along the circle the event belongs, why not add the event to the image at *every* point along the circle? When these are combined with circles from neighboring traces, one can expect in certain places the amplitudes to interfere constructively and build an event. In particular, an event should form at the true reflector location. Figure 2.9 illustrates the process. An arc passing through the apparent reflector location is tangent to the true reflector location. Arcs from neighboring traces build constructively near the points of tangency. Away from these points, the energy is dispersed.

The arc-swinging concept is in fact theoretically identical to the hyperbola-summing concept. Algorithmically, they differ in that one takes the input data trace by trace and disperses it, while the other builds the output image point by point by

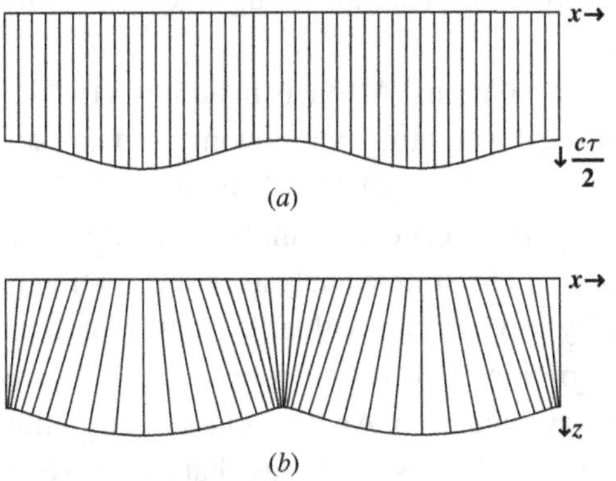

Figure 2.8 Image of a curved surface before and after migration. The upper image is that of a curved reflector prior to migration, the vertical lines measuring the distance or renormalized traveltime to the reflector from points x on the Earth's surface. The lower image is the reflector after migration, the near-vertical lines representing rays from points on the Earth's surface to points on the reflector. The migration was performed by swinging each vertical line up dip until the tangent of the unmigrated image dip equals the sine of the new image dip. Assuming the migrated image is correct, then prior to migration, the anticlines appear broader and the synclines appear narrower than they really are.

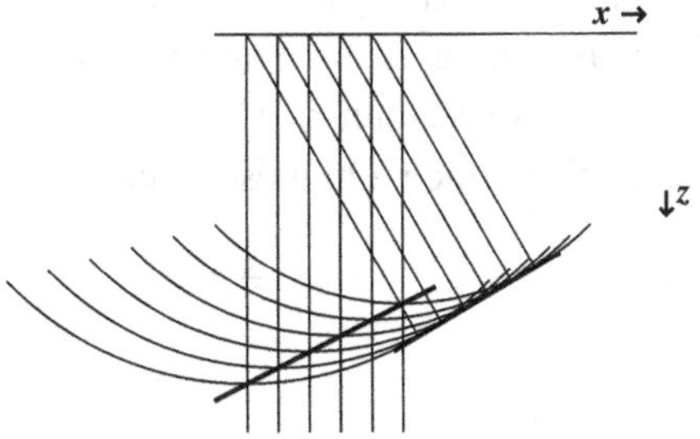

Figure 2.9 Migration by superposition of the data in Figure 2.5. The data are broadcast along arcs of constant traveltime, allowing them to build constructively at the true reflector location.

summing data into it. Data flow is very different in the two processes, but in the end, the same computation is performed.

Either concept presents migration as a wave-theoretical process. As algorithms, they are attractively simple. They can be implemented post- or prestack and in variable-velocity media where traveltimes between any two points can be computed. While the geometrical concept requires that events be identified and their dips measured a priori, the superposition concept requires no a-priori information except velocity: just take every non-zero amplitude, broadcast it along an arc, and see what adds up. Modern "Kirchhoff migration" algorithms are based on the superposition principle.

There are, of course, limitations. Artifacts in addition to "real" events are easy to produce, though intelligent weighting, windowing, and anti-aliasing along the arcs are helpful. For complex velocity structures where there may be multiple raypaths between two points, the method becomes more difficult to implement and at some point becomes intractable. The simple analysis in this section reveals nothing about amplitudes, though they can be dealt with at least asymptotically.

Ray-theoretical migration algorithms rely on a high-frequency asymptotic approximation to the wave equation. Another class of migration algorithms operate from the full wave equation. This is not to say that full wave-equation methods are devoid of approximation, but they can circumvent some of the limitations of ray theory.

2.2.5 Migration phase

It is time to pay a little attention to the phase generated by the migration operation. A basic 2-D Kirchhoff algorithm, whether by broadcast or by summation, imposes a phase shift on the data of about 45°. We can see this by examining Figure 2.10, depicting migration by broadcast of a flat reflector located at depth z_r. The broadcast operation creates an image along the tangent line $z = z_r$, but it also broadcasts a non-zero amplitude into depths less than z_r, along circles of radius z_r centered at points $(x_r, 0)$ along the Earth's surface. On these circles,

$$z^2 + (x - x_r)^2 = z_r^2, \tag{2.18}$$

or

$$|x - x_r| = \sqrt{z_r^2 - z^2}. \tag{2.19}$$

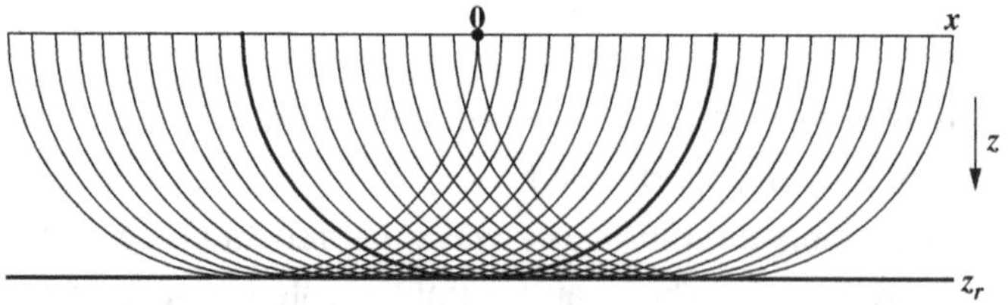

Figure 2.10 Migration by broadcast of a flat reflector at depth z_r. No data are broadcast into the migrated image at depths below z_r, but at depths z above z_r, data are broadcast into the migrated image with amplitude proportional to $(z_r - z)^{-1}$.

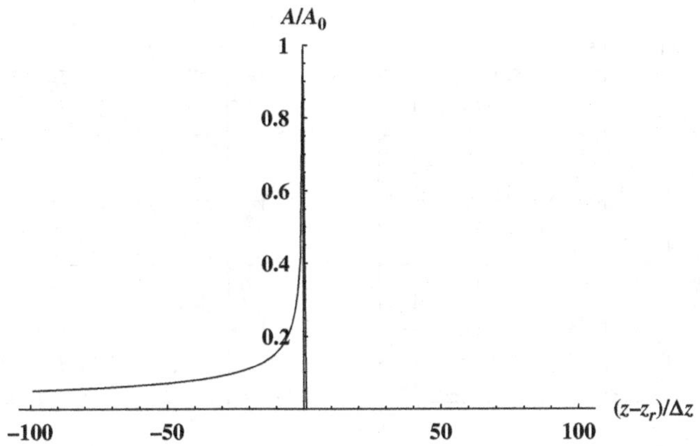

Figure 2.11 Kirchhoff migration phase. Kirchhoff migration, unless corrected, imposes a phased wavelet on the migrated image. For 2-D post-stack migration, the wavelet is proportional to $(z_r - z)^{-1/2}$, or the anticausal half-integral filter.

The total amplitude A broadcast into depth z above z_r will be inversely proportional to the slope of the circles at depth z; i.e.,

$$A(z) \propto \left|\frac{dx}{dz}\right| = \frac{z}{\sqrt{{z_r}^2 - z^2}} \simeq \frac{\sqrt{z_r/2}}{\sqrt{z_r - z}} \propto \frac{1}{\sqrt{z_r - z}}. \tag{2.20}$$

To lowest order in $z_r - z$, the amplitude of the broadcast at depths less than z_r falls off as $(z_r - z)^{-1/2}$, as depicted in Figure 2.11.

Of course we would like the migrated image of the reflector to be a symmetrical impulse centered on z_r. To get there we must add a further operation to the migration algorithm, which compresses the response into an impulse. Fortunately, this is easy to do. The wavelet imposed upon the image is recognizable as the time-reversed or anticausal half-integral filter $\Theta_{h-}(z - z_r)$, as defined in Section E.6 of Appendix E. In the frequency domain, this filter is just $1/\sqrt{i\omega}$. Its inverse is $\sqrt{i\omega}$, or the time-reversed half-derivative filter $d_{h-}(z - z_r) = -1/\sqrt{4\pi(z_r - z)^3}$, defined in Appendix E.5. Because of the $\sqrt{i}$, these filters are said to have 45° of phase. Convolution of the migrated data with the anticausal half-derivative will produce a sharp symmetrical image of reflectors at the correct location.

2.2.6 Migration of diffractions

There is more to seismic data than the specular reflections admitted by Snell's law (Keller, 1953, 1978; Born & Wolf, 1980). As a simple example, examine Figure 2.12. The upper left panel shows a simulated zero-offset seismic response to a

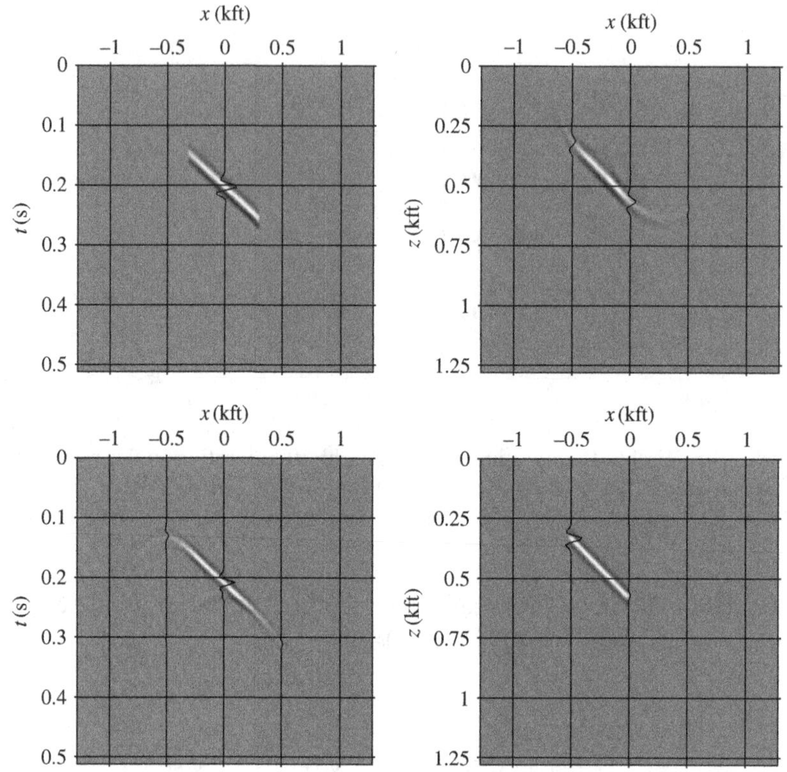

Figure 2.12 An illustration of the role diffractive energy plays in migration. The upper left panel is a simulated zero-offset data set for a dipping reflector of slope 1/2 and ends at −500 ft and 0 ft. Only the specular reflection predicted by Snell's law appears in this simulation. The upper right panel shows a migration of the upper left panel, which moves the main body of the reflection to its true location but smears and distorts the reflector end points. The lower left panel is a more complete simulation including diffractions from the reflector end points, constructed via the diffraction integral discussed in later chapters. The lower right panel is a migration of the lower left. This migration puts the reflector in its proper place, and accurately depicts the terminations.

dipping reflector of slope 1/2 defined for $-500 < x < 0$ ft. As discussed in Section 2.2.2, the data contain an image of the reflector downslope of its true lateral location. This simulation contains only the specular reflection, so the data terminate abruptly at the reflector ends. As a minor concession to reality, the reflector image has been bandlimited by convolution with a Ricker wavelet (Appendix E).

The upper right panel shows a migration of the data on the upper left. The main body of the reflector has been moved to its true location, but the reflector ends are smeared and distorted by migration "smiles". If one were trying to locate a fault accurately, this migration would not be helpful.

The lower left panel shows another simulation of the seismic data, this time allowing diffractive energy from the reflector ends. The lower right panel is a migration of these data. In this case, the reflector assumes its true location and its ends are clearly and accurately depicted.

In real seismic data, much (some might claim all) of the energy comes from non-specular diffractions. The diffractive energy carries information necessary to image complex geometries accurately. A good migration algorithm must image diffractive as well as specular energy. Both wave- and ray-theoretical algorithms are generally capable of doing this. While ray-theoretical methods adhere to Snell's law for propagated waves, they are not so restrictive for reflections. As was pointed out in Section 2.1.3, an algorithm which can migrate lines or planes can also image point diffractions. However, some algorithms attempt to measure and respond to a dominant dip at every point. Such algorithms may lose diffractive information retained by more open-minded algorithms.

2.3 Downward continuation

There are other ways to look at migration, including so-called *full wave-equation* methods, which depend less on the far-field asymptotic approximations inherent in ray theory. Central to full wave-equation methods is the concept of *downward continuation* (Claerbout, 1971, 1976; Berkhout, 1981).

The deeper a reflecting point lies within the Earth, the more spread out its unmigrated image. Suppose one could calculate what a data set would look like were the sources and receivers placed at a finite depth within the Earth. Closer to the sources and receivers, point reflectors should occur earlier in time and look more pointlike, as depicted in Figure 2.13. As the source and receiver become infinitesimally close to the reflecting point, its response is a point infinitesimally close to time zero.

This suggests another way to look at migration. In this view (Claerbout, 1971) the migrated image at depth z is the time-zero component of the data wavefield downward-continued to depth z. For 2-D, zero-offset data, the mapping takes place in a 3-D space with coordinates x (horizontal position), z (source–receiver depth), and t (time). As illustrated in Figure 2.14, unmigrated data lie on the x–t plane ($z = 0$), and the migrated image lies on the x–z plane ($t = 0$). Downward-continued data occupy the interior of the volume.

There are complications. First of all, while the zero-time downward-continued data should be related to the reflectivity, it probably will not *be* the reflectivity. The image function certainly will differ in amplitude, and quite possibly phase, from a true reflectivity function.

Another complication: strictly, continuation of sources and receivers must be handled separately. Each requires its own wave equation, even if sources and

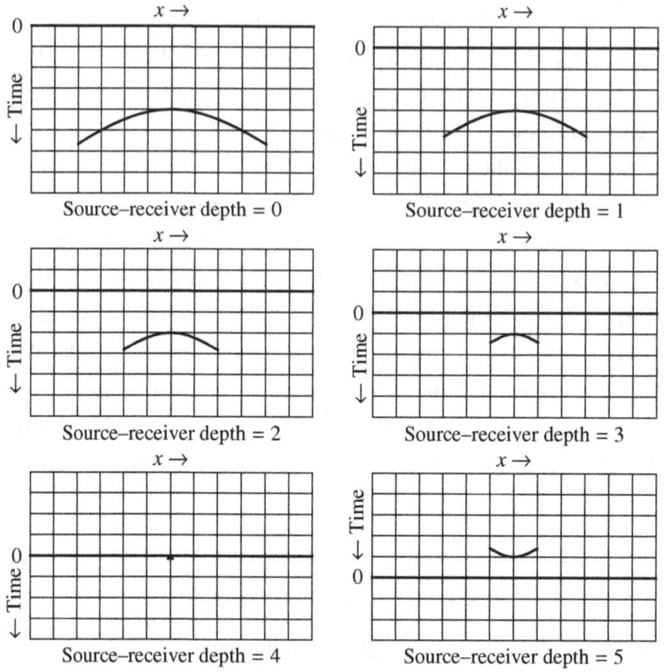

Figure 2.13 Downward continuation of a point response. A depiction of the evolution of the response of a point reflector at depth = 4 as the source–receiver plane is moved downward and closer to the reflector. The traveltime to the event becomes smaller and the event shrinks, becoming a point at zero time when the source–receiver plane reaches the reflector. If the source–receiver plane is moved past the reflector, the event reappears as an inverse image in negative time.

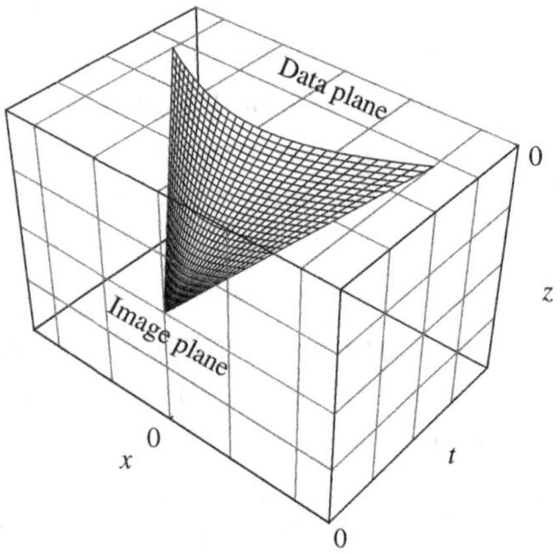

Figure 2.14 Migration as a mapping from the x–t data plane ($z = 0$) to the x–z image plane ($t = 0$). In this illustration, the data is a diffraction hyperbola on the top surface and the image is a point on the front surface. The interior of the volume consists of data with source–receiver location downward-continued to various depths.

receivers are continued from a common location to a common location. Less strictly, under some conditions it is possible to downward-continue zero-offset data with a single wave equation for the source–receiver location. This is actually implicit in the ray-theoretical migration concepts discussed above, and follows from the assumption that the raypath from source to reflector is identical to the return raypath from reflector to receiver.

If one is unconcerned about amplitudes, one can further simplify the physics by invoking the *exploding-reflector* (Claerbout, 1971) concept. The idea is to relate the zero-offset data set to an experiment in which sources are placed in the Earth at every reflector location, all timed to fire at time $t = 0$. The resulting raypath from any reflecting point to any receiver is the same as the zero-offset raypath to that receiver, except of course it is traversed twice in the zero-offset experiment versus once in the exploding reflector experiment. Thus, one starts with zero-offset data, divides time (or velocity) by two, invokes the exploding-reflector model, downward-continues receivers only, and forms the migrated image as the zero-time downward-continued data. It is understood that the exploding-reflector model will not get amplitudes right, but in the absence of multiple raypaths between reflectors and source–receiver points, it should form an image at the correct location.

2.3.1 Constant-velocity plane-wave extrapolation

Suppose we have a plane wave $P_{\hat{k}}(\tilde{x}, t)$ traveling with constant velocity c in the direction $\hat{\boldsymbol{k}}$. If we know the value of this wave anywhere (say, at $\tilde{x}_0$), we know it everywhere:

$$P_{\hat{k}}(\tilde{x}, t) = P_{\hat{k}}\left(\tilde{x}_0, t - \frac{\hat{\boldsymbol{k}}}{c} \cdot (\tilde{x} - \tilde{x}_0)\right). \tag{2.21}$$

Extrapolation of a plane wave in any direction is just a matter of applying the appropriate (wavelength–frequency) time shift. Some types of migration (e.g. *f–k*, phase shift, and Radon transform) work by decomposing data into plane-wave components. For each component, equation (2.21) can be used to perform the downward continuation.

2.3.2 Difference-equation wave extrapolation

The wave equation being second order, in order to extrapolate into a region from its bounding surface we would expect to have to know both the wave and its normal derivative. In practice, we seldom have such luxury, and must make do with what we have. Fortunately, there are workarounds. For the plane wave (2.21) it was not necessary to know derivatives of $P_{\hat{k}}$ at $\tilde{x}_0$ because $P_{\hat{k}}$ and its derivatives are not independent. Generally, if we know that the wave P_u we wish to extrapolate is "upgoing", a relation between P_u and its derivatives is implied. Where velocity is locally constant, we can even write a first-order "one-way" wave equation which can extrapolate from the starting boundary given only P_u. Where a "one-way" description is inadequate, we can often impose boundary conditions that give us an effective relation between the wavefield and its normal derivative on the starting boundary.

2.3.3 Green's theorem continuation

A general way to formulate continuation into an area or volume is as a boundary value problem using Green's theorem. Consider a solution $P(\tilde{\boldsymbol{x}}, \omega)$ to the 2-D scalar wave equation at $\tilde{\boldsymbol{x}} = (x, z)$, i.e. satisfying

$$\left(\tilde{\nabla}^2 + \frac{\omega^2}{c(\tilde{\boldsymbol{x}})^2}\right) P(\tilde{\boldsymbol{x}}, \omega) = 0 \tag{2.22}$$

within an area $\mathcal{V}$ with closed boundary $\mathcal{S}$, as per Figure 2.15.

Suppose we have a solution $G(\tilde{\boldsymbol{x}}'|\tilde{\boldsymbol{x}};\omega)$ to the inhomogeneous scalar wave equation

$$\left(\tilde{\nabla}^2 + \frac{\omega^2}{c(\tilde{\boldsymbol{x}})^2}\right) G\left(\tilde{\boldsymbol{x}}'|\tilde{\boldsymbol{x}}; \omega\right) = -\delta\left(\tilde{\boldsymbol{x}} - \tilde{\boldsymbol{x}}'\right), \tag{2.23}$$

known for arbitrary $\tilde{\boldsymbol{x}}$ and $\tilde{\boldsymbol{x}}'$ within $\mathcal{V}$. (In this notation, the coordinates to the left of the vertical bar represent a measurement or receiver location, while coordinates to the right of the bar but left of the semicolon specify a source location. To the right of the semicolon are variables – in this case, frequency – that do not depend uniquely on source or receiver.) Such a solution, called a Green's function, is not unique, and may be chosen to be causal, anticausal, or a combination of the two. It may also be chosen to accommodate convenient boundary conditions. For more discussion of Green's functions, see Chapter 6.

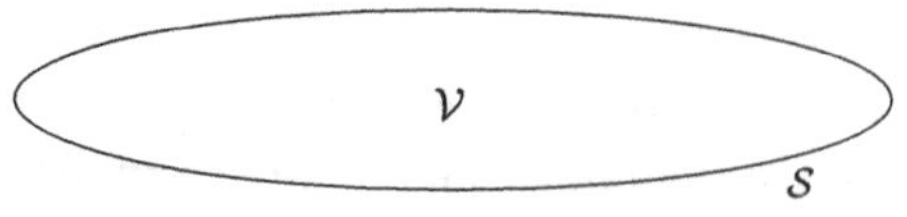

Figure 2.15 A region $\mathcal{V}$ bounded by $\mathcal{S}$.

Given P and G within $\mathcal{V}$, we invoke *Green's second identity* (Jackson, 1962), or

$$\int_{\mathcal{V}} \left(P\left(\tilde{x}', \omega\right) \tilde{\nabla}_{\tilde{x}'}{}^{2} G\left(\tilde{x}|\tilde{x}'; \omega\right) - G\left(\tilde{x}|\tilde{x}'; \omega\right) \tilde{\nabla}_{\tilde{x}'}{}^{2} P\left(\tilde{x}', \omega\right) \right) d\tilde{x}'$$
$$= \oint_{\mathcal{S}} \left(P\left(\tilde{x}', \omega\right) \frac{\partial G\left(\tilde{x}|\tilde{x}'; \omega\right)}{\partial n'} - G\left(\tilde{x}|\tilde{x}'; \omega\right) \frac{\partial P\left(\tilde{x}', \omega\right)}{\partial n'} \right) ds'. \quad (2.24)$$

Invoking (2.22) and (2.23), (2.24) becomes

$$P\left(\tilde{x}, \omega\right) = -\oint_{\mathcal{S}} \left(P\left(\tilde{x}', \omega\right) \frac{\partial G\left(\tilde{x}|\tilde{x}'; \omega\right)}{\partial n'} - G\left(\tilde{x}|\tilde{x}'; \omega\right) \frac{\partial P\left(\tilde{x}', \omega\right)}{\partial n'} \right) ds'. \quad (2.25)$$

Equation (2.25) expresses the field P at any point inside $\mathcal{V}$ in terms of the value of P and its normal derivative on the boundary $\mathcal{S}$, as illustrated in Figure 2.16. That is, the formula continues the field from the boundary $\mathcal{S}$ to the interior of $\mathcal{V}$. This differs from simple downward continuation in two respects: (1) downward continuation proceeds from an open boundary $z = 0$, not a closed boundary $\mathcal{S}$; and (2) downward continuation only uses the field on the surface, while inward continuation (2.25) requires both the field and its normal derivative on $\mathcal{S}$.

These differences can be made to go away by choosing the right Green's function. Consider a boundary with top $\mathcal{S}_1$ at $z = 0$, bottom $\mathcal{S}_2$ at $z = z_2$, and sides at $\pm$ infinity, as depicted in Figure 2.17. Placing the sides an infinite distance

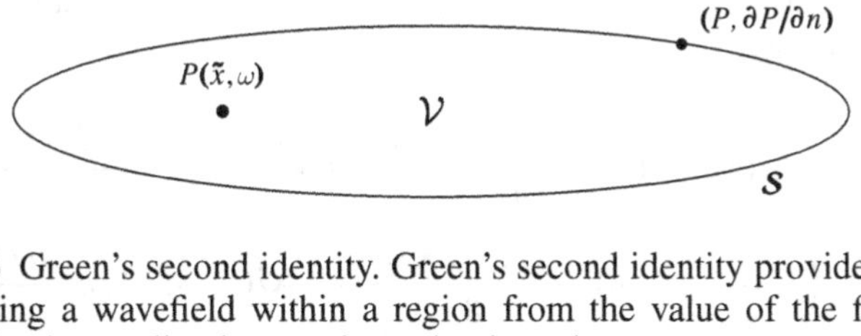

Figure 2.16 Green's second identity. Green's second identity provides a formula for computing a wavefield within a region from the value of the field and its normal derivative at all points on the region boundary.

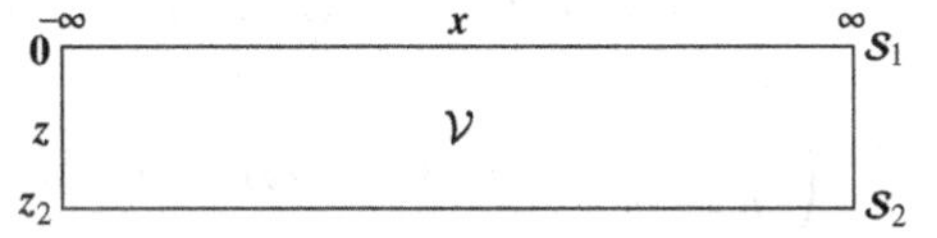

Figure 2.17 A rectangular region. The region $\mathcal{V}$ extends from $-\infty$ to $+\infty$ in the x-direction, and from 0 to z_2 in the z-direction. The upper boundary is labeled $\mathcal{S}_1$ and the lower boundary $\mathcal{S}_2$.

away prevents them from contributing to (2.25) at finite times. We are then left with

$$P(\tilde{\boldsymbol{x}}, \omega) = \int_{\mathcal{S}_1} \left(P(x', 0, \omega) \frac{\partial G(\tilde{\boldsymbol{x}} | x', 0; \omega)}{\partial z'} - G(\tilde{\boldsymbol{x}} | x', 0; \omega) \frac{\partial P(x', 0, \omega)}{\partial z'} \right) dx'$$
$$- \int_{\mathcal{S}_2} \left(P(x', z_2, \omega) \frac{\partial G(\tilde{\boldsymbol{x}} | x', z_2; \omega)}{\partial z'} - G(\tilde{\boldsymbol{x}} | x', z_2; \omega) \frac{\partial P(x', z_2, \omega)}{\partial z'} \right) dx'. \tag{2.26}$$

Let us simplify as much as possible by looking at the case of a constant velocity $c = c_0$. Then, any relevant Green's function is invariant under lateral translations, i.e. may be written

$$G(x, z|x', z'; \omega) = \mathcal{G}(z|z'; x - x', \omega). \tag{2.27}$$

Equation (2.26) then becomes the convolution

$$P(x, z, \omega)$$
$$= \int_{\mathcal{S}_1} \left(P(x', 0, \omega) \frac{\partial \mathcal{G}(z|0; x - x', \omega)}{\partial z'} - \mathcal{G}(z|0; x - x', \omega) \frac{\partial P(x', 0; \omega)}{\partial z'} \right) dx'$$
$$- \int_{\mathcal{S}_2} \left(P(x', z_2, \omega) \frac{\partial \mathcal{G}(z|z_2; x - x', \omega)}{\partial z'} - \mathcal{G}(z|z_2; x - x', \omega) \frac{\partial P(x', z_2, \omega)}{\partial z'} \right) dx'. \tag{2.28}$$

By taking a Fourier transform from x to k_x, the integrals can be replaced by simple multiplications:

$$P(k_x, z, \omega) = \left(P(k_x, 0, \omega) \frac{\partial \mathcal{G}(z|0; k_x, \omega)}{\partial z'} - \mathcal{G}(z|0; k_x, \omega) \frac{\partial P(k_x, 0, \omega)}{\partial z'} \right)$$
$$- \left(P(k_x, z_2, \omega) \frac{\partial \mathcal{G}(z|z_2; k_x, \omega)}{\partial z'} - \mathcal{G}(z|z_2; k_x, \omega) \frac{\partial P(k_x, z_2, \omega)}{\partial z'} \right). \tag{2.29}$$

The problem is now one-dimensional. P and $\mathcal{G}$ satisfy 1-D wave equations

$$\left(\frac{d^2}{dz^2} + {k_z}^2 \right) P(k_x, z, \omega) = 0, \tag{2.30}$$

$$\left(\frac{d^2}{dz^2} + {k_z}^2 \right) \mathcal{G}(z|z'; k_x, \omega) = -\delta(z - z'), \tag{2.31}$$

with vertical wavenumber k_z given by

$$k_z = \frac{\omega}{c_0}\sqrt{1 - \frac{{k_x}^2{c_0}^2}{\omega^2}}. \tag{2.32}$$

Equation (2.30) has solutions of the form

$$P(k_x, z, \omega) = \mathcal{A}_+ e^{ik_z z} + \mathcal{A}_- e^{-ik_z z}. \tag{2.33}$$

With the sign conventions used here for the Fourier transform (see Appendix C), the first term in expression (2.33) represents a downgoing wave, and the second term an upgoing wave.

The free-space solutions to (2.31) are (see Chapter 6)

$$\mathcal{G}_\pm(z|z'; k_x, \omega) = \mp\frac{e^{\pm ik_z|z - z'|}}{2ik_z}. \tag{2.34}$$

The solution $\mathcal{G}_+$ is the causal Green's function, or impulse response. In the time domain, it is zero for negative times, popping into existence at time $t = 0$, and expanding outward from the point $z = z'$ as time increases. The solution $\mathcal{G}_-$ is the anticausal or imploding Green's function. In the time domain, it exists only for negative times, contracting to the point $z = z'$ and popping out of existence at $t = 0$.

If $\mathcal{G}$ is required to satisfy (2.31) only within the region $\mathcal{V}$, then additional solutions are possible. Equation (2.31) admits a single source point, at $z = z'$. However, one can add more source points outside the region $\mathcal{V}$ without violating (2.31) within $\mathcal{V}$. This, specifically, allows construction of Green's functions that satisfy Dirichlet ($\mathcal{G} = 0$) or Neumann ($\partial\mathcal{G}/\partial n = 0$) boundary conditions on the boundary $\mathcal{S}$ of $\mathcal{V}$. To impose Neumann or Dirichlet boundary conditions on the top surface $\mathcal{S}_1$, one can use the method of images to add a single virtual source point above $\mathcal{S}_1$. Depending on the sign of the virtual source, the result is $\mathcal{G} = 0$ or $\partial\mathcal{G}/\partial n = 0$ on $\mathcal{S}_1$. To impose Neumann or Dirichlet boundary conditions on both $\mathcal{S}_1$ and $\mathcal{S}_2$ requires an infinite set of virtual source points, but is otherwise straightforward. One can perform the same construction for either the causal or the anticausal Green's functions, or for a linear combination thereof.

With Dirichlet boundary conditions imposed, equation (2.29) determines P within $\mathcal{V}$ from P on $\mathcal{S}_1$ and $\mathcal{S}_2$:

$$P(k_x, z, \omega) = P(k_x, 0, \omega)\frac{\partial\mathcal{G}(z|0; k_x, \omega)}{\partial z'} - P(k_x, z_2, \omega)\frac{\partial\mathcal{G}(z|z_2; k_x, \omega)}{\partial z'}. \tag{2.35}$$

That is, knowledge of the normal derivative of P on $\mathcal{S}_1$ and $\mathcal{S}_2$ is not required by this formula. This expression still involves data from two surfaces, while the simple downward continuation pictured in Figure 2.14 requires knowledge of the data only at $z = 0$. Equation (2.35) requires data on both surfaces because it is written for a general solution (2.33) of (1.3.30), whereas Figure 2.14 applies to upgoing waves only.

To focus on upgoing waves, let us try an anticausal Green's function satisfying the Dirichlet boundary condition on $\mathcal{S}_1$, i.e.,

$$\mathcal{G}(z|z'; k_x, \omega) = \frac{e^{-ik_z|z-z'|}}{2ik_z} - \frac{e^{-ik_z|z+z'|}}{2ik_z}. \tag{2.36}$$

At $z' = 0$, $\mathcal{G} = 0$, and (since $z > 0$ within $\mathcal{V}$)

$$\frac{\partial \mathcal{G}(z|0; k_x, \omega)}{\partial z'} = e^{-ik_z z}. \tag{2.37}$$

On the lower boundary,

$$\mathcal{G}(z|z_2; k_x, \omega) = +\frac{e^{-ik_z(z_2-z)}}{2ik_z} - \frac{e^{-ik_z(z_2+z)}}{2ik_z}, \tag{2.38}$$

and

$$\frac{\partial \mathcal{G}(z|z_2, k_x, \omega)}{\partial z'} = -ik_z \mathcal{G}(z|z_2, k_x, \omega). \tag{2.39}$$

Thus, the expression (2.29) for P within $\mathcal{V}$ becomes

$$P(k_x, z, \omega) = P(k_x, 0, \omega)\, e^{-ik_z z} + \left(P(k_x, z_2, \omega)\, ik_z + \frac{\partial P(k_x, z_2, \omega)}{\partial z'} \right) \mathcal{G}(z|z_2, k_x, \omega). \tag{2.40}$$

For an upgoing wave,

$$\frac{\partial P(k_x, z_2, \omega)}{\partial z'} = -ik_z P(k_x, z_2, \omega), \tag{2.41}$$

and the second term in (2.40) vanishes, leaving

$$P(k_x, z, \omega) = P(k_x, 0, \omega)\, e^{-ik_z z}. \tag{2.42}$$

Thus, an upgoing wave can be continued using information only from the upper surface. For a downgoing wave, the lower surface contributes to the result. For an upgoing wave, the anticausal Green's function removes the contribution of the lower surface from the Green's theorem expression. This makes sense, in that the anticausal Green's function extrapolates backward in time, which, for an upgoing wave, means downward in depth. Similarly, a downgoing wave can be extrapolated from the upper surface only, provided the causal Green's function is used. These observations extend to the case of a variable velocity, to the extent that up- and downgoing waves remain well defined.

2.4 Reverse time migration

Perhaps the most straightforward (and expensive) downward continuation method is *reverse time* migration (Whitmore, 1983; Baysal *et al.*, 1983; Zhang & Zhang, 2009), which forms a difference equation from the full, two-way wave equation to propagate data backward in time and downward in depth to form the $t = 0$ image at depth z. Consulting Figure 2.14, we expect the data to vary rapidly in both time and depth, so a reverse time algorithm will require dense sampling and high accuracy in both z and t.

In two dimensions, invoking the exploding-reflector model, seismic wave propagation is assumed to be governed by the scalar wave equation

$$\left(\frac{\partial^2}{\partial x^2} + \frac{\partial^2}{\partial z^2} - \frac{4}{c(x,z)^2}\frac{\partial^2}{\partial t^2}\right) P(x,z,t) = 0. \tag{2.43}$$

To perform reverse time migration, we convert the wave equation into a difference equation. The simplest way to do this is to replace the second derivatives with second difference operators, leading to the reverse time extrapolation equation (actually, it can go backward or forward)

$$\begin{aligned} P(i_x, i_z, i_t - 1) = {} & \left[P(i_x, i_z + 1, i_t) + P(i_x, i_z - 1, i_t)\right] a_z \\ & + \left[P(i_x + 1, i_z, i_t) + P(i_x - 1, i_z, i_t)\right] a_x \\ & + P(i_x, i_z, i_t)\, 2\,(1 - a_x - a_z) - P(i_x, i_z, i_t + 1), \end{aligned} \tag{2.44}$$

where

$$a_z = a_z(i_x, i_z) = \left(\frac{c(i_x, i_z)\,\Delta t}{\Delta z}\right)^2, \tag{2.45}$$

and

$$a_x = a_x(i_x, i_z) = \left(\frac{c(i_x, i_z)\,\Delta t}{\Delta x}\right)^2. \tag{2.46}$$

For stability, the factor $1 - a_x - a_z$ must be negative, which requires

$$\sqrt{\frac{1}{\Delta x^2} + \frac{1}{\Delta z^2}} < \frac{1}{c_{\max}\Delta t}. \tag{2.47}$$

To perform a reverse time migration, we set the wavefield $P(i_x, 0, i_t)$ at zero depth to the data $D(i_x, i_t)$. We begin the algorithm at the largest time (say, n_t) in the data and work back to earlier times, stopping when we reach $i_t = 0$. $P(i_x, i_z, 0)$ becomes the migrated image.

Actually, more initial and boundary data are needed. The algorithm asks for the wavefield at the boundaries $i_z = -1$ and at $i_z = n_z + 1$ (or equivalently, the normal derivative of the data at the upper and lower boundaries). It also wants the times at $i_t = n_t + 1$ and at $n_t + 2$, and, if 1 and n_x are the data limits in the x-direction, at the boundaries $i_x = 0$ and $i_x = n_x + 1$. The simplest thing to do about the missing boundary and initial data is to set it all to zero. The actual values, had we measured them, were certainly non-zero, but the information they carried is lost to us. However, setting the boundary values to zero will produce artificial reflections at the boundaries, which are likely to contaminate the results. Such reflections can be reduced by employing "non-reflecting" boundary conditions, which pass outgoing waves but inhibit incoming waves.

As the algorithm works backward in time, upward-propagating waves work their way down into the Earth, while downward-propagating waves migrate to shallower depths. If we can assume that the wavefield at zero depth is purely upgoing, it initially continues downward. There are, however, opportunities for waves to change direction, and the reverse time algorithm can follow such movements. Thus, a reverse-time extrapolation may not always be strictly equivalent to downward continuation.

A simple example is shown in Figures 2.18–20. A faulted boundary, smoothed with a Ricker wavelet (Appendix E, Section E.3), is imbedded (see Figure 2.18) in a medium where velocity varies vertically as $c(z) = 5000 + 1.875z$ feet per second. The velocity gradient has been chosen large enough to produce turning waves in the resulting data. A synthetic data set generated from this model by running the extrapolator forward in time is shown in Figure 2.19. A reflecting boundary condition was employed at $z = 0$. This produces downward traveling waves which reflect from the bottom of the depth window and would eventually interfere with the primary reflection data, had we not chosen the time window to avoid it. Some phase is apparent in the data, likely owing to the simplistic boundary conditions imposed. Reverse time migration of this data is shown in Figure 2.20.

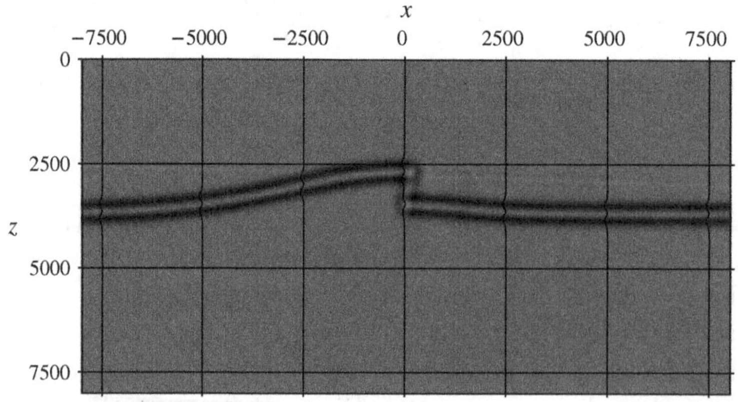

Figure 2.18 Model of a faulted boundary, smoothed with a Ricker wavelet. Depths are in feet.

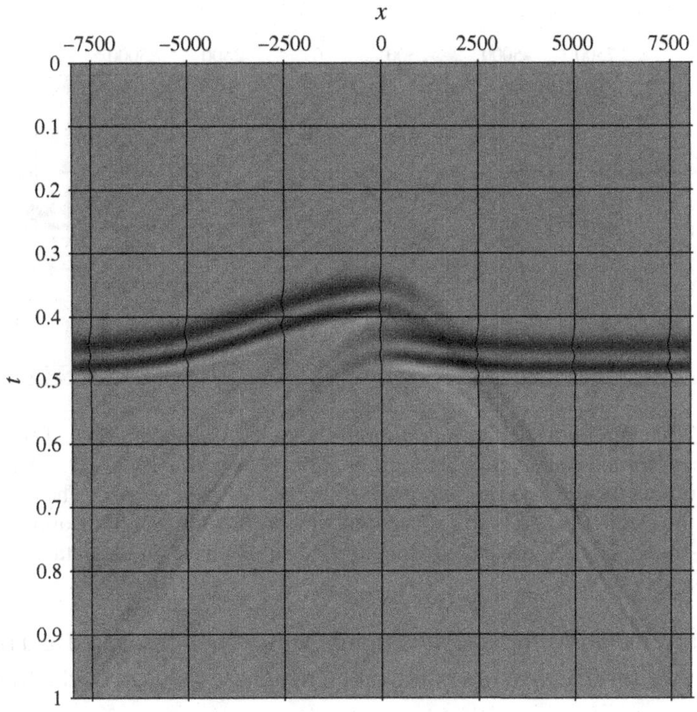

Figure 2.19 Synthetic seismic section for the faulted boundary model, made by running the reverse-time extrapolation equation (2.44) forward in time. The flat portions of the reflecting boundary do not appear to be exactly zero phase, suggesting some phase has been induced by the imposed boundary conditions. Distances are in feet and time in seconds.

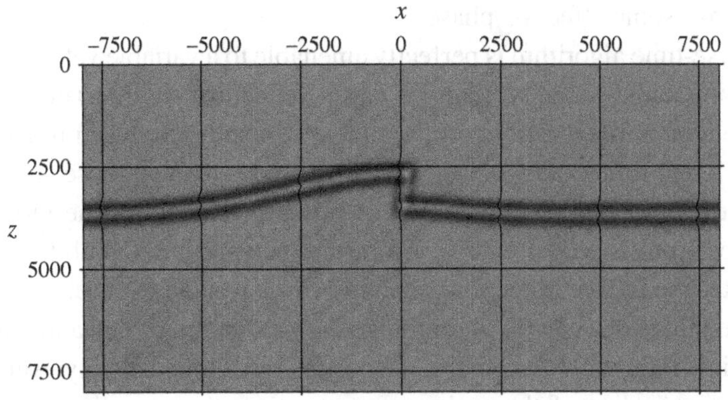

Figure 2.20 Reverse time migration of the faulted boundary data. The structure of the boundary, including the steeply dipping fault, is well recovered. The boundary image is zero phase – the phase induced in the image by the forward time operator has been removed by the reverse time operator, not too surprising in that the two operators used the same boundary conditions.

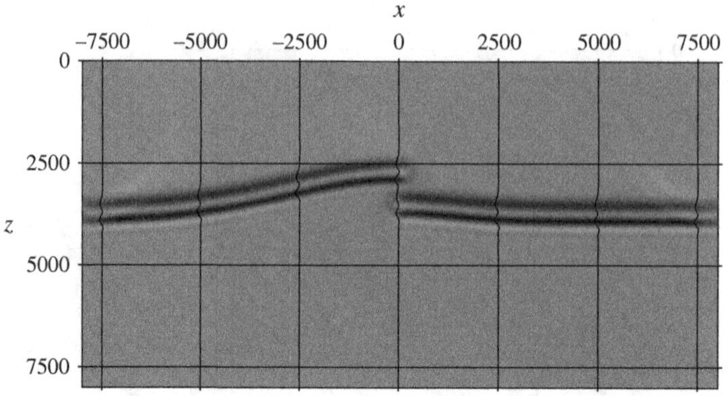

Figure 2.21 Kirchhoff depth migration of the faulted boundary data. The boundary, except for the steeply dipping fault, is well recovered. Although the Kirchhoff migration can handle large angles, it stops short of 90° and beyond. The phase visible on the synthetic data is still present on the migrated image, indicating that the reverse-time modeling algorithm has imposed some phase on the data.

The boundary, including the steeply dipping fault, is well recovered and phase-free. Apparently, the reverse time operation undid the phase induced by the forward time operation.

For comparison, Figure 2.21 shows a Kirchhoff migration of the same data. The Kirchhoff migration is able to recover the reflecting boundary except for the steeply dipping fault, which cannot be imaged without turning waves. The Kirchhoff migration retains the phase seen in the synthetic data, again suggesting that the boundary conditions employed by the forward-time modeling and reverse-time migration have some effect on phase.

The reverse-time algorithm is perfectly amenable to a variable velocity – perhaps too much so. Rapid velocity changes cause reflections. If the velocity used to migrate the data is not exactly correct, these artificially produced reflections will occur at different times and with different amplitudes than do those actually in the data. Moreover, for ordinary migration of primary reflections, the incident wave must be free from reflected energy. From this perspective, the ability of reverse time migration to induce reflections can only cause trouble for itself.

There are things one can do to minimize the production of unwanted reflections in reverse time migration. An obvious step is to smooth the velocity. Another technique is to use, in lieu of the scalar wave equation, the acoustic wave equation, which allows changes in both velocity and density. If density is constrained so that acoustic impedance is constant, then reflection strength for waves normal to a reflecting boundary is zero. Of course, waves at other angles will still produce reflections.

The oversimplified boundary conditions employed in the above example did have some effect on the results, and it is easy to see where improvements could be made. However, recent work by Weglein *et al.* (2011a, b) suggests intrinsic problems with reverse-time boundary conditions as traditionally employed. Specifically, one would like to think that the data can be accurately extrapolated backward in time from surface measurements only. Unfortunately, measurements from the sides and bottom of the propagating region may be required too. To avoid this requirement, Weglein *et al.* (2011a, b) propose a Green's theorem-based reverse-time algorithm using a special Green's function designed so that both the Green's function and its normal derivative disappear on the side and lower boundaries.

In summary, the reverse-time algorithm, though not without its problems, appears amenable to post-stack, exploding- reflector migration. The little example shown above highlights (in absentia) one of the principles of good code construction: do not test migration code only on data synthesized by the same method.

2.5 Claerbout one-way difference-equation migration

In Figure 2.14, the data can be seen to vary relatively slowly along the direction $z = ct/2$. Faster algorithms can be formed by reformulating the wave equation in a moving coordinate system, in which upward-traveling waves are well described. The resulting equation is *one-way*, in that downward-traveling waves are lost. As long as the data fit the model well, one-way difference-equation methods can accurately continue data with much less effort than reverse-time methods. Where the data do not fit the model, one-way methods can fail completely.

By their nature, one-way difference-equation methods tend to have an upper bound in the propagation angles they can faithfully describe. We illustrate this with a simple, 2-D difference equation based on a "15°" approximation to the exploding-reflector wave equation (Claerbout, 1976). Imagine a wave $P(x, z, t)$ generated by exploding reflectors at depth, traveling with velocity $c/2$, and measured at the surface $z = 0$. Between the reflector and the surface, the wave will be governed by the wave equation (2.43).

This equation varies rapidly in both space and time. For waves traveling upward in a near-vertical direction, we would like to slow the wave equation down. We can do this by choosing a new coordinate system which travels along with the wave, at least approximately. One such coordinate system is

$$x' = x, \quad z' = z, \quad t' = t + 2z/c. \tag{2.48}$$

In these coordinates, the wave equation (2.43) becomes

$$\left(\frac{\partial^2}{\partial x'^2} + \frac{\partial^2}{\partial z'^2} + \frac{4}{c}\frac{\partial^2}{\partial t' \partial z'}\right) P(x', z', t') = 0. \tag{2.49}$$

Provided the wave is changing slowly with depth in the new coordinates, the second derivative of P with depth should be small relative to the other terms. Discarding it, we are left with the approximate wave equation

$$\left(\frac{\partial^2}{\partial x'^2} + \frac{4}{c}\frac{\partial^2}{\partial t' \partial z'}\right) P(x', z', t') \simeq 0. \tag{2.50}$$

To see how good this approximation is, suppose P is an upward-traveling plane wave of temporal frequency ω:

$$P(x, z, t) = e^{-i(\omega t - k_z z - k_x x)}, \tag{2.51}$$

or, in terms of the new coordinates,

$$P(x', z', t') = e^{-i\left(\omega t' - (2\omega/c + k_z)\, z' - k_x x'\right)}. \tag{2.52}$$

Evaluating the approximate wave equation (2.50) for the plane wave (2.52), we find that

$$k_z \simeq -\frac{2\omega}{c} + \frac{{k_x}^2 c}{4\omega}. \tag{2.53}$$

Compared to the dispersion relation for the exact equation (2.43),

$$k_z = -\frac{2\omega}{c}\sqrt{1 - \frac{{k_x}^2 c^2}{4\omega^2}}, \tag{2.54}$$

the approximate form (2.53) appears to be the first two terms in an expansion of the square root:

$$k_z = -\frac{2\omega}{c}\left(1 - \frac{{k_x}^2 c^2}{8\omega^2} - \frac{{k_x}^4 c^4}{128\omega^4} + \cdots\right). \tag{2.55}$$

The direction of propagation φ, relative to the vertical, is defined by the ratio k_x/k_z:

$$\tan\varphi = -k_x/k_z, \tag{2.56}$$

from which it follows

$$\sin\varphi = \frac{k_x c}{2\omega}, \quad \cos\varphi = -\frac{k_z c}{2\omega}. \tag{2.57}$$

Equation (2.53) amounts to approximating $1 - \cos\varphi$ with $0.5\sin^2\varphi$. For directions near the vertical, the approximation is pretty good, 98% accurate at 15°. Past 15°, it starts to deteriorate.

We wish to use (2.50) to extrapolate the data to deeper depths and smaller times, starting from $z' = 0$, $t' > 0$, and ending at $z' > 0$, $t' = 2z'/c$. To make a difference equation out of (2.50), start by discretizing the depth and time coordinates. Postpone discretization in x for now:

$$z' \to z_j = j\Delta z, t' = t_k = k\Delta t, P(x', z', t') \to P_{jk}(x'). \tag{2.58}$$

2.5.1 Implicit and explicit continuation formulas

In the simplest case, we make from (2.50) a recursive formula for $P_{j+1,k}$ given $P_{j,k}$, $P_{j,k+1}$, and $P_{j+1,k+1}$. A simple discrete approximation to the time–depth derivative term is

$$\begin{aligned} &\frac{4}{c}\frac{\partial^2}{\partial t' \partial z'} P(x', \Delta z(j+0.5), \Delta t(k+0.5)) \\ &\simeq \frac{4}{c\Delta z \Delta t}\left(P_{j+1k+1} - P_{j+1k} - P_{jk+1} + P_{jk}\right). \end{aligned} \tag{2.59}$$

To maintain consistency, the x-derivative term must also be evaluated at $j+0.5$, $k+0.5$. A straightforward approximation of the data at this point is

$$P(x', \Delta z(j+0.5), \Delta t(k+0.5)) \simeq \frac{1}{4}\left(P_{j+1k+1} + P_{j+1k} + P_{jk+1} + P_{jk}\right). \tag{2.60}$$

Thus, (2.50) becomes

$$\frac{\partial^2}{\partial x'^2}\left(P_{j+1k+1}+P_{j+1k}+P_{jk+1}+P_{jk}\right) + \frac{16}{c\Delta z \Delta t}\left(P_{j+1k+1}-P_{j+1k}-P_{jk+1} + P_{jk}\right) \simeq 0, \tag{2.61}$$

or, setting

$$T_{xx} \equiv -\frac{c\Delta z \Delta t}{8}\frac{\partial^2}{\partial x'^2}, \tag{2.62}$$

we have

$$(1 + T_{xx}/2)\, P_{j+1k} = (1 - T_{xx}/2)\left(P_{j+1k+1} + P_{jk}\right) - (1 + T_{xx}/2)\, P_{jk+1}. \tag{2.63}$$

This is a formula for P_{j+1k} in terms of the data at smaller depths and later times, as depicted diagrammatically in Figure 2.22. It is *implicit* in the sense that instead of solving directly for P_{j+1k}, it solves for a differential operator (in x) times P_{j+1k}. To complete the solution, this operator must be inverted. If the second x-derivative is approximated by a difference operator, then a banded matrix must be inverted to find P_{j+1k}.

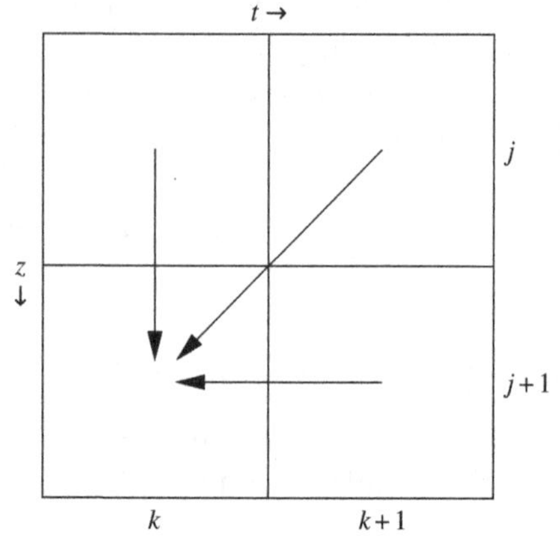

Figure 2.22 Flow diagram in depth and time for the 15° finite difference downward continuation operator. The operation in the third dimension x is not depicted here. As the calculated field proceeds downward, the operation works backward in time. This is consistent with the notion that migration moves data up dip.

It is possible to write an explicit formula for P_{j+1k}, which circumvents the matrix inversion. Instead of (2.60), approximate the data at $j + 0.5$, $k + 0.5$ by

$$P(x', \Delta z(j + 0.5), \Delta t(k + 0.5)) \simeq \frac{1}{2}\left(P_{j+1k+1} + P_{jk}\right) . \tag{2.64}$$

Then, (2.50) becomes

$$\frac{\partial^2}{\partial x'^2}\left(P_{j+1k+1} + P_{jk}\right) + \frac{8}{c\Delta z\Delta t}\left(P_{j+1k+1} - P_{j+1k} - P_{jk+1} + P_{jk}\right) \simeq 0, \tag{2.65}$$

or

$$P_{j+1k} = (1 - T_{xx})\left(P_{j+1k+1} + P_{jk}\right) - P_{jk+1}. \tag{2.66}$$

No matrix inversion is required in the explicit formula.

2.5.2 Stability

We can get some feel for the accuracy and stability of (2.63) and (2.66) by attempting a plane-wave solution. Substituting the waveform (2.52) into the implicit formula (2.63), we obtain

$$\begin{aligned}\left(1 + T_{xx}/\,2\right) e^{i\left(\hat{k}_z+2\omega/c\right)\Delta z} &= \left(1 - T_{xx}/\,2\right)\left(e^{i\left(\hat{k}_z+2\omega/c\right)\Delta z} e^{-i\omega\Delta t} + 1\right)\\ &\quad - \left(1 + T_{xx}/\,2\right) e^{-i\omega\Delta t},\end{aligned} \tag{2.67}$$

with

$$T_{xx} = \frac{c\Delta z \Delta t}{8} {k_x}^2. \tag{2.68}$$

In (2.67), we have placed a "hat" over the vertical wavenumber k_z to indicate that it is a numerically approximate form satisfying (2.67) and differing from the exact form (2.54) or the 15° form (2.53).

Solving for $\hat{k}_z$,

$$e^{i\hat{k}_z \Delta z} = \frac{(1 - T_{xx}/2) - (1 + T_{xx}/2)\, e^{-i\omega\Delta t}}{(1 + T_{xx}/2) - (1 - T_{xx}/2)\, e^{-i\omega\Delta t}} e^{-i2\omega\Delta z/c}. \tag{2.69}$$

The first question to ask about this expression is whether it defines a real k_z, since a $k_z$with a negative imaginary component will certainly be unstable. Since T_{xx} is real, the magnitude of the numerator and denominator in (2.67) are the same, hence the expression is a pure phase and k_z is real. The implicit formula would appear to be unconditionally stable.

For the explicit form (2.66), the analog to (2.69) is

$$e^{i\hat{k}_z \Delta z} = \frac{1 - T_{xx} - e^{-i\omega\Delta t}}{1 - (1 - T_{xx})\, e^{-i\omega\Delta t}} e^{-i2\omega\Delta z/c}. \tag{2.70}$$

The magnitudes of the numerator and denominator again appear equal. However, for the explicit formula there is an opportunity for the denominator and numerator to be zero. This can happen at Nyquist frequency, where $e^{i\omega\Delta t} = -1$, at any point where $T_{xx} = +2$. To assure stability for the explicit formula, the magnitude of T_{xx} must be kept < 2. If k_x is bounded by a Nyquist wavenumber $k_{\max} = \pi/\Delta x$, then

$$|T_{xx}| < \frac{c\Delta z \Delta t}{8} {k_{\max}}^2. \tag{2.71}$$

Typically, Δt and $k_{\max}$ are determined by the data. To ensure stability, the depth step Δz for the explicit formula must be chosen

$$\Delta z < \frac{16}{c\Delta t {k_{\max}}^2}. \tag{2.72}$$

2.5.3 Accuracy

Rewrite the implicit form (2.69) as

$$e^{i\gamma_{\text{impl}}} = \frac{(1 + T_{xx}/2) - (1 - T_{xx}/2)\, e^{-i\omega\Delta t}}{(1 - T_{xx}/2) - (1 + T_{xx}/2)\, e^{-i\omega\Delta t}}, \tag{2.73}$$

where

$$\hat{\gamma}_{\text{impl}} \equiv \Delta z \left(\frac{2\omega}{c} + \hat{k}_z \right) \tag{2.74}$$

is an approximation to the phase

$$\gamma \equiv \Delta z \left(\frac{2\omega}{c} + k_z \right) = \frac{2\omega \Delta z}{c}(1 - \cos\varphi), \tag{2.75}$$

or to its 15° approximation

$$\gamma_{15} \equiv \frac{\omega \Delta z}{c} \sin^2 \varphi. \tag{2.76}$$

These phases as a function of angle are shown in Figure 2.23a and b.

It is not obvious that (2.73), in its current form, yields a good approximation to γ. We can fix that by solving (2.73) for T_{xx}:

$$T_{xx} = -2 \tan \frac{\hat{\gamma}_{\text{impl}}}{2} \tan \frac{\omega \Delta t}{2}, \tag{2.77}$$

and in turn solving this equation for $\hat{\gamma}_{\text{impl}}$:

$$\hat{\gamma}_{\text{impl}} = -2 \arctan \left(\frac{T_{xx}}{2 \tan \frac{\omega \Delta t}{2}} \right). \tag{2.78}$$

Accuracy will depend in part on how the second x-derivative is implemented. Since our algorithm is based on a 15° approximation to the wave equation, we would not expect to gain a great deal with a high-order approximation to the x-derivative. Consequently, we choose to approximate the second derivative with a second difference operator. Assume that $x' \to x_i = i\Delta x$. Then

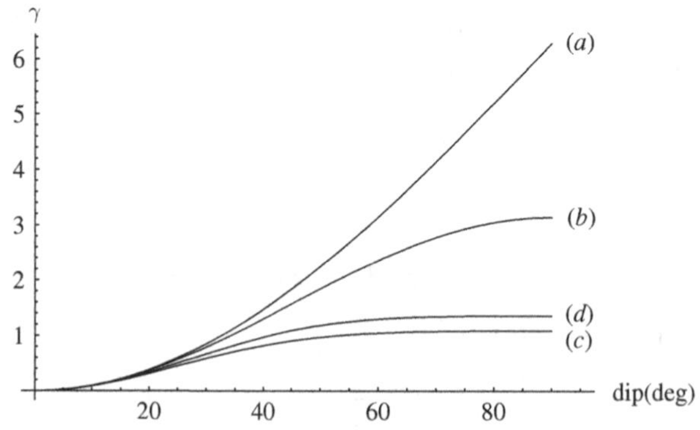

Figure 2.23 Plots of (a) the phase γ, as defined in equation (2.75), (b) its 15° approximation (2.76), (c) the implicit approximation (2.83), and (d) the explicit approximation (2.86). In this example, we have chosen $\Delta t = 0.004$ s, $c = 10\,000$ fps, $\Delta x = 80'$, $\Delta z = 160'$, and $\omega \Delta t = \pi/4$.

$$\frac{\partial^2}{\partial x'^2} P(x', z', t') \to \Delta_{xx} P(x', z', t')$$
$$= \frac{P\left(x_{i+1}, z', t'\right) + P\left(x_{i-1}, z', t'\right) - 2P\left(x_i, z', t'\right)}{\Delta x^2}. \tag{2.79}$$

This becomes for a plane wave of wavenumber k_x

$$\Delta_{xx} P = -\frac{4}{\Delta x^2} \sin^2 \frac{k_x \Delta x}{2} P. \tag{2.80}$$

With the second difference approximation in x, operator T_{xx} in (2.6.2) becomes

$$T_{xx} \equiv \frac{c \Delta z \Delta t}{2 \Delta x^2} \sin^2 \left(\frac{k_x \Delta x}{2} \right), \tag{2.81}$$

or, in terms of propagation angle φ,

$$T_{xx} \equiv \frac{c \Delta z \Delta t}{2 \Delta x^2} \sin^2 \left(\frac{\omega \Delta x \sin \varphi}{c} \right). \tag{2.82}$$

Thus, equation (2.78) becomes

$$\hat{\gamma}_{\text{impl}} = 2 \arctan \left(\frac{c \Delta z \Delta t \sin^2 \left(\frac{\omega \Delta x \sin \varphi}{c} \right)}{4 \Delta x^2 \tan \frac{\omega \Delta t}{2}} \right). \tag{2.83}$$

In the limit of small ω, this expression reduces to

$$\hat{\gamma}_{\text{impl}} \to \frac{2\omega}{c} \Delta z \frac{\sin^2 \varphi}{2}, \tag{2.84}$$

which is the 15° approximation to γ. At finite frequencies, $\hat{\gamma}$ differs from the 15° approximation as seen in Figure 2.23c.

For the explicit algorithm, we have

$$e^{i \hat{\gamma}_{\text{expl}}} = \frac{1 - (1 - T_{xx}) e^{-i\omega \Delta t}}{1 - T_{xx} - e^{-i\omega \Delta t}}, \tag{2.85}$$

leading to

$$\hat{\gamma}_{\text{expl}} = \arctan \left(\frac{- \sin(\omega \Delta t) T_{xx} \left(1 - T_{xx}/2\right)}{(1 - T_{xx}) - \left(1 - T_{xx} + T_{xx}{}^2/2\right) \cos(\omega \Delta t)} \right). \tag{2.86}$$

With equation (2.82) for T_{xx}, $\hat{\gamma}_{\text{expl}}$ reduces in the limit of small ω to the 15° approximation (2.76). At finite frequencies, the phase differs from the 15° approximation as shown in Figure 2.23d.

The magnitude of the second difference approximation (2.80) is bounded by $4/\Delta x^2$, a slightly smaller number than the $\pi^2/\Delta x^2$ for the exact second derivative.

This allows the explicit formula to remain stable for slightly larger depth steps ($k_{\max} = 2/\Delta x$ versus $k_{\max} = \pi/\Delta x$) than would otherwise be the case.

In practice, it is often possible to adjust the coefficients of the difference operator to increase its accuracy for a target range of frequencies and dips, using, in effect, one source of error to compensate for another. When making such adjustments, one should keep in mind that it is really the exact phase γ one is trying to emulate, not the 15° approximation.

2.5.4 Higher-order approximations

Higher-order approximations to the wave equation are easy enough to make, though of course more expensive to implement than the simple 15° formulas shown here. Generally, finite difference operators will have by their nature some limitations in dip, frequency, and wavenumber.

2.6 Frequency–wavenumber migration methods

Where velocity can be considered constant, Fourier methods may be used. Invoking the exploding-reflector model, we start with the data $D(x,t) = P(x, 0, t)$ *at* $z = 0$, then Fourier transform over *x* and *t*, as per Appendix C:

$$P\left(k_x, 0, \omega\right) = D\left(k_x, \omega\right) = \int dx \int dt D(x, t) e^{-i(k_x x - \omega t)}. \tag{2.87}$$

This transformation decomposes the data into plane-wave components, with k_x the horizontal component of wavenumber and ω the angular frequency. We refer to this decomposition as transforming to the *f–k* or *frequency–wavenumber domain.* Assuming that the wavefield $P(x, z, t)$ obeys the exploding-reflector wave equation (2.43), each Fourier component $P\left(k_x, z, \omega\right)$ of the data satisfies a 1-D wave equation

$$\left(\frac{d^2}{dz^2} + \frac{4\omega^2}{c^2} - {k_x}^2\right) P\left(k_x, z, \omega\right) = 0. \tag{2.88}$$

For a plane wave, the corresponding vertical component of wavenumber is

$$k_z = k_z\left(\omega, k_x\right) = \mp\frac{2\omega}{c}\sqrt{1 - \frac{{k_x}^2 c^2}{4\omega^2}}, \tag{2.89}$$

with the minus and plus signs pertaining to upgoing and downgoing waves, respectively.

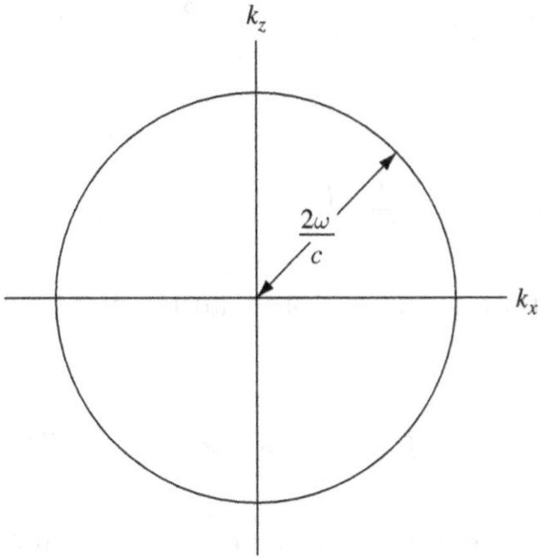

Figure 2.24 Wavenumber plot. For a plane wave physically propagating at frequency ω, the wavenumbers k_x and k_z must lie on a circle of radius $2\omega/c$.

The wavenumber $k_z(\omega, k_x)$ is real, provided the magnitude of horizontal wavenumber $|k_x| < 2|\omega|/c$. In this region, the data function at depth propagates as a plane wave. Outside this region, k_z is imaginary, and the wave damps exponentially in the direction of propagation.

For a physically propagating wave at a given frequency ω, k_x and k_z lie on the circle ${k_x}^2 + {k_z}^2 = 4\omega^2/c^2$, as depicted in Figure 2.24.

In the physically propagating region, since we are dealing with an upward-traveling plane wave, downward continuation of the data in the $k_x - \omega$ domain is very simple:

$$P(k_x, z, \omega) = P(k_x, 0, \omega)\, e^{ik_z(\omega, k_x)z}. \tag{2.90}$$

Since the wave propagates upward, the sign of k_z is opposite to that of ω. In the non-physically propagating or *evanescent* region, the downward-continued data increases exponentially in magnitude with depth. The practical effect of this is to amplify noise, so it is generally best to exclude evanescent data from the migration process.

The exploding-reflector model assumes that sources located on the reflectors fire at time zero. Equation (2.90) is technically inconsistent with this model, since it is a solution to the source-free wave equation (2.88). However, for positive times, the difference is immaterial. We desire only to extrapolate the wave back to zero time, for which task the source-free wave equation is appropriate.

Migration having been defined as the time-zero component of the downward-continued data, we have

$$R\left(k_x, z\right) = \frac{1}{2\pi} \int d\omega P\left(k_x, z, \omega\right), \tag{2.91}$$

or, using (2.90) and (2.87),

$$R\left(k_x, z\right) = \frac{1}{2\pi} \int d\omega\, D\left(k_x, \omega\right) e^{ik_z(\omega, k_x)z}. \tag{2.92}$$

To complete the operation, one performs an inverse Fourier transform from k_x back to x:

$$R(x, z) = \frac{1}{(2\pi)^2} \int dk_x \int d\omega\, D\left(k_x, \omega\right) e^{i(k_x x + k_z(\omega, k_x)z)}. \tag{2.93}$$

If the data $D\left(k_x, \omega\right)$ has a maximum frequency ω_f, then exclusion of the evanescent data results in some wavenumbers missing from the image (2.93). Specifically, only wavenumbers satisfying

$$k_x{}^2 + k_z{}^2 < 4\omega_f{}^2/c^2 \tag{2.94}$$

are included. As discussed more fully in Chapter 4, this limits the horizontal and vertical resolution of the image.

2.6.1 Phase-shift migration

The migration operation (2.93), if implemented directly, is called *phase-shift migration* (Gazdag, 1978), because it involves the phase shift $e^{ik_z(\omega, k_x)z}$. Phase-shift migration is generalizable to a (slowly) vertically varying velocity by replacing the phase shift with $e^{i \int_0^z k_z(\omega, k_x, z')\, dz'}$. With the addition of a depth-varying amplitude factor, the phase shift operator becomes the WKBJ approximation (for further information on the WKBJ method, see Chapman, 1978; Clayton & Stolt, 1981; Bleistein, 1984). Accommodating a horizontally varying velocity is more difficult, though there are methods to do so (e.g. PSPI, or phase-shift-plus-interpolation as per Gazdag & Sguazzero, 1984), at least approximately.

2.6.2 Frequency–wavelength migration

A further simplification of the migration formula (2.93) is possible (Stolt, 1978). The frequency integral can be converted to an inverse Fourier transform by a change of integration variable from ω to k_z:

$$R(x, z) = \frac{1}{(2\pi)^2} \int dk_x \int dk_z \left|\frac{d\omega}{dk_z}\right| D\left(k_x, \omega\left(k_x, k_z\right)\right) e^{i(k_x x + k_z z)}, \tag{2.95}$$

with

$$\omega\left(k_x, k_z\right) = -\frac{ck_z}{2}\sqrt{1 + \frac{k_x{}^2}{k_z{}^2}}. \tag{2.96}$$

The k_z-derivative of ω is

$$\frac{d\omega}{dk_z} = \frac{c^2}{4}\frac{k_z}{\omega} = -\frac{c}{2\sqrt{1 + k_x{}^2/k_z{}^2}}. \tag{2.97}$$

Thus,

$$R(x, z) = \frac{c}{2(2\pi)^2}\int dk_x \int dk_z \frac{D\left(k_x, \omega\right)}{\sqrt{1 + k_x{}^2/k_z{}^2}} e^{i(k_x x + k_z z)}. \tag{2.98}$$

The two wavenumber integrals are now inverse Fourier transforms. A forward Fourier transform of (2.98) from x, z to k_x, k_z leaves one with

$$R\left(k_x, k_z\right) = \frac{c}{2}\frac{D\left(k_x, \omega\right)}{\sqrt{1 + k_x{}^2/k_z{}^2}}. \tag{2.99}$$

That is, the double (space–time) Fourier transform of the data is proportional to the double (space–depth) Fourier transform of the image, with frequency and wavenumber related by the dispersion relation (2.96).

Formula (2.99) we refer to as *f–k migration*. Since the formula (2.99) has no pretensions of true amplitude, the factor of $c/2$ could be discarded, and debatably the factor of $|k_z/\omega|$ as well. With or without the factors, the attraction of *f–k* migration is that it reduces migration to three simple, computationally efficient steps: (1) forward-Fourier transform the data from (x, t) to (k_x, ω); (2) resample the data from ω to k_z; and (3) inverse-Fourier transform the data from (k_x, k_z) to (x, z). Within the sampling limitations of the data, there is no limit to the dip angle Fourier methods can recover.

2.6.3 Frequency–wavelength time migration

As written, the *f–k* migration algorithm converts data in the time domain into an image in the depth domain. If desired, the depth axis can be scaled in time units by defining vertical traveltime $\tau \equiv 2z/c$ and vertical traveltime frequency $\omega_M \equiv -k_z c/2$. In terms of these quantities,

$$\omega = \omega_M\sqrt{1 + c^2\, k_x{}^2 / 4\omega_M{}^2}, \tag{2.100}$$

$$R(x, \tau) = \frac{1}{(2\pi)^2}\int dk_x \int d\omega_M \left|\frac{\omega_M}{\omega}\right| D\left(k_x, \omega\right) e^{i(k_x x - \omega_M \tau)}, \tag{2.101}$$

and

$$R\left(k_x, \omega_M\right)=\frac{D\left(k_x, \omega\right)}{\sqrt{1+c^2\, {k_x}^2 / 4 {\omega_M}^2}}. \tag{2.102}$$

Equation (2.102), unlike (2.99), leaves the $k_x = 0$ component of the data completely unchanged:

$$R(0, \omega)= D(0, \omega). \tag{2.103}$$

Forming images in terms of vertical traveltime rather than depth is called *time migration*, and is discussed further in Section 2.8.

2.6.4 How *f–k* migration works

So how does *f–k* migration work? The forward transform separates the data into components of various dips. The data component $D\left(k_x, \omega\right)$ is the amplitude of a line dipping at an angle $\alpha_{\text{app}} = \arctan\left(\frac{k_x c}{2\omega}\right) = \arctan\left(\frac{pc}{2}\right)$ in the x–t plane. The corresponding image component $R\left(k_x, k_z\right)$ is the amplitude of a line dipping at angle $\alpha_{\text{true}} = \arctan\left(-\frac{k_x}{k_z}\right)$ in the x–z plane. If $\tan\left(\alpha_{\text{true}}\right) = -\frac{k_x}{k_z} = \frac{pc}{2\sqrt{1-p^2 c^2/4}}$, then $\tan\left(\alpha_{\text{app}}\right) = \frac{pc}{2} = \frac{k_x c}{2\omega} = \frac{-k_x/k_z}{\sqrt{1+{k_x}^2/{k_z}^2}} = \sin\left(\alpha_{\text{true}}\right)$. This identity is the geometrical migration formula (2.14). It would appear that *f–k* migration works by separating out the individual dip components of the data, applying the geometrical migration formula (2.14) to each component, then reassembling the components back into an image. Downward-continuation migration may have different conceptual roots than ray-theoretical or Kirchhoff migration, but it amounts to the same thing in the end.

Well, almost. The simplified zero-offset downward-continuation migration presented in this chapter does implicitly make ray-theoretical approximations, but they are not necessarily unavoidable. To the extent that downward-continuation migration can avoid the far-field high-frequency approximations inherent in ray theory, the approach has potentially broader applicability.

2.6.5 Relating *f–k* migration to Kirchhoff imaging

It is not difficult to convert the *f–k* migration formula to a Kirchhoff equivalent. Since the *f–k* migration formula (2.99) is not true amplitude, we generalize it slightly to

$$R\left(k_x, k_z\right)= \Omega\left(k_x / k_z\right) D\left(k_x, \omega\right), \tag{2.104}$$

with Ω some unspecified function of image function wavenumber ratio k_x/k_z. To express this formula in the space domain, perform inverse Fourier transforms of

R over k_x and k_z, while expressing D as a forward Fourier transform of its spatial components:

$$R(x,z) = \frac{1}{4\pi^2}\int\int dk_x dk_z \Omega\left(k_x/k_z\right)\int dx_s D\left(x_i, \omega\left(k_x, k_z\right)\right) e^{i(k_x(x - x_s)+k_z z)}. \tag{2.105}$$

Change integration variable from k_z to ω, and rearrange the integrals:

$$R(x,z) = -\frac{1}{\pi^2 c^2}\int dx_s \int d\omega D\left(x_s, \omega\right) \times \int dk_x \frac{\omega\Omega\left(k_x/k_z\left(\omega, k_x\right)\right)}{k_z\left(\omega, k_x\right)} e^{i(k_x(x - x_s)+k_z(\omega,k_x)z)}, \tag{2.106}$$

with

$$k_z(\omega\,, k_x) = -S(\omega)\sqrt{\frac{4\omega^2}{c^2} - {k_x}^2}. \tag{2.107}$$

The k_x integral has the phase-integral form shown in equation (F.48) in Appendix F and the stationary-phase solution (F.56), or

$$\int dk_x \frac{\omega\Omega\left(k_x/k_z\left(\omega, k_x\right)\right)}{k_z\left(\omega, k_x\right)} e^{i\left(k_x(x - x_s)-S(\omega)z\sqrt{\frac{4\omega^2}{c^2}-{k_x}^2}\right)} \simeq z\frac{\Omega\left(\underset{\wedge}{k_x}/\underset{\wedge}{k_z}\left(\omega, k_x\right)\right)}{\underset{\wedge}{k_z}\left(\omega, k_x\right)}\sqrt{\frac{32\pi i\omega^3}{c^4\tau^3}} e^{-i\omega\tau}, \tag{2.108}$$

where

$$\tau = \tau\left(x - x_s, z\right) = \frac{2}{c}\sqrt{z^2 + \left(x - x_s\right)^2}, \tag{2.109}$$

$$\underset{\wedge}{k_x} = -\frac{4\omega}{c^2}\frac{x - x_s}{\tau}, \tag{2.110}$$

and

$$\underset{\wedge}{k_z} = k_z\left(\omega, \underset{\wedge}{k_x}\right) = -\frac{4\omega}{c^2}\frac{z}{\tau}. \tag{2.111}$$

Substituting these last two expressions,

$$\int dk_x \frac{\omega\Omega\left(k_x/k_z\left(\omega, k_x\right)\right)}{k_z\left(\omega, k_x\right)} e^{i\left(k_x(x - x_s)-S(\omega)z\sqrt{\frac{4\omega^2}{c^2}-{k_x}^2}\right)} \simeq -\Omega\left(\frac{x - x_s}{z}\right)\sqrt{\frac{2\pi i\omega}{\tau}} e^{-i\omega\tau}, \tag{2.112}$$

and $R(x, z)$,

$$R(x,z) \simeq \frac{4}{\sqrt{2\pi}c^2}\int \frac{dx_s}{\sqrt{\tau}}\Omega\left(\frac{x-x_s}{z}\right)\frac{1}{2\pi}\int d\omega\sqrt{i\omega}D\left(x_s,\omega\right)e^{-i\omega\tau}. \quad (2.113)$$

The frequency integral is the inverse temporal Fourier transform of the data $D\left(x_i, \omega\right)$ multiplied by the half-derivative operator $\sqrt{i\omega}$, described in Appendix E. The half-derivative filter corrects for the phase generated by a Kirchhoff migration, as discussed in Section 2.2.5. In the time domain, the filter can be written as a convolution:

$$R(x,z) \simeq \int dx_s A\left(x,z,x_s\right)\left[d_{h-} * D\right]\left(x_s, \tau\left(x-x_s, z\right)\right), \quad (2.114)$$

with

$$A\left(x,z,x_s\right) = \frac{4}{c^2\sqrt{2\pi\tau}}\Omega\left(\frac{x-x_s}{z}\right). \quad (2.115)$$

Apart from the convolution, the formula (2.114) sums the data along the Kirchhoff migration path (2.109). This summation is amplitude-weighted by the factor $A(x, z, x_s)$. The *f–k* formula is asymptotically equivalent to a phase-corrected amplitude-weighted Kirchhoff migration. If we use in the *f–k* migration the amplitude factor in equation (2.99), or

$$\Omega\left(k_x/k_z\right) \rightarrow \frac{c}{2}\frac{1}{\sqrt{1+k_x{}^2/k_z{}^2}}, \quad (2.116)$$

then the space–time amplitude factor is

$$A\left(x,z,x_s\right) \rightarrow \frac{4z}{c^2\sqrt{2\pi\tau^3}}. \quad (2.117)$$

2.6.6 Frequency–wavelength data synthesis and its relation to Kirchhoff modeling

The *f–k* migration formula can be used to synthesize data from a reflectivity model. We need only solve (2.104) for D:

$$D\left(k_x,\omega\right) = R\left(k_x,k_z\right)/\,\Omega\left(k_x/k_z\right). \quad (2.118)$$

This operation is the inverse of a migration operation. It can be converted into a space–time inverse migration formula using the line of reasoning employed in the last section. Start with

$$D\left(x_s,t\right) = \frac{1}{4\pi^2}\int dk_x\int d\omega\Omega^{-1}\left(k_x/k_z\right)\int dx R\left(x,k_z\right)e^{-i\left(k_x\left(x-x_s\right)+\omega t\right)}. \quad (2.119)$$

With a change of integration variable from ω to k_z,

$$D(x_s, t) = \frac{c}{8\pi^2} \int dx \int dk_z R(x, k_z) \int dk_x \frac{e^{-i(k_x(x - x_s)+\omega t)}}{\sqrt{1 + k_x{}^2/k_z{}^2}\,\Omega(k_x, k_z)}. \quad (2.120)$$

The phase function in this case is $f(k_x) = -k_x(x - x_s) - \omega t$, where

$$\omega = -S(k_z)\frac{c}{2}\sqrt{k_x{}^2 + k_z{}^2}. \quad (2.121)$$

The stationary phase approximation (F.56) becomes in this case

$$\int dk_x \frac{e^{-i\left(k_x(x - x_s) - S(k_z)\frac{ct}{2}\sqrt{k_x{}^2+k_z{}^2}\right)}}{\sqrt{1 + k_x{}^2/k_z{}^2}\,\Omega(k_x/k_z)} \simeq \frac{e^{ik_z\zeta}}{\Omega}\left(\frac{x - x_s}{\zeta}\right)\sqrt{\frac{2\pi i k_z}{\zeta}},$$

where

$$\zeta = \zeta(x - x_s, t) = \sqrt{\left(\frac{ct}{2}\right)^2 - (x - x_s)^2}. \quad (2.122)$$

With the stationary phase approximation, the space–time inverse-migration expression becomes

$$\begin{aligned} D(x_s, t) &= \frac{c}{2}\int \frac{dx}{\sqrt{2\pi\zeta}} \frac{1}{\Omega\left(\frac{x-x_s}{\zeta}\right)} \frac{1}{2\pi} \int dk_z R(x, k_z)\sqrt{ik_z}e^{ik_z\zeta} \\ &= \int dx B(x, x_s, t)\left[d_{h+} * R\right](x, \zeta), \end{aligned} \quad (2.123)$$

with

$$B(x, x_s, t) = \frac{c}{2\sqrt{2\pi\zeta}}\Omega^{-1}\left(\frac{x - x_s}{\zeta}\right). \quad (2.124)$$

If we were to choose (2.116) for Ω, then

$$B(x, x_s, t) \to \frac{ct}{2\sqrt{2\pi\zeta^3}}. \quad (2.125)$$

The forward and inverse migration formulas follow the same space–time trajectories, though in opposite directions. As discussed in Section 2.2, integral-equation space–time migration can be implemented as either a gather or a scatter operation. In gather mode, to form a migrated image function R at a point (x, z), one weights and sums data D from each surface position x_s along the traveltime trajectory $t = \tau(x - x_s, z)$ (see Figure 2.25a). In scatter mode, one weights and sums data at a specific place and time (x_s, t) into R at all points along the trajectory $(x, \zeta(x - x_s, t))$, as shown in Figure 2.25b. Inverse migration can also be implemented as a gather or a scatter operation, with the roles of τ and ζ reversed.

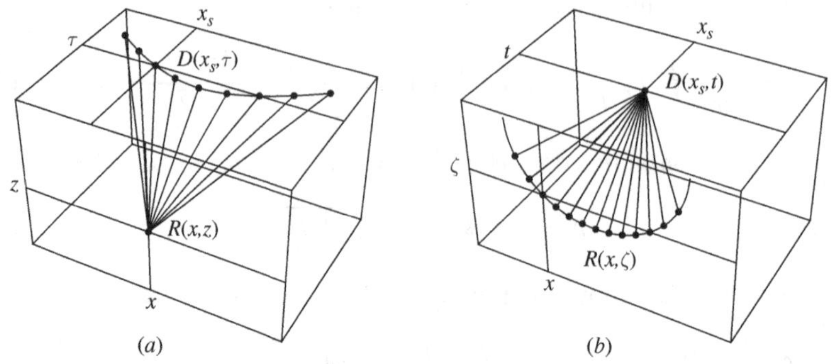

Figure 2.25 Kirchhoff migration. Kirchhoff migration can be implemented as a gather operation, as illustrated in panel (a), collecting data along the trajectory $(x_s, \tau(x - x_s, z))$ and summing into $R(x, z)$, or as a scatter operation, as shown in panel (b) broadcasting data from (x_s, t) onto the path $(x, \zeta(x - x_s, t))$, (b). Inverse migration can also be implemented either way, with the roles of the two trajectory functions reversed: panel (a) now represents broadcasting the image function at $R(x, z)$ onto the data path $(x_s, \tau(x - x_s, z))$, and panel (b) gathering R along the path $(x, \zeta(x - x_s, t))$ into D at (x_s, t).

2.7 Other methods

2.7.1 Radon-transform migration

Closely related to Fourier migration is Radon-transform or p–τ migration (Ottolini and Claerbout, 1984; Beylkin, 1985; Bisset and Durrani, 1990). Starting with a data set $D(x, t)$, perform the Radon transform (equation (C.37) in Appendix C)

$$D(p_x, \tau) = \int dx D(x, t = \tau + p_x x). \tag{2.126}$$

The Radon transform itself can be transformed to the frequency domain:

$$D(p_x, \omega) = \int d\tau D(p_x, \tau) e^{i\omega\tau}, \tag{2.127}$$

where (see equation C.42) it is equal to the double Fourier transform (2.87)

$$D(p_x, \omega) = D(k_x = \omega p_x, \omega). \tag{2.128}$$

We already know how to migrate data in f–k space. Invoking (2.93) and substituting (2.128) for $D(k_x, \omega)$,

$$R(x, z) = \frac{1}{(2\pi)^2} \int dk_x \int d\omega\, D(p_x, \omega) e^{i(k_x x + k_z z)}. \tag{2.129}$$

Changing integration variable from k_x to p_x, and noting that

$$k_z = -\frac{2\omega}{c}\sqrt{1 - \frac{{p_x}^2 c^2}{4}}, \tag{2.130}$$

the migration equation (2.129) becomes

$$R(x, z) = \frac{1}{(2\pi)^2} \int dp_x \int d\omega \rho(\omega) D(p_x, \omega)\, e^{i\omega\left(p_x x - \frac{2z}{c}\sqrt{1-\frac{{p_x}^2 c^2}{4}}\right)}, \tag{2.131}$$

with $\rho(\omega) \equiv |\omega|$ the 1-D "rho" filter. The ω-integral is just an inverse Fourier transform and a rho filter, evaluated at time $t = \frac{2z}{c}\sqrt{1 - {p_x}^2\, c^2/4} - p_x x$. This leaves

$$R(x, z) = \frac{1}{2\pi} \int dp_x \overline{D}^{(\rho)} \left(p_x, \frac{2z}{c}\sqrt{1 - \frac{{p_x}^2 c^2}{4}} - p_x x\right), \tag{2.132}$$

with $\bar{D}^{(\rho)}$ equal to the convolution $\rho * D$. This is an equation for migration of data in p–τ space. While analogous to the phase-shift migration operation (2.93), it rescales the time axis in lieu of computation of a phase shift.

An inverse Radon-transform formulation

The expression (2.132) looks a lot like an inverse Radon transform, and in fact, might be referred to as a generalized Radon transform. Computationally, it is no more difficult to perform than an inverse Radon transform. Even so, there might be some practical advantage if equation (2.132) could be converted to an inverse Radon transform, which would have the form

$$R(x, z) = \frac{1}{2\pi} \int dq_x \overline{R}^{(\rho)} (q_x, z + q_x x)\,. \tag{2.133}$$

In equation (2.132), try the change of variable from p_x to q_x, where

$$q_x = \frac{c p_x}{2\sqrt{1 - c^2\, {p_x}^2/4}}, \tag{2.134}$$

or equivalently,

$$p_x = \frac{2}{c}\frac{q_x}{\sqrt{1 + {q_x}^2}}. \tag{2.135}$$

Then (2.132) becomes

$$R(x, z) = \frac{1}{2\pi} \int dq_x \frac{2}{c}\frac{1}{\left(1 + {q_x}^2\right)^{3/2}} \overline{D}^{(\rho)} \left(\frac{2}{c}\frac{q_x}{\sqrt{1 + {q_x}^2}}, \frac{2}{c}\frac{z + q_x x}{\sqrt{1 + {q_x}^2}}\right). \tag{2.136}$$

This is the inverse Radon transform (2.133), with

$$\overline{R}^{(\rho)}(q_x, z) = \frac{2}{c}\frac{1}{\left(1+{q_x}^2\right)^{3/2}}\overline{D}^{(\rho)}\left(\frac{2}{c}\frac{q_x}{\sqrt{1+{q_x}^2}}, \frac{2}{c}\frac{z}{\sqrt{1+{q_x}^2}}\right). \tag{2.137}$$

The Radon-transform migration algorithm starts with the Radon transform $D(x,\ t) \rightarrow \bar{D}^{(\rho)}(p_x, \tau)$. Next, one rescales p_x to q_x with the formula (2.134) or (2.135), and τ to z with the formula

$$\tau = \frac{2}{c}\frac{z}{\sqrt{1+{q_x}^2}}, \tag{2.138}$$

or equivalently,

$$z = \frac{c}{2}\tau\sqrt{1+{q_x}^2}. \tag{2.139}$$

Equation (2.137) migrates the data in p–τ space, after which it may be returned to x–z space with an inverse Radon transform.

The algorithm can be described as follows. A forward Radon transform decomposes the data into dip components, with $\bar{D}^{(\rho)}(p_x, \tau)$ the amplitude of the component at the apparent angle $\alpha_{\text{app}} = \arctan\left(\frac{cp_x}{2}\right)$ with respect to the vertical, and intersecting the axis $x = 0$ at time τ. After migration, according to the migration equation (2.137), the same component maps to $\bar{R}^{(\rho)}(q_x, z)$, the amplitude of the migrated component of dip $\alpha_{\text{true}} = \arctan(q_x)$. As for the other migration concepts, $\tan\alpha_{\text{app}} = \sin\alpha_{\text{true}}$.

The Radon transform method may look a little cumbersome, but it is actually quite powerful. It handles a depth-variable velocity as well as phase-shift migration, with less tendency to accumulate error than does computation of a phase shift. It can handle a generally variable velocity to some extent, though not as capably as some difference-equation methods.

2.7.2 Hybrid methods

Different migration methods are good at different things, so it makes sense to combine them. The possibilities are too numerous to mention, but we can illustrate with one or two. The phase-shift method is good at $c(z)$, but needs help with $c(x, z)$. One can start with a phase-shift at some reference velocity, say $c_0(z)$, then use another method to handle deviations from c_0. The phase-shift takes place in the k_x domain. The other method may remain in this domain, or may return to the x-domain, where the deviations are more easily characterized. All it takes to come up with a new migration method is a little imagination.

2.8 Time migration

Migration methods and algorithms differ in their ability to handle variations in background or propagation velocity. Difference-equation methods, if properly formulated, may be able to propagate accurately through very complex media. Kirchhoff methods can handle a variable velocity, provided reasonable rays can be constructed and traveltimes defined throughout the medium. Frequency–wavenumber migration is strictly applicable only in a constant-velocity Earth.

Migration goals and objectives also differ. No matter how capable and accurate a migration method, the result can be no better than the velocity information provided to it. In fact, given imperfect information, a less accurate but more forgiving algorithm may provide the better image.

Time migration is a process that presents the image as a function of vertical traveltime instead of depth. Vertical traveltime τ, defined as

$$\tau(z) \equiv 2\int_0^z \frac{dz}{c(z)}, \tag{2.140}$$

can, within limits, be converted to depth post-migration. Its advantages are that the output is more easy to relate to unmigrated time-data, and that traveltime is less sensitive than depth to velocity errors.

2.8.1 Root mean square velocity

In a constant-velocity medium, a point diffractor at (x_0, z_0) shows up on a zero-offset data set as a hyperbola

$$\begin{aligned} t(x) &= \frac{2}{c}\sqrt{(x-x_0)^2 + z_0{}^2} \\ &= \sqrt{\frac{4\,(x-x_0)^2}{c} + t_0{}^2} \\ &\simeq t_0 + \frac{2\,(x-x_0)^2}{ct_0}. \end{aligned} \tag{2.141}$$

where

$$t_0 = \frac{2z_0}{c}. \tag{2.142}$$

Where velocity varies with depth, according to Snell's law, the curve generalizes to

$$t = 2\int_0^z \frac{dz}{c(z)\sqrt{1-p^2c^2(z)}}, \tag{2.143}$$

where the ray parameter p is a constant proportional to the sine of the propagation angle ϕ, divided by velocity:

$$p = \frac{\sin\phi(z)}{c(z)}. \tag{2.144}$$

The ray parameter is related to horizontal displacement from the diffraction point by

$$x - x_0 = p \int_0^{z_0} \frac{dz c(z)}{\sqrt{1 - p^2 c^2(z)}}. \tag{2.145}$$

For small ray parameters ($|p|\ c \ll 1$), the relation between p and displacement becomes linear:

$$x - x_0 \simeq p \int_0^{z_0} dz c(z), \tag{2.146}$$

whereas the relation between time and p becomes quadratic:

$$t \simeq 2 \int_0^{z_0} \frac{dz}{c(z)} + p^2 \int_0^{z_0} dz c(z). \tag{2.147}$$

Thus, in terms of horizontal displacement, the time curve becomes

$$t \simeq t_0 + \frac{2\,(x - x_0)^2}{c_{\mathrm{rms}}{}^2\,(z_0)\,t_0}, \tag{2.148}$$

where

$$t_0 = 2 \int_0^{z_0} \frac{dz}{c(z)}, \tag{2.149}$$

and

$$c_{\mathrm{rms}}{}^2\,(z_0) = \frac{2}{t_0} \int_0^{z_0} dz c(z). \tag{2.150}$$

The time curve from a point diffractor in a $c(z)$ medium looks, near the apex, like the time curve in a constant-velocity medium. However, there are two effective velocities involved. The traveltime at the apex is determined by an average over depth of inverse velocity or slowness, and the moveout velocity for displacements near the apex is determined by an average over depth of velocity. If we rescale the depth axis in traveltime units as per (2.140), then

$$t_0 = \tau_0 = \tau\,(z_0)\,, \tag{2.151}$$

$$t \simeq \tau_0 + \frac{2\,(x - x_0)^2}{c_{\mathrm{rms}}{}^2\,(\tau_0)\,\tau_0}, \tag{2.152}$$

with

$$c_{\mathrm{rms}}{}^2\,(\tau_0) = \frac{1}{\tau_0} \int_0^{\tau_0} d\tau c^2(\tau). \tag{2.153}$$

The moveout velocity c_{rms} is seen to be a root-mean-square traveltime average of velocity (Dix, 1955).

To first approximation, a point diffraction at a given traveltime in a depth-variable medium looks like a point diffraction in a constant-velocity medium, with effective velocity c_{rms}. Thus, to migrate an event near time τ_0, one can use a constant-velocity algorithm with velocity $c_{\text{rms}}(\tau_0)$. This velocity cannot be used to recover the correct image depth, but if the output is left in units of traveltime, that doesn't matter. While the model upon which rms time migration is based depends on depth only, moderate lateral variations in velocity are easily accommodated.

An rms time migration is clearly approximate, especially for steep dips and for strong lateral velocity variations. However, for data of moderate dip and velocity variation, it is a reasonable approximation. For integral-equation migration, each output point is calculated independently, making it very compatible with rms time migration. Difference equation methods are very flexible, and are generally amenable to time migration. Frequency–wavenumber migration wants to use a single velocity for the entire output range, which complicates things somewhat. The most straightforward implementation of *f–k* rms time migration is to generate outputs for a whole suite of velocities, then select the output with the desired rms velocity at each point.

Figure 2.26 shows a Kirchhoff time migration of the synthetic faulted boundary data introduced in Figure 2.19. Velocity in this model increases linearly ($c(z) = 5000 + 1.875z$) with depth. Compared with a Kirchhoff depth migration (see Figure 2.21), the time migration images most of the reflector equally well. However, near the fault, reflector edges are a little blurred and exhibit residual edge diffractions.

2.8.2 Time stretching

An alternative method for incorporating rms velocity changes is by stretching the time axis (Stolt, 1978) in such a way that velocity appears constant. Suppose diffractive events at points (x_0, τ_0) have the moveout curve (2.152). We would like to define a "stretched" time coordinate t_s for which diffractive events have a constant-velocity moveout curve. The rate of change of time with x in unstretched coordinates is approximately

$$\frac{dt}{dx} \simeq \frac{4(x - x_0)}{c_{\text{rms}}{}^2(t)t}. \tag{2.154}$$

In the stretched-time coordinate t_s, we desire

$$\frac{dt_s}{dx} \simeq \frac{4(x - x_0)}{c_0{}^2 t_s}. \tag{2.155}$$

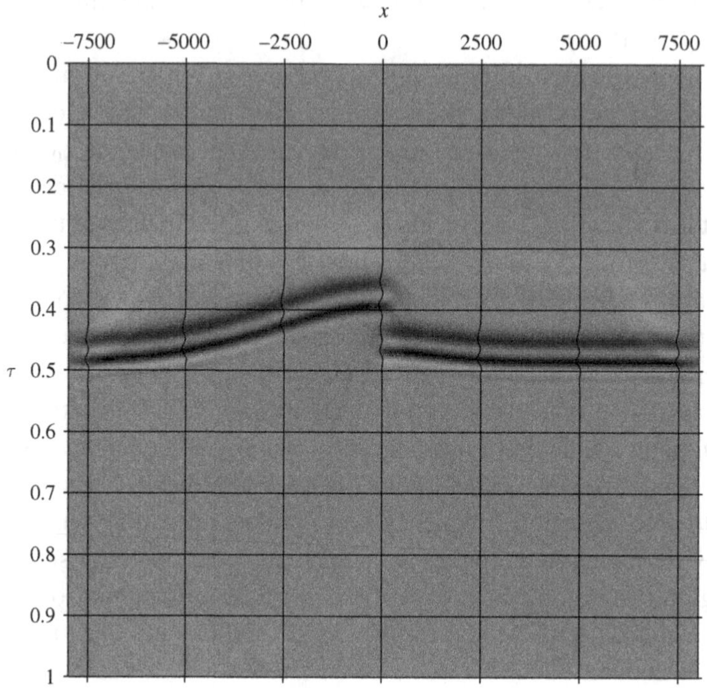

Figure 2.26 Kirchhoff time migration of the faulted boundary data (Figure 2.19). The boundary, except near the fault, is well recovered. Compared with a Kirchhoff depth migration (Figure 2.21) the boundary end points are less focused and show residual diffractions.

The stretch factor is found by taking the ratio of these two expressions:

$$\frac{dt_s}{dt} \simeq \frac{c_{\text{rms}}{}^2(t)t}{c_0{}^2 t_s}. \tag{2.156}$$

Rearranging, we have

$$t_s dt_s \simeq \frac{c_{\text{rms}}{}^2(t)}{c_0{}^2} t dt, \tag{2.157}$$

and integrating,

$$t_s{}^2 \simeq 2 \int_0^t dt' \frac{c_{\text{rms}}{}^2(t')}{c_0{}^2} t'. \tag{2.158}$$

By stretching the time axis according to this formula prior to migration, diffraction curves assume approximately the constant-velocity ($c = c_0$) form (see Figure 2.27), allowing a constant-velocity migration to be performed on the entire data

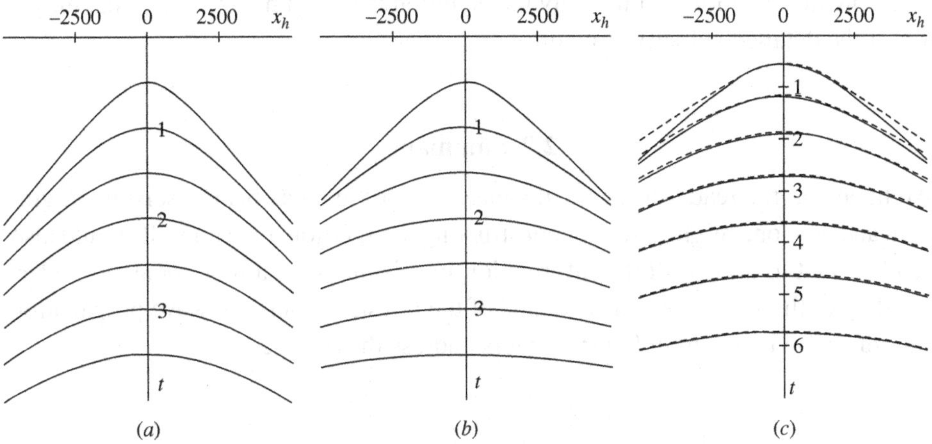

Figure 2.27 An illustration of time stretching. The left panel (a) shows diffraction curves for a constant-velocity (c = 5000 fps) medium. Seven curves are shown, spaced every half-second at the apex. The center panel (b) shows diffraction curves in a variable-velocity medium, specifically $c = 5000 + 2500t$ fps. The deeper curves, experiencing higher average velocity, have noticeably less curvature than the constant-velocity curves in panel (a). The right panel (c) shows the variable-velocity curves of panel (b), after time stretching according to formula (2.158), with constant-velocity (c = 5000 fps) curves also shown as dashed lines. Note that the time axis has been rescaled in this panel. The stretched curves overlie the constant-velocity curves except at small times and large offsets.

set. After migration, events are still in stretched time, hence must be unstretched by inverting (2.158).

Time stretching has obvious limitations. A standard rms time migration uses rms velocity at the image point, whereas time stretching uses rms velocity at the unmigrated location. Neither approximation is correct. In the presence of a moderate lateral velocity gradient, time stretching may be slightly more accurate, but the remaining error may be more difficult to estimate and remove. Since the time stretching formula was developed near the apex of diffractions, it will lose accuracy at greater dips. This deficiency can be addressed up to a point by either modifying the effective wave equation and resulting migration formula (Stolt, 1978) or by keying the stretch to points away from the diffraction apex.

2.8.3 Dealing with complexity

Time migration is not designed to accommodate steep dips or severe velocity variations. When such occur, time migration might still be helpful as a first step in assessing the situation, but the only real solution is a "depth" migration that makes use of the detailed velocity structure.

2.9 Summary

We hope that the reader has gained some feel for the ideas behind seismic migration, and how one might go about constructing a migration algorithm. It should also be clear that to access all the information in seismic data, a more detailed understanding of the seismic experiment and of the physics of seismic wave propagation and reflection is required. Later chapters address these issues.

Exercises

2.1 Write a routine to perform Kirchhoff migration when velocity is at most depth variable, assuming that the half-derivative operation has been applied to the data prior to migration. The output will be a 2-D image function *img*(*ixm, izm*). Consider as inputs a 2-D data function *data*(*ixd, it*), a 2-D traveltime function *itxz*(*izm, jx*), and a 2-D amplitude function *ampxz*(*izm, jx*), with $jx = |ixm - ixd| + 1$.

2.2 Starting from the reverse time extrapolation formula (2.44),
   (a) Write a subroutine to extrapolate the reverse time migration operator one time step.
   (b) Write a subroutine to perform a reverse time exploding-reflector migration.

2.3 Starting with the first-order (15°) approximation to the square-root equation (2.54),
   (a) in the $k_x - \omega$ domain, write a one-way 1-D differential wave-propagation operator for upgoing waves;
   (b) write a corresponding $k_x - \omega$-domain difference equation;
   (c) write the dispersion relation of the difference equation, for a depth step Δz a multiple m of $c\ \Delta t/2$, and frequency ω a fraction $1/n$ of Nyquist frequency.

2.4 Starting with a second-order approximation to the square-root equation (2.54),
   (a) write a one-way differential wave-propagation operator for upgoing waves;
   (b) write a corresponding frequency-domain difference equation;
   (c) write and compare dispersion relations using a three-point, a five-point, and a Fourier operator for the second-derivative with respect to x.

2.5 (a) Formulate the Claerbout 15° downward-continuation operator as a time-migration operator, assuming that velocity varies slowly enough that its derivatives can be ignored.
(b) Approximating the x-derivative term with a second difference operator, write an explicit difference-equation extrapolator.

2.6 Write a routine to perform a 2-D exploding-reflector *f–k* migration.

2.7 Starting with a theoretical data set from a single dipping layer,
$D(x,t) = \mathcal{R}_0 \delta\left(t - t_0 - px\right),$
calculate the migrated image function using
(a) the phase-shift migration formula (2.93),
(b) the *f–k* migration formula (2.99),
(c) the *f–k* time migration formula (2.102),
(d) the Kirchhoff migration formula (2.114) with amplitude (2.117), and
(e) the slant-stack migration formula (2.132).

3

Prestack migration

We have formulated basic migration concepts in terms of zero-offset data, even though real data are anything but zero-offset. Fortunately, the ideas are generalizable to finite-offset data.

3.1 Prestack migration concepts

Kirchhoff migration is perfectly amenable to prestack data. As long as one can calculate (or tabulate) the traveltimes from source to reflection point to receiver, one can sum into a point all the data that might have come from that point, or, alternatively, broadcast a data point into all the locations from which the data might have come. Kirchhoff methods are happy with restricted data sets such as constant offset, though in some circumstances they may produce artifacts. There are of course issues of amplitude, phase, aliasing, etc., which must be dealt with, some of which are discussed below.

A prestack difference-equation migration algorithm requires extrapolation of both sources and receivers. Given full surface coverage, one could do this by applying the wave equation separately to source and receiver coordinates, downward continuing each in turn, and taking the time-zero doubly downward-continued data to be the migrated image. To image a restricted data set, one must be more devious. For example, a constant-source set can be imaged by downward-continuing the receiver coordinates and cross-correlating at each image point with a forward extrapolation of the source impulse. The imaging principle here is also due to Claerbout (1976): *Reflectors exist at points where the arrival of the downgoing wave from the source coincides with the upgoing reflected wave*. The quantitative connection between the resulting image and reflectivity is more obscure than for double downward continuation, but with some effort it can be established.

The simplicity of downward continuation in *f–k* space makes wavenumber–frequency migration methods particularly easy to construct and explain. However, these methods are not amenable to migration of constant-offset or most other partial data sets. Fourier transforms over both source and receiver locations are required, which implies that for every source, there must be a complete set of receivers, and vice versa. By invoking asymptotic approximations, the demand for completeness can be relaxed somewhat, but the full flexibility enjoyed by other methods is elusive.

3.2 Prestack *f–k* migration in two dimensions

Given a full 2-D data set $D\left(x_g|x_s;t\right)$, where x_s is source location, x_g is receiver location, and t is time, one can perform a Fourier transform over all coordinates:

$$D\left(k_{gx}|k_{sx};\omega\right) = \int dx_g \int dx_s \int dt\, D\left(x_g|x_s;t\right) e^{i\left(k_{sx}x_s - k_{gx}x_g + \omega t\right)}. \qquad (3.1)$$

In the convention employed here, the forward time transform is performed with positive phase, as is the forward source-coordinate transform. The receiver-coordinate transform is performed with negative phase. For a discussion of these conventions, see Appendix C.

Downward continuation

To perform downward continuation, we consider the data to be the expression of a wave recorded at $\left(x_g, 0\right)$ produced by a source at $(x_s, 0)$, and reflected from multiple subsurface locations (x, z). Denoting this wave as $P\left(x_g, z_g|x_s, z_s;\omega\right)$, we have

$$D\left(x_g|x_s;t\right) = P\left(x_g, 0\,|x_s,\ 0;t\right), \qquad (3.2)$$

and, in the frequency–wavenumber domain,

$$D\left(k_{gx}|k_{sx};\omega\right) = P\left(k_{gx}, 0\,|k_{sx},\ 0;\omega\right). \qquad (3.3)$$

(While we have written these equations for 2-D data, the 3-D expression is a simple extension.) Analogous to Section 2.6, the Fourier transforms isolate a single plane-wave component of both the wave from the source to the reflectors and the wave from the reflectors to the receiver. Assuming a constant velocity c, the wavenumber vector of the (downgoing) source wave is

$$\tilde{\boldsymbol{k}}_s = \left(k_{sx}, k_{sz}\right), \qquad (3.4)$$

where

$$k_{sz} = +\frac{\omega}{c}\sqrt{1 - \frac{{k_{sx}}^2 c^2}{\omega^2}}, \qquad (3.5)$$

and that of the (upgoing) receiver wave is

$$\tilde{\boldsymbol{k}}_g = \left(k_{gx}, k_{gz}\right), \tag{3.6}$$

with

$$k_{gz} = -\frac{\omega}{c}\sqrt{1 - \frac{{k_{gx}}^2 c^2}{\omega^2}}. \tag{3.7}$$

As for poststack migration, we confine attention to the physically propagating wavenumbers, where the arguments of both square roots are positive and the vertical wavenumbers are real.

To downward-continue both the source and receiver waves to the same depth z, we simply multiply the data by the phase term $e^{i(k_{gz}-k_{sz})z}$:

$$P\left(k_{gx}, z\,|k_{sx},\ z; \omega\right) = \ D\left(k_{gx}|k_{sx}; \omega\right) e^{i(k_{gz}-k_{sz})z}. \tag{3.8}$$

Following the convention of Appendix C for Fourier transforms, the source and receiver wavenumber phases are given opposite signs. This is because downward continuation extrapolates the source wave forward and the receiver wave backwards.

Defining k_z to be the total vertical wavenumber for the downward-continued data, we have

$$k_z \equiv k_{gz} - k_{sz} \ = -\frac{\omega}{c}\left(\sqrt{1 - \frac{{k_{gx}}^2 c^2}{\omega^2}} + \sqrt{1 - \frac{{k_{sx}}^2 c^2}{\omega^2}}\right). \tag{3.9}$$

Equation (3.9) is, for obvious reasons, often called the *double-square-root* equation. As shown in Figure 3.1, vertical wavenumber k_z is real within the *propagating region* $-\omega/c < k_{gx} < \omega/c$, $-\omega/c < k_{sx} < \omega/c$. For small horizontal wavenumbers its value is near $-2\omega/c$. When both $|k_{sx}c/\omega|$ and $\left|k_{gx}c/\omega\right|$ approach 1, k_z drops to zero. When both horizontal wavenumbers exceed this limit, k_z becomes imaginary. If one wavenumber does, k_z becomes complex.

Imaging

Analogous to Section 2.6, we would like to define migration in terms of the time-zero component of the downward-continued data. The extra dimension in prestack data complicates things a little. We must set the downward-continued source and receiver locations to be equal, so that the upgoing and downgoing waves coincide. We would also like a migration operator that does not change the phase or amplitude spectrum of flat reflectors. In two dimensions, there is some ambiguity with this goal. For 2-D point (more precisely, line) sources and receivers in a 2-D Earth, the impulse response is a Hankel function (Appendix E, Section E.10), not a delta function. Hence, if the image function for a flat reflector is zero phase, the

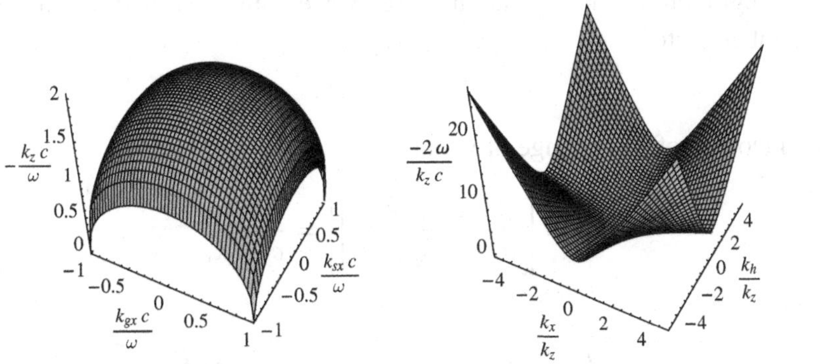

Figure 3.1 The double-square-root equation (3.9) for k_z as a function of frequency ω, and horizontal wavenumbers k_{sx} and k_{gx}, (left), and its inverse solution (3.17) for ω as a function of k_x, k_h, and k_z(right). The double-square-root equation has a real (propagating) solution for k_z only for $|k_{gx}c/\omega| < 1$ and $|k_{sx}c/\omega| < 1$. $|k_z c/\omega|$ is bounded above by 2 and below by 0. The inverse equation has a real solution for ω for all real values of k_z, k_x, and k_h. $|2\,\omega/k_z\,c|$ is bounded below by 1 but has no upper bound.

corresponding data function will be the 45° Hankel function. On the other hand, for real-Earth data, even where the Earth may be approximated as 2-D, the sources and receivers should be considered points in three dimensions. In this 2.5-dimensional setting, the impulse response is a delta function, hence for flat reflectors we would like a zero-phase image function to correspond to a zero-phase data function. To achieve the desired phase, we define the image function $R(x,z)$ to be proportional to a half-integral filtered (Appendix E, Section E.6) downward-continued wave at $x_s = x_g \equiv x$, $z_g = z_s = z$, and $t = 0$ (note that source and receiver must converge on the same point):

$$R(x, z) \equiv (\Theta_{h+} * P)\left(x_g = x, z_g = z\,|x_s = x, z_s = z; t = 0\right)$$
$$= \frac{1}{(2\pi)^3} \int \frac{d\omega}{\sqrt{-i\omega}} \int dk_{sx} \int dk_{gx} D\left(k_{gx}|k_{sx}; \omega\right) e^{i(k_z z + k_x x)}, \tag{3.10}$$

with k_z as defined in equation (3.9), and

$$k_x \equiv k_{gx} - k_{sx}. \tag{3.11}$$

with k_h the lateral offset wavenumber

$$k_h \equiv k_{gx} + k_{sx}. \tag{3.12}$$

Two of the three integrals in equation (3.10) can be converted to inverse Fourier transforms with a change of integration variable. We choose as a new set of integration variables, lateral offset wavenumber k_h and the image-point wavenumber vector

$$\tilde{\boldsymbol{k}} \equiv (k_x, k_z) . \tag{3.13}$$

The Jacobian of this change of variables is

$$\left| \text{Det} \left[\frac{\partial \left(k_x, k_h, k_z\right)}{\partial \left(k_{sx}, k_{gx}, \omega\right)} \right] \right| = \frac{2\omega k_z}{c^2 k_{gz} k_{sz}} . \tag{3.14}$$

Thus,

$$R(x, z) = \frac{c^2}{2(2\pi)^3} \int dk_z \int dk_x e^{i(k_x x + k_z z)} \int dk_h \frac{k_{gz} k_{sz}}{\sqrt{-i\omega\omega k_z}} D\left(k_{gx} | k_{sx}; \omega\right) . \tag{3.15}$$

The outer two integrals are now inverse Fourier transforms from k_z to z and k_x to x. Consequently, the double Fourier transform of R is proportional to the remaining integral over k_h:

$$R\left(k_x, k_z\right) = \frac{c^2}{4\pi} \int dk_h D\left(k_{gx} | k_{sx}; \omega\right) \frac{k_{gz} k_{sz}}{\sqrt{-i\omega\omega k_z}} . \tag{3.16}$$

Expression (3.16) is an integral over offset wavenumber k_h. Within the integral, k_x and k_z are fixed, while k_{gx}, k_{sx}, and ω vary with k_h. The relations between the variables can be expressed as follows. Source and receiver wavenumbers are $k_{gx} = (k_x + k_h) / 2$, and $k_{sx} = (k_h - k_x) / 2$. Within the propagating region, the double-square-root equation (3.9) can be solved for ω as a function of k_z, k_x, and k_h (see Figure 3.1):

$$\omega = -\frac{c k_z}{2} \sqrt{\left(1 + {k_x}^2 / {k_z}^2\right) \left(1 + {k_h}^2 / {k_z}^2\right)} . \tag{3.17}$$

The source and receiver vertical wavenumbers k_{sz} and k_{gz} can also be expressed in terms of k_z, k_x, and k_h:

$$k_{sz} = -\frac{k_z}{2} \left(1 - \frac{k_x k_h}{{k_z}^2}\right) , \tag{3.18}$$

and

$$k_{gz} = \frac{k_z}{2} \left(1 + \frac{k_x k_h}{{k_z}^2}\right) . \tag{3.19}$$

3.2.1 Constant-angle f–k migration

The expression (3.16) has no pretensions of true amplitude. In fact, it has one obvious deficiency. Depending on which physical properties are changing at a point, a realizable reflection coefficient will depend on angle of incidence γ, whereas the quantity R in (3.16) has no angle dependence. We suspect that the angle dependence has been smothered by the integration over offset wavenumber. This integration occurred because we wanted sources and receivers to be downward-continued to the same point, so that offset at the reflection point must be zero. However, something subtle was missed in the argument above: offset must go to zero *in the limit* as time approaches zero, and limits do funny things sometimes. For example, a unit step function $\Theta(x)$ has the limit zero as x approaches zero from below, but the limit one if x approaches zero from above. Similarly, reflectivity, defined as the limit of the downward-continued data as time and offset approach zero, will depend on the *direction* from which the downward-continued waves approach zero (Clayton and Stolt, 1981).

We accommodate the directional dependence as follows. Changing integration variable from k_h to angle of incidence γ, equation (3.16) becomes

$$R\left(k_x, k_z\right) = \frac{c^2}{4\pi} \int d\gamma D\left(k_{gx}|k_{sx}; \omega\right) \frac{k_{gz}k_{sz}}{\sqrt{-i\omega\omega k_z}} \left|\frac{dk_h}{d\gamma}\right|. \tag{3.20}$$

For an infinite-aperture migration, γ would range from $-\pi/2$ to $\pi/2$. We can confine attention to a specific angle of incidence γ, by writing (3.20) as

$$R\left(k_x, k_z\right) \equiv \int d\gamma r\left(k_x, k_z, \gamma\right), \tag{3.21}$$

where

$$r\left(k_x, k_z, \gamma\right) = \frac{c^2}{4\pi} D\left(k_{gx}|k_{sx}; \omega\right) \frac{k_{gz}k_{sz}}{\sqrt{-i\omega\omega k_z}} \left|\frac{dk_h}{d\gamma}\right|. \tag{3.22}$$

We have assumed a relation exists for incident angle in terms of the known parameters. To find one, define propagation angles ϕ_i and ϕ_r for the source and receiver waves relative to the vertical as

$$\tilde{\boldsymbol{k}}_s = \left(k_{sx}, k_{sz}\right) = \frac{\omega}{c}\left(-\sin\phi_i, \cos\phi_i\right), \tag{3.23}$$

$$\tilde{\boldsymbol{k}}_g = \left(k_{gx}, k_{gz}\right) = -\frac{\omega}{c}\left(\sin\phi_r, \cos\phi_r\right), \tag{3.24}$$

As defined, these two angles are positive for travel from right to left, as illustrated in Figure 3.2. For specular reflectors, these two angles can be related to dip angle α and angle of incidence γ as depicted in Figure 3.2:

$$\alpha = \left(\phi_i - \phi_r\right)/\,2, \tag{3.25}$$

$$\gamma = \left(\phi_i + \phi_r\right)/\,2. \tag{3.26}$$

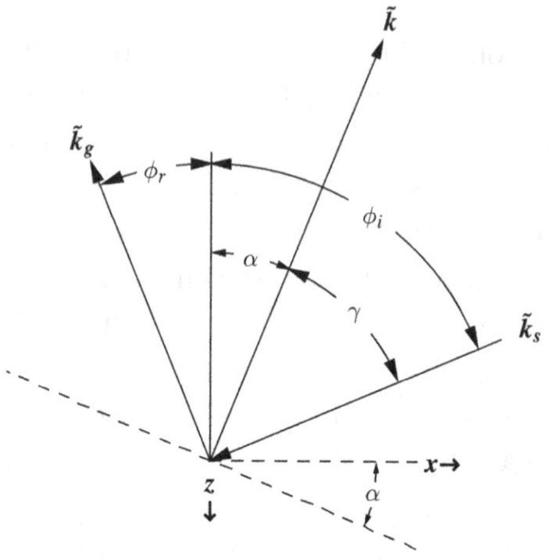

Figure 3.2 Wavenumber plot. Incoming and reflected wavenumber vectors, and associated propagation angles ϕ_i and ϕ_r relative to the vertical, chosen positive for propagation from right to left. Also shown are the midpoint vector $\tilde{\boldsymbol{k}} = \tilde{\boldsymbol{k}}_g - \tilde{\boldsymbol{k}}_s$, the angle α between $\tilde{\boldsymbol{k}}$ and the vertical, and the angle γ between $\tilde{\boldsymbol{k}}$ and $\tilde{\boldsymbol{k}}_s$ or $\tilde{\boldsymbol{k}}_g$. At a specular reflection point, the reflector (indicated by a dotted line) will be normal to the vector $\tilde{\boldsymbol{k}}$ and have dip α.

From the definitions (3.9), (3.23) and (3.24), it follows that

$$k_z = -\frac{\omega}{c}(\cos\phi_i + \cos\phi_r)\,. \tag{3.27}$$

With the relations (3.25) and (3.26), k_z can be expressed in terms of incident and dip angles:

$$k_z = -\frac{2\omega}{c}\cos\alpha\cos\gamma. \tag{3.28}$$

We also have from (3.11) an expression for midpoint wavenumber

$$\begin{aligned} k_x &= \frac{\omega}{c}(\sin\phi_i - \sin\phi_r) \\ &= \frac{2\omega}{c}\sin\alpha\cos\gamma, \end{aligned} \tag{3.29}$$

and from (3.12), for offset wavenumber,

$$\begin{aligned} k_h &= -\frac{\omega}{c}(\sin\phi_i + \sin\phi_r) \\ &= -\frac{2\omega}{c}\sin\gamma\cos\alpha. \end{aligned} \tag{3.30}$$

From these formulas it follows that dip angle α depends on the ratio k_x/k_z,

$$\tan\alpha = -k_x/k_z, \tag{3.31}$$

while angle of incidence γ depends on k_h/k_z:

$$\tan\gamma = k_h/k_z, \tag{3.32}$$

or

$$\cos\gamma = \frac{1}{\sqrt{1+k_h{}^2/k_z{}^2}}. \tag{3.33}$$

Equation (3.32) is the relation needed to evaluate $dk_h/d\gamma$. We have

$$\frac{dk_h}{d\gamma} = \frac{k_z}{\cos^2\gamma}. \tag{3.34}$$

Hence, (3.22) becomes

$$r(k_x, k_z, \gamma) = \frac{-c^2}{4\pi} D\left(k_{gx}|k_{sx};\omega\right)\frac{1}{|\omega|\sqrt{-i\omega}}\frac{k_{gz}k_{sz}}{\cos^2\gamma}. \tag{3.35}$$

Equations (3.17), (3.18), and (3.19) for ω, k_{sz}, and k_{gz} can be expressed in terms of k_x, k_z, and γ as

$$\omega = -k_z\frac{c}{2}\sqrt{1+k_x{}^2/k_z{}^2}\sec\gamma, \tag{3.36}$$

$$k_{sz} = -\frac{k_z}{2}\left(1-\frac{k_x\tan\gamma}{k_z}\right), \tag{3.37}$$

and

$$k_{gz} = \frac{k_z}{2}\left(1+\frac{k_x\tan\gamma}{k_z}\right). \tag{3.38}$$

Equation (3.35) gives a one-to-one mapping between *f–k* data points and the *k*-space migrated image at a particular angle of incidence. The amplitude factor in (3.35) should not be taken seriously. The most one can say is that, with the above amplitude, the integral over incidence angle of the angle-dependent image function $r(k_x, k_z, \gamma)$ is the image function $R(k_x, k_z)$, as defined in equation (3.16).

3.2.2 Amplitude-preserving prestack f–k migration in 2.5 dimensions

Guided by the form (3.35), let's derive an amplitude-faithful expression for 2-D prestack *f–k* migration. We seek an expression of the form

$$D(k_x, k_h, \omega) = R(k_x, k_z, \gamma)\,\mathcal{A}(\alpha, \gamma, \omega), \tag{3.39}$$

with α the dip angle, defined as

$$\tan\alpha = -k_x/k_z, \tag{3.40}$$

γ the incident angle, given by

$$\tan\gamma = k_h/k_z, \tag{3.41}$$

and frequency ω given by

$$\omega = -\frac{k_z c}{2}\sqrt{\left(1 + {k_x}^2/{k_z}^2\right)\left(1 + {k_h}^2/{k_z}^2\right)}, \tag{3.42}$$

also expressible as (3.28).

To figure out what $\mathcal{A}$ must be, look at a single dipping reflector. Invoking the principle of superposition, we claim that a formula that can migrate a dipping reflector can migrate anything. Spatially, given a reflector at $z = z_0 + x\tan\alpha_0$, the formula for R is

$$R(x, z, \gamma) = R_0(\gamma)\delta\left((z - z_0)\cos\alpha_0 - x\sin\alpha_0\right). \tag{3.43}$$

In terms of wavenumber, R is

$$\begin{aligned} R\left(k_x, k_z, \gamma\right) &= R_0(\gamma)\int\int dx dz \delta\left((z - z_0)\cos\alpha_0 - x\sin\alpha_0\right)e^{-i(k_x x + k_z z)} \\ &= 2\pi R_0(\gamma)e^{-ik_z z_0}\delta\left(k_x\cos\alpha_0 + k_z\sin\alpha_0\right). \end{aligned} \tag{3.44}$$

The delta function confirms what we should already know for a dipping layer, namely (see equation (3.31)) $\tan\alpha_0 = -k_x/k_z$. By definition, tangent of incident angle is given by (3.32), or $\tan\gamma = k_h/k_z$. Since dip is now fixed at α_0, the equation (3.42) for frequency ω reduces to

$$\omega = -\frac{k_z c}{2}\sec\alpha_0\sqrt{1 + {k_h}^2/{k_z}^2}, \tag{3.45}$$

or, inversely,

$$k_z = -\text{Sign}(\omega)\sqrt{\frac{4\omega^2}{c^2}\cos^2\alpha_0 - {k_h}^2}. \tag{3.46}$$

The space–time formula for the reflection response D can be determined from the method of images. In two dimensions, the impulse response is a Hankel function (see, e.g., Clayton & Stolt, 1981), corresponding to line sources and receivers. Typically, however, we want point sources and receivers, even if the experiment is otherwise two-dimensional. In three dimensions, the reflected impulse response is a delta function

$$D\left(x_m, x_h, t\right) = \frac{R_0(\gamma)\delta(t - r/c)}{r}, \tag{3.47}$$

with r the total raypath length (see exercise 3.1),

$$r = 2\cos\alpha_0\sqrt{{x_h}^2 + {z_m}^2}, \tag{3.48}$$

z_m the reflector depth directly below the midpoint x_m,

$$z_m \equiv z_0 + x_m \tan\alpha_0, \tag{3.49}$$

and γ the ray-theoretical angle of incidence (see exercise 3.1)

$$\gamma = -\arctan\left(\frac{x_h}{z_m}\right). \tag{3.50}$$

The amplitude decay in the impulse response of $1/r$ is referred to as spherical divergence.

A minor problem in linking R to D is that the 2-D wave equation anticipates cylindrical, not spherical divergence. We address this problem by multiplying the 3-D data by the square root of time:

$$\begin{aligned} D^{(\text{tdc})}(x_m, x_h, t) &\equiv \sqrt{t}D(x_m, x_h, t) \\ &= \frac{R_0(\gamma)\delta(t - r/c)}{\sqrt{rc}}. \end{aligned} \tag{3.51}$$

We expect to be able, at least asymptotically, to relate R and $D^{(\text{tdc})}$ (tdc standing for time divergence corrected) with a 2-D mapping of the form (3.39). Although there is also a phase difference between the 2-D Hankel function and the 3-D impulse response, we hope to include that in the coefficient $\mathcal{A}$.

To confirm the form (3.39), we must Fourier transform $D^{(\text{tdc})}$ over x_m, x_h, and t, and compare to $R(k_x, k_z, \gamma)$.

$$D^{(\text{tdc})}(k_m, k_h, \omega) = \int dx_m \int dx_h \frac{R_0(\gamma)e^{i(\omega r/c - k_h x_h - k_m x_m)}}{\sqrt{rc}}. \tag{3.52}$$

Whereas R is a function of k_x, k_z, and γ, $D^{(\text{tdc})}$ is a function of k_m, k_h, and ω.

Perform the integral over offset using the stationary phase approximation. We seek the stationary point of the phase argument

$$\begin{aligned} f(x_h) &= \omega r/c - k_h x_h \\ &= \frac{2\omega}{c}\cos\alpha_0\sqrt{{x_h}^2 + {z_m}^2} - k_h x_h, \end{aligned} \tag{3.53}$$

i.e. the point at which its derivative,

$$f'(x_h) = \frac{2\omega\cos\alpha_0 x_h}{c\sqrt{{x_h}^2 + {z_m}^2}} - k_h, \tag{3.54}$$

is zero. The stationary point is

$$\underset{\wedge}{x}_h = z_m \frac{k_h\, c\, \text{Sign}(\omega)}{\sqrt{4\omega^2 \cos^2\alpha_0/c^2 - {k_h}^2}} = -z_m \frac{k_h}{k_z}, \tag{3.55}$$

at which

$$\sqrt{{\underset{\wedge}{x}_h}^2 + {z_m}^2} = \frac{2|\omega| \cos\alpha_0 z_m}{c\sqrt{4\omega^2 \cos^2\alpha_0/c^2 - {k_h}^2}} = -\frac{2\omega\ \cos\alpha_0 z_m}{ck_z}, \tag{3.56}$$

$$\underset{\wedge}{r} = -4\cos^2\alpha_0 \frac{\omega z_m}{ck_z}, \tag{3.57}$$

and

$$f\left(\underset{\wedge}{x}_h\right) = -k_z z_m, \tag{3.58}$$

the expressions in terms of vertical wavenumber k_z following from (3.46).

Note also that the ray-theoretical expression (3.50) for angle of incidence becomes the wave-theoretical expression (3.41):

$$\underset{\wedge}{\gamma} = \underset{\wedge}{\gamma}\,(k_m, k_h, \omega) - \arctan\left(\underset{\wedge}{x}_h/z_m\right) = \arctan\left(k_h/k_z\right). \tag{3.59}$$

This allows us to write the relation between ω and k_z in terms of $\underset{\wedge}{\gamma}$:

$$k_z = -\frac{2\omega}{c}\cos\alpha_0 \cos\underset{\wedge}{\gamma}. \tag{3.60}$$

The second derivative of the phase argument is

$$f''\left(x_h\right) = \frac{2\omega\cos\alpha_0 {z_m}^2}{c\,\left({x_h}^2 + {z_m}^2\right)^{3/2}}, \tag{3.61}$$

with stationary value

$$f''\left(\underset{\wedge}{x}_h\right) = \frac{2\omega\cos\alpha_0\cos^3\underset{\wedge}{\gamma}}{cz_m} = -\frac{k_z\cos^2\underset{\wedge}{\gamma}}{z_m}. \tag{3.62}$$

The stationary phase approximation to the offset integral thus yields

$$\begin{aligned} D^{(\text{tdc})}\left(k_m, k_h, \omega\right) &= \int dx_m e^{-ik_m x_m} \sqrt{\frac{2\pi i}{\underset{\wedge}{r} c f''\left(\underset{\wedge}{x}_h\right)}} R_0\left(\underset{\wedge}{\gamma}\right) e^{if\left(\underset{\wedge}{x}_h\right)} \\ &= \int dx_m \sqrt{\frac{\pi i}{2\omega}} \frac{R_0\left(\underset{\wedge}{\gamma}\right)}{\cos\underset{\wedge}{\gamma}\cos\alpha_0} e^{-i\left(k_z z_m + k_m x_m\right)}. \end{aligned} \tag{3.63}$$

Now perform the Fourier transform over midpoint. Remembering the definition (3.49) of z_m,

$$
\begin{aligned}
D^{(\mathrm{tdc})}(k_m, k_h, \omega) &= e^{-ik_z z_0}\sqrt{\frac{\pi i}{2\omega}}\frac{R_0(\gamma)}{\cos\gamma\cos\alpha_0}\int dx_m e^{-ix_m(k_z\tan\alpha_0+k_m)} \\
&= 2\pi e^{-ik_z z_0}\sqrt{\frac{\pi i}{2\omega}}R_0(\gamma)\frac{\delta(k_z\tan\alpha_0+k_m)}{\cos\gamma\cos\alpha} \\
&= 2\pi e^{-ik_z z_0}\sqrt{\frac{\pi i}{2\omega}}R_0(\gamma)\frac{\delta(k_z\sin\alpha_0+k_m\cos\alpha_0)}{\cos\gamma}. \qquad (3.64)
\end{aligned}
$$

Comparing with equation (3.44), we invoke the principle of superposition to assert that for R a general function of k_m, k_z, and γ,

$$D^{(\mathrm{tdc})}(k_x, k_h, \omega) = \sqrt{\frac{\pi i}{2\omega}}\frac{R(k_m, k_z, \gamma)}{\cos\gamma}. \qquad (3.65)$$

This is of the form (3.39), with

$$\mathcal{A}^{(\mathrm{tdc})}(\alpha, \gamma, \omega) \equiv \frac{1}{\cos\gamma}\sqrt{\frac{\pi i}{2\omega}}. \qquad (3.66)$$

This expression should (at least asymptotically) preserve reflection amplitude and phase. It does have the same phase as the more naive expression (3.35), though of course a different amplitude. An equivalent 2.5-D expression is derived in Chapter 9 via scattering theory. It differs from (3.65) in that there, the 2.5-D divergence correction (3.51) is performed after the migration. The difference is superficial: the divergence correction may be applied either before migration as $\sqrt{t}$, or after migration as $\sqrt{z}$ with a slightly altered amplitude coefficient.

3.3 Prestack slant-stack migration in 2.5 dimensions

Expressing prestack migration in terms of slant stacks or Radon transforms is surprisingly easy. Start with the amplitude-preserving 2.5-D formula (3.65), with the k-space representation of R replaced by

$$R^{(2.5\mathrm{D})}(q_x, q_h, k_z) = R^{(2.5\mathrm{D})}(k_x, k_z, \gamma), \qquad (3.67)$$

with $q_x = k_x/k_z = -\tan\alpha$ and $q_h = k_h/k_z = \tan\gamma$. In this representation, R is the Fourier–Radon transform

$$R^{(2.5\mathrm{D})}(q_x, q_h, k_z) = \int\int dx_m dz R^{(2.5\mathrm{D})}(x_m, z - q_x x_m, \gamma)\, e^{-ik_z z}. \qquad (3.68)$$

Equation (3.65) becomes

$$R^{(2.5\mathrm{D})}\left(q_x, q_h, k_z\right) = D^{(2.5\mathrm{D})}\left(k_x, k_h, \omega\right)\sqrt{\frac{-2i\omega}{\pi\left(1+{q_h}^2\right)}}. \tag{3.69}$$

In terms of the new parameters, the relation between k_z and ω is

$$\omega = -\frac{k_z c}{2}\sqrt{\left(1+{q_x}^2\right)\left(1+{q_h}^2\right)}. \tag{3.70}$$

Expanding the data into space-domain components,

$$R^{(2.5\mathrm{D})}\left(q_x, q_h, k_z\right) = \sqrt{\frac{-2i\omega}{\pi\left(1+{q_h}^2\right)}}\int\int dx_m dx_h D^{(2.5\mathrm{D})}\left(x_m, x_h, \omega\right) e^{-i\left(k_x x_m + k_h x_h\right)}, \tag{3.71}$$

and performing the inverse Fourier transform from k_z to z,

$$R^{(2.5\mathrm{D})}\left(q_x, q_h, z\right) = \frac{1}{2\pi}\int dk_z\sqrt{\frac{-2i\omega}{\pi\left(1+{q_h}^2\right)}}\int\int dx_m dx_h \times D^{(2.5\mathrm{D})}\left(x_m, x_h, \omega\right) e^{ik_z\left(z - q_x x_m - q_h x_h\right)}. \tag{3.72}$$

Changing integration variable from k_z to ω,

$$R^{(2.5D)}\left(q_x, q_h, z\right) = \frac{2}{c}\sqrt{\frac{2}{\pi}}\frac{1}{\sqrt{\left(1+{q_x}^2\right)\left(1+{q_h}^2\right)}}\int\int dx_m dx_h \frac{1}{2\pi}\int d\omega\sqrt{-i\omega}D^{(2.5\mathrm{D})}\left(x_m, x_h, \omega\right) e^{-i\frac{2\omega}{c}\frac{\left(z-q_x x_m - q_h x_h\right)}{\sqrt{\left(1+{q_x}^2\right)\left(1+{q_h}^2\right)}}}. \tag{3.73}$$

The frequency integral is an inverse Fourier transform back to time of the data convolved with a half-derivative operator and evaluated at time

$$\tau = \frac{2}{c}\frac{\left(z - q_x x_m - q_h x_h\right)}{\sqrt{\left(1+{q_x}^2\right)\left(1+{q_h}^2\right)}}, \tag{3.74}$$

so that

$$R^{(2.5\mathrm{D})}\left(q_x, q_h, z\right) = \frac{2}{c}\sqrt{\frac{2}{\pi}}\frac{1}{\sqrt{\left(1+{q_x}^2\right)\left(1+{q_h}^2\right)}} \int\int dx_m dx_h \left(d_{h+} * D^{(2.5\mathrm{D})}\right)\left(x_m, x_h, \frac{2}{c}\frac{\left(z - q_x x_m - q_h x_h\right)}{\sqrt{\left(1+{q_x}^2\right)\left(1+{q_h}^2\right)}}\right). \tag{3.75}$$

This operation is a 2-D generalized slant stack, in which the time coordinate is not only shifted but compressed.

3.4 Prestack ray-theoretical migration

Given almost any source–receiver configuration, it is possible to perform a migration using ray-theoretical concepts. If rays can be computed from source and receiver locations to arbitrary points in the subsurface, then for a given source and receiver, a map can be generated of traveltimes from source to all possible reflection points to receiver. An event at traveltime τ can be broadcast to all possible reflection points consistent with traveltime τ using the method of Section 2.2.4. Combining the data from many sets of sources and receivers, we can expect an event to build constructively along actual reflector locations, with destructive interference attenuating the response at locations where no reflector exists. The principal requirement, other than the availability of rays, is that the source–receiver configuration result in dense coverage of the image space. Typically, ray-theoretical migration uses a common-offset data set, though other configurations are certainly possible.

For constant velocity, if amplitude is not an issue we can use the amplitude-challenged 2-D *f–k* formula (3.10) to derive an asymptotically equivalent Kirchhoff migration algorithm. We return the data to space–time, obtaining (after some rearranging)

$$R(x,z) = \frac{1}{(2\pi)^3}\int dx_g \int dx_s \int dt D\left(x_g|x_s;t\right)\int d\omega \frac{e^{i\omega t}}{\sqrt{-i\omega}} \times \int dk_{sx}\, e^{-i(k_{sz}z+k_{sx}(x-x_s))}\int dk_{gx} e^{i\left(k_{gz}z+k_{gx}\left(x-x_g\right)\right)}. \tag{3.76}$$

Since k_{gz} is a function of k_{gx} and ω, and k_{sz} is a function of k_{sx} and ω (see equations (3.5) and (3.7)), the phases in the k_{gx} and k_{sx} integrals are nonlinear in their arguments. Exact solutions to these integrals are possible, but it is more convenient to seek an asymptotic solution through the stationary phase approximation. Invoking Appendix F, the k_{gx} and k_{sx} integrals can be approximately evaluated to be

$$\int dk_{sx}\, e^{-i(k_{sz}z+k_{sx}(x-x_s))} \simeq e^{-i\omega r_s/c}\sqrt{\frac{2\pi i\omega z^2}{c{r_s}^3}}, \tag{3.77}$$

and

$$\int dk_{gx}\, e^{i\left(k_{gz}z+k_{gx}\left(x-x_g\right)\right)} \simeq e^{-i\omega r_g/c}\sqrt{\frac{2\pi i\omega z^2}{c{r_g}^3}}, \tag{3.78}$$

with r_s the distance from source $(x_s, 0)$ to image point $(x,\ z)$

$$r_s = \sqrt{z^2 + (x - x_s)^2} \tag{3.79}$$

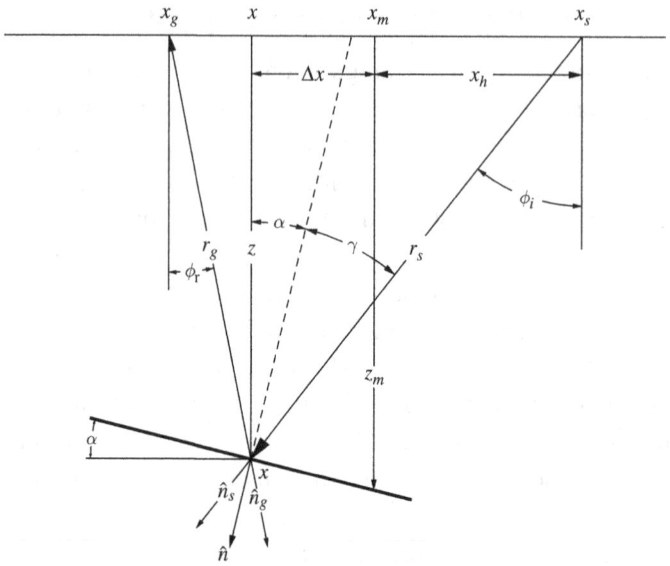

Figure 3.3 Ray-theoretical reflection from a locally planar surface, shown as two-dimensional for simplicity. The ray starts at point x_s, and travels along the raypath $\boldsymbol{r}_s$ to the reflection point $x = (x, z)$, making angle ϕ_i with the vertical. At $\boldsymbol{x}$, the ray encounters a reflector dipping at angle α relative to the horizontal, and is reflected to the point x_g along the ray $\boldsymbol{r}_g$, which travels at angle ϕ_r relative to the vertical. Defining unit vectors $\hat{\boldsymbol{n}}_s$ aligned with $\boldsymbol{r}_s$ and $\hat{\boldsymbol{n}}_g$ aligned opposite to $\boldsymbol{r}_g$, Snell's law demands that their sum $\hat{\boldsymbol{n}}_s + \hat{\boldsymbol{n}}_g$ point in the direction $\hat{\boldsymbol{n}}$ normal to the reflector at $\boldsymbol{x}$. This vector makes angle α with the vertical and angle γ with either the reflected or incident ray. Defining horizontal displacement between reflection position x and source–receiver midpoint x_m to be Δx, and half-offset x_h as the horizontal distance between midpoint x_m and either source position x_s or receiver position x_r, the dip and incident angles α and γ can be related to $\Delta x/ct$ and x_h/ct, as per equations (4.84) and (4.85). To lowest order, $2\ \Delta x/ct \sim \sin\alpha$, and $2\ x_h \big/ ct \sim \sin\gamma$. (If these relations were perfectly true, the dashed line normal to the reflector from the reflection point would intersect the Earth's surface exactly at x_m.) Consequently, the maximum aperture in a migration algorithm determines the maximum dip that can be imaged, while a maximum offset in the data determines the maximum incident angle that can be imaged, within the limitations of ray theory.

and r_g the distance from image point (x, z) to receiver $\left(x_g, 0\right)$

$$r_g = \sqrt{z^2 + \left(x - x_g\right)^2}, \tag{3.80}$$

as depicted in Figure 3.3. Thus, setting

$$r = r_g + r_s, \tag{3.81}$$

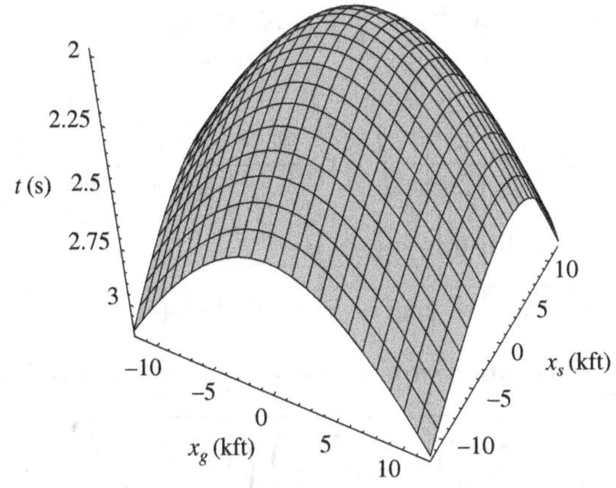

Figure 3.4 Two-dimensional trajectory function $t\left(x_s, x_g\right) = r/c$.

the image function R becomes

$$R(x,z) = \frac{z^2}{(2\pi)^2 c}\int dx_g \int dx_s \int dt \frac{D\left(x_g|x_s;t\right)}{\left(r_s r_g\right)^{3/2}} \int d\omega\sqrt{-i\omega}e^{i\omega(t-r/c)}. \tag{3.82}$$

The frequency integral can be evaluated as the half-derivative filter

$$\int d\omega\sqrt{-i\omega}e^{i\omega(t-r/c)} = 2\pi d_{h+}(t-r/c). \tag{3.83}$$

The time integral is a convolution of d_{h+} with the data function:

$$R(x,z) = \frac{z^2}{(2\pi)c}\int dx_g \int dx_s \frac{\left[d_{h+} * D\right]\left(x_g|x_s;t=r/c\right)}{\left(r_s r_g\right)^{3/2}}. \tag{3.84}$$

As was the case for the *f–k* formula from which this expression is derived, one should not take either the amplitude or the phase too seriously.

According to this equation, full Kirchhoff migration is a weighted summation of the data over all source and receiver positions, along a trajectory of traveltimes corresponding to raypaths from the source position to the image point to the receiver point. This trajectory function is plotted as a function of x_g and x_s in Figure 3.4.

The formula (3.84) does not imply that the summation *must* be over all source and receiver positions. One can pluck out a subset of the complete data set, corresponding to a line on the surface depicted in Figure 3.4. An example would be a constant-offset section. With $x_h = \left(x_g - x_s\right)/2$, and $x_m = \left(x_g + x_s\right)/2$, we can write (3.84) as

$$R(\tilde{\boldsymbol{x}}) = \int dx_h \mathcal{R}(\tilde{\boldsymbol{x}}, x_h), \tag{3.85}$$

with

$$\mathcal{R}(\tilde{\boldsymbol{x}}, x_h) = \frac{z^2}{(2\pi)c} \int dx_m \frac{d_{h+} * D\left(x_g | x_s; t = r/c\right)}{\left(r_s r_g\right)^{3/2}}. \tag{3.86}$$

The trajectory function for this subset is

$$ct = r = \sqrt{z^2 + (\Delta x - x_h)^2} + \sqrt{z^2 + (\Delta x + x_h)^2}, \tag{3.87}$$

with $x_m - x \equiv \Delta x$. This expression can be solved for z, obtaining

$$z = \frac{ct}{2} \sqrt{\left(1 - 4\frac{\Delta x^2}{c^2 t^2}\right)\left(1 - 4\frac{x_h{}^2}{c^2 t^2}\right)}. \tag{3.88}$$

Substituting this expression for z, the ray lengths r_s and r_g can be rewritten as

$$r_s = \frac{ct}{2} - 2\frac{\Delta x x_h}{ct}, \tag{3.89}$$

$$r_g = \frac{ct}{2} + 2\frac{\Delta x x_h}{ct}. \tag{3.90}$$

The angles made by the incident and reflected rays with the vertical are given by the formulas

$$\sin\phi_i = \frac{x_h - \Delta x}{r_s}, \quad \cos\phi_i = \frac{z}{r_s}, \tag{3.91}$$

and

$$\sin\phi_r = \frac{x_h + \Delta x}{r_g}, \quad \cos\phi_r = \frac{z}{r_g}. \tag{3.92}$$

We have defined these angles to be positive when $x_g > x$ and $x_s < x$. Angle of incidence γ is defined to be $\gamma \equiv (\phi_i + \phi_r)/2$, so that

$$\begin{aligned} \cos 2\gamma &= \cos(\phi_i + \phi_r) = \cos\phi_i \cos\phi_r - \sin\phi_i \sin\phi_r \\ &= \frac{z^2 + \Delta x^2 - x_h{}^2}{r_s r_g} = \frac{1 - 8\frac{x_h{}^2}{c^2 t^2} + 16\frac{\Delta x^2 x_h{}^2}{c^4 t^4}}{1 - 16\frac{\Delta x^2 x_h{}^2}{c^4 t^4}}, \end{aligned} \tag{3.93}$$

or

$$\cos\gamma = \sqrt{\frac{1 + \cos 2\gamma}{2}} = \sqrt{\frac{1 - 4\frac{x_h{}^2}{c^2 t^2}}{1 - 16\frac{\Delta x^2 x_h{}^2}{c^4 t^4}}}, \tag{3.94}$$

$$\sin\gamma = \frac{2x_h}{ct} \sqrt{\frac{1 - 4\frac{\Delta x^2}{c^2 t^2}}{1 - 16\frac{\Delta x^2 x_h{}^2}{c^4 t^4}}}, \tag{3.95}$$

and

$$\tan\gamma = \frac{2x_h}{ct}\sqrt{\frac{1-4\frac{\Delta x^2}{c^2t^2}}{1-4\frac{x_h{}^2}{c^2t^2}}}. \tag{3.96}$$

This can also be written as

$$\tan\gamma = \frac{x_h}{z}\left(1-4\frac{\Delta x^2}{c^2t^2}\right). \tag{3.97}$$

Dip angle is $\alpha = (\phi_r - \phi_i)/2$, or

$$\begin{aligned}\cos 2\alpha &= \cos(\phi_r + \phi_i) = \cos\phi_i\cos\phi_r + \sin\phi_i\sin\phi_r \\ &= \frac{z^2 - \Delta x^2 + x_h{}^2}{r_s r_g} = \frac{1-8\frac{\Delta x^2}{c^2t^2}+16\frac{\Delta x^2 x_h{}^2}{c^4t^4}}{1-16\frac{\Delta x^2 x_h{}^2}{c^4t^4}},\end{aligned} \tag{3.98}$$

$$\cos\alpha = \sqrt{\frac{1-4\frac{\Delta x^2}{c^2t^2}}{1-16\frac{\Delta x^2 x_h{}^2}{c^4t^4}}}, \tag{3.99}$$

$$\sin\alpha = \frac{2\Delta x}{ct}\sqrt{\frac{1-4\frac{x_h{}^2}{c^2t^2}}{1-16\frac{\Delta x^2 x_h{}^2}{c^4t^4}}}, \tag{3.100}$$

and

$$\tan\alpha = \frac{2\Delta x}{ct}\sqrt{\frac{1-4\frac{x_h{}^2}{c^2t^2}}{1-4\frac{\Delta x^2}{c^2t^2}}}, \tag{3.101}$$

or, in mixed notation,

$$\tan\alpha = \frac{\Delta x}{z}\left(1-4\frac{x_h{}^2}{c^2t^2}\right). \tag{3.102}$$

Other subsets (such as constant source or receiver location) could similarly be extracted.

Either the full migration formula or its subsets can be generalized to variable velocity. If one is not too concerned with amplitudes, all one need do is replace the straight-ray distances with those for variable-velocity rays. Constant or variable, the Kirchhoff formula lies in asymptopia.

3.5 Prestack *f–k* migration in three dimensions

Migration in *f–k* space of a 3-D data set is a straightforward extension of 2-D migration. Without regard to the question as to whether a full, 5-D seismic experiment is practical, we want a migration operator of the form

$$R(\boldsymbol{k}, \gamma) = D\left(\tilde{\boldsymbol{k}}_g | \tilde{\boldsymbol{k}}_s; \omega\right) \mathcal{A}\left(\tilde{\boldsymbol{k}}_g | \tilde{\boldsymbol{k}}_s; \omega\right), \tag{3.103}$$

where D is a full 3-D surface data set, R is the desired image function, and $\mathcal{A}$ is an amplitude-phase term. Data are a function of the receiver wavenumber 2-vector $\tilde{\boldsymbol{k}}_g = \left(k_{gx}, k_{gy}\right)$, the source wavenumber 2-vector $\tilde{\boldsymbol{k}}_s = \left(k_{sx}, k_{sy}\right)$, and frequency ω. The wavenumbers $\tilde{\boldsymbol{k}}_g$ and $\tilde{\boldsymbol{k}}_s$ are the Fourier conjugates of receiver position $\tilde{\boldsymbol{x}}_g = \left(x_g, y_g\right)$, and source position $\tilde{\boldsymbol{x}}_s = (x_s, y_s)$. The image function depends on image wavenumber $\boldsymbol{k}$ (the Fourier conjugate of image position $\boldsymbol{x}$), and also possibly on incident angle γ. (Conceivably, reflectivity might also depend on source–receiver azimuthal angle ψ, an additional parameter not present in the 2-D case. It appears in three dimensions because sources and receivers are separated in two dimensions, and may assume many different relative orientations. With azimuth included, both the data and image functions are 5-D, though for the models of reflection and transmission considered here, R is supposed to be independent of azimuth.) The amplitude-phase term $\mathcal{A}$ is also 5-D, though it is assumed to vary slowly with the spatial wavenumbers. We could determine $\mathcal{A}$ in the same manner as in Section 3.2.1 or 3.2.2, but for now we will leave it open.

3.5.1 Vertical wavenumbers

The data are presumed to obey the wave equation, so source and receiver vertical wavenumber can be inferred from the horizontal wavenumbers. The wave from source to reflector travels in the positive z direction, so the source vertical wavenumber k_{sz} will have the same sign as frequency ω:

$$k_{sz} = +\frac{\omega}{c}\sqrt{1 - \frac{\left(k_{sx}{}^2 + k_{sy}{}^2\right) c^2}{\omega^2}}. \tag{3.104}$$

Waves from reflector to receiver travel in the negative z direction, so k_{gz} must have sign opposite to ω:

$$k_{gz} = -\frac{\omega}{c}\sqrt{1 - \frac{\left(k_{gx}{}^2 + k_{gy}{}^2\right) c^2}{\omega^2}}. \tag{3.105}$$

The vertical wavenumbers k_{gz} and k_{sz} can be considered the last components of the 3-vectors

$$\boldsymbol{k}_s = \left(k_{sx}, k_{sy}, k_{sz}\right) \tag{3.106}$$

and

$$\boldsymbol{k}_g = \left(k_{gx}, k_{gy}, k_{gz}\right). \tag{3.107}$$

These vectors both have magnitude

$$k_g = k_s = \frac{\omega}{c}. \tag{3.108}$$

3.5.2 Image function wavenumber

We next define the 3-vector $\boldsymbol{k}$ to be

$$\boldsymbol{k} = \left(k_x, k_y, k_z\right) = \boldsymbol{k}_g - \boldsymbol{k}_s. \tag{3.109}$$

The wavenumber vector $\boldsymbol{k}$ points upward and bisects the source and receiver wavenumber vectors $\boldsymbol{k}_s$ and $\boldsymbol{k}_g$. If a specular reflector exists at a reflection point, $\boldsymbol{k}$ is normal to it (see Figure H.1). This occurs because the Fourier transforms have organized the data into components of particular dips. The element $D\left(\tilde{\boldsymbol{k}}_g|\tilde{\boldsymbol{k}}_s;\omega\right)$ contains all the data pertaining to reflectors normal to $\boldsymbol{k} = \boldsymbol{k}_g - \boldsymbol{k}_s$. A Fourier transform of the image function $R(\boldsymbol{x}, \gamma)$ to $R(\boldsymbol{k}, \gamma)$ performs the same reflector dip decomposition on the image, so we are justified in identifying the $\boldsymbol{k}$ of equation (3.109) with the image function wavenumber in (3.103).

The vertical component of image function wavenumber honors a 3-D double-square-root equation:

$$k_z \equiv k_{gz} - k_{sz} = -\frac{\omega}{c}\left(\sqrt{1 - \frac{\left(k_{sx}{}^2 + k_{sy}{}^2\right)c^2}{\omega^2}} + \sqrt{1 - \frac{\left(k_{gx}{}^2 + k_{gy}{}^2\right)c^2}{\omega^2}}\right). \tag{3.110}$$

3.5.3 Midpoint and offset wavenumbers

We can, if desired, convert source–receiver coordinates to midpoint–offset coordinates. Midpoint and offset are defined as

$$\begin{aligned}\tilde{\boldsymbol{x}}_m &= \left(\tilde{\boldsymbol{x}}_g + \tilde{\boldsymbol{x}}_s\right)/2,\\ \tilde{\boldsymbol{x}}_h &= \left(\tilde{\boldsymbol{x}}_g - \tilde{\boldsymbol{x}}_s\right)/2,\end{aligned} \tag{3.111}$$

with corresponding midpoint and offset wavenumbers

$$\begin{aligned}\tilde{\boldsymbol{k}}_m &= \tilde{\boldsymbol{k}}_g - \tilde{\boldsymbol{k}}_s,\\ \tilde{\boldsymbol{k}}_h &= \tilde{\boldsymbol{k}}_g + \tilde{\boldsymbol{k}}_s.\end{aligned} \tag{3.112}$$

Note that the midpoint 2-vector comprises the first two coordinates of the image function wavenumber $\boldsymbol{k}$:

$$\boldsymbol{k} = \left(\tilde{\boldsymbol{k}}_m, k_z\right) = \left(k_x, k_y, k_z\right). \tag{3.113}$$

Equation (3.110) can be solved for frequency as a function of $\tilde{\boldsymbol{k}}_m$, $\tilde{\boldsymbol{k}}_h$, and k_z:

$$\omega = -\frac{k_z c}{2}\sqrt{1 + \tilde{k}_m{}^2/k_z{}^2 + \tilde{k}_h{}^2/k_z{}^2 + \left(\tilde{\boldsymbol{k}}_m \cdot \tilde{\boldsymbol{k}}_h\right)^2/k_z{}^4}. \tag{3.114}$$

Using this relation, the vertical wavenumbers k_{gz} and k_{sz} can also be expressed in terms of $\tilde{\boldsymbol{k}}_m$, $\tilde{\boldsymbol{k}}_h$, and k_z:

$$k_g z = +\frac{k_z}{2}\left(1 - \frac{\tilde{\boldsymbol{k}}_m \cdot \tilde{\boldsymbol{k}}_h}{k_z^2}\right), \tag{3.115}$$

$$k_s z = -\frac{k_z}{2}\left(1 + \frac{\tilde{\boldsymbol{k}}_m \cdot \tilde{\boldsymbol{k}}_h}{k_z^2}\right). \tag{3.116}$$

The offset wavenumber can also be extended into a 3-vector:

$$\boldsymbol{k}_h = \boldsymbol{k}_g + \boldsymbol{k}_s. \tag{3.117}$$

The offset 3-vector is normal to $\boldsymbol{k}$, hence lies in the plane of specular reflectors. Its z-component is

$$k_h z = -\frac{\tilde{\boldsymbol{k}}_m \cdot \tilde{\boldsymbol{k}}_h}{2k_z}. \tag{3.118}$$

3.5.4 Dip and incident angle

Since $\boldsymbol{k}$ is normal to specular reflectors, the ratio of horizontal to vertical image function wavenumbers determines the reflector dip:

$$\left|\tilde{\boldsymbol{k}}_m / k_z\right| \equiv |\tan\alpha|. \tag{3.119}$$

In two dimensions, the ratio of offset wavenumber k_h to vertical wavenumber k_z determines incident angle γ. In three dimensions, the relation is not quite so simple. The angle between the incoming wavenumber $\boldsymbol{k}_s$ and the outgoing wavenumber $\boldsymbol{k}_g$ is twice the angle of incidence:

$$\cos 2\gamma = -\frac{c^2}{\omega^2}\boldsymbol{k}_g \cdot \boldsymbol{k}_s. \tag{3.120}$$

In terms of $\tilde{\boldsymbol{k}}_m$, $\tilde{\boldsymbol{k}}_h$, and k_z,

$$\cos 2\gamma = 2\cos^2\gamma - 1 = -\frac{\left(\tilde{k}_{h^2} - \tilde{k}_{m^2} - k_{z^2} + (\tilde{\boldsymbol{k}}_m \cdot \tilde{\boldsymbol{k}}_h)^2/k_{z^2}\right)}{k_{z^2} + \tilde{k}_{m^2} + \tilde{k}_{h^2} + (\tilde{\boldsymbol{k}}_m \cdot \tilde{\boldsymbol{k}}_h)^2/k_{z^2}}, \tag{3.121}$$

or

$$\cos^2\gamma = \frac{\tilde{\boldsymbol{k}}_m{}^2 + k_z{}^2}{k_z{}^2 + \tilde{k}_m{}^2 + \tilde{k}_h{}^2 + (\tilde{\boldsymbol{k}}_m \cdot \tilde{\boldsymbol{k}}_h)^2/k_z{}^2}. \tag{3.122}$$

Continuing,

$$\sin^2\gamma = \frac{\tilde{k}_{h^2} + (\tilde{\boldsymbol{k}}_m \cdot \tilde{\boldsymbol{k}}_h)^2/k_z{}^2}{k_{z^2} + \tilde{k}_{m^2} + \tilde{k}_{h^2} + (\tilde{\boldsymbol{k}}_m \cdot \tilde{\boldsymbol{k}}_h)^2/k_{z^2}}, \tag{3.123}$$

and

$$\tan^2\gamma = \frac{\tilde{k}_h{}^2}{k_z{}^2}\frac{1 + (\tilde{\boldsymbol{k}}_m \cdot \tilde{\boldsymbol{k}}_h)^2/\,\tilde{k}_h{}^2 k_z{}^2}{1 + \tilde{\boldsymbol{k}}_m{}^2/k_z{}^2}. \tag{3.124}$$

If $\tilde{\boldsymbol{k}}_m$ and $\tilde{\boldsymbol{k}}_h$ are aligned, then the expression for $\tan^2\gamma$ reduces to the 2-D relation (3.32).

Frequency ω can be expressed in terms of $\tilde{\boldsymbol{k}}_m$, k_z, and γ:

$$\omega = -\frac{k_z c}{2\cos\gamma}\sqrt{1 + \tilde{k}_m{}^2/k_z{}^2}, \tag{3.125}$$

Magnitudes of $\boldsymbol{k}$ and $\boldsymbol{k}_h$ are

$$k = |\boldsymbol{k}_g - \boldsymbol{k}_s| = \sqrt{2\left(\frac{\omega^2}{c^2} - \boldsymbol{k}_g \cdot \boldsymbol{k}_s\right)} = \sqrt{2\frac{\omega^2}{c^2}(1 + \cos 2\gamma)} = 2\frac{|\omega\cos\gamma|}{c}, \tag{3.126}$$

$$k_h = |\boldsymbol{k}_g + \boldsymbol{k}_s| = \sqrt{2\left(\frac{\omega^2}{c^2} + \boldsymbol{k}_g \cdot \boldsymbol{k}_s\right)} = \sqrt{2\frac{\omega^2}{c^2}(1 - \cos 2\gamma)} = 2\frac{|\omega\sin\gamma|}{c}. \tag{3.127}$$

3.5.5 Image amplitude and phase

We now turn attention to the amplitude-phase function $\mathcal{A}$. In the spirit of Section 3.2.2, we choose $\mathcal{A}$ to give us the "right" data function in the case of a single dipping reflector, and use the principle of superposition to claim we have found the right answer for an arbitrary image function. We choose the x-axis to align with reflector dip, so that

$$R(x,\gamma) = R_0(\gamma)\delta\left((z - z_0)\cos\alpha - x\sin\alpha\right). \tag{3.128}$$

The space–time formula for the reflection response D can be determined from the method of images. The desired response is a delta function

$$D\left(\tilde{\boldsymbol{x}}_g|\tilde{\boldsymbol{x}}_s;t\right) = \frac{R_0(\gamma)\delta(t-2r/c)}{2r}, \tag{3.129}$$

where r is half the total distance from source to reflector to receiver. The amplitude decay of $1/r$ in the impulse response is referred to as spherical divergence. This is an asymptotic formula in that we have inserted a plane-wave specular reflection coefficient $R_0(\gamma)$ into the formula. Consequently, we will be content with an asymptotic relation between (3.128) and (3.129), or its frequency-domain equivalent

$$D\left(\tilde{\boldsymbol{x}}_g|\tilde{\boldsymbol{x}}_s;\omega\right) = e^{i2\omega r/c}\frac{R_0(\gamma)}{2r}. \tag{3.130}$$

In terms of wavenumber, R is

$$\begin{aligned} R(k,\gamma) &= R_0(\gamma)\int\int\int dx\,dy\,dz\,\delta\left((z-z_0)\cos\alpha - x\sin\alpha\right)e^{-i(k_x x+k_y y+k_z z)} \\ &= (2\pi)^2 R_0(\gamma)e^{-ik_z z_0}\delta\left(k_y\right)\delta\left(k_x\cos\alpha+k_z\sin\alpha\right). \end{aligned} \tag{3.131}$$

Invoking (3.103), the frequency–wavenumber reflection response D must be

$$D\left(\tilde{\boldsymbol{k}}_g|\tilde{\boldsymbol{k}}_s;\omega\right) = \frac{(2\pi)^2\,R_0(\gamma)}{\mathcal{A}\left(\tilde{\boldsymbol{k}}_g|\tilde{\boldsymbol{k}}_s;\omega\right)}e^{-ik_z z_0}\delta\left(k_y\right)\delta\left(k_x\cos\alpha+k_z\sin\alpha\right). \tag{3.132}$$

The delta functions in D are satisfied when

$$\begin{aligned} k_y &= 0, \\ k_x &= -k_z\tan\alpha, \\ 1+\tilde{\boldsymbol{k}}_m{}^2/k_z{}^2 &= \sec^2\alpha, \\ \tan^2\gamma &= \frac{k_{hx}{}^2+k_{hy}{}^2\cos^2\alpha}{k_z{}^2}. \end{aligned} \tag{3.133}$$

In space-frequency, the data function becomes

$$\begin{aligned} D\left(\tilde{\boldsymbol{x}}_g|\tilde{\boldsymbol{x}}_s;\omega\right) = \frac{1}{(2\pi)^2}\int\int d^2k_g d^2k_s \frac{R_0(\gamma)}{\mathcal{A}\left(\tilde{\boldsymbol{k}}_g|\tilde{\boldsymbol{k}}_s;\omega\right)}e^{i\left(\tilde{\boldsymbol{k}}_g\cdot\tilde{\boldsymbol{x}}_g-\tilde{\boldsymbol{k}}_s\cdot\tilde{\boldsymbol{x}}_s-k_z z_0\right)} \\ \times\delta\left(k_y\right)\delta\left(k_x\cos\alpha+k_z\sin\alpha\right). \end{aligned} \tag{3.134}$$

Make a change of integration variables to midpoint and offset wavenumber:

$$\begin{aligned} D\left(\tilde{\boldsymbol{x}}_g|\tilde{\boldsymbol{x}}_s;\omega\right) &= \frac{1}{4(2\pi)^2}\int\int d^2k_m d^2k_h \, \frac{R_0(\gamma)}{\mathcal{A}\left(\tilde{\boldsymbol{k}}_g|\tilde{\boldsymbol{k}}_s;\omega\right)} e^{i\left(\tilde{\boldsymbol{k}}_g\cdot\tilde{\boldsymbol{x}}_g-\tilde{\boldsymbol{k}}_s\cdot\tilde{\boldsymbol{x}}_s-k_z z_0\right)} \\ &\quad\times \delta\left(k_y\right)\delta\left(k_x\cos\alpha+k_z\sin\alpha\right) \\ &= \frac{1}{16\pi^2}\int d^2k_h \frac{1}{\cos\alpha}\frac{R_0(\gamma)}{\mathcal{A}\left(\tilde{\boldsymbol{k}}_g|\tilde{\boldsymbol{k}}_s;\omega\right)} e^{i\left(\tilde{\boldsymbol{k}}_h\cdot\tilde{\boldsymbol{x}}_h-k_z z_0+x_m\tan\alpha\right)}. \end{aligned} \quad (3.135)$$

Given that $k_y = 0$ and $k_x = -k_z\tan\alpha$, equation (3.114) can be solved for k_z as a function of ω and $\tilde{\boldsymbol{k}}_h$:

$$k_z = -\mathrm{Sign}(\omega)\sqrt{\left(\frac{4\omega^2}{c^2} - k_{hy}{}^2\right)\cos^2\alpha - k_{hx}{}^2}. \quad (3.136)$$

In the far field, the remaining integrals can be evaluated via stationary phase. The result is

$$D\left(\tilde{\boldsymbol{x}}_g|\tilde{\boldsymbol{x}}_s;\omega\right) \simeq -\frac{i\omega z_m\cos\alpha}{4\pi c r^2}\frac{R_0(\underset{\wedge}{\gamma})}{\mathcal{A}\left(\underset{\wedge}{\tilde{\boldsymbol{k}}_g}|\underset{\wedge}{\tilde{\boldsymbol{k}}_s};\omega\right)} e^{2i\omega r/c}, \quad (3.137)$$

with $\underset{\wedge}{\tilde{\boldsymbol{k}}_g}$ and $\underset{\wedge}{\tilde{\boldsymbol{k}}_s}$ the stationary values of the source and receiver wavenumbers, and $\underset{\wedge}{\gamma}$ the stationary value of angle of incidence.

Comparing equation (3.137) to the dipping-layer formula (3.130), we discover that the amplitude-phase function must be

$$\mathcal{A}\left(\underset{\wedge}{\tilde{\boldsymbol{k}}_g}|\underset{\wedge}{\tilde{\boldsymbol{k}}_s};\omega\right) \equiv -\frac{i\omega z_m\cos\alpha}{2\pi c r}. \quad (3.138)$$

The stationary value of angle of incidence is given by the relation

$$\cos\underset{\wedge}{\gamma} = \frac{z_m\cos\alpha}{r}, \quad (3.139)$$

The stationary values of the various wavenumbers are

$$\underset{\wedge}{k_{hx}} = \frac{2\omega}{c}\frac{\cos^2\alpha\, x_h}{r}, \qquad \underset{\wedge}{k_{hy}} = \frac{2\omega}{c}\frac{y_h}{r}, \quad (3.140)$$

and

$$\underset{\wedge}{k_z} = -\frac{2\omega}{c}\frac{\cos^2\alpha\, z_m}{r}. \quad (3.141)$$

With this last relation, we can write the amplitude-phase function (3.138) in terms of vertical wavenumber as

$$\mathcal{A}\left(\underset{\wedge}{\tilde{\boldsymbol{k}}_g}|\underset{\wedge}{\tilde{\boldsymbol{k}}_s};\,\omega\right)=\frac{i\underset{\wedge}{k_z}}{4\pi\,\cos\alpha}. \tag{3.142}$$

The dip angle α is also expressible in terms of wavenumbers as

$$\cos\alpha=\frac{1}{\sqrt{1+\underset{\wedge}{\tilde{k}_m}^2/\underset{\wedge}{k_z}^2}}, \tag{3.143}$$

which allows us to write the dip-independent expression

$$\mathcal{A}\left(\tilde{\boldsymbol{k}}_g|\tilde{\boldsymbol{k}}_s;\,\omega\right)=\frac{ik_z\sqrt{1+\tilde{k}_m{}^2/k_z{}^2}}{4\pi}. \tag{3.144}$$

With this choice for $\mathcal{A}$, the operator (3.103) will (asymptotically, anyway) yield a true-amplitude reflectivity function for reflectors of any dip or orientation, which is to say, it will convert any superposition of dipping-layer data functions of the form (3.129) into a superposition of true-amplitude reflections.

3.6 Prestack reverse-time migration

One can implement prestack reverse-time migration by continuing reflection data backward in time while continuing forward in time a wave from the source, forming two wavefields $P_s(x, y, z, t)$ and $P_r(x, y, z, t)$. We know that these two wavefields are related through the reflectivity, and would like to infer from them the reflectivity function. Try the following, and assume that at any point (x, y, z) in space,

$$P_r(x, y, z, t) = R(x, y, z)P_i(x, y, z, t) + n(x, y, z, t). \tag{3.145}$$

That is, at every point in space and time, the reflected wave has a component proportional to reflectivity times the incident wave at the same point in space and time. The reflected wave may have additional components, but we hope they will not correlate with the incident wave at (x, y, z, t). This suggests a weighted integral over time:

$$\int_{-\infty}^{\infty} dt\,P_i(x, y, z, t)P_r(x, y, z, t) = R(x, y, z)\int_{-\infty}^{\infty} dt\,P_i(x, y, z, t)^2 + \int_{-\infty}^{\infty} dt\,P_i(x, y, z, t)n(x, y, z, t). \tag{3.146}$$

In the hope that the last term is negligible, we have

$$R(x, y, z) \simeq \frac{\int_{-\infty}^{\infty} dt\, P_i(x, y, z, t) P_r(x, y, z, t)}{\int_{-\infty}^{\infty} dt\, P_i(x, y, z, t)^2}. \tag{3.147}$$

By neglecting the last term we open the door to artifacts. However, it is reasonable to hope for decent results. To test the basic concept, look at a simple case for which we can produce analytic results, namely, a constant-velocity layered medium. Put a single source at position $(x_s, y_s, 0)$, and endow it with a Gaussian wavelet (Appendix E, Section E.3) of characteristic width t_w. The forward-continued response to this source is

$$P_i(x, y, z, t) = \frac{1}{\sqrt{\pi} t_w \tau_i} e^{-(t-\tau_i)^2/{t_w}^2}, \tag{3.148}$$

with

$$\tau_i = \sqrt{(x - x_s)^2 + (y - y_s)^2 + z^2}/c. \tag{3.149}$$

With the given normalization, the integral over all time of P_i is $1/r_i$. In the limit of small t_w it approaches the delta function form

$$P_i(x, y, z, t) \underset{t_w \to 0}{\to} \delta(t - \tau_i)/\tau_i. \tag{3.150}$$

The forward-continued wave is oblivious to reflectors in the medium.

Suppose the medium does have flat reflectors at depths a_j. If reflectivity is small enough that transmission effects are negligible, then the reflection response at zero depth is

$$P_r(x, y, 0, t) = \sum_j \frac{R_j}{\sqrt{\pi} t_w \tau_j(x,y,0)} e^{-(t-\tau_j(x,y,0))^2/{t_w}^2}, \tag{3.151}$$

where

$$\tau_j(x, y, 0) = \sqrt{(x - x_s)^2 + (y - y_s)^2 + 4{a_j}^2}/c. \tag{3.152}$$

If we backward-propagate the reflected wave, we obtain

$$P_r(x, y, z, t) = \sum_j \frac{R_j}{\sqrt{\pi} t_w \tau_j(x, y, z)} e^{-(t-\tau_j(x, y, z))^2/{t_w}^2}, \tag{3.153}$$

where

$$\tau_j(x, y, z) = \sqrt{(x - x_s)^2 + (y - y_s)^2 + \left(2a_j - z\right)^2}/c. \tag{3.154}$$

This expression holds both above and beneath the reflectors because the backward-propagated field, unlike the actual reflected wavefield, does not see the reflectors.

Now we try the reflectivity formula (3.147). The denominator in this case is

$$\int_{-\infty}^{\infty} dt\, P_i(x,y,z,t)^2 = \frac{1}{\pi t_w{}^2 \tau_i{}^2}\int_{-\infty}^{\infty} dt e^{-2(t-\tau_i)^2/t_w{}^2} = \frac{1}{\sqrt{2\pi}\, t_w \tau_i{}^2}. \tag{3.155}$$

The numerator evaluates to

$$\int_{-\infty}^{\infty} dt\, P_i(x,y,z,t) P_r(x,y,z,t) = \sum_j \frac{R_j}{\pi t_w{}^2 \tau_i \tau_j}\int_{-\infty}^{\infty} dt e^{-[(t-\tau_i)^2+(t-\tau_j)^2]/t_w{}^2} = \frac{1}{\sqrt{2\pi}\, t_w \tau_i}\sum_j \frac{R_j}{\tau_j} e^{-(\tau_i-\tau_j)^2/2t_w{}^2}. \tag{3.156}$$

Combining numerator and denominator,

$$R(x,y,z) \simeq \sum_j \frac{\tau_i}{\tau_j} R_j e^{-(\tau_i-\tau_j)^2/2t_w{}^2}. \tag{3.157}$$

Since t_w is small, the exponential functions in the reflectivity estimate should be sharply peaked around points where $\tau_i = \tau_j$. At the jth such point, the peak occurs where

$$\sqrt{(x-x_s)^2 + (y-y_s)^2 + z^2} = \sqrt{(x-x_s)^2 + (y-y_s)^2 + (2a_j - z)^2}, \tag{3.158}$$

the obvious solution to which is $z = a_j$. For z near a_j, define $\epsilon = z - a_j$, in terms of which

$$\tau_i - \tau_j = \frac{1}{c}\left(\sqrt{(x-x_s)^2 + (y-y_s)^2 + (a_j+\epsilon)^2} - \sqrt{(x-x_s)^2 + (y-y_s)^2 + (a_j-\epsilon)^2}\right) = \frac{2a_j\epsilon}{c\sqrt{(x-x_s)^2 + (y-y_s)^2 + a_j{}^2}} + O\left(\epsilon^3\right). \tag{3.159}$$

For t_w sufficiently small, only the term lowest order in ϵ is significant, whence (replacing the ratio τ_i/τ_j by 1),

$$R(x,y,z) \simeq \sum_j R_j \exp\left[-\frac{2\left(z-a_j\right)^2}{c^2 t_w{}^2\left(1 + \left((x-x_s)^2 + (y-y_s)^2\right)/a_j{}^2\right)}\right]. \tag{3.160}$$

In this instance, the formula (3.147) gives a bandlimited reflectivity estimate, with responses proportional to the reflection coefficients R_j centered at the true reflector depths a_j. The characteristic width of the migrated response increases with distance between source and observation point, and decreases with reflector depth a_j. Summing images from multiple source points would mitigate this effect, and could also help with any artifacts.

Equation (3.147) made the assumption that reflectivity depends on spatial position only, and not on incoming and reflected angles. Unlike asymptotic methods, in which waves know the direction they are traveling, a finite-difference reverse-time operator does not so easily distill directional information. This is not to say that an amplitude-preserving reverse-time migration algorithm is impossible, but it would require a more sophisticated imaging algorithm than (3.147).

Amplitude and artifact issues aside, the ability of reverse-time migration to handle two-way wave propagation through complex structure enhances its appeal. A 3-D prestack reverse-time migration will be computationally very intensive and will consume large amounts of computer memory and bandwidth. Even so, as computer power continues to increase, interest in reverse-time migration continues to grow.

Exercises

3.1 Verify geometrically that formulas (3.48) and (3.81) for r are equivalent, and that the ray-theoretical incident angle is given by (3.50).

3.2 Derive an amplitude-preserving migration formula analogous to equation (3.65) in which the 2.5-D divergence correction is performed after migration.

3.3 Using the formula (3.65), derive the 2.5-D response to a 2-D point image.

3.4 Evaluate the integral

$$D(\tilde{\boldsymbol{x}}_g|\tilde{\boldsymbol{x}}_s;\omega) = \frac{1}{16\pi^2}\int d^2k_h \frac{\mathcal{A}\left(\tilde{\boldsymbol{k}}_g|\tilde{\boldsymbol{k}}_s;\omega\right)}{\cos\alpha} R_0(\gamma) e^{i(\tilde{\boldsymbol{k}}_h\cdot\tilde{\boldsymbol{x}}_h - k_z(z_0 + x_m\tan\alpha))}$$

using the stationary phase approximation, under the assumption that $\mathcal{A}$ and R_0 are slowly varying in the wavenumbers, and that the dipping-layer conditions (3.133) and (3.136) are met.

3.5 Starting from the 2.5-D *f–k* formula (3.65), use the stationary phase approximation to derive space–time modeling and migration equations.

4
Migration limitations

4.1 Perfect and imperfect migrations

A perfect 2-D migration should image a reflecting point as a point and a reflecting surface as a line. In practice, we know that limitations in the data and in the processing algorithms prevent this from happening. Among the things that contribute to imperfections in the seismic image are limited bandwidth and coarse sampling intervals in the data, missing data, and limitations to processing aperture and reflector dip in the processing algorithm. Migration with the wrong velocity likewise results in an imperfect image.

In this chapter we look at the effect of some of these limitations on the seismic image. We confine the analysis to a constant-velocity propagation model. Results in many cases can be generalized to variable velocity under the assumption of "locally constant velocities", wherein propagation velocity can be considered constant in some neighborhood of each image point.

4.2 Effects of finite bandwidth

One of the simplest and easiest to analyze situations is the effect of finite bandwidth (Berkhout, 1984). In what follows, we fix the medium velocity to the constant value c. Suppose the data, otherwise white, have a minimum angular frequency ω_i and a maximum angular frequency ω_f. Frequency is related to spatial wavenumber through the dispersion relation of the medium.

4.2.1 Zero-offset two-dimensional data

For zero-offset data in two dimensions, the relation between frequency and horizontal and vertical wavenumbers k_x and k_z is

$$\omega = -\frac{k_z c}{2}\sqrt{1 + {k_x}^2/{k_z}^2}. \tag{4.1}$$

For a given value of ω, equation (4.1) describes a circle in wavenumber space. If absolute frequency is limited to the region $|\omega| \in \left(\omega_i, \omega_f\right)$, then wavenumbers are confined to the region

$$\frac{2\,\omega_i}{c} < \sqrt{{k_x}^2 + {k_z}^2} < \frac{2\omega_f}{c}, \tag{4.2}$$

which describes a "doughnut" shape (Figure 4.1).

Migration to a bandlimited point

Given bandlimited data, a point reflector at (x_0, z_0) will image, not to a point delta function, but to the bandlimited point

$$\mathcal{R}_p\left(x, z; x_0, z_0; \omega_i, \omega_f\right) = \frac{R}{(2\pi)^2} \int\int_{\mathcal{D}(\omega_i,\omega_f)} dk_x dk_z e^{i(k_x(x-x_0)+k_z(z-z_0))}, \tag{4.3}$$

where $\mathcal{D}\left(\omega_i, \omega_f\right)$ is the doughnut-shaped region described by the relation (4.2) and illustrated in Figure 4.1. Defining wavenumber and distance vectors $\tilde{\boldsymbol{k}} = (k_x, k_z)$ and $\tilde{\boldsymbol{\rho}} = (x - x_0, z - z_0)$, with magnitudes

$$k = \sqrt{{k_x}^2 + {k_z}^2}, \tag{4.4}$$

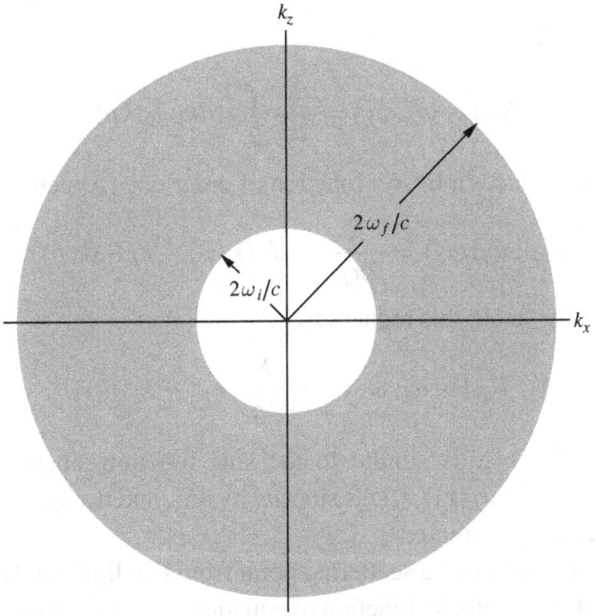

Figure 4.1 Frequency–wavenumber plot. Where the seismic data are limited to frequencies in the range $\left(\omega_i, \omega_f\right)$, the migrated data are limited in wavenumber to the doughnut-shaped region described by the relation (4.2).

and

$$\rho = \sqrt{(x - x_0)^2 + (z - z_0)^2}, \tag{4.5}$$

the integral (4.3) can be written

$$\begin{aligned} \mathcal{R}_p\left(x, z; x_0, z_0; \omega_i, \omega_f\right) &= \mathcal{R}_p\left(\rho; \omega_i, \omega_f\right) \\ &= \frac{R}{(2\pi)^2} \int_{k_i}^{k_f} dkk \int_{-\pi}^{\pi} d\theta e^{i\tilde{k}\cdot\tilde{\rho}} \\ &= \frac{R}{(2\pi)^2} \int_{k_i}^{k_f} dkk \int_{-\pi}^{\pi} d\theta e^{i\tilde{k}\rho\cos\theta}, \end{aligned} \tag{4.6}$$

with

$$k_i = \frac{2\omega_i}{c}, \tag{4.7}$$

and

$$k_f = \frac{2\omega_f}{c}. \tag{4.8}$$

The θ-integral is recognizable as a Bessel function of zero order (see Appendix E.10):

$$J_0(k\rho) \equiv \frac{1}{2\pi} \int_{-\pi}^{\pi} d\theta e^{ik\rho\cos\theta}, \tag{4.9}$$

so that

$$\mathcal{R}_p\left(\rho; \omega_i, \omega_f\right) = \frac{R}{2\pi} \int_{k_i}^{k_f} dkk J_0(k\rho). \tag{4.10}$$

The k-integral produces a Bessel function of order one (Appendix E.10):

$$\mathcal{R}_p\left(\rho; \omega_i, \omega_f\right) = \frac{R}{2\pi\rho} \left(k_f J_1\left(k_f\rho\right) - k_i J_1\left(k_i\rho\right)\right). \tag{4.11}$$

For ρ sufficiently small, this reduces to

$$\mathcal{R}_p\left(\rho; \omega_i, \omega_f\right) \underset{\rho\to 0}{\rightarrow} \frac{R}{4\pi} \left(k_f^2 - k_i^2\right). \tag{4.12}$$

The function $J_1(x)/x$ is similar to the sinc function sin(x)/x, as shown in Figure 4.2. For large x, $J_1(x)/x$ lags sin(x)/x by 45°, and decays faster by a factor of $\sqrt{2/\pi x}$, as seen in the figure.

A perfect 2-D migration of a scattering point from bandlimited data is thus not a point but rather the Bessel-sinc function of equation (4.11), as shown in Figure 4.3. The peak amplitude of the bandlimited reflectivity is given by (4.12). The resolution of the point is the same in all directions. That is, the point is equally well resolved in the x-direction as in the z-direction, or all directions in between. This

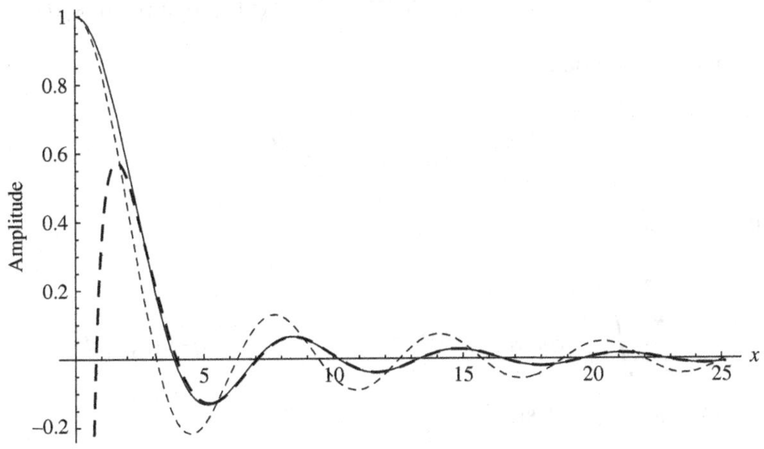

Figure 4.2 Bessel-sinc function. The solid line is $2\,J_1(x)/\,x$ (where J_1 is a Bessel function of the first kind of order 1. The finely dashed line is $\sin(x)/x$. The coarsely dashed line is $\sqrt{\frac{8}{\pi x^3}}\,\sin(x-\pi/4)$, the large-$x$ asymptotic form for $2\,J_1(x)/\,x$.

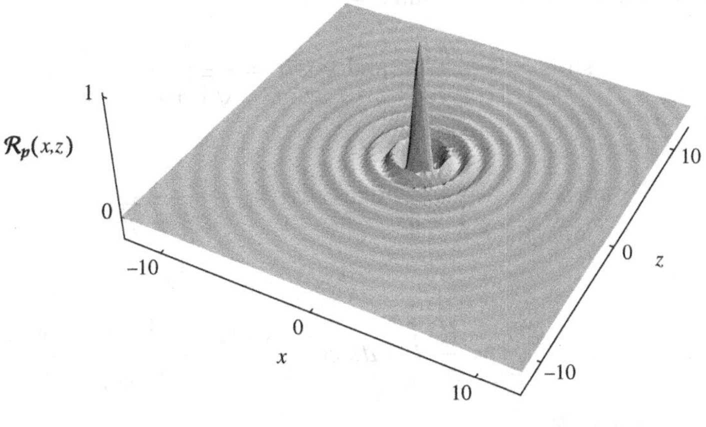

Figure 4.3 Bandlimited point reflection. A "perfect" 2-D bandlimited migration of a point reflection, resulting in a directionally invariant Bessel-sinc function (4.12). For this figure, k_i and k_f were chosen as 1 and 5 in units of inverse length, respectively, and reflection strength normalized to a peak amplitude of 1.

result presupposes that the data is adequately sampled in the x-direction, which is usually not the case.

Migration to a bandlimited line

A 2-D line image with dip tangent $q = \tan\alpha$ passing through the point (x_r, z_r) has the k-space form

$$\mathcal{R}_S\left(k_x, k_z; x_r, z_r; q, \omega_i, \omega_f\right) = 2\pi Re^{-i(k_x x_r + k_z z_r)}\sqrt{1+q^2}\delta\left(k_x + k_z q\right), \quad (4.13)$$

where, for bandlimited data,

$$\frac{2\omega_i}{c} < \sqrt{{k_x}^2 + {k_z}^2} < \frac{2\omega_f}{c}. \quad (4.14)$$

In x and z, we have

$$\mathcal{R}_S\left(x, z; x_r, z_r; q, \omega_i, \omega_f\right) = \frac{R}{2\pi}\int\int_{\mathcal{D}} dk_x dk_z e^{i(k_x(x-x_r)+k_z(z-z_r))}\sqrt{1+q^2}\delta\left(k_x + k_z q\right), \quad (4.15)$$

or, if $k_z = -k\cos\alpha$, $k_x = k\sin\alpha$,

$$\mathcal{R}_S\left(x, z; x_r, z_r; p, \omega_i, \omega_f\right) = \frac{R}{2\pi}\int_{k_i}^{k_f} dk \int_{-\pi}^{\pi} d\alpha\sqrt{1+q^2}\cdot e^{ik((x-x_r)\sin\alpha-(z-z_r)\cos\alpha)}\delta(\sin\alpha - \cos\alpha q). \quad (4.16)$$

The delta function is satisfied at $\tan\alpha = q$, i.e.

$$\sin\alpha = \frac{q}{\sqrt{1+q^2}} \text{ and } \cos\alpha = \frac{1}{\sqrt{1+q^2}}, \quad (4.17)$$

or

$$\sin\alpha = -\frac{q}{\sqrt{1+q^2}} \text{ and } \cos\alpha = -\frac{1}{\sqrt{1+q^2}}. \quad (4.18)$$

Thus,

$$\begin{aligned}\mathcal{R}_S\left(x, z; x_r, z_r; q, \omega_i, \omega_f\right) &= \frac{R}{\pi}\int_{k_i}^{k_f} dk\cos\left(\frac{k\left((z-z_r)-(x-x_r)q\right)}{\sqrt{1+q^2}}\right)\\ &= \frac{\sqrt{1+q^2}R}{\pi\left((z-z_r)-(x-x_r)q\right)}\\ &\times\left(\sin\left(\frac{k_f\left((z-z_r)-(x-x_r)q\right)}{\sqrt{1+q^2}}\right)-\sin\left(\frac{k_i\left((z-z_r)-(x-x_r)q\right)}{\sqrt{1+q^2}}\right)\right).\end{aligned} \quad (4.19)$$

As $z \to z_r + q(x - x_r)$, the reflectivity function approaches its peak amplitude

$$\mathcal{R}_S \to \frac{R}{\pi}\left(k_f - k_i\right). \quad (4.20)$$

Since migration has the effect of shifting vertical wavelengths to lower numbers, one might reasonably ask whether migration has affected resolution. For zero dip, the answer is clearly no. At steep dips, *vertical* resolution or k_z is certainly affected,

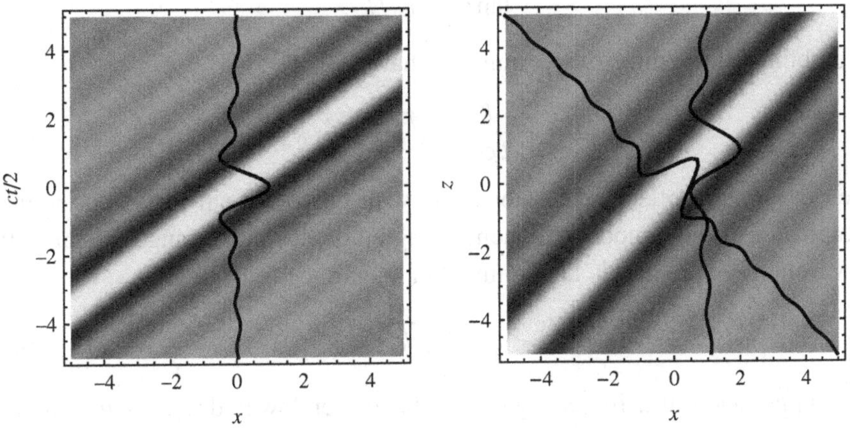

Figure 4.4 The effect of migration on frequency bandwidth. On the left is an unmigrated reflector, of dip 45°, with data bandwidth $\omega\epsilon\ (c/2\lambda_0, 5c/2\lambda_0)$, with λ_0 a characteristic wavelet distance. Rescaled to dimensionless depth $z = ct/2\lambda_0$, a vertical "wiggle trace" sees a sinc function $(\sin(5z) - \sin(z))\ /\ (4z)$ times the peak amplitude (4.20). On the right is the migrated reflector. A vertical wiggle trace passing through the reflector sees the lower-frequency sinc function $(\sin(5z\cos(\pi/4) - \sin(z\cos(\pi/4)))/(4z\cos(\pi/4))$. However, a wiggle trace passing through the reflector normal to it sees the original bandwidth.

extended on the low end and curtailed on the high end. However, wavelength *magnitude* k is not affected by migration, which means that resolution in the direction normal to dip is not affected by zero-offset migration. Vertical measurements, such as on a migrated seismic trace, may appear to be lower frequency when dipping layers are encountered, but this is because the layers are being encountered at an angle. Plot the image along lines normal to the dipping layer, and resolution is untainted by migration, as illustrated in Figure 4.4.

4.2.2 Effects of finite bandwidth on finite-offset data

For non-zero offsets, the situation is slightly more complicated. Data in k–ω space are of the form $D(k_m, k_h, \omega)$, with k_m the midpoint wavenumber, k_h the offset wavenumber, and ω the temporal angular frequency. The image has the form $R(k_x, k_z, \gamma)$, with k_x and k_z the horizontal and vertical wavenumbers, and γ the angle of incidence. The relations between the data and image parameters are

$$k_x = k_m, \tag{4.21}$$

$$\omega = -\frac{k_z c}{2}\sqrt{\left(1 + k_m{}^2/k_z{}^2\right)\left(1 + k_h{}^2/k_z{}^2\right)}, \tag{4.22}$$

$$\tan\gamma = k_h/k_z. \tag{4.23}$$

Combining these three expressions, we have

$$\omega = -\frac{k_z c}{2\cos\gamma}\sqrt{1 + k_x{}^2/k_z{}^2}, \tag{4.24}$$

which is the zero-offset expression (4.1) with an additional factor cos γ in the denominator. If the temporal frequency magnitude lies in the range (ω_i, ω_f), then the wavenumbers k_x and k_z are confined in magnitude to the region

$$k_i(\gamma) = \frac{2\omega_i\cos\gamma}{c} < \sqrt{k_x{}^2 + k_z{}^2} < k_f(\gamma) = \frac{2\omega_f\cos\gamma}{c}. \tag{4.25}$$

The upper and lower frequency bounds have been lowered relative to zero angle by a factor of cosine of angle of incidence.

Point reflectors

Repeating the analysis of equations (4.3) through (4.11), the bandlimited image of a point is seen to be

$$\begin{aligned}\mathcal{R}_p\left(x, z; x_0, z_0; \omega_i, \omega_f; \gamma\right) &= \mathcal{R}_p\left(\rho; \omega_i, \omega_f; \gamma\right)\\ &= \frac{R}{2\pi\rho}\left(k_f(\gamma)J_1\left(k_f\rho\right) - k_i(\gamma)J_1\left(k_i\rho\right)\right), \end{aligned}\tag{4.26}$$

where

$$\rho = \sqrt{(x - x_0)^2 + (z - z_0)^2}. \tag{4.27}$$

Relative to the imaged reflectivity at zero dip,

$$\mathcal{R}_p\left(\rho; \omega_i, \omega_f; \gamma\right) = \cos^2\gamma\mathcal{R}_p\left(\rho\cos\gamma; \omega_i, \omega_f; 0\right). \tag{4.28}$$

The bandlimited point image is the same isotropic Bessel-sinc function as found for the zero-offset case, but with the wavelet stretched in all directions by a factor of cos γ. The wavenumber lowering is related to the phenomenon of NMO or normal moveout stretch (Tygel *et al.*, 1994), and represents a real loss of resolution at non-zero offsets. It occurs because, at non-zero angles of incidence, reflecting layers are being encountered at an angle, lengthening the perceived thickness of the reflector. For an isolated reflection, the original bandwidth can be recovered by compressing the reflector image by the factor cos γ. In practical situations, however, isolating individual reflections may be difficult.

The peak amplitude of the stretched wavelet is reduced by a factor of $\cos^2\gamma$ relative to a zero-offset image:

$$\mathcal{R}_p \underset{\rho\to 0}{\to} \frac{R}{4\pi}\left(k_f{}^2 - k_i{}^2\right)\cos^2\gamma. \tag{4.29}$$

The reduction in amplitude is just sufficient to preserve area under the wavelet. If one picks any point $\rho_f(\gamma) = \rho_f(0)\cos\gamma$ on the wavelet, then the area of the wavelet out to that point is

$$\mathcal{A}\left(\rho_f(0); \omega_i, \omega_f; \gamma\right) = 2\pi \int_0^{\rho_f(\gamma)} d\rho\rho\mathcal{R}_p\left(\rho; \omega_i, \omega_f; \gamma\right). \tag{4.30}$$

Invoking (4.28), and changing integration variable to $\rho'=\rho\cos\gamma$,

$$\mathcal{A}\left(\rho_f(0); \omega_i, \omega_f; \gamma\right) = 2\pi \int_0^{\rho_f(0)} d\rho'\rho'\mathcal{R}_p\left(\rho'; \omega_i, \omega_f; 0\right). \tag{4.31}$$

That is, the area under the wavelet out to any fixed point (e.g. a zero crossing, half-amplitude or half-power point) is independent of incidence angle.

Line reflectors

The image bandwidth of a line reflector is also shifted downward by a factor of $\cos\gamma$:

$$\begin{aligned}\mathcal{R}_S\left(x, z; x_r, z_r; q, \omega_i, \omega_f, \gamma\right) &= \frac{R\sqrt{1+q^2}}{\pi\left((z-z_r)-(x-x_r)q\right)} \\ &\times \left(\sin\left(\frac{2\omega_f\cos\gamma\left((z-z_r)-(x-x_r)q\right)}{c\sqrt{1+q^2}}\right)\right. \\ &\left. -\sin\left(\frac{2\omega_i\cos\gamma\left((z-z_r)-(x-x_r)q\right)}{c\sqrt{1+q^2}}\right)\right).\end{aligned} \tag{4.32}$$

The peak amplitude of this function is

$$\mathcal{R}_S\left(x_r, z_r; x_r, z_r; q, \omega_i, \omega_f\right) = \frac{2}{c}\frac{R}{\pi}(\omega_f - \omega_i)\cos\gamma, \tag{4.33}$$

which is also reduced by the factor $\cos\gamma$ over that of the zero-offset image. The amplitude reduction of a bandlimited line reflector is less than the reduction for a point reflector, because the point-reflector wavelet stretches outward in two dimensions, whereas the wavelet of a line reflector expands only in one.

Define coordinates (μ, ν), respectively parallel and normal to the reflector, such that $\nu = 0$ on the reflector. Then $\mathcal{R}_S$ depends on ν but not μ, and we have the relation

$$\begin{aligned}\mathcal{R}_S(\nu, \gamma) &= \frac{R}{\pi\nu}\left(\sin\left(\frac{2\omega_f\cos\gamma\nu}{c}\right) - \sin\left(\frac{2\omega_i\cos\gamma\nu}{c}\right)\right) \\ &= \cos\gamma\mathcal{R}_S(\nu\cos\gamma, 0).\end{aligned} \tag{4.34}$$

At finite γ, $\mathcal{R}_S$ loses both resolution and amplitude as $\cos\gamma$. However, an integral over ν between any two points fixed on the wavelet (zero crossings, half-power points, etc.) is unchanged with γ.

4.2.3 Effects of finite bandwidth on zero-offset 3-D data

In three dimensions, the idea is much the same. The relation between frequency and wavenumber is

$$\omega = -\frac{k_z c}{2}\sqrt{1 + k_x{}^2/k_z{}^2 + k_y{}^2/k_z{}^2}, \tag{4.35}$$

so that for $|\omega| \in (\omega_i, \omega_f)$,

$$k_i \equiv \frac{2\omega_i}{c} < k \equiv \sqrt{k_x{}^2 + k_y{}^2 + k_z{}^2} < k_f \equiv \frac{2\omega_f}{c}. \tag{4.36}$$

Point reflectors

The bandlimited image of a point reflector at (x_0, y_0, z_0) is

$$\mathcal{R}_p\left(\boldsymbol{x}; \boldsymbol{x}_0; \omega_i, \omega_f\right) = \frac{1}{(2\pi)^3}\int_{k_i}^{k_f} dk k^2 \int_{-\pi}^{\pi} d\theta \int_0^{\pi} d\phi \sin\phi e^{ikr\cos\phi}, \tag{4.37}$$

where

$$r = \sqrt{(x - x_0)^2 + (y - y_0)^2 + (z - z_0)^2}. \tag{4.38}$$

The θ-integral in (4.37) evaluates to 2π. Changing integration variable from ϕ to $u = \cos\phi$, (4.37) becomes

$$\mathcal{R}_p\left(\boldsymbol{x}; \boldsymbol{x}_0; \omega_i, \omega_f\right) = \frac{1}{(2\pi)^2}\int_{k_i}^{k_f} dk k^2 \int_{-1}^{1} du e^{ikru}, \tag{4.39}$$

Evaluating the u-integral,

$$\mathcal{R}_p\left(\boldsymbol{x}; \boldsymbol{x}_0; \omega_i, \omega_f\right) = \frac{2}{(2\pi)^2 r}\int_{k_i}^{k_f} dk k \sin(kr). \tag{4.40}$$

Evaluating the last integral,

$$\begin{aligned}\mathcal{R}_p\left(\boldsymbol{x}; \boldsymbol{x}_0; \omega_i, \omega_f\right) = \frac{2}{(2\pi)^2 r^3}&\left(\sin\left(k_f r\right) - k_f r\cos\left(k_f r\right)\right.\\ &\left.- \sin\left(k_i r\right) + k_i r\cos\left(k_i r\right)\right).\end{aligned} \tag{4.41}$$

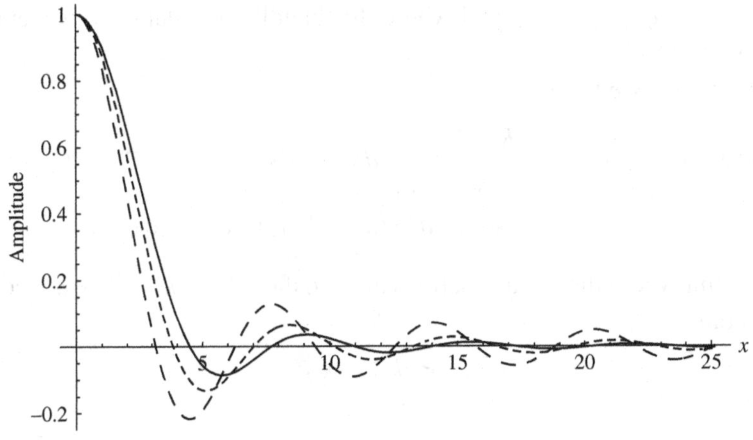

Figure 4.5 The spherical Bessel-sinc function. Comparison of the 3-D spherical Bessel-sinc function $3j_1(x)/x$ (the solid line) with the 2-D Bessel-sinc function $2J_1(x)/x$ (the finely dashed line), and the 1-D sinc function $\sin(x)/x$ (the coarsely dashed line).

This expression is related to the spherical Bessel function of order 1:

$$j_1(x) \equiv \frac{\sin(x) - x\cos(x)}{x^2}, \tag{4.42}$$

in terms of which

$$\mathcal{R}_p\left(\boldsymbol{x}; \boldsymbol{x}_0; \omega_i, \omega_f\right) = \frac{1}{2\pi^2 r}\left({k_f}^2 j_1\left(k_f r\right) - {k_i}^2 j_1\left(k_i r\right)\right). \tag{4.43}$$

We find that the bandlimited point image in three dimensions is an isotropic spherical Bessel-sinc function. Figure 4.5 illustrates the form of $j_1(x)/x$, as compared with an ordinary sinc function. At large x, the spherical Bessel-sinc function has a phase delay of $\pi/2$ relative to the sinc function, and decreases as $1/x^2$ instead of as $1/x$.

As $\boldsymbol{x} \to \mathbf{x}_0$, $\mathcal{R}_p$ approaches its peak value

$$\mathcal{R}_p\left(\boldsymbol{x}_0; \boldsymbol{x}_0; \omega_i, \omega_f\right) = \frac{\left({k_f}^3 - {k_i}^3\right)}{6\pi^2}. \tag{4.44}$$

Reflecting planes

In three dimensions, the k-space image of a reflecting plane is

$$\mathcal{R}_S\left(\boldsymbol{k}; \boldsymbol{x}_r; \boldsymbol{q}, \omega_i, \omega_f\right) = (2\pi)^2 R e^{-i\boldsymbol{k}\cdot\boldsymbol{x}_r}\sqrt{1+q^2}\,\delta\left(k_x + k_z q_x\right)\delta\left(k_y + k_z q_y\right), \tag{4.45}$$

where $\boldsymbol{q} = (q_x, q_y)$, $q = |\boldsymbol{q}|$, and where, for bandlimited data, the wavenumbers must obey (4.36).

In x–y–z space, we have

$$\mathcal{R}_S(\boldsymbol{x}; \boldsymbol{x}_r; \boldsymbol{q}, \omega_i, \omega_f) = \frac{R}{2\pi} \int\int\int_{\mathcal{D}} dk_x dk_y dk_z e^{i(k_x(x-x_r)+k_y(y-y_r)+k_z(z-z_r))} \times\sqrt{1+q^2}\delta(k_x + k_z q_x)\,\delta(k_y + k_z q_y), \tag{4.46}$$

At the point where the delta functions are satisfied, $k = |k_z|\sqrt{1+q^2}$, condition (4.36) becomes

$$k_{zi} < |k_z| < k_{zf}, \tag{4.47}$$

with

$$k_{zi} = \frac{2\omega_i}{c\sqrt{1+q^2}} \qquad k_{zf} = \frac{2\omega_f}{c\sqrt{1+q^2}}. \tag{4.48}$$

Integrating over k_x and k_y, reflectivity becomes

$$\mathcal{R}_S(\boldsymbol{x}; \boldsymbol{x}_r; \boldsymbol{q}, \omega_i, \omega_f) = \frac{R\sqrt{1+q^2}}{\pi} \int_{k_{zi}}^{k_{zf}} dk_z \cos\left(\sqrt{1+q^2}k_z\nu\right), \tag{4.49}$$

where ν is the cartesian coordinate normal to the reflecting surface:

$$\nu = \frac{(z-z_r) - q_x(x-x_r) - q_y(y-y_r)}{\sqrt{1+q^2}}. \tag{4.50}$$

Evaluating the k_z-integral,

$$\mathcal{R}_S(\boldsymbol{x}; \boldsymbol{x}_r; \boldsymbol{q}, \omega_i, \omega_f) = \frac{R}{\pi\nu}\left(\sin(k_f\nu) - \sin(k_i\nu)\right). \tag{4.51}$$

As $z \to z_r + q_x(x-x_r) + q_y(y-y_r)$, the reflectivity function approaches the same peak amplitude as its 2-D counterpart:

$$\mathcal{R}_S \to \frac{2R}{c\pi}\left(\omega_f - \omega_i\right). \tag{4.52}$$

4.2.4 Effects of finite bandwidth on finite-offset 3-D data

For finite offset, the data parameters are $(k_{mx}, k_{my}) \equiv \tilde{\boldsymbol{k}}_m$, $(k_{hx}, k_{hy}) \equiv \tilde{\boldsymbol{k}}_h$, and ω, while the image parameters are k_x, k_y, k_z, and γ. The relations between the two sets of parameters are (see Section 3.5)

$$k_x = k_{mx}, \tag{4.53}$$

$$k_y = k_{my}, \tag{4.54}$$

$$\tan\gamma = \frac{\tilde{k}_h}{k_z}\sqrt{\frac{1+\left(\tilde{\boldsymbol{k}}_m\cdot\tilde{\boldsymbol{k}}_h\right)^2/\left(k_z\tilde{k}_h\right)^2}{1+\tilde{k}_m{}^2/k_z{}^2}}, \tag{4.55}$$

and

$$\omega = -\frac{k_z c}{2}\sqrt{1+\tilde{k}_m{}^2/k_z{}^2+\tilde{k}_h{}^2/k_z{}^2+\left(\tilde{\boldsymbol{k}}_m\cdot\tilde{\boldsymbol{k}}_h\right)^2/k_z{}^4}. \tag{4.56}$$

Equation (4.56) can be rewritten as

$$\omega = -\frac{k_z c}{2\cos\gamma}\sqrt{1+\left(k_x{}^2+k_y{}^2\right)/k_z{}^2}, \tag{4.57}$$

which, as was the case in two dimensions, is the zero-offset form with an additional factor of $\cos\gamma$ in the denominator. Thus, the zero-offset analysis still holds, with the wavenumber limits lowered by the $\cos\gamma$ factor:

$$k_i \rightarrow \frac{2\omega_i\cos\gamma}{c}, \tag{4.58}$$

$$k_f \rightarrow \frac{2\omega_i\cos\gamma}{c}. \tag{4.59}$$

Point reflectors

The finite-offset bandlimited image point is the isotropic spherical Bessel-sinc function (4.43), with the wavenumber magnitude limits (4.58) and (4.59). As for the 2-D case, the wavenumber passband is shifted downward at non-normal incidence because the raypaths encounter the reflecting layers at an angle, stretching the distance between layers. We have the relation

$$\mathcal{R}_p\left(r;\omega_i,\omega_f,\gamma\right) = \frac{8\cos^2\gamma}{(2\pi)^2c^2r}\left(\omega_f{}^2 j_1\left(\frac{2\omega_f r\cos\gamma}{c}\right)-\omega_i{}^2 j_1\left(\frac{2\omega_i r\cos\gamma}{c}\right)\right), \tag{4.60}$$

from which it follows that

$$\mathcal{R}_p\left(r;\omega_i,\omega_f,\gamma\right) = \cos^3\gamma\,\mathcal{R}_p\left(r\cos\gamma;\omega_i,\omega_f,0\right). \tag{4.61}$$

For finite γ, amplitude is reduced by $\cos^3\gamma$, which implies that, analogously to the 2-D case, volumes under the wavelet (for example, the volume between zero crossings) are independent of γ.

Planar reflectors

The bandlimited image of a reflecting plane at incident angle γ is

$$\begin{aligned}\mathcal{R}_S(\nu,\gamma) &= \frac{R}{\pi\nu}\left(\sin\left(\frac{2\omega_f\cos\gamma\,\nu}{c}\right) - \sin\left(\frac{2\omega_i\cos\gamma\,\nu}{c}\right)\right)\\ &= \cos\gamma\,\mathcal{R}_S(\nu\cos\gamma,0),\end{aligned} \tag{4.62}$$

with ν a cartesian coordinate normal to the reflecting plane. The wavelet is stretched by the factor cos γ in one direction only, so that if $\Delta\nu$ is a measure of the width of the wavelet, then $\Delta\nu(\gamma) = \Delta\nu(0)/\cos\gamma$. Amplitude is reduced by cos γ also, so that an integral of $\mathcal{R}_S$ over the width $\Delta\nu$ does not vary with γ.

4.3 Space-domain image amplitude renormalization

We noted in the last section that peak image amplitude in *x–y–z*-space is reduced when angle of incidence (or source–receiver offset) is non-zero. Some authors (e.g. Black *et al.*, 1993) have argued that the imaged data should be renormalized by dividing amplitude by cos γ. As seen above, this would make the peak amplitude of flat specular reflectors independent of γ.

One can argue on either side of this proposal. It is just an attempt to compensate for one of the effects of finite data bandwidth, and is neither right nor wrong. In favor of the argument is that as normal moveout (NMO) is usually applied, the result is a wavelet stretched but retaining its peak amplitude.

On the other hand, while dividing by cos γ may preserve the peak amplitude of specular reflectors, we saw above that it does not preserve peak amplitude of point reflectors. It is also true that if further processing of the image is to be done, renormalization may have some unintended consequences. From the do-not-renormalize point of view, if current NMO algorithms do not reduce the amplitude of stretched wavelets, they should.

In this text, the amplitudes of imaged data will not be renormalized. Consequently, true-amplitude migration formulas in this book may differ from some formulas in the literature. To use the formulas in this book correctly, the amplitude of an event should not be measured by its peak, but rather by an integral of the associated wavelet over a characteristic width of the wavelet.

4.4 Effects of spatial sampling

Typically, data are undersampled in the horizontal direction. In *k*-space, this reduces the maximum horizontal wavenumber in the data. This clips the sides of the doughnut-shaped pass region, as illustrated in Figure 4.6.

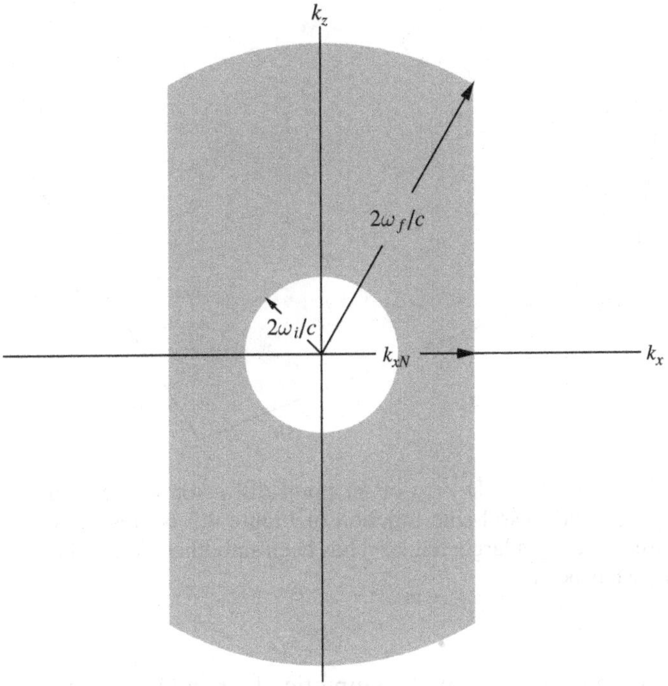

Figure 4.6 The image wavenumber pass region when data is undersampled in x. In this case, the Nyquist wavenumber k_{xN} in the x-direction is one-half the maximum image wavenumber $k_f = 2\,\omega_f / c$.

Needless to say, this destroys the symmetry of the spatial image. For directions near the vertical, the image is unaffected. At larger angles, the image is damaged, as illustrated in Figure 4.7.

4.5 Effects of maximum dip

Processing algorithms are often dip limited. If dip angle α is limited to some value $\alpha_{\max}$, in wavenumber space, this limits the ratio $\tan\alpha = k_x / k_z$ to values less than $\tan\alpha_{\max}$. Figure 4.8 shows the wavenumber pass region.

The spatial image takes the form of Figure 4.9. As one would expect, x–z symmetry in not preserved. Moreover, an X-shaped artifact is superimposed on the point image, the bars of the X following the maximum dip angle.

4.6 Effects of finite spatial aperture

Practical data sets are limited in the range of obtainable offsets and midpoints, and practical processing algorithms may be limited in the distance between input

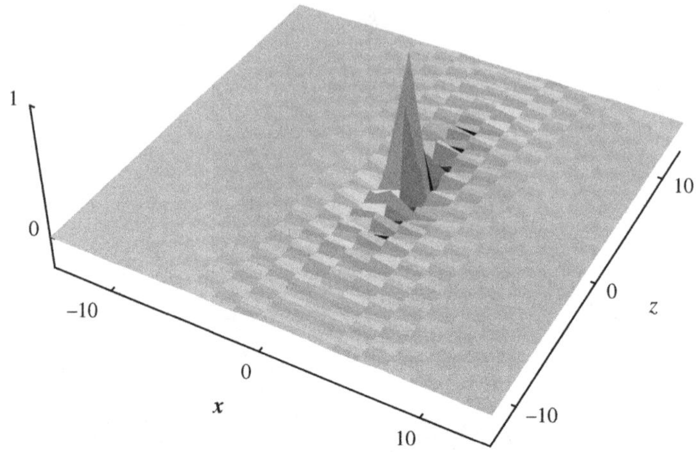

Figure 4.7 Image of a 2-D zero-offset point diffractor where data is spatially undersampled. The Bessel-sinc function of Figure 4.3 is preserved in the near-vertical direction, but at larger angles it has been smoothed away. The x–z isotropy of the image has been lost.

midpoint and output image point. Asymptotically, at a given image depth, a maximum offset translates into a maximum incidence angle, and a maximum distance between image coordinate and data midpoint translates into a maximum dip that can be imaged. The exact formulas depend on the detailed propagation model. We present the problem here in the simplest case of a 2-D medium of constant velocity.

4.6.1 The finite aperture filter

Equation (3.43) relates an angle-dependent reflectivity function to an f–k-transformed data function. To form the f–k transform perfectly, one must have data from all offsets and midpoints. Suppose, however, that data are available only for a finite range of offsets, say $-h_{\max} < x_h < h_{\max}$. In the k_h-domain, this limitation effectively convolves the data with the filter

$$f(k_h) = \int_{-h_{\max}}^{h_{\max}} e^{-ik_h x_h}\, dx_h = \frac{e^{ik_h h_{\max}} - e^{-ik_h h_{\max}}}{ik_h} = 2\frac{\sin(k_h h_{\max})}{k_h}. \tag{4.63}$$

Convolution with this filter smooths the data over an interval of order $\pi / h_{\max}$. We expect the data to already be smooth and slowly varying over small changes in incident angle, the tangent of which is given by equation (3.32). Consequently, at large k_z, the offset limitation should have little effect. To be faithful at smaller k_z, however, one will need to worry about the missing data.

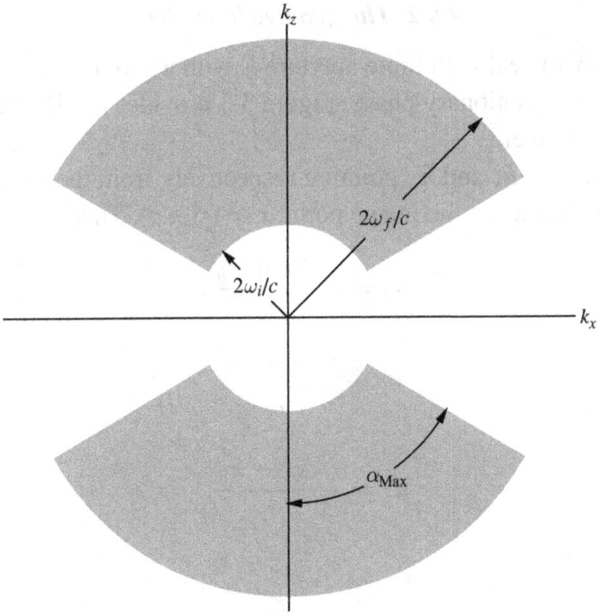

Figure 4.8 The image wavenumber pass region when data is dip limited. In this case, the maximum dip α_{max} is 60°.

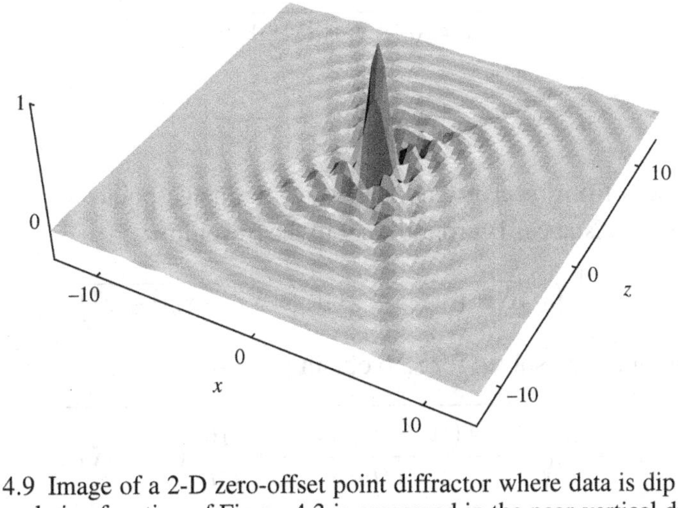

Figure 4.9 Image of a 2-D zero-offset point diffractor where data is dip limited. The Bessel-sinc function of Figure 4.3 is preserved in the near-vertical direction, but at larger angles it has been smoothed away. The x–z isotropy of the image has been lost. In addition an X has been superimposed on the image. The bars of the X follow the maximum dip angle.

4.6.2 The geometric model

The simplest way to deal with finite aperture is with a geometrical approach, more or less equivalent to stationary phase. Figure 3.3 provides a 2-D illustration of the geometry of the problem.

Define unit vectors $\hat{\boldsymbol{n}}_s$ and $\hat{\boldsymbol{n}}_g$ pointing respectively from the source position $\boldsymbol{x}_s$ and receiver position $\boldsymbol{x}_g$ to the image point $\boldsymbol{x} = (x, y, z)$. Then

$$\tilde{\boldsymbol{n}}_g = -\frac{\Delta\tilde{\boldsymbol{x}} + \tilde{\boldsymbol{x}}_h}{r_g}, \tag{4.64}$$

$$n_{gz} = \frac{z}{r_g}, \tag{4.65}$$

$$\tilde{\boldsymbol{n}}_s = -\frac{\Delta\tilde{\boldsymbol{x}} - \tilde{\boldsymbol{x}}_h}{r_s}, \tag{4.66}$$

$$n_{sz} = \frac{z}{r_s}, \tag{4.67}$$

where $\tilde{\boldsymbol{n}}_g$ and $\tilde{\boldsymbol{n}}_s$ are 2-vectors containing the x and y components of $\hat{\boldsymbol{n}}_g$ and $\hat{\boldsymbol{n}}_s$, $\Delta\tilde{\boldsymbol{x}} = (\Delta x, \Delta y) = (x_m - x, y_m - y)$ contains the horizontal distances from the image point to the data midpoint in the x and y directions, $\tilde{\boldsymbol{x}}_h = (x_h, y_h)$ contains the x and y components of half-offset, and z is the depth to the image point. The quantities

$$r_g = \sqrt{|\Delta\tilde{\boldsymbol{x}} + \tilde{\boldsymbol{x}}_h|^2 + z^2} \tag{4.68}$$

and

$$r_s = \sqrt{|\Delta\tilde{\boldsymbol{x}} - \tilde{\boldsymbol{x}}_h|^2 + z^2} \tag{4.69}$$

are the distances from receiver and source to the image point, respectively. The sum of these two distances is proportional to the traveltime:

$$t = \frac{1}{c}\left(\sqrt{|\Delta\tilde{\boldsymbol{x}} + \tilde{\boldsymbol{x}}_h|^2 + z^2} + \sqrt{|\Delta\tilde{\boldsymbol{x}} - \tilde{\boldsymbol{x}}_h|^2 + z^2}\right). \tag{4.70}$$

This equation can be solved for z, to obtain

$$z \to \frac{ct}{2}\sqrt{1 - \frac{4\left(|\tilde{\boldsymbol{x}}_h|^2 + |\Delta\tilde{\boldsymbol{x}}|^2\right)}{c^2t^2} + \frac{16}{c^4t^4}\left(\tilde{\boldsymbol{x}}_h \cdot \Delta\tilde{\boldsymbol{x}}\right)^2}. \tag{4.71}$$

Substituting this expression for z into the definitions of r_g and r_s, we obtain expressions without a square root:

$$r_g = \frac{ct}{2} + \frac{2\tilde{\boldsymbol{x}}_h \cdot \Delta\tilde{\boldsymbol{x}}}{ct}, \tag{4.72}$$

$$r_s = \frac{ct}{2} - \frac{2\tilde{\boldsymbol{x}}_h \cdot \Delta\tilde{\boldsymbol{x}}}{ct}. \tag{4.73}$$

Now, for Snell's law to hold, the dip at the image point must be the angle that the sum of $\boldsymbol{n}_g$ and $\boldsymbol{n}_s$ makes with the vertical.That is,

$$\tan^2\alpha = \frac{|\tilde{\boldsymbol{n}}_g + \tilde{\boldsymbol{n}}_s|^2}{\left(n_{gz} + n_{sz}\right)^2}. \tag{4.74}$$

Substituting (4.64)–(4.67) into (4.74),

$$\tan^2\alpha = \frac{\left|\Delta\tilde{\boldsymbol{x}}\left(r_g + r_s\right) + \tilde{\boldsymbol{x}}_h\left(r_s - r_g\right)\right|^2}{z^2\left(r_g + r_s\right)^2}. \tag{4.75}$$

With the expressions (4.72) and (4.73), this becomes

$$\begin{aligned} \tan^2\alpha &= \left(\left|\Delta\tilde{\boldsymbol{x}} - 4\tilde{\boldsymbol{x}}_h \frac{\tilde{\boldsymbol{x}}_h \cdot \Delta\tilde{\boldsymbol{x}}}{c^2t^2}\right|\right)^2 / z^2 \\ &= \frac{|\Delta\tilde{\boldsymbol{x}}|^2}{z^2} - 8\frac{(\tilde{\boldsymbol{x}}_h \cdot \Delta\tilde{\boldsymbol{x}})^2}{c^2t^2z^2} + 16\frac{|\tilde{\boldsymbol{x}}_h|^2}{z^2}\frac{(\tilde{\boldsymbol{x}}_h \cdot \Delta\tilde{\boldsymbol{x}})^2}{c^4t^4}, \end{aligned} \tag{4.76}$$

or

$$|\tan\alpha| = \frac{|\Delta\tilde{\boldsymbol{x}}|}{z}\sqrt{1 - 8\frac{(\tilde{\boldsymbol{x}}_h \cdot \Delta\tilde{\boldsymbol{x}})^2}{|\Delta\tilde{\boldsymbol{x}}|^2 c^2t^2} + 16\frac{|\tilde{\boldsymbol{x}}_h|^2}{|\Delta\tilde{\boldsymbol{x}}|^2}\frac{(\tilde{\boldsymbol{x}}_h \cdot \Delta\tilde{\boldsymbol{x}})^2}{c^4t^4}}. \tag{4.77}$$

At small offsets, tangent of maximum dip is just the aperture magnitude $\Delta\tilde{x}$ divided by image depth. At larger offsets, maximum dip also depends on offset.

Using equation (4.71), depth may be removed from this equation in favor of traveltime. The result is

$$|\sin\alpha| = \frac{2|\Delta\tilde{\boldsymbol{x}}|}{ct}\sqrt{\frac{1 - 8\frac{(\tilde{\boldsymbol{x}}_h \cdot \Delta\tilde{\boldsymbol{x}})^2}{|\Delta\tilde{\boldsymbol{x}}|^2c^2t^2} + 16\frac{|\tilde{\boldsymbol{x}}_h|^2}{|\Delta\tilde{\boldsymbol{x}}|^2}\frac{(\tilde{\boldsymbol{x}}_h \cdot \Delta\tilde{\boldsymbol{x}})^2}{c^4t^4}}{\left(1 - \frac{4|\tilde{\boldsymbol{x}}_h|^2}{c^2t^2}\right)\left(1 - \frac{16(\tilde{\boldsymbol{x}}_h \cdot \Delta\tilde{\boldsymbol{x}})^2}{c^4t^4}\right)}}. \tag{4.78}$$

Maximum incidence angle is found the same way. Tangent squared of incident angle is given by

$$\tan^2\gamma = \frac{|\tilde{\boldsymbol{n}}_g - \tilde{\boldsymbol{n}}_s|^2}{\left(n_{gz} + n_{sz}\right)^2} = \frac{\left(\left|\tilde{\boldsymbol{x}}_h - 4\Delta\tilde{\boldsymbol{x}}\frac{\tilde{\boldsymbol{x}}_h \cdot \Delta\tilde{\boldsymbol{x}}}{c^2t^2}\right|\right)^2}{z^2}, \tag{4.79}$$

or

$$|\tan\gamma| = \frac{|\tilde{\boldsymbol{x}}_h|}{z}\sqrt{1 - 8\frac{(\tilde{\boldsymbol{x}}_h \cdot \Delta\tilde{\boldsymbol{x}})^2}{|\tilde{\boldsymbol{x}}_h|^2 c^2t^2} + 16\frac{|\Delta x|^2}{|\tilde{\boldsymbol{x}}_h|^2}\frac{(\tilde{\boldsymbol{x}}_h \cdot \Delta\tilde{\boldsymbol{x}})^2}{c^4t^4}}, \tag{4.80}$$

At small dips, the maximum incidence angle is maximum offset divided by z. For larger dip, the maximum incidence angle is diminished by the square root factor of (4.80).

As for dips, depth may be replaced by traveltime, resulting in

$$|\sin\gamma| = \frac{2\,|\tilde{\boldsymbol{x}}_h|}{ct}\sqrt{\frac{1-8\frac{(\tilde{x}_h\cdot\Delta\tilde{x})^2}{|\tilde{x}_h|^2c^2t^2}+16\frac{|\Delta x|^2}{|\tilde{x}_h|^2}\frac{(\tilde{x}_h\cdot\Delta\tilde{x})^2}{c^4t^4}}{\left(1-\frac{4|\Delta\tilde{x}|^2}{c^2t^2}\right)\left(1-\frac{16(\tilde{x}_h\cdot\Delta\tilde{x})^2}{c^4t^4}\right)}}. \tag{4.81}$$

In the case where $\Delta\tilde{x}$ and $\tilde{x}_h$ are aligned (or in two dimensions, where they are always aligned), the expressions (4.77) and (4.80) for tan α and tan γ simplify to

$$|\tan\alpha| \to \frac{|\Delta\tilde{\boldsymbol{x}}|}{z}\left(1-4\frac{|\tilde{\boldsymbol{x}}_h|^2}{c^2t^2}\right), \tag{4.82}$$

and

$$|\tan\gamma| \to \frac{|\tilde{\boldsymbol{x}}_h|}{z}\left(1-4\frac{|\Delta\tilde{\boldsymbol{x}}|^2}{c^2t^2}\right). \tag{4.83}$$

Similarly, with depth replaced by traveltime,

$$|\sin\alpha| \to \frac{2\,|\Delta\tilde{\boldsymbol{x}}|}{ct}\sqrt{\frac{1-\frac{4|\tilde{x}_h|^2}{c^2t^2}}{1-1-\frac{16|\tilde{x}_h||\Delta\tilde{x}|^2}{c^4t^4}}}, \tag{4.84}$$

and

$$|\sin\gamma| \to \frac{2\,|\tilde{\boldsymbol{x}}_h|}{ct}\sqrt{\frac{1-\frac{4|\Delta\tilde{x}|^2}{c^2t^2}}{1-1-\frac{16|\tilde{x}_h||\Delta\tilde{x}|^2}{c^4t^4}}}. \tag{4.85}$$

Limited midpoint aperture imposes a maximum dip that can be imaged, and limited offset range imposes a maximum angle of incidence. These maxima generally get smaller with image depth. Maximum dip also gets smaller as angle of incidence increases, and maximum angle of incidence gets smaller as dip increases. Figure 4.10 shows maximum angle of incidence versus depth and dip angle, given a maximum offset of $x_h = h_{\max}$.

4.6.3 A test of the geometric model

We can test the geometrical or asymptotic assignments of dip and incidence angle with a 2-D prestack *f–k* migration. We will use the 2.5-D amplitude-preserving formula (3.65). To keep things as simple as possible, we choose R to correspond

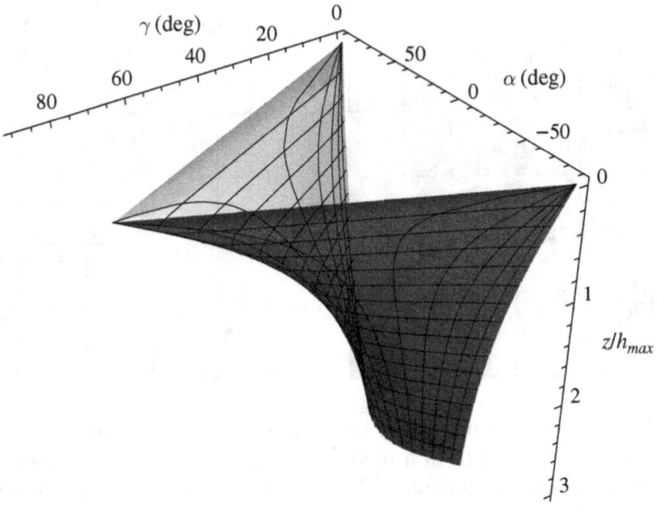

Figure 4.10 Maximum angle of incidence. Asymptotic formula for maximum angle of incidence versus depth and dip angle, given a maximum offset of $x_h = h_{\max}$. The maximum angle of incidence that can be imaged decreases with depth and dip angle.

to a single flat reflector at depth $z_0 = 1$, with reflection coefficient R_0 independent of γ. Then

$$D^{(2.5\mathrm{D})}(k_x, k_h, \omega) = R_0 \frac{\delta(k_x)}{\cos\gamma} \sqrt{\frac{\pi}{2}\frac{i}{\omega}}. \tag{4.86}$$

In space–time, the data assume the form (3.51), or

$$D^{(2.5\mathrm{D})}(x_m, x_h, t) \equiv \sqrt{t}\, D^{(3\mathrm{D})}(x_m, x_h, t) = \frac{R_0(\gamma)\delta(t - 2r/c)}{\sqrt{2rc}}, \tag{4.87}$$

where $r = \sqrt{{z_0}^2 + {x_h}^2}$, and $c = 10$.

The space–time data function is depicted on the left-hand side of Figure 4.11. For this data set, offset has been limited to $|x_h| < 1$, which geometrically illuminates the reflector for incident angles less than 45°. An *f–k* migration of this data set is shown on the right-hand of Figure 4.11. The image shows a constant-amplitude reflection for small angles. Near 45°, the amplitude begins to fail, dropping to about half-amplitude at 45°, and fading rapidly at angles larger than 45°. Also present is a diffraction pattern characteristic of an edge effect, conceding that a sharp edge

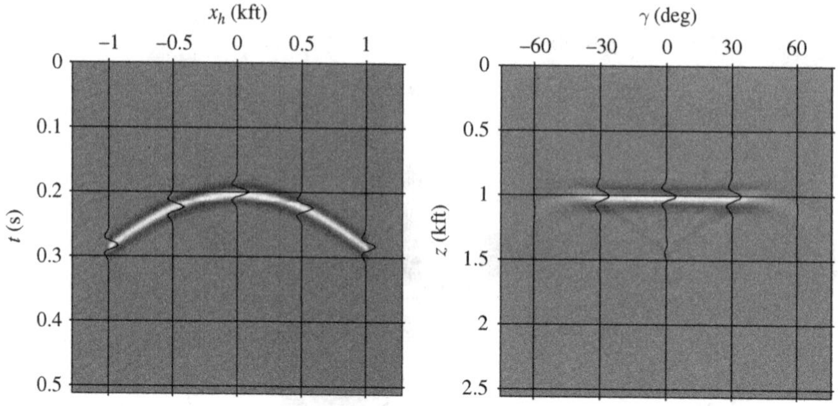

Figure 4.11 Prestack *f–k* migration of a flat reflector from offset-limited data. The reflector is at unit depth in a medium of velocity $c = 10$. Half-offset x_h is limited to magnitudes less than one. The data set is shown on the left as a function of offset and time; its migration is shown on the right as a function of incident angle and depth. Geometrically, incident angles less than 45° are illuminated. This is approximately confirmed by the *f–k* migration, though the geometrical model fails to account for edge-diffraction artifacts generated by the sharp data cutoff at maximum offset.

in offset does not exactly correspond to a sharp edge in incident angle. This simple test confirms, up to a point, the geometric model, while also illuminating its limitations.

We can analyze this result in more detail using the expression (F.34) for the stationary-phase approximation to an integral with boundaries. Solving the *f–k* expression (3.65) for *R*,

$$R(\gamma, k_z) = \sqrt{\frac{-2i\omega}{\pi}} \cos\gamma\, D^{(2.5\mathrm{D})}(k_h, \omega)\,, \tag{4.88}$$

we write

$$R(\gamma, z) = \frac{1}{\sqrt{2\pi}^3} \int dk_z e^{ik_z z} \sqrt{-i\omega} \cos\gamma \int dx_h e^{-ik_h x_h} D^{(2.5\mathrm{D})}(x_h, \omega)\,. \tag{4.89}$$

In these expressions, k_h and ω are related to γ and k_z as $k_h = k_z \tan\gamma$, and $\omega = -k_z c/(2\cos\gamma)$. Invoking (4.87), and expressing ω and k_h in terms of k_z and γ,

$$R(\gamma, z) = \frac{\cos\gamma\, R_0}{2\sqrt{\pi}^3} \int dk_z e^{ik_z z} \sqrt{\frac{ik_z}{2\cos\gamma}} \int dx_h \frac{e^{-i\frac{k_z}{\cos\gamma}(r - x_h \sin\gamma)}}{\sqrt{r}}\,. \tag{4.90}$$

The integral over offset can be evaluated via stationary phase. In the absence of offset boundaries, we have

$$\int dx_h \frac{e^{-i\frac{k_z}{\cos\gamma}(r-x_h\sin\gamma)}}{\sqrt{r}} \simeq \sqrt{\frac{-2\pi i}{k_z\cos\gamma}}e^{-ik_z z_0}, \tag{4.91}$$

thus

$$R(\gamma, z) \simeq \frac{1}{2\pi}\int dk_z e^{ik_z(z-z_0)} = R_0\delta\left(z-z_0\right). \tag{4.92}$$

If, however, offset is limited to $h_{\max}$, then, according to equation (F.34) in Appendix F, the stationary phase approximation becomes

$$\int_{-h_{\max}}^{h_{\max}} dx_h \frac{e^{-i\frac{k_z}{\cos\gamma}(r-x_h\sin\gamma)}}{\sqrt{r}} \simeq \sqrt{\frac{-2\pi i}{k_z\cos\gamma}}e^{-ik_z z_0} + i\frac{\cos\gamma\sqrt{r_{\max}}}{k_z} \left(\frac{e^{-i\frac{k_z}{\cos\gamma}(r_{\max}+h_{\max}\sin\gamma)}}{h_{\max}+r_{\max}\sin\gamma} + \frac{e^{-i\frac{k_z}{\cos\gamma}(r_{\max}-h_{\max}\sin\gamma)}}{h_{\max}-r_{\max}\sin\gamma}\right), \tag{4.93}$$

where $r_{\max} = \sqrt{{z_0}^2 + {h_{\max}}^2}$.

A naive estimate $\hat{R}$ of R in this case would be

$$\hat{R}(\gamma, z) = \frac{\cos\gamma R_0}{2\sqrt{\pi^3}}\int dk_z e^{ik_z z}\sqrt{\frac{ik_z}{2\cos\gamma}}\int_{-h_{\max}}^{h_{\max}} dx_h \frac{e^{-i\frac{k_z}{\cos\gamma}(r-x_h\sin\gamma)}}{\sqrt{r}}, \tag{4.94}$$

which, according to (4.93), is

$$\begin{aligned}
\hat{R}(\gamma, z) &= \frac{R_0}{2\pi}\int dk_z e^{ik_z z} \\
&\times \left(e^{-ik_z z_0} + \sqrt{\frac{-i r_{\max}\cos^3\gamma}{2\pi k_z}}\left(\frac{e^{-i\frac{k_z}{\cos\gamma}(r_{\max}+h_{\max}\sin\gamma)}}{h_{\max}+r_{\max}\sin\gamma} + \frac{e^{-i\frac{k_z}{\cos\gamma}(r_{\max}-h_{\max}\sin\gamma)}}{h_{\max}-r_{\max}\sin\gamma}\right)\right) \\
&= R_0\delta\left(z-z_0\right) + \frac{R_0\sqrt{r_{\max}\cos^3\gamma}}{(2\pi)^{3/2}h_{\max}+r_{\max}\sin\gamma}\int dk_z\sqrt{\frac{-i}{k_z}}e^{i\,k_z\left((z-\frac{1}{\cos\gamma}(r_{\max}+h_{\max}\sin\gamma)\right)} \\
&+ \frac{R_0\sqrt{r_{\max}\cos^3\gamma}}{(2\pi)^{3/2}h_{\max}-r_{\max}\sin\gamma}\int dk_z\sqrt{\frac{-i}{k_z}}e^{i\,k_z\left((z-\frac{1}{\cos\gamma}(r_{\max}-h_{\max}\sin\gamma)\right)}.
\end{aligned} \tag{4.95}$$

The k_z- integrals are recognizable as the half-integral operator in equation (E.48):

$$\hat{R}(\gamma,z)=R_0\delta\left(z-z_0\right)+\frac{R_0\sqrt{r_{\max}\cos^3\gamma}}{(2\pi)^{3/2}\left(h_{\max}+r_{\max}\sin\gamma\right)}\Theta_{h+}\left(z-\frac{r_{\max}}{\cos\gamma}-h_{\max}\tan\gamma\right)$$
$$+\frac{R_0\sqrt{r_{\max}\cos^3\gamma}}{(2\pi)^{3/2}\left(h_{\max}-r_{\max}\sin\gamma\right)}\Theta_{h+}\left(z-\frac{r_{\max}}{\cos\gamma}+h_{\max}\tan\gamma\right). \quad (4.96)$$

The boundary correction predicts a 45° diffraction pattern centered on the edges of the geometrical model.

4.6.4 The stationary phase model

According to the geometric model, a finite range of offsets translates into a finite range of offset wavenumbers. The offset wavenumber range, however, depends on data frequency, dip and depth. This variability will, unless compensated for, distort the amplitude of the migrated image. Most of the formulas presented so far have made no pretensions of true amplitude, so we could, if we wish, feign indifference. However, acknowledging formulas that do attempt to honor amplitudes, we ought not to dismiss the matter.

Confining ourselves to two dimensions, we start with an equation relating a migrated image in the (k_x, k_z) domain to data in the $\left(k_{gx}, k_{sx}, \omega\right)$ domain. We write a simplified version of (3.39), namely

$$D\left(k_{gx}|k_{sx};\omega\right)=s(\omega)A(\gamma,\alpha)a\left(k_x,k_z\right), \quad (4.97)$$

with α and γ dip and incidence angle as given by equations (3.31) and (3.32). The amplitude function A is slowly varying in dip and in incidence angle. The incident-angle-dependent reflectivity function of equation (3.39) is replaced by an angle-independent potential kernel $a\left(k_x, k_z\right)$, so that all the dependence on incident angle is relegated to the coefficient A. We also assume that a function of frequency $s(\omega)$ is separable from the other terms. The wavenumbers k_x, k_h, and k_z are given in terms of k_{gx}, k_{sx}, and ω by equations (3.11), (3.12), and (3.9). We shall see in later chapters that this formula has the form predicted by linearized scattering theory for a scalar wave equation.

To solve for the potential kernel a given the data D, we can write a general formula

$$a\left(k_m,k_z\right)=\int dk_h D\left(\frac{k_x+k_h}{2}\left|\frac{k_h-k_x}{2};\ \omega\right.\right)L(\alpha,\gamma)s^{-1}(\omega), \quad (4.98)$$

where L is a slowly varying function, arbitrary except for the constraint

$$\int dk_h A(\gamma, \alpha) L(\alpha, \gamma) = 1. \tag{4.99}$$

The potential function has the spatial (x, z) representation

$$a(x, z) = \frac{1}{(2\pi)^2} \int dk_x \int dk_h \int dk_z e^{i(k_x x + k_z z)} \times D\left(\frac{k_x + k_h}{2} \middle| \frac{k_h - k_x}{2}; \omega\right) L(\alpha, \gamma) s^{-1}(\omega). \tag{4.100}$$

If we change integration variables to k_{gx}, k_{sx}, and ω, we get

$$a(x, z) = \frac{1}{(2\pi)^2} \frac{2}{c^2} \int d\omega s^{-1}(\omega) \int dk_{gx} e^{i(k_{gx} x + k_{gz} z)} \int dk_{sx} e^{-i(k_{sx} x + k_{sz} z)} \times D\left(k_{gx} | k_{sx}; \omega\right) L(\alpha, \gamma) \left| \frac{\omega k_z}{k_{gz} k_{sz}} \right|. \tag{4.101}$$

Return the data to the space–frequency domain:

$$a(x, z) = \frac{1}{(2\pi)^2} \frac{2}{c^2} \int dx_g \int dx_s \int d\omega s^{-1}(\omega) D\left(x_g | x_s; \omega\right) \times \int dk_{gx} e^{i(k_{gx}(x - x_g) + k_{gz} z)} \int dk_{sx} e^{-i(k_{sx}(x - x_s) + k_{sz} z)} L(\alpha, \gamma) \left| \frac{\omega k_z}{k_{gz} k_{sz}} \right|. \tag{4.102}$$

4.7 Extra and missing data

Realizable data are never perfect. Additive noise is always present, and often there are gaps or missing areas in the data. Low levels of random noise do not present a great problem since in the migration process the noise tends to interfere destructively and cancel. High-amplitude isolated noise spikes are another matter. In two dimensions and zero offset, a spike at data point (x_0, t_0) creates an artifact along the trajectory

$$z(x) = \frac{c t_0}{2} \sqrt{1 - \frac{4 (x - x_0)^2}{c^2 t_0^2}}, \tag{4.103}$$

as seen in Figure 4.12. Such artifacts can be a nuisance where they occur. They are best removed before migration.

Missing data also create artifacts. Figure 4.13 migrates a data set with a gap of several traces. Migration fills in the missing data with an anticlinal artifact. It may seem unlikely that anyone would mistake such an artifact for a real anticline, but stranger things have happened.

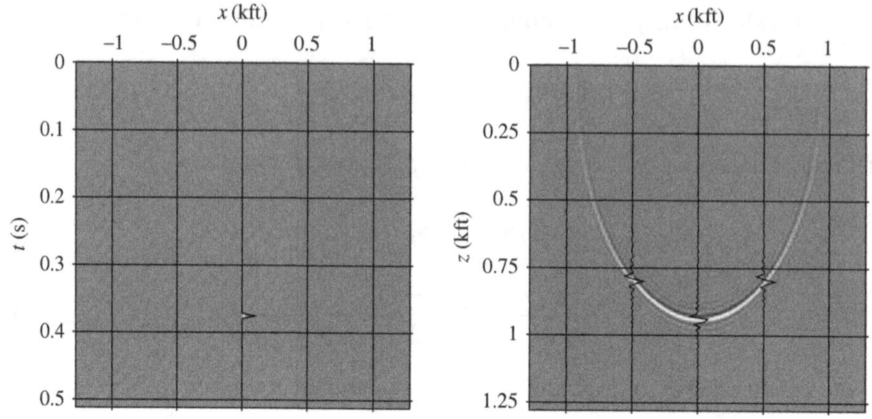

Figure 4.12 Migration of a noise spike in zero-offset data. An isolated data point migrates into a hemispherical event.

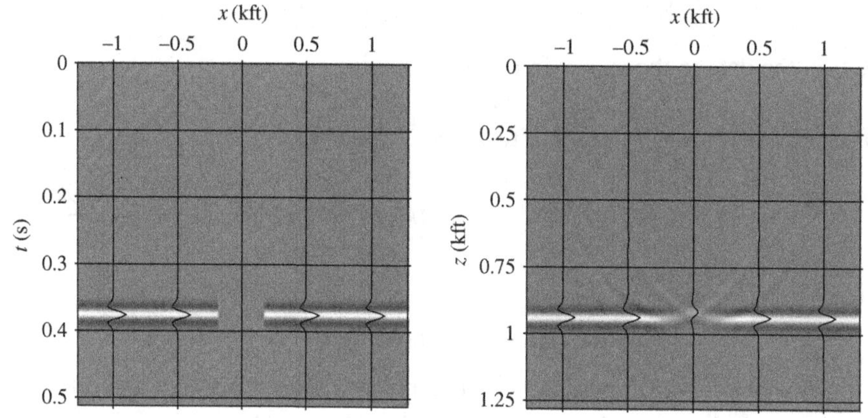

Figure 4.13 Migration of a 2-D zero-offset data set with missing data. Migration fills in the gap with an anticlinal artifact.

4.8 Effects of a finite time window

Seismic data are generally collected over a limited time interval, say $t = (0, T_f)$. Since migration generally moves data from later towards earlier times, a finite time window may result in missing data, which may affect the completeness and quality of the migrated image. When this occurs, the only cure is to collect more data over a larger time window.

Another, less obvious effect of a finite time window may occur when Fourier transforms are employed. A Fourier transform over the time window $(0, T_f)$ results in data at discrete frequency intervals $\Delta\omega = 2\pi / T_f$, the transform implicitly assuming that outside the time window, the time-domain data field actually repeats itself at intervals of T_f. A problem occurs when interpolation in the frequency domain is required. In particular, *f–k* migration maps from frequency ω to k_z, with intervals $\Delta k_z = 2\pi / Z_f$. This requires resampling or interpolating the frequency-domain data at irregular frequency intervals. The result will depend upon the method of interpolation.

As a simple example, the upper left panel of Figure 4.14 depicts data from a pair of point reflectors in two dimensions.

Performing an *f–k* migration on this data with a simple nearest-neighbor frequency-domain interpolation, the result is the upper right panel in Figure 4.14. While the migration does image the two points, it also produces numerous artifacts. This is because the algorithm sees the data extending to infinite positive and negative time, repeating every 0.512 seconds. Only the first interval $(0, T_f)$ is imaged correctly; for all the other intervals, the migration algorithm disperses the data along trajectories of different curvature.

If we try a linear frequency-domain interpolation, the result is the lower left panel in Figure 4.14. While there are noticeably fewer artifacts than produced by the nearest-neighbor interpolation, they have not been eliminated. A linear frequency-domain interpolation still assumes the data to be non-zero outside the time window, though attenuated.

4.8.1 Zero padding the time function

To work perfectly, *f–k* migration needs the data to be zero outside the time window $(0, T_f)$. One way to accomplish this is to simply pad the time-domain data with additional zeros before taking the Fourier transform. If one lengthens the time window this way by a factor of M, then the frequency interval is shortened by the same factor. The image wavenumber interval Δk_z is not affected. For a given k_z value, one can pick from a denser sample the closest corresponding ω-value. The lower right panel in Figure 4.14 shows the result of extending the time window by adding zeros to a length of 8 T_f. Artifacts are well attenuated, except for some cross-shapes attributable to the steep-dip attenuation present in the data before migration.

4.8.2 Interpolation in the frequency domain

Equivalent to time-domain zero-padding is using a sinc function to interpolate the data in the frequency domain. Specifically, for time-domain data $d_i, i = 0, \ldots, N-1$, sampled at intervals Δt, we have the discrete Fourier transform

$$D_j^{(N)} = \Delta t \sum_{i=0}^{N-1} d_i e^{i \frac{2\pi}{N} ij}, \; j = 0, \; \ldots, \; N-1, \tag{4.104}$$

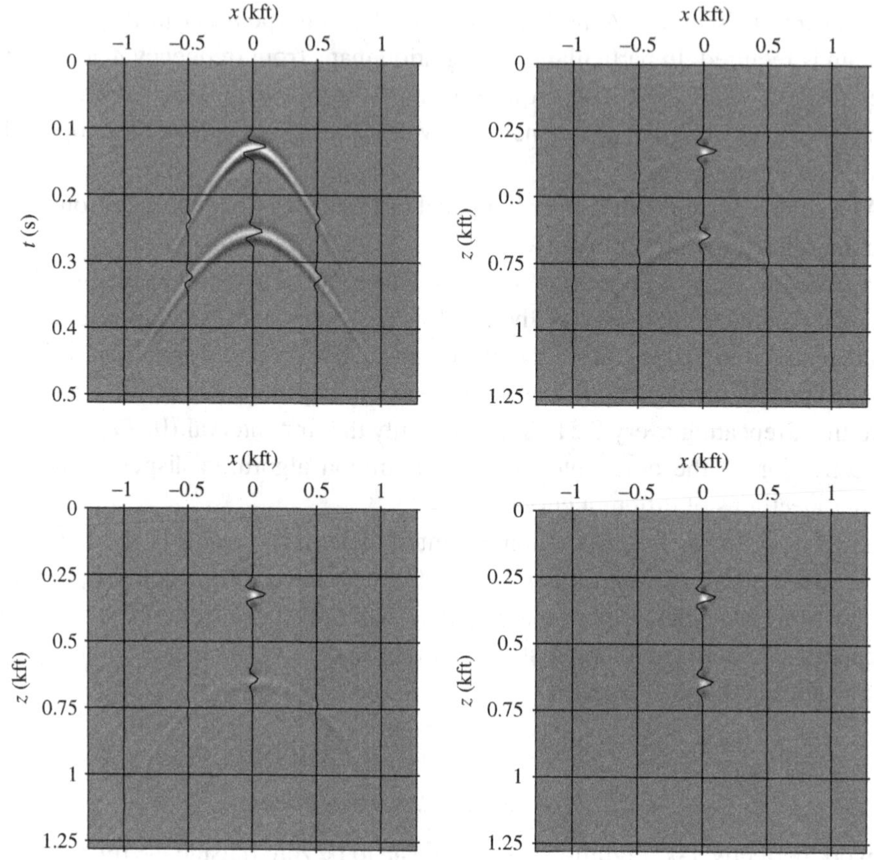

Figure 4.14 Migration of two 2-D point diffractors. Upper left is the data set, consisting of two diffractions convolved with a Ricker wavelet. The diffractions exhibit 45° of phase. Upper right is an *f–k* migration of the data with no zero-padding of the data and nearest-neighbor interpolation from ω to k_z. Numerous artifacts are apparent in this migration. Since the discrete Fourier transform assumes that the data set repeats itself in time, the algorithm is confused about where in time the data belong, and combines many incorrect migrations with the correct one. Lower left is the result of a linear interpolation between ω and k_z. Artifacts are reduced but not eliminated. Lower right results from padding the time data with seven time-window's-worth of zeros (which reduces the interval $\Delta\omega$ between frequencies by a factor of 8), then performing a nearest-neighbor interpolation to k_z. To the eye, this panel produces an image of the two diffracting points free of artifacts.

sampled at frequency intervals $\Delta\omega^{(N)}=\frac{2\pi}{N\Delta t}$. The inverse transform is

$$d_i = \frac{\Delta\omega^{(N)}}{2\pi} \sum_{j=0}^{N-1} D_j{}^{(N)} e^{+i\frac{2\pi}{N}ij}. \tag{4.105}$$

If we append zeros to d_i out to length M times N, then the discrete Fourier transform becomes

$$D_l^{(MN)} = \Delta t \sum_{i=0}^{N-1} d_i e^{i \frac{2\pi}{MN} il}, \quad l = 0, \ldots, MN-1, \tag{4.106}$$

sampled at intervals $\Delta\omega^{(MN)} = \frac{2\pi}{MN\Delta t} = \Delta\omega^{(N)} / M$. The length-$N$ transform is a subset of the length M times N transform:

$$D_{Mj}{}^{(MN)} = D_j{}^{(N)}. \tag{4.107}$$

Remaining terms in the length M times N transform can be expressed as linear combinations of the terms in the length N transform. For the general coefficient $l = M\, j_0 + k$, where $j_0 = 0, \ldots, N-1,\ k = 0, \ldots, M-1$, we can combine (4.105) and (4.106) to obtain

$$D_{Mj_0+k}{}^{(MN)} = \Delta t \frac{\Delta\omega^{(N)}}{2\pi} \sum_{i=0}^{N-1} \sum_{j=0}^{N-1} D_j{}^{(N)} e^{-i\frac{2\pi}{N} ij} e^{i\frac{2\pi}{MN} i(Mj_0 + k)}, \tag{4.108}$$

or, changing the order of the sums,

$$D_{Mj_0+k}{}^{(MN)} = \frac{1}{N} \sum_{j=0}^{N-1} D_j{}^{(N)} \sum_{i=0}^{N-1} e^{-i\frac{2\pi}{N} i(j - j_0 - \frac{k}{M})}. \tag{4.109}$$

The sum over i is recognizable as a geometric progression, which evaluates to

$$\begin{aligned} \sum_{i=0}^{N-1} e^{i\frac{2\pi}{N} i(j_0 - j + \frac{k}{M})} &= \frac{e^{i2\pi(j_0 - j + \frac{k}{M})} - 1}{e^{i\frac{2\pi}{N}(j_0 - j + \frac{k}{M})} - 1} \\ &= e^{i\pi\frac{N-1}{N}(j_0 - j + \frac{k}{M})} \frac{\sin\left(\pi\left(j_0 - j + \frac{k}{M}\right)\right)}{\sin\left(\frac{\pi}{N}\left(j_0 - j + \frac{k}{M}\right)\right)} \\ &= \frac{(-1)^{j_0 - j} e^{i\pi\frac{N-1}{N}(j_0 - j + \frac{k}{M})} \sin\left(\pi\frac{k}{M}\right)}{\sin\left(\frac{\pi}{N}\left(j_0 - j + \frac{k}{M}\right)\right)}. \end{aligned} \tag{4.110}$$

With this information, the terms in the length M times N transform are expressible as

$$D_{Mj_0+k}{}^{(MN)} = \frac{e^{i\frac{\pi}{N}((N-1)\frac{k}{M} - j_0)} \sin\left(\pi\frac{k}{M}\right)}{N} \sum_{j=0}^{N-1} \frac{D_j{}^{(N)} e^{i\pi\frac{j}{N}}}{\sin\left(\frac{\pi}{N}\left(j_0 - j + \frac{k}{M}\right)\right)}. \tag{4.111}$$

The finely spaced ω-space data field is a phase-skewed sinc-function interpolation of the original ω-space data. It is skewed because the original time-domain data exists only in positive time (i.e. $0 < t < T_f$). Were it equally distributed over positive and negative times (that is, $-T_f / 2 < t < T_f / 2$), the interpolator would

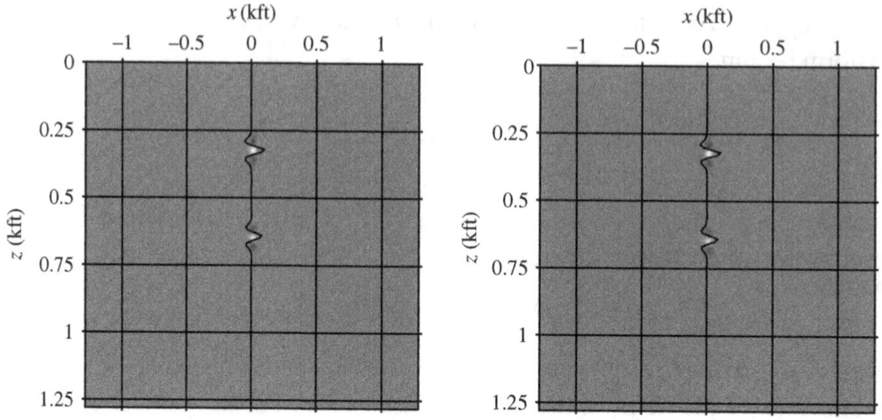

Figure 4.15 zero-padded versus sinc-function migration. A comparison of zero-padded *f*–*k* migration (left) and a migration using a sinc-function interpolator in the frequency domain (right). An operator length of 16 was used to resample the data by a factor of 8. The result is very similar to zero-padding.

be real. The above interpolator is an alternative to zero padding the time-domain data before the transform. As written, the length of the interpolation operator is the length of the entire frequency-domain data set, which would be expensive to implement. In practice, the operator can be shortened to a reasonable length. The most important thing is to get the phase right. For example, an interpolation operator of length L is

$$D_{M j_0+k}{}^{(Mn_t)} = \sum_{j=j_0-L/2+1}^{j_0+L/2} D_j{}^{(n_t)} C_{M(j_0-j)+k}, \tag{4.112}$$

with

$$C_r = \frac{e^{i\pi r/M} \sin(\pi r/M)}{L \tan(\pi r/ML)}. \tag{4.113}$$

The result, shown in Figure 4.15, is very similar to the zero-padded result.

4.9 Undermigration and overmigration

In Section 2.3, migration is presented as the construction of a downward-continued wavefield at time $t = 0$. In the exploding-reflector model, at time zero sources explode at every reflector location. The time-zero wavefield at depth should image the sources, hence the reflectors. For a more sophisticated reflectivity model, the same concept still applies. If one downward-continues both sources and receivers to a point in the subsurface, the wavefield at time zero should be proportional to reflectivity at that point.

What happens, however, if the depth of the downward-continued wavefield is miscalculated? One might think the data have been continued to a certain depth, when in fact the continuation has only gone part way. This could easily happen if the velocity chosen for the migration is slower than the true velocity. In this case the result is similar to one of the partially downward-continued panels depicted in Figure 2.13: a point reflector is seen not as a point but as a downward-curving hyperbola on the migrated image. We can see this from the 2-D exploding-reflector migration formula (2.99). If $R(x, z)$ is a point reflector at (x_0, z_0),

$$R(x, z) = \delta(x - x_0)\,\delta(z - z_0)\,, \tag{4.114}$$

then, from (2.99),

$$D(k_x, \omega) = -\frac{2\omega}{ck_z} e^{-i(k_x x_0 + k_z z_0)}, \tag{4.115}$$

with

$$k_z = -\frac{2\omega}{c}\sqrt{1 - {k_x}^2\, c^2 \big/ 4\omega^2}. \tag{4.116}$$

It is convenient in what follows to express depth in units of traveltime; i.e. $\tau = 2z/c$, so that reflectivity, as a function of x and τ, is (ignoring the change in scale)

$$R(x, \tau) = \delta(x - x_0)\,\delta(\tau - \tau_0)\,, \tag{4.117}$$

and its double Fourier transform is (note that a time transform is given the opposite sign of a space transform)

$$R(k_x, \omega_m) = e^{-i(k_x x_0 - \omega_m \tau_0)}, \tag{4.118}$$

with $\omega_m = -k_z c/2$. We do this because the vertical traveltime position of an event is unaffected by an error in velocity, whereas the apparent depth of an event will vary with velocity. In terms of ω_m, the migration formula becomes the time-migration expression (2.102), or

$$R(k_x, \omega_m) = \frac{\omega_m}{\omega} D(k_x, \omega)\,, \tag{4.119}$$

with

$$\omega_m = \omega\sqrt{1 - {k_x}^2\, c^2 \big/ 4\omega^2}. \tag{4.120}$$

In terms of ω_m, the data (4.115) are

$$D(k_x, \omega) = \frac{\omega}{\omega_m} e^{-i(k_x x_0 - \omega_m \tau_0)}. \tag{4.121}$$

Back in x–τ, this data set is known to take the form of a hyperbola with apex at τ_0 and curvature determined by the velocity c.

If one migrates the data (4.121) using the formula (4.119) but with the wrong velocity c_0, one has the result

$$R_0\left(k_x, \omega_0\right) = \frac{\omega_0}{\omega_m} e^{-i\left(k_x x_0 - \omega_m \tau_0\right)}, \tag{4.122}$$

with

$$\omega_0 = \omega\sqrt{1 - k_x{}^2\, c_0{}^2 \big/ 4\omega^2}. \tag{4.123}$$

It is clear that R_0 is not the double Fourier transform of a point. We could perform an inverse Fourier transform on R_0 and observe that in x–τ space it remains a hyperbola. However, let us stay in the Fourier domain for now.

Solving for ω_m as a function of ω_0,

$$\omega_m = \omega_0\sqrt{1 - \frac{k_x{}^2}{4\omega_0{}^2}\left(c^2 - c_0{}^2\right)}. \tag{4.124}$$

The expression (4.122) for R_0 has the same form as (4.121) for data D, with c^2 replaced by $c^2 - c_0{}^2$. Thus, the x–τ representation of R_0 must also be a hyperbola with apex at τ_0, with curvature governed by the effective residual velocity $c_r = \sqrt{c^2 - c_0{}^2}$.

We have seen that migration with too small a velocity images a reflecting point into a hyperbolic "frown", essentially the diffraction pattern expected in a medium of velocity $c_r = \sqrt{c^2 - c_0{}^2}$. What happens, however, if we migrate with too large a velocity? In this case,

$$\omega_m = \omega_0\sqrt{1 + \frac{k_x{}^2}{4\omega_0{}^2}\left(c_0{}^2 - c^2\right)}. \tag{4.125}$$

In x–τ, we still observe a diffraction pattern, but now the "frown" has become a "smile". We have in this case downward-continued too far, and the diffraction has turned upside down.

One might wonder, given the exploding-reflector model, why, when we downward-continue past an event's true location, it does not simply "unexplode" out of existence. It would, if the downward-continuation operator knew about the source at the reflector location. However, downward continuation is done with the source-free wave equation, which knows nothing about sources or reflections, hence thinks the wave has always existed. If we downward-continue the point-source response back to its place and time of origin, the wave does not extinguish, but passes through a focal point at the source location, and continues downward in negative time (see Figure 2.13).

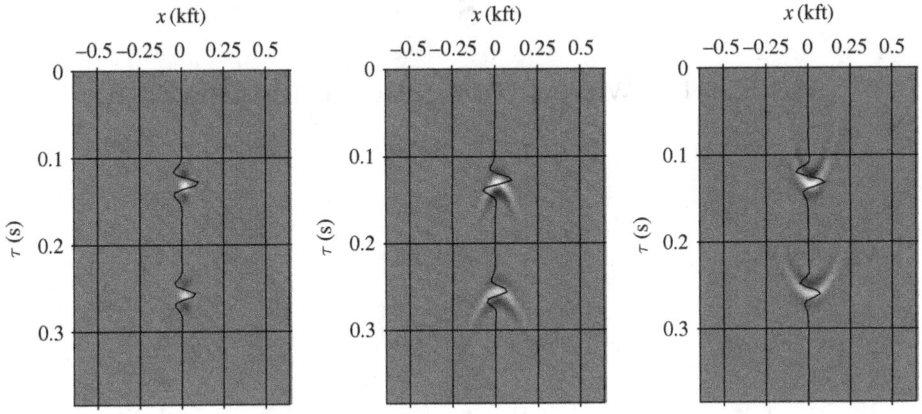

Figure 4.16 Undermigration and overmigration of point diffractions. The left panel shows the point diffractors of Figure 4.14, migrated at the correct velocity. The second panel is migrated with a velocity 10% too low, and the third panel is migrated with a velocity that is 10% too high.

It follows from the above that "undermigrated" data will have residual downward-curving diffractions, while "overmigrated" data will have residual upward-curving diffractions, as shown in Figure 4.16. Where the curvature of these diffractions can be measured, an estimate of deviation of migration velocity from true velocity can be made, and an improved migration performed.

Exercises

4.1 Starting with the differential equation

$$x^2\frac{d^2J_m(x)}{dx^2} + x\frac{dJ_m(x)}{dx} + \left(x^2 - m^2\right) J_m(x) = 0,$$

and the requirement that

$$J_m(x) \underset{x\to 0}{\rightarrow} \frac{1}{m!}\left(\frac{x}{2}\right)^m,$$

derive a power-series solution for the Bessel function J_m.

4.2 From the power-series expansion of the Bessel function, verify that

$$\frac{d(xJ_1(x))}{dx} = xJ_0(x),$$

and that (4.11) follows from (4.10).

4.3 Using (4.77) and (4.71), derive (4.78).

4.4 In the case where $\Delta\tilde{x}$ and $\tilde{x}_h$ are aligned, show that the expressions (4.77) and (4.80) for $\tan\alpha$ and $\tan\gamma$ simplify to (4.82) and (4.83).

5

Models for wave propagation and reflection

5.1 The need for models

So far, we have presented seismic migration in terms of highly simplified models of wave propagation and reflection. To proceed, we need to acknowledge that seismic waves are not scalar, and that reflections are tied to rapid changes in any of a number of Earth properties. That is, we need to incorporate more realistic models of both wave propagation and reflection.

We are fortunate to have been able to come this far without a complete Earth model, though there is a reason for our success. Look at a simple 1-D Earth, with reflectors at z_1 and z_2, which we probe at time zero with an impulsive plane wave. We posit that the wave travels in each layer without attenuation or dispersion with a well-defined velocity. In the absence of multiple reflections, we expect to see an impulsive response from each reflector, at times $t_1 = 2z_1/c_1$ and $t_2 = 2z_1/c_1 + 2\left(z_2 - z_1\right)/c_2$; i.e.

$$D(t) = A_1\delta\left(t - t_1\right) + A_2\delta\left(t - t_2\right) . \tag{5.1}$$

The quantities A_1 and A_2 are related to the reflection coefficients R_1 and R_2 at the two boundaries:

$$A_1 = R_1, \tag{5.2}$$

and, if T_{12} and T_{21} are forward and backward transmission coefficients between the two layers,

$$A_2 = T_{12}R_2T_{21}. \tag{5.3}$$

In the frequency domain,

$$D(\omega) = A_1e^{i\omega t_1} + A_2e^{i\omega t_2} . \tag{5.4}$$

If we know the velocities, a simple migration of these data can be effected by downward continuation followed by restriction to time zero. The downward-continued data field is

$$\begin{aligned} P(\omega, z) &= D(\omega)e^{-2i\omega z/c_1}, & z \leq z_1, \\ &= D(\omega)e^{-2i\omega(z_1/c_1+(z-z_1)/c_2)}, & z > z_1. \end{aligned} \tag{5.5}$$

Restricting to time zero, the migrated data field is

$$\begin{aligned} R(z) &= \frac{1}{2\pi}\int d\omega P(\omega, z) \qquad (5.6) \\ &= \frac{1}{2\pi}\int d\omega \left(A_1 e^{i\omega(t_1-2z/c_1)} + A_2 e^{i\omega(t_2-2z/c_1)}\right), \quad z \leq z_1; \\ &= \frac{1}{2\pi}\int d\omega \left(A_1 e^{i\omega(t_1-2z_1/c_1-2(z-z_1)/c_2)} + A_2 e^{i\omega(t_2-2z_1/c_1-2(z-z_1)/c_2)}\right), \quad z > z_1. \end{aligned}$$

Evaluating the frequency integrals,

$$\begin{aligned} R(z) &= A_1\delta\left(t_1 - 2z/c_1\right) + A_2\delta\left(t_2 - 2z/c_1\right), \quad z \leq z_1; \qquad (5.7) \\ &= A_1\delta\left(t_1 - 2z_1/c_1 - 2\left(z - z_1\right)/c_2\right) \\ &+ A_2\delta\left(t_2 - 2z_1/c_1 - 2\left(z - z_1\right)/c_2\right), \qquad z > z_1. \end{aligned}$$

Of the four delta functions, the second $\delta\left(t_2 - 2z/c_1\right)$ cannot be satisfied for $z \leq z_1$, and the third $\delta\left(t_1 - 2z_1/c_1 - 2\left(z - z_1\right)/c_2\right)$ cannot be satisfied for $z > z_1$. This leaves

$$R(z) = \frac{c_1}{2}A_1\delta\left(z_1 - z\right) + \frac{c_2}{2}A_2\delta\left(z_2 - z\right). \tag{5.8}$$

Given the correct velocity information, this simple imaging algorithm has correctly located the two reflectors. Until we become interested in the reflection amplitudes, the velocity information is all we need. This observation continues to hold for 2-D and 3-D geometries.

For the simple 1-D example above, migration has also provided some amplitude information in the form of $c_1A_1/2$ and $c_2A_2/2$. This information, however, cannot be deciphered without formulas for the reflection and transmission coefficients. Once reflection amplitudes become of interest, we need to know more about reflection and propagation in the medium.

5.2 Wave equations

We generally assume that seismic wave propagation is governed by a wave equation, but which one? Wave propagation in the real Earth is more complex than any model we are likely to devise, but that does not mean that a tractable model cannot, in the right circumstances, be useful. We look here at scalar, acoustic, and elastic waves.

5.2.1 The scalar wave equation

Simplest is the scalar wave equation

$$\left(\nabla^2 - \frac{1}{c(\boldsymbol{x})^2}\frac{\partial^2}{\partial t^2}\right)\psi_{\text{scalar}}(\boldsymbol{x}, t) = 0, \tag{5.9}$$

with the scalar wavefield at $(\boldsymbol{x}, t)$ for velocity $c(\boldsymbol{x})$ represented by $\psi(\boldsymbol{x}, t)$. In the case of constant velocity, equation (5.9) has plane-wave solutions

$$\psi_{\text{scalar}}(\boldsymbol{x}, t; \boldsymbol{k}) = e^{i(\boldsymbol{k}\cdot\boldsymbol{x} - \omega t)}, \tag{5.10}$$

with frequency ω equal to wavenumber magnitude times velocity:

$$\omega = kc. \tag{5.11}$$

Consider a reflecting boundary at which velocity changes from c_0 to c_1, and a plane wave approaching the boundary at incident angle γ_0. This wave will produce reflected and transmitted waves. The wave equation (5.9) requires that the total wave amplitude must be continuous at the boundary, which implies that the transmission angle γ_1 is given by Snell's law:

$$\frac{\sin\gamma_1}{c_1} = \frac{\sin\gamma_0}{c_0}. \tag{5.12}$$

Equation (5.9) also implies that the derivative of the total wavefield is continuous across the boundary, from which it follows that the reflection coefficient, or ratio of incoming to reflected amplitude, is

$$R_{\text{scalar}}(\gamma) = \frac{c_1\cos\gamma_0 - c_0\cos\gamma_1}{c_1\cos\gamma_0 + c_0\cos\gamma_1}. \tag{5.13}$$

Similarly, the transmission coefficient is

$$T_{\text{scalar}}(\gamma) = \frac{2c_1\cos\gamma_0}{c_1\cos\gamma_0 + c_0\cos\gamma_1}. \tag{5.14}$$

The scalar wave equation would appear to be an imperfect choice for modeling seismic energy. For one thing, the seismic wavefield is not scalar. For another, there is no reason to suppose that real rocks would produce reflection coefficients with the angular dependence found in (5.13). Even so, the scalar wave equation is the most common basis for a propagation model. It is also used to model reflections more often than one might suppose, in applications where the angular dependence of reflection amplitudes is not critically important.

5.2.2 The acoustic wave equation

Slightly less simple is the acoustic wave equation

$$\left(\nabla\cdot\frac{1}{\rho(\boldsymbol{x})}\nabla - \frac{1}{\rho(\boldsymbol{x})c(\boldsymbol{x})^2}\frac{\partial^2}{\partial t^2}\right)\psi_a(\boldsymbol{x}, t) = 0, \tag{5.15}$$

where ρ is density, c is velocity, and ψ_a is now the acoustic pressure. To satisfy (5.15) at a boundary, ψ_a must be continuous, which implies that Snell's law (5.12) holds also for acoustic waves. The quantity $\rho^{-1}\nabla\psi_a$ must also remain continuous across a boundary, which leads to the reflection and transmission coefficients

$$R_a(\gamma) = \frac{\rho_1 c_1 \cos\gamma_0 - \rho_0 c_0 \cos\gamma_1}{\rho_1 c_1 \cos\gamma_0 + \rho_0 c_0 \cos\gamma_1}, \tag{5.16}$$

$$T_a(\gamma) = \frac{2\rho_1 c_1 \cos\gamma_0}{\rho_1 c_1 \cos\gamma_0 + \rho_0 c_0 \cos\gamma_1}. \tag{5.17}$$

Real rocks are no more acoustic than they are scalar, so one might ask whether the acoustic wave equation is at all useful for imaging. Actually, it provides a better propagation model than does the scalar wave equation, because it will adjust wave amplitude as density varies. As a reflection model, it does produce realistic-looking amplitude variations with angle, even though the physical causes attributed to the variations are not correct. Even so, special cases can be useful:

Constant reflectivity: When density but not velocity changes across a boundary, the resulting reflection coefficient reduces to

$$R_a(\gamma) \to \frac{\rho' - \rho}{\rho' + \rho}, \tag{5.18}$$

which is independent of angle of incidence. An imaging algorithm based on an imperfect reflection model will generally leave residual amplitude variations with angle. If the acoustic constant-velocity model is used, the residual variations will be untainted by any bias of the imaging model, hence can be analyzed, post-migration, by a more realistic reflection model. This brings into play three separate physical models: a propagation model, a pre-migration reflection model, and a post-migration reflection model. Theoretically, this may be unsatisfying. Practically, it is hard to argue with approximations that work, until they begin to fail and more rigorous approaches become necessary.

Zero reflectivity: If density varies inversely with velocity, then impedance ρc is constant across a boundary, and at normal incidence ($\gamma_0 = 0$), reflection coefficient is zero. Because a constant-impedance model attenuates reflections, it is useful to describe propagation in a variable-velocity medium.

5.2.3 The elastic wave equation and beyond

Intuitively, one might expect that the elastic wave equation should provide a more realistic model than the acoustic or scalar wave equations. Of course, the real Earth isn't elastic either, but at least one can hope to describe both compressional and shear wave propagation under the same umbrella (Marfurt, 1978; Schleicher *et al.*, 2007).

For an isotropic, elastic Earth, the governing wave equation is (in the frequency domain)

$$\mathcal{L}_E(\boldsymbol{x}, \omega)\boldsymbol{u}(\boldsymbol{x}, \omega) = 0, \tag{5.19}$$

where $\boldsymbol{u}$ is the displacement vector, and $\mathcal{L}_E$ the 3×3 matrix

$$\mathcal{L}_E(\boldsymbol{x}, \omega) = \begin{pmatrix} L_{xx} & L_{xy} & L_{xz} \\ L_{yx} & L_{yy} & L_{yz} \\ L_{zx} & L_{zy} & L_{zz} \end{pmatrix}, \tag{5.20}$$

with elements

$$\begin{aligned} L_{ii} &= \partial_i(\lambda + 2\mu)\partial_i + \sum_{j\neq i} \partial_j \mu \partial_j + \rho\omega^2, \; i, j = x, \; y, \; z; \\ L_{ij} &= \partial_i \lambda \partial_j + \partial_j \mu \partial_i, \; j \neq i. \end{aligned} \tag{5.21}$$

It is understood in this equation that the partial derivatives ∂_i and ∂_j operate on everything to their right. The quantities λ and μ are the Lamé parameters of the medium, and ρ is density.

Where derivatives of the parameters can be neglected, the matrix $\mathcal{L}_E$ is diagonalizable into P- and S-wave components. Define a matrix of partial derivatives $\boldsymbol{\Pi}$ as

$$\begin{aligned} \boldsymbol{\Pi} &\equiv \begin{pmatrix} i\alpha/\omega & 0 & 0 & 0 \\ 0 & i\beta/\omega & 0 & 0 \\ 0 & 0 & i\beta/\omega & 0 \\ 0 & 0 & 0 & i\beta/\omega \end{pmatrix} \begin{pmatrix} \partial_x & \partial_y & \partial_z \\ 0 & -\partial_z & \partial_y \\ \partial_z & 0 & -\partial_x \\ -\partial_y & \partial_x & 0 \end{pmatrix} \\ &= \frac{i}{\omega} \begin{pmatrix} \alpha\partial_x & \alpha\partial_y & \alpha\partial_z \\ 0 & -\beta\partial_z & \beta\partial_y \\ \beta\partial_z & 0 & -\beta\partial_x \\ -\beta\partial_y & \beta\partial_x & 0 \end{pmatrix}. \end{aligned} \tag{5.22}$$

Acting on a displacement vector $\boldsymbol{u}$, the first row of $\boldsymbol{\Pi}$ is proportional to the divergence of $\boldsymbol{u}$, and the remaining rows the curl of $\boldsymbol{u}$ (see Appendix B):

$$\boldsymbol{\Psi} \equiv \begin{pmatrix} \psi \\ A \end{pmatrix} = \boldsymbol{\Pi}\boldsymbol{u} = \frac{i}{\omega} \begin{pmatrix} \alpha\partial_x u_x + \alpha\partial_y u_y + \alpha\partial_z u_z \\ \beta\partial_y u_z - \beta\partial_z u_y \\ \beta\partial_z u_x - \beta\partial_x u_z \\ \beta\partial_x u_y - \beta\partial_y u_x \end{pmatrix} = \frac{i}{\omega} \begin{pmatrix} \alpha\boldsymbol{\nabla}\cdot\boldsymbol{u} \\ \beta(\boldsymbol{\nabla}\times\boldsymbol{u})_x \\ \beta(\boldsymbol{\nabla}\times\boldsymbol{u})_y \\ \beta(\boldsymbol{\nabla}\times\boldsymbol{u})_z \end{pmatrix}. \tag{5.23}$$

The transformation operator $\boldsymbol{\Pi}$ has been normalized as suggested by Yanglei Zou of MOSRP. With this normalization, the P and S wave functions ψ and A retain the dimension of displacement, and plane-wave reflections assume the familiar Zoeppritz form as presented in Aki and Richards (1980). Though $\boldsymbol{\Pi}$ maps a 3-D vector into a 4-D space, the space is restricted by the constraint $\boldsymbol{\nabla}\cdot(\boldsymbol{\nabla}\times\boldsymbol{u}) = \boldsymbol{\nabla}\cdot A = 0$ (see equation (B.34)). Thus, there are only three independent dimensions present.

If $\mathbf{\Pi}$ is postmultiplied by its transpose, the result is a 4×4 matrix:

$$\mathbf{\Pi}\mathbf{\Pi}^T \equiv \frac{-1}{\omega^2}\begin{pmatrix} \alpha^2\nabla^2 & 0 & 0 & 0 \\ 0 & \beta^2\nabla^2 - \beta^2\partial_x\partial_x & -\beta^2\partial_x\partial_y & -\beta^2\partial_x\partial_z \\ 0 & -\beta^2\partial_x\partial_y & \beta^2\nabla^2 - \beta^2\partial_y\partial_y & -\beta^2\partial_y\partial_z \\ 0 & -\beta^2\partial_x\partial_z & -\beta^2\partial_y\partial_z & \beta^2\nabla^2 - \beta^2\partial_z\partial_z \end{pmatrix}. \tag{5.24}$$

Operating on a 4-vector $(\psi, \boldsymbol{A})$, we have

$$\mathbf{\Pi}\mathbf{\Pi}^T \begin{pmatrix} \psi \\ \boldsymbol{A} \end{pmatrix} = \frac{\nabla^2}{\omega^2}\begin{pmatrix} \alpha^2 & \psi \\ \beta^2 & \boldsymbol{A} \end{pmatrix} + \frac{1}{\omega^2}\begin{pmatrix} 0 \\ \beta^2\nabla\,(\nabla\cdot\boldsymbol{A}) \end{pmatrix}. \tag{5.25}$$

For the subspace of vectors where $\nabla\cdot\boldsymbol{A} = 0$,

$$\mathbf{\Pi}\mathbf{\Pi}^T \begin{pmatrix} \psi \\ \boldsymbol{A} \end{pmatrix} = \mathcal{J}\begin{pmatrix} \psi \\ \boldsymbol{A} \end{pmatrix}, \tag{5.26}$$

where

$$\mathcal{J} = -\frac{\nabla^2}{\omega^2}\begin{pmatrix} \alpha^2 & 0 & 0 & 0 \\ 0 & \beta^2 & 0 & 0 \\ 0 & 0 & \beta^2 & 0 \\ 0 & 0 & 0 & \beta^2 \end{pmatrix}. \tag{5.27}$$

The matrix $\mathcal{J}$ can be formally inverted to obtain:

$$\mathcal{J}^{-1} = -\nabla^{-2}\omega^2\begin{pmatrix} \alpha^{-2} & 0 & 0 & 0 \\ 0 & \beta^{-2} & 0 & 0 \\ 0 & 0 & \beta^{-2} & 0 \\ 0 & 0 & 0 & \beta^{-2} \end{pmatrix}. \tag{5.28}$$

Provided derivatives of λ, μ, and ρ can be neglected, the elastic wave operator $\boldsymbol{\mathcal{L}}_E$ can be diagonalized with the operator Π:

$$\boldsymbol{\mathcal{L}}_D = \mathbf{\Pi}\boldsymbol{\mathcal{L}}_E\mathbf{\Pi}^T\mathcal{J}^{-1} = \begin{pmatrix} \mathcal{L}_P & 0 & 0 & 0 \\ 0 & \mathcal{L}_S & 0 & 0 \\ 0 & 0 & \mathcal{L}_S & 0 \\ 0 & 0 & 0 & \mathcal{L}_S \end{pmatrix}, \tag{5.29}$$

valid, as for $\mathbf{\Pi}\mathbf{\Pi}^T$, in the subspace $\nabla\cdot\boldsymbol{A} = 0$, and where $\mathcal{L}_P$ and $\mathcal{L}_S$ are P- and S-wave operators

$$\mathcal{L}_P(\boldsymbol{x}, t) = \rho\alpha^2\left(\nabla^2 + \frac{\omega^2}{\alpha^2}\right), \tag{5.30}$$

$$\mathcal{L}_S(\boldsymbol{x}, t) = \rho\beta^2\left(\nabla^2 + \frac{\omega^2}{\beta^2}\right), \tag{5.31}$$

and where α and β are P- and S-wave velocities, related to the original parameters as

$$\alpha = \sqrt{\frac{\lambda + 2\mu}{\rho}}, \tag{5.32}$$

and

$$\beta = \sqrt{\frac{\mu}{\rho}}. \tag{5.33}$$

We note that the matrix $\mathcal{J} = \boldsymbol{\Pi}\boldsymbol{\Pi}^T$, acting on a solution $\boldsymbol{\Psi}$ to the wave equation $\boldsymbol{L}_D\boldsymbol{\Psi} = 0$, yields the unit matrix.

With the 4-vector $\boldsymbol{\Psi} \equiv \boldsymbol{\Pi u}$ the P–S decomposition of $\boldsymbol{u}$ (as per equation (5.23)), the elastic wave equation becomes

$$\boldsymbol{\mathcal{L}}_D\boldsymbol{\Psi} = \begin{pmatrix} \mathcal{L}_P\psi \\ \mathcal{L}_S\boldsymbol{A} \end{pmatrix} = 0. \tag{5.34}$$

Where the seismic parameters vary too rapidly for their derivatives to be neglected, the above transformation does not diagonalize the wave equation. Often, however, the interpretation of the elements of $\boldsymbol{\Psi}$ as P- and S-waves is still reasonable, with off-diagonal terms in $\boldsymbol{\mathcal{L}}_D$ representing conversions of one wave type to another. Abrupt changes in seismic properties produce reflections as described by the Zoeppritz equations (see e.g. Aki and Richards, 1980).

It will prove useful to perform an additional rotation of the shear components in the wave operator (5.33) into SH, SV, and the null space of longitudinal shear waves. To do so, we need a reference plane, which is not yet defined.

Although one cannot help but feel that the elastic wave equation is a step closer to reality, it is not clear just how big that step is. The real Earth is not isotropic or elastic. While the elastic wave equation (5.19) can be generalized to accommodate anisotropy, the number of required parameters rises significantly, and how close one can come to the real Earth with a tractable model is an open discussion (Aki and Richards, 1980; Chapman and Coates, 1994; de Hoop and Bleistein, 1997).

The real Earth also exhibits attenuation, affecting both propagation and reflection. At least approximately, these effects can be incorporated into propagation and reflection models. But even having done so, are we there yet? The answer is, not likely.

5.3 Building reflections

To build a reflected wave, one needs to combine a propagation model with a reflection model. For plane waves and planar reflectors, this is a trivial matter. Imagine a downgoing plane wave traveling with velocity c_i in the direction of its wavenumber vector $\boldsymbol{k}_i = \omega\hat{\boldsymbol{k}}_i/c_i$, incident on a reflecting surface. Reflected from the surface will be a plane wave traveling with velocity c_r in the direction of the wavenumber $\boldsymbol{k}_r = \omega\hat{\boldsymbol{k}}_r/c_r$. We have allowed for the possibility that $c_i \neq c_r$ to accommodate converted waves.

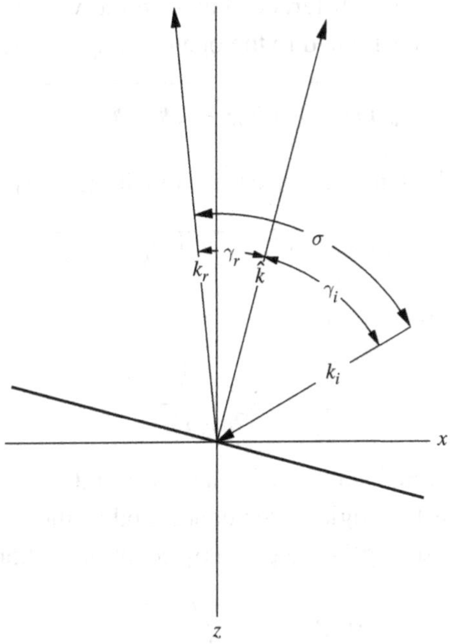

Figure 5.1 Illustration of a planar reflection. Confined to two dimensions for simplicity. A plane wave traveling with velocity c_i incident on a sloping reflecting surface has direction vector $\hat{\boldsymbol{k}}_i$. The reflected plane wave has direction vector $\hat{\boldsymbol{k}}_r$, traveling with velocity c_r. In this illustration, $c_i = 2$, and $c_r = 1$. Relative to the normal $\hat{\boldsymbol{k}}$ of the reflecting surface, the incident wave makes angle γ_i and the reflected wave γ_r. The sum of the incident and reflected angles is the opening angle σ. According to Snell's law, $\sin(\gamma_i)/c_i = \sin(\gamma_r)/c_r = \sin(\sigma)\Big/\sqrt{c_i^2 + c_r^2 + 2c_i c_r \cos(\sigma)}$.

The dot product of the incident and reflected wavenumbers is minus the cosine of the opening angle between them:

$$\hat{\boldsymbol{k}}_i \cdot \hat{\boldsymbol{k}}_r = -\cos\sigma. \tag{5.35}$$

Moreover, according to Snell's law (see Figure 5.1 and Appendix H, Figure H.1), the components of $\boldsymbol{k}_r$ and $\boldsymbol{k}_i$ in the plane of the reflector must be equal. Consequently, the difference $\boldsymbol{k} = \boldsymbol{k}_r - \boldsymbol{k}_i$ between the incident and reflected wavenumbers points in the direction $\hat{\boldsymbol{k}}$ normal to the reflecting surface:

$$\hat{\boldsymbol{k}} = \frac{\boldsymbol{k}_r - \boldsymbol{k}_i}{|\boldsymbol{k}_i - \boldsymbol{k}_r|} = \frac{\hat{\boldsymbol{k}}_r/c_r - \hat{\boldsymbol{k}}_i/c_i}{\left|\hat{\boldsymbol{k}}_i/c_i - \hat{\boldsymbol{k}}_r/c_r\right|}. \tag{5.36}$$

Now, the magnitude of the difference between the velocity-weighted incoming and reflected unit vectors is related to the opening angle σ between them:

$$\left|\hat{\boldsymbol{k}}_i/c_i - \hat{\boldsymbol{k}}_r/c_r\right| = \sqrt{1/c_i{}^2 + 1/c_r{}^2 - 2\hat{\boldsymbol{k}}_i \cdot \hat{\boldsymbol{k}}_r/c_i c_r} = \frac{2c_e(\sigma)}{c_i c_r}, \tag{5.37}$$

where $c_e(\sigma)$ is the angle-dependent average velocity (see Appendix H)

$$c_e(\sigma) = \frac{1}{2}\sqrt{c_i{}^2 + c_r{}^2 + 2c_i c_r \cos\sigma}. \tag{5.38}$$

Consequently, $\hat{\boldsymbol{k}}$ can be written

$$\hat{\boldsymbol{k}} = \frac{\hat{\boldsymbol{k}}_r\, c_i - \hat{\boldsymbol{k}}_i\, c_r}{2c_e(\sigma)}. \tag{5.39}$$

The incident and reflected waves make angles γ_i and γ_r with respect to $\hat{\boldsymbol{k}}$, where γ_i may be referred to as the angle of incidence, and γ_r the angle of reflection. The cosines and sines of these angles are given by equations (H.56) through (H.59), or

$$\cos\gamma_i = \frac{(c_r + c_i \cos\sigma)}{2c_e(\sigma)}, \tag{5.40}$$

$$\cos\gamma_r = \frac{(c_i + c_r \cos\sigma)}{2c_e(\sigma)}, \tag{5.41}$$

and

$$\frac{\sin\gamma_i}{c_i} = \frac{\sin\gamma_r}{c_r} = \frac{\sin\sigma}{2c_e(\sigma)}. \tag{5.42}$$

In the absence of converted waves, $c_i = c_r = c$, $c_e = c\cos(\sigma/2)$, and the formula (5.42) reduces to $\gamma_i = \gamma_r = \gamma = \sigma/2$.

Given the input direction $\hat{\boldsymbol{k}}_i$ and the reflector normal $\hat{\boldsymbol{k}}$, the output direction $\hat{\boldsymbol{k}}_r$ is

$$\hat{\boldsymbol{k}}_r = \frac{\hat{\boldsymbol{k}}_i\, c_r + 2\,\hat{\boldsymbol{k}}\, c_e(\sigma)}{c_i}. \tag{5.43}$$

If $c_r = c_i$, this expression simplifies to

$$\hat{\boldsymbol{k}}_r = \hat{\boldsymbol{k}}_i + 2\,\hat{\boldsymbol{k}}\cos(\sigma/2). \tag{5.44}$$

If $\mathcal{A}_s$ is the amplitude of the incoming wave at a point on the reflector, and $R(\sigma)$ is the reflection coefficient, then the reflected wave at that point has amplitude

$$\mathcal{A}_g = \mathcal{A}_s R(\sigma). \tag{5.45}$$

In the real world, waves and reflecting surfaces are not perfectly planar. Where one can assume that waves and boundaries are locally planar, one can construct reflected waves using specular concepts. Alternatively, one can invoke scattering theory, characterizing reflections in terms of scattering potentials. The scattering-theory approach is more general in that it allows construction of reflected waves where specular reflectors do not exist. It is also better grounded physically, and ultimately easier to understand. First-order scattering theory produces a linearized reflectivity which may become inaccurate for large reflections or large incident angles, but this deficiency is remediable.

5.4 A scattering-theory model for reflection data

5.4.1 Exact and background wave operators

The scattering-theory approach begins with a wave equation operator and a Green's function. The wave equation operator is a differential operator, which we represent abstractly by $\mathcal{L}$. A solution Ψ to the homogeneous wave equation obeys

$$\mathcal{L}\Psi = 0. \tag{5.46}$$

Examples of this equation include the scalar, acoustic, and elastic wave equations (5.9), (5.15), and (5.19). The Green's function G obeys a similar equation

$$\mathcal{L}G = -I, \tag{5.47}$$

which means that formally, the Green's function is an inverse to the wave operator $-\mathcal{L}$:

$$G = -\mathcal{L}^{-1}. \tag{5.48}$$

The Green's function, or impulse response, carries all the results of a scattering experiment, including a direct arrival, simple and multiple reflections, and all other phenomena allowed by the wave equation operator $\mathcal{L}$.

In terms of space and frequency, equation (5.48) is expressible as

$$\mathcal{L}(\boldsymbol{x}, \omega)G\left(\boldsymbol{x}|\boldsymbol{x}_s, \omega\right) = -\delta\left(\boldsymbol{x} - \boldsymbol{x}_s\right), \tag{5.49}$$

where $\boldsymbol{x}_s = (x_s, y_s, z_s)$ is the source coordinate, and $\boldsymbol{x} = (x, y, z)$ the observation point. Without loss of generality, we consider only positive ω. The Green's function solution to this equation as written has infinite bandwidth. In contrast, a realizable experiment has a finite bandwidth, limited in both low and high temporal frequency. In one respect, finite bandwidth is an advantage, leading to useful approximations. One can divide physical changes into three regimes. The first, or "low-frequency", regime consists of trends or changes that are slow relative to the wavelengths in the data. These changes do not produce observable reflections, but do affect the direction, amplitude, velocity, and phase of wavefronts. The second "high-frequency" regime produces the reflections we see on a seismic record. The third "ultra-high-frequency" regime produces complex behavior that may not be directly observable but impacts effective physical properties. Our theory deals with the effective physical properties as measured in the high-frequency regime.

The existence of a low-frequency regime allows construction of an approximate or background wave operator $\mathcal{L}_0$ and corresponding Green's function G_0. The physical parameters of the background wave operator are assumed to change slowly enough that observable reflections are not generated. Even though the full wave equation may entail a vector wavefield, because the approximate wavefields do not reflect, one can expect them to separate into scalar components (see e.g. Section 5.2.3). Each scalar component of the approximate Green's function will obey

$$G_0 = -\mathcal{L}_0^{-1}, \tag{5.50}$$

or, more explicitly, in the space–frequency domain,

$$\mathcal{L}_0(\boldsymbol{x}, \omega) G_0 (\boldsymbol{x}|\boldsymbol{x}_s, \omega) = -\delta(\boldsymbol{x} - \boldsymbol{x}_s). \tag{5.51}$$

The difference between the exact and approximate wave operators is the scattering potential:

$$\mathcal{V} = \mathcal{L} - \mathcal{L}_0. \tag{5.52}$$

5.4.2 The Lippmann–Schwinger equation and the Born approximation

The full and approximate Green's functions are related via the Lippmann–Schwinger equation (Newton, 1966; Taylor, 1972)

$$G = G_0 + G_0 \mathcal{V} G, \tag{5.53}$$

as is easily confirmed by substituting the defining equation (5.52) for $\mathcal{V}$ into this equation, and invoking the inverse relations (5.48) and (5.50).

An alternate formulation of the Lippmann–Schwinger equation is

$$G = (I - G_0 \mathcal{V})^{-1} G_0, \tag{5.54}$$

with I the identity or unit operator. Either form can be expanded as a power series in the quantity $G_0\mathcal{V}$:

$$G = G_0 + G_0\mathcal{V}G_0 + G_0\mathcal{V}G_0\mathcal{V}G_0 + G_0\mathcal{V}G_0\mathcal{V}G_0\mathcal{V}G_0 + \cdots. \tag{5.55}$$

Justification for the Lippmann–Schwinger equation or its expansion requires that the potential $\mathcal{V}$ be "small", or more precisely, that the norm of the operator $G_0\mathcal{V}$ be less than one. This poses an apparent problem for seismic data, where velocity increases systematically with depth, and hard-rock seismic velocities may be several times the velocity of sound in water. Variations from a constant background are neither small nor localized, and the operator $G_0\mathcal{V}$ most certainly has a norm greater than 1. Fortunately, it is mostly at low frequencies that $\boldsymbol{G}_0\boldsymbol{\mathcal{V}}$ becomes large. Within the seismic bandwidth, it is relatively well behaved. Thus, use of the Lippmann–Schwinger equation or its expansion on seismic data carries an implicit high-frequency assumption.

For modeling or migration, we can approximate the Lippmann–Schwinger equation with the Born approximation, which replaces the exact Green's function on the right-hand side of (5.53) with the approximate one (Taylor, 1972; Cohen & Bleistein, 1977; Raz, 1981; Clayton & Stolt, 1981; Stolt & Weglein, 1985; Bleistein *et al.*, 1985; Cohen *et al.*, 1986)

$$\boldsymbol{G} \simeq \boldsymbol{G}_0 + \boldsymbol{G}_0\boldsymbol{\mathcal{V}}\boldsymbol{G}_0. \tag{5.56}$$

The Born approximation can be viewed as the linear term in an expansion of the Lippmann–Schwinger equation in powers of the scattering potential. It becomes exact in the limit of small $\boldsymbol{\mathcal{V}}$. Since the approximate Green's function $\boldsymbol{G}_0$ has been engineered so as not to contain reflected energy, the Born approximation contains only primary reflections. One can enhance the ability of the Born approximation to handle large primary reflections by modifying the potential term $\boldsymbol{\mathcal{V}}$, though we will not worry about that here.

Since reflected energy is confined to the second term in (5.56), we can define a reflected-wave data set $\boldsymbol{D}$ to be

$$\boldsymbol{D} \simeq \mathfrak{w}\boldsymbol{G}_0\boldsymbol{\mathcal{V}}\boldsymbol{G}_0. \tag{5.57}$$

Because the Green's functions $\boldsymbol{G}_0$ have infinite bandwidth, while a realizable data set $\boldsymbol{D}$ will have finite bandwidth, we have included in $\boldsymbol{D}$ a finite-bandwidth wavelet $\mathfrak{w}$, which for simplicity we will take to be independent of source and receiver location and direction.

The potential term $\boldsymbol{\mathcal{V}}$ is a pseudo-local operator in space–time or space–frequency. Like $\boldsymbol{\mathcal{L}}$ and $\boldsymbol{\mathcal{L}}_0$, it may contain partial derivatives with respect to any or all of the space–time variables. The Born approximation, expressed abstractly in (5.57), can be written as a space–frequency integral equation

$$D\left(\boldsymbol{x}_g|\boldsymbol{x}_s;\omega\right) = \mathfrak{w}(\omega)\int\!\!\int\!\!\int d^3x G_0\left(\boldsymbol{x}_g\middle|\,\boldsymbol{x};\omega\right)\mathcal{V}(\boldsymbol{x},\omega)G_0\left(\boldsymbol{x}|\boldsymbol{x}_s;\omega\right), \tag{5.58}$$

with $\boldsymbol{x} = (x, y, z)$ the reflector position, $\boldsymbol{x}_s = (x_s, y_s, z_s)$ the source position, and $\boldsymbol{x}_g = (x_g, y_g, z_g)$ the receiver position. For surface reflection data, sources and receivers are located at zero depth, i.e. at $z_g = z_s = 0$. In this case, the 3-vectors $\boldsymbol{x}_s$ and $\boldsymbol{x}_g$ can be replaced by 2-vectors $\tilde{\boldsymbol{x}}_s = (x_s, y_s)$ and $\tilde{\boldsymbol{x}}_g = \left(x_g, y_g\right)$, and we can write

$$D\left(\tilde{\boldsymbol{x}}_g|\tilde{\boldsymbol{x}}_s;\omega\right) = \mathfrak{w}(\omega)\int\!\!\int\!\!\int d^3x G_0\left(\tilde{\boldsymbol{x}}_g\middle|\,\boldsymbol{x};\omega\right)\mathcal{V}(\boldsymbol{x},\omega)G_0\left(\boldsymbol{x}|\tilde{\boldsymbol{x}}_s;\omega\right). \tag{5.59}$$

The Born approximation (5.59) provides a simple but powerful way to calculate seismic reflection data. The formula is expressed in terms of a scattering potential which is a point property of the reflecting volume, hence does not require or rely on the existence of specular reflecting surfaces. We shall see, however, that the scattering potential, and hence the reflected data, can be expressed in terms of a pointwise reflectivity function, if desired.

The great appeal of scattering theory is its solid foundation in terms of measurable physical parameters. Scattering theory forms reflections from a scattering potential, in turn defined from intrinsic physical Earth properties. Contrasted with the scattering potential, reflectivity seems a vague and insubstantial concept – viewed from a distance, it appears to be well defined, but the closer one looks, the less appears to be there.

On the other hand, reflectivity is a concept more familiar to geoscientists, and therefore carries a higher comfort factor. We will not abandon either concept in what follows, but make no secret where our preference lies.

Exercises

5.1 Given a reflecting boundary at $z = 0$, derive the reflection and transmission coefficients for (a) a scalar wave, and (b) an acoustic wave.

5.2 (a) Express the scalar reflection coefficient (5.13) as a function of incident angle γ_0 and the velocity perturbation $a_c = 1 - c_0{}^2/c_1{}^2$. (b) Express the acoustic reflection coefficient (5.16) in terms of γ_0, a_c, and the density perturbation $a_\rho = 1 - \rho_0/\rho_1$.

5.3 Derive the Lippmann–Schwinger equation.

5.4 Write the second-order term in the Born expansion of the Lippmann–Schwinger equation as a space–frequency domain integral equation.

6

Green's functions

6.1 The general Green's function

When describing wave propagation, the fundamental entity is the Green's function, or impulse response. If $\mathcal{L}(\boldsymbol{x},\ t)$ represents the wave operator, so that the source-free wave equation is

$$\mathcal{L}(\boldsymbol{x}, t)\Psi(\boldsymbol{x}, t) = 0, \tag{6.1}$$

then the time-domain Green's function is a solution to the point-source wave equation

$$\mathcal{L}(\boldsymbol{x}, t)G\left(\boldsymbol{x}\,|\boldsymbol{x}_s\,, t - t_s\right) = -\delta\left(t - t_s\right)\delta\left(\boldsymbol{x} - \boldsymbol{x}_s\right). \tag{6.2}$$

Viewed as an operator, the Green's function $\boldsymbol{G}$ is the negative of an inverse to the wave operator $\boldsymbol{\mathcal{L}}$.

There are many solutions to the point-source wave equation, all of which are considered Green's functions. The *impulse response* or *causal Green's function* $\boldsymbol{G}_+$ is that particular solution to (6.2) or (5.47) corresponding to a wave emanating from point $\boldsymbol{x}_s$ at time t_s. Other possible solutions include an *imploding* or *anticausal* Green's function $\boldsymbol{G}_-$ in which a wave converges upon point $\boldsymbol{x}_s$, vanishing after time t_s. Mathematically, $\boldsymbol{G}_+$ and $\boldsymbol{G}_-$ are adjoints, so to know one is to know the other. Unless otherwise stated, the Green's function of interest to us will be the impulse response.

The impulse response as written has infinite bandwidth, hence is not physically realizable. A realizable response would be the convolution of $\boldsymbol{G}$ with a bandlimited source function, but we will not worry about that at this point.

There is also an issue of scaling. Equation (6.2) describes the response to a "unit" impulse. If one defines a new wave operator $\epsilon\boldsymbol{\mathcal{L}}$ proportional to $\boldsymbol{\mathcal{L}}$, then the unit response to this operator is scaled by ϵ^{-1}. Thus there would appear to be some degree of arbitrariness in the scaling or normalization of the impulse response. The scaling also depends upon the physical attribute represented by the wave function

$\Psi(\boldsymbol{x}, t)$. Possible choices include displacement, velocity, acceleration, pressure, and so on, each of which leads to a different scale factor for the wave operator. The choice of physical attribute is actually determined by the data. More important than the choice itself is to remain consistent with the choice after it is made.

Part of the power of the Green's function is its ability to describe the response to arbitrary source configurations. Suppose, given a distributed source function $\rho(\boldsymbol{x}, t)$, we desire the resulting wavefunction solution $P(\boldsymbol{x}, t)$, such that

$$\mathcal{L}(\boldsymbol{x}, t) P(\boldsymbol{x}, t) = -\rho(\boldsymbol{x}, t). \tag{6.3}$$

If we know the Green's function satisfying (6.1.2), then an immediate solution is

$$P(\boldsymbol{x}, t) = \int d^3x_s \int_{-\infty}^{t} dt_s \rho\left(\boldsymbol{x}_s, t_s\right) G_+\left(\boldsymbol{x} \left| \boldsymbol{x}_s\right., t - t_s\right). \tag{6.4}$$

The volume integral is over all space. Since G_+ is causal, it is non-zero only if $t > t_s$, so the time integral is from $-\infty$ to t.

If the source distribution is known for all space and time, then all that is needed to determine P is the impulse response G_+. No knowledge of spatial or temporal boundary conditions is required. If, on the other hand, the source is known only within a closed region of space and time, then boundary conditions enter in.

To determine the boundary conditions involved, we need to invoke the *fundamental theorem of integral calculus*, which states that if $F(x)$ and $f(x)$ are functions such that $dF/dx = f$, then

$$\int_a^b f(x)dx = F(b) - F(a). \tag{6.5}$$

In three dimensions, this theorem becomes the *divergence theorem*:

$$\int_{\mathcal{V}} \nabla \cdot \boldsymbol{A}\, d^3x = \oint_{\mathcal{S}} \boldsymbol{A} \cdot \hat{\boldsymbol{n}} d^2s, \tag{6.6}$$

with $\boldsymbol{A}$ a vector function defined within a volume $\mathcal{V}$ bounded by the surface $\mathcal{S}$, and $\hat{\boldsymbol{n}}$ a unit vector pointing outwardly normal to $\mathcal{S}$.

From the divergence theorem, a short step leads to *Green's second identity* or *Green's theorem.* Choosing for $\boldsymbol{A}$ the quantity $\phi\nabla\psi - \psi\nabla\phi$, equation (6.6) becomes

$$\int_{\mathcal{V}} (\phi\nabla^2\psi - \psi\nabla^2\phi)\, d^3x = \oint_{\mathcal{S}} (\phi\nabla\psi - \psi\nabla\phi) \cdot \hat{\boldsymbol{n}}\, d^2s. \tag{6.7}$$

This relation ties wavefunction values within a volume to values on the boundary of the volume.

6.2 The scalar Green's function

The scalar Green's function satisfies

$$\left(\nabla^2 - \frac{1}{c^2}\frac{\partial^2}{\partial t^2}\right) G_{\text{scalar}}\left(\boldsymbol{x}\,|\boldsymbol{x}_s, t - t_s\right) = -\delta\left(t - t_s\right)\delta\left(\boldsymbol{x} - \boldsymbol{x}_s\right). \tag{6.8}$$

6.2.1 Constant velocity

Where velocity c is constant, equation (6.8) has the causal solution

$$G_{\text{scalar}}\left(\boldsymbol{x}\,|\boldsymbol{x}_s, t - t_s\right) = \frac{\delta\left(t - t_s - r/c\right)}{4\pi r}, \tag{6.9}$$

where r is the distance between source and observation point:

$$r = |\boldsymbol{x} - \boldsymbol{x}_s| = \sqrt{(x_s - x)^2 + (y_s - y)^2 + (z_s - z)^2}. \tag{6.10}$$

The impulse response is seen to be an spherical shell, expanding outward from the source position $\boldsymbol{x}_s$ in all directions with velocity c.

The Green's function (6.9) obeys reciprocity, which is to say that $G_{\text{scalar}}(\boldsymbol{x}|\,\boldsymbol{x}_s, t) = G_{\text{scalar}}(\boldsymbol{x}_s|\,\boldsymbol{x}, t)$. That is, G_{scalar} is invariant under an interchange of source and receiver.

The constant-velocity Green's function is translationally invariant; that is, it depends only on the difference between the source and receiver positions $\boldsymbol{x} - \boldsymbol{x}_s$. This means that we can write a simplified Green's function $\mathcal{G}_{\text{scalar}}(\Delta\boldsymbol{x}, t)$ as

$$\mathcal{G}_{\text{scalar}}(\Delta\boldsymbol{x}, t) \equiv G_{\text{scalar}}\left(\boldsymbol{x}_s + \Delta\boldsymbol{x}|\boldsymbol{x}_s, t\right) = \frac{\delta(t - |\Delta\boldsymbol{x}|/c)}{4\pi\,|\Delta\boldsymbol{x}|}. \tag{6.11}$$

Expression in terms of $\boldsymbol{k_x}$, $\boldsymbol{k_y}$, $\boldsymbol{z}$, $\boldsymbol{\omega}$ for the constant-velocity Green's function

If one takes the Fourier transforms over x, y, and t of the constant-velocity point-source wave equation (6.8), one obtains

$$\left(\frac{d^2}{dz^2} - {k_x}^2 - {k_y}^2 + \frac{\omega^2}{c^2}\right) \mathcal{G}_{\text{scalar}}(\tilde{\boldsymbol{k}}, z{-}z_s, \omega) = -\delta\left(z{-}z_s\right), \tag{6.12}$$

with frequency ω the Fourier conjugate of t, and the 2-vector wavenumber $\tilde{\boldsymbol{k}} = \left(k_x, k_y\right)$ the Fourier conjugate of $\tilde{\boldsymbol{x}} = (x, y)$. While the space–time expression of the Green's function is real, its representation in $\tilde{\boldsymbol{k}}$–ω space is necessarily complex. Because the space–time function is real, we have

$$\mathcal{G}_{\text{scalar}}(-\tilde{\boldsymbol{k}}, z, -\omega) = \mathcal{G}_{\text{scalar}}(\tilde{\boldsymbol{k}}, z, \omega)^*, \tag{6.13}$$

where the * denotes the complex conjugate. Exploiting this relation, we confine attention to positive ω, knowing that negative frequencies can be recovered using (6.13).

Equation (6.12) is an ordinary differential equation in z, with causal solution

$$\mathcal{G}_{\text{scalar}}(\tilde{\boldsymbol{k}}, z - z_s, \omega) = -\frac{e^{ik_z|z-z_s|}}{2ik_z}, \tag{6.14}$$

where k_z is the vertical wavenumber

$$k_z = \frac{\omega}{c}\sqrt{1 - \frac{\left(k_x{}^2 + k_y{}^2\right)c^2}{\omega^2}}. \tag{6.15}$$

Note that since ω has been constrained positive, k_z is necessarily positive (or at least non-negative) also. The sign of the exponent of (6.14) produces a wave moving outward in time from the point $z = 0$. The choice of a negative sign in the exponent produces a wave imploding on the point $z = 0$ at time $t = 0$; i.e. the anticausal Green's function.

With a little effort, one can verify that equation (6.14) is the triple Fourier transform of the space–time Green's function (6.11).

6.2.2 Depth-variable velocity

If velocity varies only with depth, one can Fourier-transform out the x, y, and t coordinates in (6.2.8), resulting in a slight generalization of (6.12):

$$\left(\frac{d^2}{dz^2} + k_z{}^2(z)\right)\mathcal{G}_{\text{scalar}}(\tilde{\boldsymbol{k}}, z, z_s, \omega) = -\delta\,(z - z_s)\,, \tag{6.16}$$

with

$$k_z(z) = \frac{\omega}{c(z)}\sqrt{1 - \frac{c(z)^2}{\omega^2}\left(k_x{}^2 + k_y{}^2\right)}. \tag{6.17}$$

Try a solution of the form

$$\mathcal{G}_{\text{scalar}}(\tilde{\boldsymbol{k}}, z, z_s, \omega) = g_{\text{scalar}}\,(z\,|z_s\,)\,e^{i\int_{z_<}^{z_>} dz' k_z(z')}, \tag{6.18}$$

where $z_<$ and $z_>$ are the smaller and larger of z and z_s, respectively. Substituting into (6.16), we have for $z \neq z_s$,

$$\frac{d^2}{dz^2}g_{\text{scalar}} + \frac{i}{g_{\text{scalar}}}\frac{d}{dz}\left(g_{\text{scalar}}{}^2 k_z\right) = 0. \tag{6.19}$$

At this point we make a high-frequency (or high- k_z) approximation, and assume we can ignore the second derivative of g. We can then infer that

$$g_{\text{scalar}}\,(z\,|z_s\,) \propto \frac{1}{\sqrt{|k_z(z)|}}. \tag{6.20}$$

As $z \to z_s$, we demand that $\mathcal{G}_{\text{scalar}}$ approach the constant-velocity form (6.14). This implies that $g_{\text{scalar}}\,(z\,|z_s\,)$ must be

$$g_{\text{scalar}}\left(z\,|z_s\right) = \frac{-1}{2i\sqrt{k_z(z)k_z\left(z_s\right)}}, \tag{6.21}$$

and $\mathcal{G}_{\text{scalar}}$ must be

$$\mathcal{G}_{\text{scalar}}(\tilde{\boldsymbol{k}}, z, z_s, \omega) = -\frac{e^{i\int_{z_<}^{z_>} dz' k_z(z')}}{2i\sqrt{k_z(z)k_z\left(z_s\right)}}. \tag{6.22}$$

This expression also exhibits reciprocity; that is, it is invariant under an exchange of z and z_s.

6.2.3 Generally variable velocity

For generally variable velocity, symmetry is lost and expressions are not quite so simple. Under the assumption that propagation velocity varies slowly relative to seismic wavelengths, some useful approximations are available. In the frequency domain, the scalar Green's function obeys

$$\left(\nabla^2 + \frac{\omega^2}{c(\boldsymbol{x})^2}\right) G_{\text{scalar}}\left(\boldsymbol{x}\,|\boldsymbol{x}_s, \omega\right) = -\delta\left(\boldsymbol{x} - \boldsymbol{x}_s\right). \tag{6.23}$$

We try a solution of the asymptotic form

$$G_{\text{scalar}}\left(\boldsymbol{x}\,|\boldsymbol{x}_s, \omega\right) \simeq g_{\text{scalar}}\left(\boldsymbol{x}\,|\boldsymbol{x}_s\right) e^{i\omega\tau\left(\boldsymbol{x}|\boldsymbol{x}_s\right)}. \tag{6.24}$$

In this expression, an exponential term is multiplied by an amplitude function g_{scalar}. The phase of the exponential term is determined by a traveltime function τ. Substituting (6.24) into (6.23), we obtain for $\boldsymbol{x} \neq \boldsymbol{x}_s$,

$$\left(\nabla^2 g_{\text{scalar}} + i\omega\left(2\nabla g_{\text{scalar}} \cdot \nabla\tau + g_{\text{scalar}}\nabla^2\tau\right) + \omega^2\left(\frac{1}{c(\boldsymbol{x})^2} - |\nabla\tau|^2\right)\right) e^{-i\omega\tau} = 0. \tag{6.25}$$

Under the assumption that g_{scalar} and τ do not depend on frequency, the coefficients of ω^2, ω, and ω^0 must each be zero. This implies

$$\frac{1}{c(\boldsymbol{x})} = |\nabla\tau|, \tag{6.26}$$

$$\nabla \cdot \left({g_{\text{scalar}}}^2\nabla\tau\right) = 0, \tag{6.27}$$

and

$$\nabla^2 g_{\text{scalar}} = 0. \tag{6.28}$$

The first of these equations is called the Eikonal equation, which controls how traveltime unfolds. The second term is the transport equation, which determines how the amplitude function varies spatially. The third equation generally cannot be satisfied along with the first two. However, under the assumption of a slowly varying velocity, the third term should be negligible relative to the first two.

For a constant velocity, one can compare the asymptotic form (6.24) to the exact solution, found by a Fourier transform of (6.9) from time to frequency:

$$G_{\text{scalar}}(\boldsymbol{x}\,|\boldsymbol{x}_s\,,\omega) \rightarrow \frac{e^{i\omega r/c}}{4\pi r}. \tag{6.29}$$

Thus, in the constant-velocity case,

$$g_{\text{scalar}}(\boldsymbol{x}\,|\boldsymbol{x}_s) \rightarrow \frac{1}{4\pi r}, \tag{6.30}$$

and

$$\tau(\boldsymbol{x}\,|\boldsymbol{x}_s) \rightarrow \frac{r}{c}. \tag{6.31}$$

One can easily check that all three equations (6.26), (6.27), and (6.28) are satisfied exactly for constant velocity provided $r \neq 0$.

For a variable velocity, the situation is of course more complicated. However, as $\boldsymbol{x}$ approaches $\boldsymbol{x}_s$, we invoke the approximation of a locally constant velocity, so when source and receiver are near enough, the Green's function approaches the constant-velocity form (6.29). Thus the amplitude function approaches (6.30),

$$g_{\text{scalar}}(\boldsymbol{x}\,|\boldsymbol{x}_s) \rightarrow \frac{1}{4\pi r} \text{ as } \boldsymbol{x} \rightarrow \boldsymbol{x}_s, \tag{6.32}$$

and the traveltime function approaches (6.31):

$$\tau(\boldsymbol{x}|\boldsymbol{x}_s) \rightarrow \frac{r}{c(\boldsymbol{x}_s)} \text{ as } \boldsymbol{x} \rightarrow \boldsymbol{x}_s. \tag{6.33}$$

Using (6.32) and (6.33) as a starting point, the Eikonal and transport equations provide the tools needed to construct an approximate Green's function in a (slowly) variable velocity medium. This function is asymptotic in frequency. For observation points near to the source, it reduces to the constant-velocity Green's function. Construction methodology is the subject of numerous publications (e.g. Beydoun & Keho, 1987; Cerveny, 2001; Schleicher *et al.*, 2007), and will not be duplicated here.

6.3 The acoustic Green's function

For acoustic waves, the equation for the Green's function is

$$\left(\nabla \cdot \frac{1}{\rho(\boldsymbol{x})}\nabla - \frac{1}{\rho(\boldsymbol{x})c(\boldsymbol{x})^2}\frac{\partial^2}{\partial t^2}\right) G_a(\boldsymbol{x}\,|\boldsymbol{x}_s\,, t - t_s) = -\delta(t - t_s)\,\delta(\boldsymbol{x} - \boldsymbol{x}_s). \tag{6.34}$$

6.3.1 Constant velocity and density

Where velocity and density are constant, (6.34) reduces to

$$\left(\nabla^2 - \frac{1}{c^2}\frac{\partial^2}{\partial t^2}\right) G_a(\boldsymbol{x}|\boldsymbol{x}_s, t - t_s) = -\rho\delta\,(t - t_s)\,\delta\,(\boldsymbol{x} - \boldsymbol{x}_s)\,. \tag{6.35}$$

This is proportional to the corresponding scalar wave equation. We can immediately write down the impulse-response solution, differing from the scalar Green's function by a factor of density:

$$G_a\,(\boldsymbol{x}\,|\boldsymbol{x}_s\,, t - t_s) = \mathcal{G}_a\,(\boldsymbol{x} - \boldsymbol{x}_s, t - t_s) = \frac{\delta\,(t - t_s - r/c)}{\mathcal{C}_a{}^2 4\pi r}, \tag{6.36}$$

with

$$\mathcal{C}_a = \rho^{-1/2}. \tag{6.37}$$

Since the constant-velocity Green's function is translationally invariant, we can write a simplified function $\mathcal{G}_a$ that is a function of the difference between the source and observation point just as for the scalar case.

The corresponding *k*–*ω*–*z*-domain expression is

$$\mathcal{G}_a(\tilde{\boldsymbol{k}}, z, \omega) \;=\; -\frac{e^{ik_z|z|}}{\mathcal{C}_a{}^2 2ik_z}, \tag{6.38}$$

with expression (6.15) for k_z still valid.

6.3.2 Depth-variable velocity and density

The acoustic expression for the depth-variable asymptotic Green's function is also similar to the scalar expression:

$$\mathcal{G}_a(\tilde{\boldsymbol{k}},\; z,\; z_s, \omega) \;=\; \frac{-1}{2i}\frac{e^{i\int_{z<}^{z>} dz' k_z(z')}}{\sqrt{k_z(z)k_z\,(z_s)}}\frac{1}{\mathcal{C}_a(z)\mathcal{C}_a\,(z_s)}, \tag{6.39}$$

with $\mathcal{C}_a(z) = \rho^{-1/2}(z)$ in this case depth-dependent. Like the scalar expression, the acoustic Green's function exhibits reciprocity.

6.3.3 Fully variable velocity and density

If we assume the asymptotic form (6.24) for the acoustic Green's function, and expand the wave equation into powers of ω as in (6.25), we discover that the Eikonal equation (6.26) is the same, while the transport equation becomes

$$\nabla \cdot \left(g_a{}^2\mathcal{C}_a{}^2\nabla\tau\right) = 0, \tag{6.40}$$

with $C_a(\boldsymbol{x}) = \rho^{-1/2}(\boldsymbol{x})$. Thus, the acoustic Green's function has the same traveltime function as the scalar Green's function, whereas the amplitude function is slightly different. In the neighborhood of the source point $\boldsymbol{x}_s$, the acoustic amplitude function goes to

$$g_a(\boldsymbol{x}\,|\boldsymbol{x}_s) \to \frac{1}{C_a{}^2(\boldsymbol{x}_s)\,4\pi r} \text{ as } \boldsymbol{x} \to \boldsymbol{x}_s. \tag{6.41}$$

In the presence of multipathing and caustics, the acoustic Green's function generalizes in the same manner as the scalar Green's function.

6.3.4 Constant velocity, variable density

For the special case of a constant velocity and variable density, the solution of the Eikonal equation is just the scalar relation

$$\tau(\boldsymbol{x}\,|\boldsymbol{x}_s) = \frac{r}{c}, \tag{6.42}$$

with r the distance between $\boldsymbol{x}$ and $\boldsymbol{x}_s$. The transport equation reduces to

$$\nabla \cdot \left(g_a{}^2 C_a{}^2 \nabla r\right) = 0, \tag{6.43}$$

which has solutions

$$g_a(\boldsymbol{x}\,|\boldsymbol{x}_s) \propto \frac{1}{C_a(\boldsymbol{x})r}. \tag{6.44}$$

As $r \to 0$, g_a must approach the form (6.41). This fixes the proportionality constant, whence in this case

$$g_a(\boldsymbol{x}\,|\boldsymbol{x}_s) = \frac{1}{4\pi r C_a(\boldsymbol{x}) C_a(\boldsymbol{x}_s)}. \tag{6.45}$$

In this case also the Green's function is seen to observe reciprocity.

6.4 The elastic Green's function

6.4.1 Constant α, β, ρ

The elastic Green's functions have an additional degree of complexity, since they describe a vector wavefield. Drawing from Section 5.2.3, for the constant-parameter case we can define P- and S-wave Green's functions that are inverses to the P- and S-wave operators (5.30) and (5.31):

$$G_P(\boldsymbol{x}\,|\boldsymbol{x}_s,\, t - t_s) = \frac{\delta(t - t_s - r/\alpha)}{4\pi r C_P{}^2}, \tag{6.46}$$

$$G_{Si}(\boldsymbol{x}\,|\boldsymbol{x}_s, t - t_s) = \frac{\delta\,(t - t_s - r/\beta)}{4\pi r \mathcal{C}_S{}^2}. \tag{6.47}$$

Here, G_P is a Green's function for a compressional wave, and G_{Si}, $i = 1, 2, 3$, represent shear-wave impulse responses in the x, y, and z directions. In a physical situation, arbitrary combinations of the three S-wave impulse responses are not allowed – only those combinations with zero divergence are realizable. The factors $\mathcal{C}_P$ and $\mathcal{C}_S$ are

$$\mathcal{C}_P = \sqrt{\rho}\alpha, \quad \mathcal{C}_S = \sqrt{\rho}\beta. \tag{6.48}$$

6.4.2 Depth-variable α, β, ρ

When parameters vary with depth only, the results are very similar to the scalar and acoustic cases. We have the high-frequency approximate Green's functions

$$\mathcal{G}_P(\tilde{\boldsymbol{k}}, z, z_s, \omega) = -\frac{e^{i\int_{z<}^{z>} dz' k_{zP}(z')}}{2i\mathcal{C}_P(z)\mathcal{C}_P(z_s)\sqrt{k_{zP}(z)k_{zP}(z_s)}}, \tag{6.49}$$

and

$$\mathcal{G}_{Si}(\tilde{\boldsymbol{k}}, z, z_s, \omega) = -\frac{e^{i\int_{z<}^{z>} dz' k_{zS}(z')}}{2i\mathcal{C}_S(z)\mathcal{C}_S(z_s)\sqrt{k_{zS}(z)k_{zS}(z_s)}}, \tag{6.50}$$

with $\tilde{\boldsymbol{k}}$ the Fourier conjugate of $\tilde{\boldsymbol{x}} = (x, y)$, and $z_<$ ($z_>$) the smaller (larger) of z and z_s. The vertical wavenumbers k_{zP} and k_{zS} are

$$k_{zP} = \frac{\omega}{\alpha}\sqrt{1 - \frac{\left(k_x{}^2 + k_y{}^2\right)\alpha^2}{\omega^2}},$$

$$k_{zS} = \frac{\omega}{\beta}\sqrt{1 - \frac{\left(k_x{}^2 + k_y{}^2\right)\beta^2}{\omega^2}}. \tag{6.51}$$

6.4.3 Generally variable α, β, ρ

For variable parameters we can, at least under the high-frequency approximation employed here, examine P- and S-wave components separately. In the frequency domain, we approximate a compressional Green's function as the form (6.24):

$$G_P(\boldsymbol{x}\,|\boldsymbol{x}_s, \omega) \simeq g_P(\boldsymbol{x}\,|\boldsymbol{x}_s)\, e^{i\omega\tau_P(\boldsymbol{x}|\boldsymbol{x}_s)}. \tag{6.52}$$

The Eikonal equation for this Green's function is

$$|\nabla\tau_P| = \frac{1}{\alpha(\boldsymbol{x})}, \tag{6.53}$$

and the transport equation is

$$\nabla \cdot \left(C_P{}^2 g_P{}^2 \nabla \tau_P\right) = 0. \tag{6.54}$$

The asymptotic P-wave Green's function assumes the constant-velocity form (6.46) near the source point $\boldsymbol{x}_s$. Away from $\boldsymbol{x}_s$, the Green's function is extrapolated using the Eikonal and transport equations (6.53) and (6.54).

The shear-wave asymptotic form can be written as a vector

$$\boldsymbol{G}_S\left(\boldsymbol{x}\,|\boldsymbol{x}_s,\omega\right) = \ \boldsymbol{g}_S\left(\boldsymbol{x}\,|\boldsymbol{x}_s\right) e^{i\omega\tau_S(\boldsymbol{x}|\boldsymbol{x}_s)}, \tag{6.55}$$

with the constraint

$$\boldsymbol{g}_S \cdot \nabla\tau_S = 0. \tag{6.56}$$

If one forms a local cartesian coordinate system with the third coordinate in the direction of $\nabla\tau_S$, then in this system g_{S3} is not physical. In the next chapter, we perform a rotation into this system, effectively removing the third component of the shear-wave Green's function.

The Eikonal equation for the S-wave is

$$|\nabla\tau_S| = \frac{1}{\beta(\boldsymbol{x})}, \tag{6.57}$$

and the transport equation

$$\nabla \cdot \left(C_S{}^2 |\boldsymbol{g}_S|^2 \nabla \tau_P\right) = 0. \tag{6.58}$$

As for the P-wave, the asymptotic S-wave Green's function assumes the constant-velocity form near the source, and can be extrapolated away from the source using the Eikonal and transport equations.

6.5 Local wavenumbers

Suppose we have a Green's function representing a ray from $\boldsymbol{x}_s$ to $\boldsymbol{x}$ of the form

$$G\left(\boldsymbol{x}_g|\boldsymbol{x}_s,\omega\right) \simeq \ g\left(\boldsymbol{x}_g|\boldsymbol{x}_s\right) e^{i\omega\tau\left(\boldsymbol{x}_g|\boldsymbol{x}_s\right)} \tag{6.59}$$

with associated traveltime which obeys the Eikonal equation

$$\left|\nabla_{\boldsymbol{x}_g}\tau\left(\boldsymbol{x}_g|\boldsymbol{x}_s\right)\right| = \frac{1}{c\left(\boldsymbol{x}_g\right)}, \quad \left|\nabla_{\boldsymbol{x}_s}\tau\left(\boldsymbol{x}_g|\boldsymbol{x}_s\right)\right| = \frac{1}{c\left(\boldsymbol{x}_s\right)}. \tag{6.60}$$

The gradient of τ at any point along the ray points in the direction of maximum increase in τ. We can define local unit vectors $\hat{\boldsymbol{k}}_g$ and $\hat{\boldsymbol{k}}_s$ tangent to the ray at $\boldsymbol{x}_g$ and $\boldsymbol{x}_s$:

$$\begin{aligned} \hat{\boldsymbol{k}}_g\left(\boldsymbol{x}_g\right) &= +c\left(\boldsymbol{x}_g\right)\nabla_{\boldsymbol{x}_g}\tau\left(\boldsymbol{x}_g|\boldsymbol{x}_s\right), \\ \hat{\boldsymbol{k}}_s\left(\boldsymbol{x}_s\right) &= -c\left(\boldsymbol{x}_s\right)\nabla_{\boldsymbol{x}_s}\tau\left(\boldsymbol{x}_g|\boldsymbol{x}_s\right). \end{aligned} \tag{6.61}$$

With this sign convention, the unit vectors on both ends of the ray point in the direction of travel (from $\boldsymbol{x}_s$ to $\boldsymbol{x}_g$) along the ray. We can also define local wavenumber vectors pointing in the same direction as the unit vectors. The magnitude of the wavenumber vectors must be frequency divided by local velocity, i.e.

$$k_g(\boldsymbol{x}_g) = \frac{\omega}{c(\boldsymbol{x}_g)},$$
$$k_s(\boldsymbol{x}_s) = \frac{\omega}{c(\boldsymbol{x}_s)}. \tag{6.62}$$

so that

$$\boldsymbol{k}_g(\boldsymbol{x}_g) = +\omega \nabla_{\boldsymbol{x}_g} \tau(\boldsymbol{x}_g|\boldsymbol{x}_s), \quad \boldsymbol{k}_s(\boldsymbol{x}_s) = -\omega \nabla_{\boldsymbol{x}_s} \tau(\boldsymbol{x}_g|\boldsymbol{x}_s). \tag{6.63}$$

In terms of the Green's function, we can write

$$\nabla_{\boldsymbol{x}_g} G_0(\boldsymbol{x}_g|\boldsymbol{x}_s;\omega) \simeq +i\boldsymbol{k}_g(\boldsymbol{x}_g) G_0(\boldsymbol{x}_g|\boldsymbol{x}_s;\omega),$$
$$\nabla_{\boldsymbol{x}_s} G_0(\boldsymbol{x}_g|\boldsymbol{x}_s;\omega) \simeq -i\boldsymbol{k}_s(\boldsymbol{x}_s) G_0(\boldsymbol{x}_g|\boldsymbol{x}_s;\omega). \tag{6.64}$$

That is, the gradient of the Green's function is proportional to $\pm i$ times the local wavenumber. The plus sign obtains at the end of the ray, and the minus sign at the beginning of the ray.

6.6 Multipath Green's functions

In the absence of multipathing, the amplitude function g is real and finite. Where multipathing exists, the Green's function becomes a sum of terms

$$G(\boldsymbol{x}_g|\boldsymbol{x}_s,\omega) \simeq \sum_m g_m(\boldsymbol{x}_g|\boldsymbol{x}_s) e^{i\omega\tau_m(\boldsymbol{x}_g|\boldsymbol{x}_s)}, \tag{6.65}$$

each term representing an individual ray. If a ray goes through one or more caustic points, corresponding to a path where traveltime is a local minimum, maximum, or saddle point, then for the corresponding term,

$$g_m(\boldsymbol{x}_g|\boldsymbol{x}_s) = |g_m(\boldsymbol{x}_g|\boldsymbol{x}_s)| e^{i\frac{\pi}{2}\alpha_m(\boldsymbol{x}_g|\boldsymbol{x}_s)}, \tag{6.66}$$

with $\alpha_m(\boldsymbol{x}_g|\boldsymbol{x}_s)$ the KMAH index, or a count of the number of caustics the ray has passed through between $\boldsymbol{x}_s$ and $\boldsymbol{x}$.

Where multiple raypaths exist, the total Green's function may also contain additional energy that does not correspond to a classical (i.e. extremal time) raypath. Such *diffractive* energy can be described by additional events corresponding to inflection points or discontinuities in traveltime, as described in Appendix G. The Green's function component for such an event has the form

$$g_m\left(\boldsymbol{x}_g|\boldsymbol{x}_s\right) = \left|g_m\left(\boldsymbol{x}_g|\boldsymbol{x}_s\right)\right| f_m\left(\omega; \boldsymbol{x}_g|\boldsymbol{x}_s\right), \tag{6.67}$$

with f_m a filter whose form depends upon the type of event.

6.7 Green's functions in a layered medium

6.7.1 Two-layer medium

Consider the case of scalar wave propagation of an impulse across a layer boundary. For propagation within the ith layer, assume a Green's function of the (frequency domain) form

$$G_i(\boldsymbol{x}_g|\boldsymbol{x}_s; \omega) = g_i\left(\boldsymbol{x}_g|\boldsymbol{x}_s\right)e^{i\omega\tau_i(\boldsymbol{x}_g|\boldsymbol{x}_s)}. \tag{6.68}$$

If the layers are of constant velocity c_i, then $\tau_i \to r/c_i$, and $g_i \to 1/r$, where $r = \left|\boldsymbol{x}_g - \boldsymbol{x}_s\right|$. More generally, velocity may vary within the layer and the raypath may curve, but we will assume that there is but one raypath between any two points within a layer.

Suppose we are interested in transmission of an impulse at point $\boldsymbol{x}_1$ in layer 1 to point $\boldsymbol{x}_2$ in layer 2. Propagation from $\boldsymbol{x}_1$ to points $\boldsymbol{x}_b$ on the boundary between the two layers is described by the Green's function $G_1 = g_1\left(\boldsymbol{x}_1|\boldsymbol{x}_b\right) e^{i\omega\tau_1(\boldsymbol{x}_1|\boldsymbol{x}_b)}$, while propagation from $\boldsymbol{x}_b$ to $\boldsymbol{x}_2$ in layer 2 is described by $G_2 = g_2\left(\boldsymbol{x}_b|\boldsymbol{x}_2\right) e^{i\omega\tau_2(\boldsymbol{x}_b|\boldsymbol{x}_2)}$. In the far field, at each point on the boundary, G_1 has a well-defined direction denoted by $\hat{\boldsymbol{p}}_{1b}$, a unit vector pointing in the direction of the gradient of G_1. Likewise, on the boundary G_2 has a well-defined direction denoted by $\hat{\boldsymbol{p}}_{2b}$. One can define an incident angle γ_1 between the direction of G_1 and the (upward pointing) boundary normal $\hat{\boldsymbol{n}}_b$ as

$$\cos\gamma_1 = -\hat{\boldsymbol{p}}_{1b} \cdot \hat{\boldsymbol{n}}_b. \tag{6.69}$$

Immediately across the boundary, in layer 2, the transmitted wave is

$$G_t\left(\boldsymbol{x}_b|\boldsymbol{x}_1; \omega\right) = G_1\left(\boldsymbol{x}_b|\boldsymbol{x}_1; \omega\right) T_b\left(\gamma_1\right), \tag{6.70}$$

with $T_b\left(\gamma_1\right)$ a transmission coefficient. For scalar waves, the transmission coefficient at the boundary is

$$T_b\left(\gamma_1\right) = \left(2c_{2b}\cos\gamma_1\right) / \left(c_{2b}\cos\gamma_1 + c_{1b}\cos\gamma_t\right), \tag{6.71}$$

where c_{1b} and c_{2b} are the velocities on each side of the boundary, and γ_t is the Snell refraction angle, given by

$$\cos\gamma_t = \sqrt{1 - \frac{{c_{2b}}^2}{{c_{1b}}^2}\sin^2\gamma_1}. \tag{6.72}$$

The transmitted wave at the boundary also has a well-defined propagation direction $\hat{\boldsymbol{p}}_{tb}$, related to the Snell refraction angle as

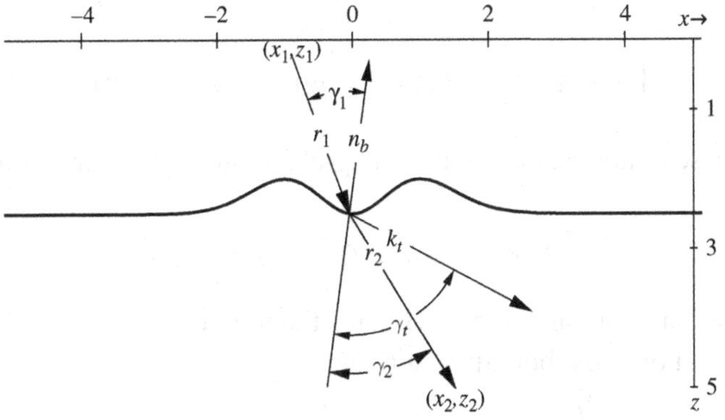

Figure 6.1 Transmission through two layers. Illustrated for simplicity in two dimensions. Velocity in the upper layer is half that in the lower layer. The boundary between the two layers is shown as a heavy line. Ray $\boldsymbol{r}_1$is shown from the source location (x_1, z_1) to a point on the boundary, and ray $\boldsymbol{r}_2$ from that point to a receiver location (x_2, z_2). Also shown is a vector $\boldsymbol{k}_t$ pointing in the direction of the Snell refraction of the ray $\boldsymbol{r}_1$. The ray $\boldsymbol{r}_1$ makes angle γ_1 with the boundary normal $\boldsymbol{n}_b$, while $\boldsymbol{r}_2$ makes angle γ_2 and $\boldsymbol{k}_t$ makes angle γ_t. The angles γ_2 and γ_t are different except at certain points on the boundary where the total traveltime from source to receiver is a minimum or maximum.

$$\cos \gamma_t = -\hat{\boldsymbol{p}}_{tb} \cdot \hat{\boldsymbol{n}}_b. \tag{6.73}$$

These angles are illustrated in Figure 6.1. Except at special points on the boundary (corresponding to a minimum or maximum in total traveltime $\tau_1 + \tau_2$), the direction $\hat{\boldsymbol{p}}_{tb}$ of the transmitted wave is different than that ($\hat{\boldsymbol{p}}_{2b}$) of the second-layer Green's function G_2. According to ray theory, the extremal points are the only ones that count. Since, however, our interest includes multipath Green's functions, we need to extend the analysis somewhat.

As the wave propagates into layer 2, the propagation direction is less easy to determine. At point $\boldsymbol{x}_2$, the wave will be a superposition of contributions from all points on the boundary. To determine the superposition, one can invoke Green's theorem (6.7). Start with an integral over the volume $\mathcal{V}_2$ of layer 2 with the Green's functions G_t and G_2 replacing ϕ and ψ:

$$I(\boldsymbol{x}_2|\boldsymbol{x}_1;\omega) = \int_{\mathcal{V}_2} d^3\boldsymbol{x} \left(G_2(\boldsymbol{x}_2|\boldsymbol{x};\omega)\, \nabla_{\boldsymbol{x}}{}^2 G_t(\boldsymbol{x}\,|\boldsymbol{x}_1;\ \omega) \right. \\ \left. - G_t(\boldsymbol{x}\,|\boldsymbol{x}_1;\ \omega)\, \nabla_{\boldsymbol{x}}{}^2 G_2(\boldsymbol{x}_2|\,\boldsymbol{x};\omega) \right). \tag{6.74}$$

Now, within $\mathcal{V}_2$, G_t has no sources,

$$\left(\nabla_{\boldsymbol{x}}{}^2 + \frac{\omega^2}{c_2{}^2} \right) G_t(\boldsymbol{x}\,|\boldsymbol{x}_1;\ \omega) = 0, \tag{6.75}$$

whereas G_2 does:

$$\left(\nabla_x{}^2 + \frac{\omega^2}{c_2{}^2}\right) G_2\left(\boldsymbol{x}_2 | \boldsymbol{x}; \omega\right) = -\delta\left(\boldsymbol{x}_2 - \boldsymbol{x}\right). \tag{6.76}$$

Invoking these relations, the volume integral $\mathfrak{I}$ reduces to the transmitted Green's function at $\boldsymbol{x}_2$,

$$\mathfrak{I}\left(\boldsymbol{x}_2|\boldsymbol{x}_1; \omega\right) = G_t\left(\boldsymbol{x}_2|\boldsymbol{x}_1; \omega\right). \tag{6.77}$$

On the other hand, application of Green's theorem converts the volume integral $\mathfrak{I}$ into an integral over the boundary surface:

$$\begin{aligned}\mathfrak{I}\left(\boldsymbol{x}_2|\boldsymbol{x}_1; \omega\right) = \int_{S_b} d^2\boldsymbol{x}_b \big(&G_2\left(\boldsymbol{x}_2|\boldsymbol{x}_b; \omega\right) \hat{\boldsymbol{n}}_b \cdot \nabla_b G_t\left(\boldsymbol{x}_b|\boldsymbol{x}_1; \omega\right) \\ &- G_t\left(\boldsymbol{x}_b|\boldsymbol{x}_1; \omega\right) \hat{\boldsymbol{n}}_b \cdot \nabla_b G_2\left(\boldsymbol{x}_2|\boldsymbol{x}_b; \omega\right)\big).\end{aligned} \tag{6.78}$$

The gradients of G_t and G_2 are (as per equation (6.64))

$$\hat{\boldsymbol{n}}_b \cdot \nabla_b G_t\left(\boldsymbol{x}_b|\boldsymbol{x}_1; \omega\right) = -i\frac{\omega}{c_{2b}} \cos\gamma_t G_t\left(\boldsymbol{x}_b|\boldsymbol{x}_1; \omega\right), \tag{6.79}$$

and

$$\hat{\boldsymbol{n}}_b \cdot \nabla_b G_2\left(\boldsymbol{x}_2|\boldsymbol{x}_b; \omega\right) = i\frac{\omega}{c_{2b}} \cos\gamma_2 G_2\left(\boldsymbol{x}_2|\boldsymbol{x}_b; \omega\right), \tag{6.80}$$

where γ_2 is the angle, with respect to the surface normal, of the ray from $\boldsymbol{x}_b$ on the boundary surface to $\boldsymbol{x}_2$. We have noted above that this angle is in general different from the Snell refraction angle γ_t. Thus, with (6.70), (6.77), (6.79), and (6.80) into (6.78),

$$\begin{aligned}G_t\left(\boldsymbol{x}_2|\boldsymbol{x}_1; \omega\right) = -i\omega \int_{S_b} &d^2\boldsymbol{x}_b G_2\left(\boldsymbol{x}_2|\,\boldsymbol{x}; \omega\right) G_1\left(\boldsymbol{x}\,|\boldsymbol{x}_1\,; \omega\right) \\ &\times \frac{T_b\left(\gamma_1\right)}{c_{2b}}\left(\cos\gamma_t + \cos\gamma_2\right).\end{aligned} \tag{6.81}$$

Invoking the asymptotic form for the Green's functions, the transmitted impulse from $\boldsymbol{x}_1$ to $\boldsymbol{x}_2$ becomes

$$G_t\left(\boldsymbol{x}_2|\boldsymbol{x}_1; \omega\right) = -i\omega \int_{S_b} d^2\boldsymbol{x}_b C\left(\boldsymbol{x}_2\,|\boldsymbol{x}_b|\,\boldsymbol{x}_1\right) e^{i\omega\tau\left(\boldsymbol{x}_2|\boldsymbol{x}_b|\boldsymbol{x}_1\right)}, \tag{6.82}$$

with amplitude term

$$C\left(\boldsymbol{x}_2\,|\boldsymbol{x}_b|\,\boldsymbol{x}_1\right) = g_2\left(\boldsymbol{x}_2|\boldsymbol{x}_b\right) g_1\left(\boldsymbol{x}_b|\boldsymbol{x}_1\right) \frac{T_b\left(\gamma_1\right)}{c_{2b}}\left(\cos\gamma_t + \cos\gamma_2\right), \tag{6.83}$$

and phase term

$$\tau\left(\boldsymbol{x}_2\,|\boldsymbol{x}_b|\,\boldsymbol{x}_1\right) = \tau_1\left(\boldsymbol{x}_2|\boldsymbol{x}_b\right) + \tau_2\left(\boldsymbol{x}_b|\boldsymbol{x}_1\right). \tag{6.84}$$

In the time domain, this expression becomes

$$G_t\left(\boldsymbol{x}_2|\boldsymbol{x}_1;t\right)=\int_{S_b} d^2\boldsymbol{x}_b C\left(\boldsymbol{x}_2\left|\boldsymbol{x}_b\right|\boldsymbol{x}_1\right)\dot{\delta}\left(t-\tau\left(\boldsymbol{x}_2\left|\boldsymbol{x}_b\right|\boldsymbol{x}_1\right)\right). \tag{6.85}$$

Equation (6.85) is of the diffraction integral form (G.1), with the operator f in this case the differentiation operator. According to Appendix G, the integration (6.85) should produce a finite number of discrete events, centered at times where traveltime goes through local extrema and inflection points. The complete Green's function may be a sum of discrete events, or

$$G_t\left(\boldsymbol{x}_2|\boldsymbol{x}_1;t\right)\simeq\sum_e g_{te}\left(\boldsymbol{x}_2|\boldsymbol{x}_1\right)f_e\left(t-\tau_e\left(\boldsymbol{x}_2|\boldsymbol{x}_1\right)\right), \tag{6.86}$$

with f_e the filter characteristic of the type of event.

6.7.2 Transmission through multiple layers

Construction of a multiple-layer Green's function is straightforward. In a three-layer medium, first calculate the two-layer Green's function at the bottom of the second layer. For each event in this Green's function, use Snell's law to calculate the transmitted wave at the top of the third layer and apply Green's theorem to calculate the transmitted wave at points within the third layer. The number of events may increase, but will remain finite. Given the total Green's function at the bottom of layer three, we can extrapolate through additional layers repeating the analysis. The result will remain in the form (6.86), though in a complex medium the number of events may become prohibitive.

6.7.3 Transmission in 2.5 dimensions

Suppose the layering is y-invariant, as in the example shown in Figure 6.1.

To calculate an impulse response for a source at (x_1, y_1, z_1) and a receiver at (x_2, y_2, z_2), one would use the diffraction integral (6.85). We have chosen a surface invariant in the y-direction, so the y-integral can be performed independently. The y-integral is in fact a 2-D diffraction integral of the form (G.5). Keeping the starting and end points y_1 and y_2 on the y-axis ($y_1 = y_2 = 0$), traveltime goes through a single minimum in the y-direction at $y_b = 0$. If

$$\tau_1\left(\boldsymbol{x}_2|\boldsymbol{x}_b\right)=\frac{r_1}{c_1},\quad \tau_2\left(\boldsymbol{x}_b|\boldsymbol{x}_2\right)=\frac{r_2}{c_2}, \tag{6.87}$$

where, on the y-axis,

$$C\to C_0=C\left(\boldsymbol{x}_2\left|(x_b,0,z_b)\right|\boldsymbol{x}_1\right), \tag{6.88}$$

$$
\begin{aligned}
r_1 \rightarrow r_{10} &= \sqrt{(x_1 - x_b)^2 + (z_1 - z_b)^2}, \\
r_2 \rightarrow r_{20} &= \sqrt{(x_2 - x_b)^2 + (z_2 - z_b)^2},
\end{aligned} \tag{6.89}
$$

and

$$
\tau \rightarrow \tau_0 = \frac{r_{1,0}}{c_1} + \frac{r_{2,0}}{c_2}. \tag{6.90}
$$

The second derivative of traveltime at its minimum is

$$
\tau_{yy0} = \frac{1}{r_{1,0} c_1} + \frac{1}{r_{2,0} c_2}. \tag{6.91}
$$

In the frequency domain,

$$
G_t\left(\boldsymbol{x}_2 | \boldsymbol{x}_1; \omega\right) \simeq -i\omega \int_{\mathcal{L}_b} d\ell_b \int dy_b C\left(\boldsymbol{x}_2 \left| \boldsymbol{x}_b \right| \boldsymbol{x}_1\right) e^{i\omega\left(\tau_0 + \tau_{yy0}\, {y_b}^2/2\right)}, \tag{6.92}
$$

where $\mathcal{L}_b$ follows the line of the boundary in the x–z plane.

The y-integral can be evaluated using the stationary-phase formula (F.11), leading to

$$
G_t\left(\boldsymbol{x}_2 | \boldsymbol{x}_1; \omega\right) \simeq \sqrt{-2\pi i\omega} \int_{\mathcal{L}_b} d\ell_b \frac{C_0}{\sqrt{\tau_{yy0}}} e^{i\omega\tau_0}. \tag{6.93}
$$

Recalling that $\sqrt{-i\omega}$ is the frequency-domain expression for the half-derivative operator d_{h+} (see equation (E.42)), we can write

$$
G_t\left(\boldsymbol{x}_2 | \boldsymbol{x}_1; \omega\right) \simeq \sqrt{2\pi} d_{h+}(\omega) \int_{\mathcal{L}_b} d\ell_b \frac{C_0}{\sqrt{\tau_{yy0}}} e^{i\omega\tau_0}. \tag{6.94}
$$

Alternatively, starting from the time-domain expression (6.85), we can invoke the diffraction-integral formula (G.14) for 2-D minima, obtaining

$$
G_t\left(\boldsymbol{x}_2 | \boldsymbol{x}_1; t\right) = \sqrt{2\pi} \int_{\mathcal{L}_b} d\ell_b \frac{C_0}{\sqrt{\tau_{yy0}}} \left[\Theta_{h+} * \dot{\delta}\right](t - \tau_0), \tag{6.95}
$$

with $\mathcal{L}_b$ the $y = 0$ line on the boundary. The half-integral operator Θ_{h+} combined with the derivative operator produces a half-derivative operator, or

$$
G_t\left(\boldsymbol{x}_2 | \boldsymbol{x}_1; t\right) = \sqrt{2\pi} \int_{\mathcal{L}_b} d\ell_b \frac{C_0}{\sqrt{\tau_{yy0}}} d_{h+}(t - \tau_0), \tag{6.96}
$$

which is the time-domain equivalent to (6.94).

The remaining integral is also a 2-D diffraction integral, which, owing to the complexity of the boundary, is less simple to evaluate. For given $\boldsymbol{x}_1$ and $\boldsymbol{x}_2$, there may be multiple events, including traveltime minima, maxima, and inflection points.

Traveltime minima. For a traveltime minimum, the diffraction integral (6.96) evaluates to a weighted delta function

$$G_{\min}\left(\boldsymbol{x}_2|\boldsymbol{x}_1;t\right)=2\pi\frac{C_{\min}}{\sqrt{\left|\tau_{yy\min}\tau_{\ell\ell\min}\right|}}\delta\left(t-\tau_{\min}\right), \tag{6.97}$$

with

$$\tau_{\min}=\tau\left(\boldsymbol{x}_2\left|\boldsymbol{x}_{\min}\right|\boldsymbol{x}_1\right), \tag{6.98}$$

and $\boldsymbol{x}_{\min}=(x_{\min},0,z_{\min})$ the location on the boundary of the traveltime minimum.

Traveltime maxima. For a maximum, the event is a Hilbert transform:

$$G_{\max}\left(\boldsymbol{x}_2|\boldsymbol{x}_1;t\right)=-2\pi\frac{C_{\max}}{\sqrt{\left|\tau_{yy\max}\tau_{\ell\ell\max}\right|}}H\left(t-\tau_{\max}\right). \tag{6.99}$$

Traveltime inflections. At an inflection point, the second derivative $\tau_{\ell\ell}=0$, and the signs of the first derivative τ_ℓ and the third derivative $\tau_{\ell\ell\ell}$ are the same. Defining the *time dilation factor* at the inflection point to be

$$\tau_d=\frac{2}{3}\sqrt{\frac{2\tau_\ell{}^3}{\tau_{\ell\ell\ell}}}, \tag{6.100}$$

the contribution to the Green's function from an inflection event is

$$G_{\text{infl}}\left(\boldsymbol{x}_2|\boldsymbol{x}_1;t\right)=\sqrt{2\pi}\frac{C_{\text{infl}}}{\sqrt{\left|\tau_{yy}\right|}}\frac{\sqrt{\tau_d}}{\left|\tau_\ell\right|}w_{h+}\left(\tau_d,t-\tau_{\text{infl}}\right), \tag{6.101}$$

with $w_{h+}(\tau_d,t)$ the half-differentiated inflection wavelet depicted in Figure E.17.

Figure 6.2 shows the raypaths for events between surface points and the subsurface point (0, 5) kft. Traveltime minima are shown in light gray, maxima in darker gray, and inflections in black. The minima are best behaved, in that surface position increases with ray angle at (0, 5). The maxima are seen to pass through a focal point, or caustic, on their way to the surface. Thus, for a maximum, surface position decreases with subsurface-point ray angle. The inflections appear to cluster near a single initial ray angle. then diffract into a wide range of surface positions. Initial ray angle for the inflections appears to fall into a gap between angles corresponding to minima and those corresponding to maxima. Put another way, if φ_2 is the angle a ray makes with the vertical at subsurface point (0, 5), and x_1 the point at which the ray intersects the wavy boundary, then minima are found in the regions $\partial\varphi_2/\partial x_1>0$, maxima in the regions $\partial\varphi_2/\partial x_1<0$, while inflections occupy the vicinity of the point $\partial\varphi_2/\partial x_1=0$.

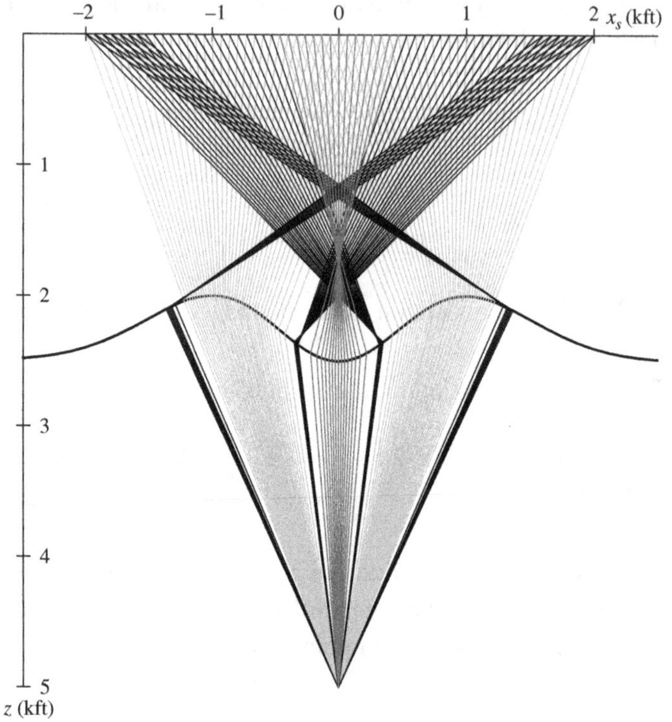

Figure 6.2 Raypath plot from a subsurface point. Raypath plot from the point (0, 5) to various surface points through a wavy boundary. Snell raypaths corresponding to traveltime minima and maxima are shown in light and darker gray, respectively. The black events are diffractions not predicted by Snell's law. They correspond to inflection points in traveltime. For the minimum-time raypaths, surface position is seen to increase with initial ray angle. For the maximum-time events, surface position decreases with ray angle. The apparent gap between the minimum and maximum time raypaths appears because position on the boundary is changing very slowly with ray angle. The inflection rays appear in the angular gap between the minimum and maximum time rays.

Also shown in this figure are a set of inflections at a large subsurface ray angle and an even larger surface ray angle. If we look at the traveltime curve $\tau(\ell)$ for the largest surface position $\boldsymbol{x}_s = (2, 0)$ (Figure 6.3), we see a barely discernible inflection located at about boundary locations $x_b = \pm 1.4$. For this event, the magnitude of the traveltime derivative is relatively large, while that of the third derivative is very small. This makes the dilation factor τ_d for this event to be large, giving the event a low-frequency character. We therefore expect the event to be inconsequential.

Figure 6.4 shows the computed Green's function corresponding to the raypaths in Figure 6.2. Whereas a constant-velocity Green's function would be a hyperbola,

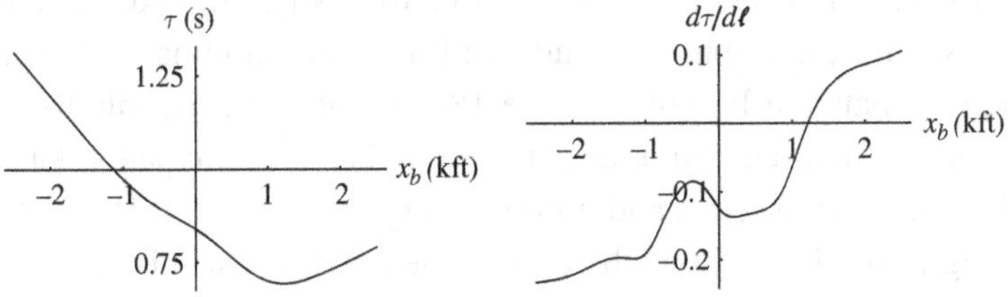

Figure 6.3 Traveltime curve and its derivative. Left: Traveltime curve between points (0, 5) and (2, 0), showing a minimum and a weak inflection. Right: Derivative of the traveltime curve.

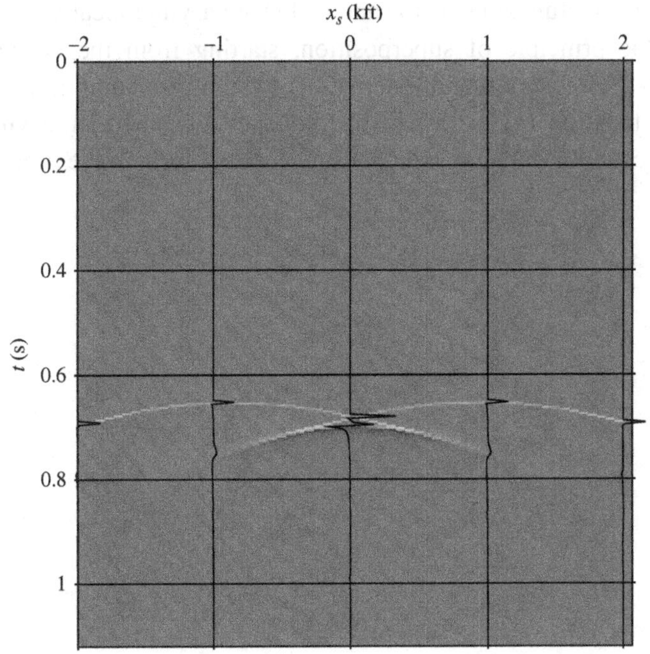

Figure 6.4 A multipath one-way Green's function for rays between the $z = 0$ surface and the point (0, 5), plotted as a function of surface position x_s and time t. The shallow, zero-phase events correspond to minimum-time raypaths. The event in the middle of the plot with a phase of 180° is a traveltime maximum. The minimum and maximum events merge at $x_s \simeq 0.4$. Past this point, Snell's law would predict a void. In fact, a diffractive event, corresponding to an inflection point in traveltime, takes over from the merged minimum and maximum. As one moves further from the Snell region, the diffraction assumes a lower frequency and eventually disappears from view.

this Green's function is retarded near $x_s = 0$, with a bowtie-like structure in the middle. The shallow, zero-phase events are traveltime minima, corresponding to the light gray raypaths in Figure 6.2. The 180° events near the middle of the plot are traveltime maxima, corresponding to the darker gray raypaths. Classical ray theory ends where the maxima and minima merge near $x = 0.4$. Diffractive events corresponding to the black raypaths extend the bowtie past this point, becoming lower and lower frequency with distance from the Snell region. As predicted, the diffractive events from the regions $x_b \simeq \pm 1.4$ are not discernible.

Exercises

6.1 Using the stationary phase approximation, show that the constant-velocity scalar Green's functions (6.14) and (6.11) are asymptotically equivalent.

6.2 Using the principle of superposition, starting from the constant-velocity scalar 3-D Green's function, derive (*a*) the corresponding time-domain 2-D Green's function, (*b*) its form in the frequency domain, (*c*) its asymptotic form in the frequency domain, and (*d*) its asymptotic form in the time domain.

7
The scattering potential

To describe reflections, we have taken the fundamental entity to be the scattering potential, defined in Chapter 5 to be the difference between a "true" and an approximate or background wave operator. Embedded in the scattering potential are perturbations from their background values of perhaps several physical parameters. Scattering or reflection is considered to be caused by the perturbations of these physical quantities.

7.1 The scattering potential as a function of angle

Given the desired wave equation, using (5.52) one can write down the appropriate scattering potential. A vector wave equation requires a diagonalization of the background wave equation into components each with a single background velocity function. The same transformation that diagonalizes the background wave equation separates the scattering potential into components corresponding to the different wave types. Diagonal components of the transformed scattering potential describe reflections that preserve wave type. Off-diagonal elements describe reflections that convert one wave type into another. Each component of the transformed scattering potential will contain perturbation terms sandwiched between coefficients which may contain differential operators; i.e. $\mathcal{V}$ is of the form

$$\mathcal{V}(\boldsymbol{x}, \omega) = \sum_i A_{Li}(\boldsymbol{x}, \omega) a_i(\boldsymbol{x}) A_{Ri}(\boldsymbol{x}, \omega), \tag{7.1}$$

with each a_i the perturbation of some physical parameter, and A_{Li} and A_{Ri} the associated coefficients. Where the coefficients contain differential operators, they can only be evaluated in the context of a seismic experiment, in which incoming and reflected waves have a well-defined directionality. Fortunately, the potential normally resides inside the Born approximation integral (5.59), or

$$D\left(\tilde{\boldsymbol{x}}_g|\tilde{\boldsymbol{x}}_s;\omega\right) = \mathfrak{w}(\omega)\int\int\int d^3x G_0\left(\tilde{\boldsymbol{x}}_g\middle|\boldsymbol{x};\omega\right)\mathcal{V}(\boldsymbol{x},\omega)G_0\left(\boldsymbol{x}\,|\tilde{\boldsymbol{x}}_s\,;\omega\right), \quad (7.2)$$

with $G_0\left(\boldsymbol{x}|\tilde{\boldsymbol{x}}_s;\,\omega\right)$ the Green's function for the incoming wave type, and $G_0\left(\tilde{\boldsymbol{x}}_g\middle|\,\boldsymbol{x};\,\omega\right)$ the Green's function for the reflected wave type. The wavelet $\mathfrak{w}(\omega)$ confines the reflection response to a realizable frequency band.

Derivatives, if any, in A_{Ri} apply to the incoming Green's function. Derivatives in A_{Li} can be applied to the reflected Green's function through integration by parts, resulting in a change in sign. Defining $\bar{A}_{Li}$ by changing the sign of derivative operators in A_{Li}, we can write

$$D\left(\tilde{\boldsymbol{x}}_g|\tilde{\boldsymbol{x}}_s;\omega\right)$$
$$= \mathfrak{w}(\omega)\sum_i\int\int\int d^3x\left(\overline{A}_{Li}G_0\right)\left(\tilde{\boldsymbol{x}}_g\middle|\,\boldsymbol{x};\omega\right)a_i(\boldsymbol{x})\left(A_{Ri}G_0\right)\left(\boldsymbol{x}\,|\tilde{\boldsymbol{x}}_s;\;\omega\right). \quad (7.3)$$

If the Green's functions are plane waves or local approximations to plane waves, then (see equation (6.64))

$$\nabla_x G_0\left(\boldsymbol{x}\,|\tilde{\boldsymbol{x}}_s;\;\omega\right) \simeq +i\boldsymbol{k}_i(\boldsymbol{x})G_0\left(\boldsymbol{x}\,|\tilde{\boldsymbol{x}}_s;\;\omega\right),$$
$$\nabla_x G_0\left(\tilde{\boldsymbol{x}}_g\middle|\,\boldsymbol{x};\omega\right) \simeq -i\boldsymbol{k}_r(\boldsymbol{x})G_0\left(\tilde{\boldsymbol{x}}_g\middle|\,\boldsymbol{x};\omega\right), \quad (7.4)$$

where $\boldsymbol{k}_i$ is the wavenumber vector of the incoming wave, and $\boldsymbol{k}_r$ is the wavenumber vector of the reflected wave. For true plane waves, these conditions are exactly met; for WKBJ asymptotic waves of the form (6.24), they are approximately met, though it is not unlikely that the complete Green's function is a superposition of several approximate plane waves. For our purposes, we can assume a single, approximate plane-wave form for the incoming and reflected waves, so that (7.4) is satisfied. The magnitudes of the wavenumber vectors, as determined by the wave equation, are

$$|\boldsymbol{k}_i| = \omega\,/c_i\,, \quad (7.5)$$

and

$$|\boldsymbol{k}_r| = \omega\,/c_r\,, \quad (7.6)$$

where c_i and c_r are the velocities appropriate to the incoming and reflected waves at $\boldsymbol{x}$. If we define unit direction vectors $\hat{\boldsymbol{k}}_i$ and $\hat{\boldsymbol{k}}_r$ pointing in the direction of wave propagation, then

$$\boldsymbol{k}_i = \frac{\omega}{c_i}\hat{\boldsymbol{k}}_i, \quad (7.7)$$

and

$$\boldsymbol{k}_r = \frac{\omega}{c_r}\hat{\boldsymbol{k}}_r. \quad (7.8)$$

The angle between the incident and reflected waves at the reflector is called the opening angle σ. In terms of the unit vectors,

$$\hat{\boldsymbol{k}}_r \cdot \hat{\boldsymbol{k}}_i = -\cos\sigma. \tag{7.9}$$

Equation (7.4) allows replacement of ∇ in A_{Ri} with $i\boldsymbol{k}_i = i\omega\hat{\boldsymbol{k}}_i(\boldsymbol{x})/c_i$, and ∇ in $\bar{A}_{Li}$ with $-i\boldsymbol{k}_r$=-i$\omega\hat{\boldsymbol{k}}_r(\boldsymbol{x})\,/c_r$. Since we have already changed signs of derivatives in the transition from A_{Li} to $\bar{A}_{Li}$, this amounts to replacing ∇ in A_{Li} with $i\boldsymbol{k}_r$= $i\omega\hat{\boldsymbol{k}}_r(\boldsymbol{x})\,/c_r$. Wavenumber magnitudes are proportional to frequency. Since the underlying wave equation is a second-order differential equation, every term in the potential is proportional to ω^2. For an isotropic potential, the net result does not depend upon the absolute directions of the incoming and outgoing wavenumbers, but only on the angle between them. This gives an expression for the scattering potential component within the Born integral in terms of the opening angle σ between the incoming and reflected waves, which we write as

$$\mathcal{V}(\boldsymbol{x},\omega) \to \omega^2\mathbb{V}(\boldsymbol{x},\sigma) = \omega^2\sum_i a_i(\boldsymbol{x})\mathbb{A}_i(\boldsymbol{x},\sigma), \tag{7.10}$$

where $\mathbb{V}$ is the *frequency-reduced* potential, and where $\mathbb{A}_i$ is formed from $A_{Li}A_{Ri}$, substituting $i\hat{\boldsymbol{k}}_r(\boldsymbol{x})\,/c_r$ for ∇ in A_{Li}, and $i\hat{\boldsymbol{k}}_i(\boldsymbol{x})/c_i$ for ∇ in A_{Ri}. If one considers anisotropic wave equations, then the functional dependence of the potential must be generalized.

7.2 The scalar scattering potential

Consider the scalar wave equation. With a Fourier transform from time to frequency, equation (6.8) is

$$\mathcal{L}_{\text{scalar}}(\boldsymbol{x},\omega)G\left(\boldsymbol{x}\,|\boldsymbol{x}_s,\ \omega\right) = -\delta\left(\boldsymbol{x}-\boldsymbol{x}_s\right), \tag{7.11}$$

where

$$\mathcal{L}_{\text{scalar}}(\boldsymbol{x},\omega) = \nabla^2 + \frac{\omega^2}{c(\boldsymbol{x})^2} \tag{7.12}$$

is the scalar wave operator. The scattering potential $\mathcal{V}$ is the difference between this operator and the background wave operator

$$\mathcal{L}_{0\text{scalar}}(\boldsymbol{x},\omega) = \nabla^2 + \frac{\omega^2}{c_0(\boldsymbol{x})^2}. \tag{7.13}$$

That is,

$$\begin{aligned}\mathcal{V}_{\text{scalar}}(\boldsymbol{x},\omega) &\equiv \mathcal{L}_{\text{scalar}}(\boldsymbol{x},\omega) - \mathcal{L}_{0\text{scalar}}(\boldsymbol{x},\omega)\\ &= \frac{\omega^2}{c_0(\boldsymbol{x})^2}\left(\frac{c_0{}^2}{c^2}-1\right) \equiv -\frac{\omega^2}{c_0(\boldsymbol{x})^2}a_c(\boldsymbol{x}),\end{aligned} \tag{7.14}$$

where

$$a_c(\boldsymbol{x}) = \left(1 - \frac{c_0{}^2}{c^2}\right). \tag{7.15}$$

Viewed as a perturbation, a reflecting surface appears as a step function in the parameters a_c (or more precisely, because of the finite bandwidth of seismic data, a bandlimited step function).

The scattering potential for the scalar wave equation contains no derivatives. Integration by parts is not required to express it in the form (7.10), as it has that form from the outset.

The frequency dependence of $\mathcal{V}_{\text{scalar}}$ is pretty simple (i.e. $\mathcal{V}_{\text{scalar}} \propto \omega^2$). In keeping with (7.10), we can introduce a frequency-independent reduced potential as

$$\mathbb{V}_{\text{scalar}}(\boldsymbol{x}) \equiv \frac{\mathcal{V}_{\text{scalar}}(\boldsymbol{x}, \omega)}{\omega^2} = -\frac{a_c(\boldsymbol{x})}{c_0(\boldsymbol{x})^2}. \tag{7.16}$$

This notation may be overkill for scalar waves, but will prove convenient when dealing with other wave equations.

7.3 The acoustic scattering potential

The full and approximate wave operators for the acoustic case are

$$\mathcal{L}_a(\boldsymbol{x}, \omega) = \boldsymbol{\nabla} \cdot \frac{1}{\rho} \boldsymbol{\nabla} + \frac{\omega^2}{\rho c^2}, \tag{7.17}$$

and

$$\mathcal{L}_{a0}(\boldsymbol{x}, \omega) = \boldsymbol{\nabla} \cdot \frac{1}{\rho_0} \boldsymbol{\nabla} + \frac{\omega^2}{\rho_0 c_0{}^2}. \tag{7.18}$$

The quantities ρ and c may vary rapidly with $\boldsymbol{x}$, while ρ_0 and c_0 may only vary slowly. In these expressions, it is understood that the $\boldsymbol{\nabla}$ operators operate on whatever is on their right. The scattering potential is

$$\mathcal{V}_a(\boldsymbol{x}, \omega) = \mathcal{L}_a(x, \omega) - \mathcal{L}_{a0}(x, \omega) = \boldsymbol{\nabla} \cdot \left(\frac{1}{\rho} - \frac{1}{\rho_0}\right) \boldsymbol{\nabla} + \omega^2 \left(\frac{1}{\rho c^2} - \frac{1}{\rho_0 c_0{}^2}\right). \tag{7.19}$$

Define the perturbations

$$a_\rho(\boldsymbol{x}) = 1 - \frac{\rho_0(\boldsymbol{x})}{\rho(\boldsymbol{x})}, \quad a_c(\boldsymbol{x}) = 1 - \frac{c_0{}^2(\boldsymbol{x})}{c^2(\boldsymbol{x})}. \tag{7.20}$$

The velocity perturbation a_c is the same as that found for the scalar wave equation perturbation equation (7.15). Consistent with the spirit of the Born approximation, we retain only terms linear in the perturbations. With this approximation,

$$\mathcal{V}_a(\boldsymbol{x},\,\omega) \simeq -\nabla \cdot \left(\frac{a_\rho}{\rho_0}\right)\nabla - \frac{\omega^2}{\rho_0 {c_0}^2}\left(a_\rho + a_c\right). \tag{7.21}$$

The derivative operators in the acoustic scattering potential mark a significant difference from the scalar potential. It is possible, however, to evaluate the derivatives within the context of the Born approximation (5.59), where the potential exists inside a volume integral sandwiched between two Green's functions. Following the argument of section 7.1, to the right of a_ρ/ρ_0, $\nabla \to i\boldsymbol{k}_i = i\omega\hat{\boldsymbol{k}}_i/c_0$, while to the left of a_ρ/ρ_0, $\nabla \to i\boldsymbol{k}_r = i\omega\hat{\boldsymbol{k}}_r/c_0$. Thus,

$$\begin{aligned}\mathcal{V}_a(\boldsymbol{x},\,\omega) &\to \boldsymbol{k}_i \cdot \boldsymbol{k}_r\left(\frac{a_\rho}{\rho_0}\right) - \frac{\omega^2}{\rho_0 {c_0}^2}\left(a_\rho + a_c\right)\\ &= \frac{\omega^2}{\rho_0 {c_0}^2}\left(\hat{\boldsymbol{k}}_i \cdot \hat{\boldsymbol{k}}_r a_\rho - \left(a_\rho + a_c\right)\right).\end{aligned} \tag{7.22}$$

In the presence of a Snell specular reflecting surface, the angle between the two wavenumber vectors would be twice the angle of incidence (or equivalently, the angle of reflection). Generally, a specular reflector is not present, but the opening angle between the two wavenumber vectors is still a useful construct. From the relation (7.9),

$$\mathcal{V}_a(\boldsymbol{x},\,\omega) \to -\frac{\omega^2}{\rho_0 {c_0}^2}\left((\cos\sigma + 1)a_\rho + a_c\right). \tag{7.23}$$

Choosing a new symbol $\mathbb{V}_a(x,\sigma)$ to represent the frequency-reduced potential as a function of σ, we have

$$\begin{aligned}\frac{\mathcal{V}_a(\boldsymbol{x},\,\omega)}{\omega^2} &\to \mathbb{V}_a(\boldsymbol{x},\sigma) \equiv \mathbb{A}_c{}^{(a)}a_c + \mathbb{A}_\rho{}^{(a)}a_\rho\\ &= -\frac{{\mathcal{C}_a}^2}{{c_0}^2}\left(a_\rho(1+\cos\sigma) + a_c\right),\end{aligned} \tag{7.24}$$

with $\mathcal{C}_a$ the acoustic scale factor ${\rho_0}^{-1/2}$. Within the Born integral (5.59), we are free to replace the differential form (7.21) of the scattering potential with the angle-dependent form (7.24), the coefficients of which are shown in Figure 7.1.

7.4 The elastic scattering potential

The elastic wave operator can be written in matrix form as (5.20), with elements (5.21). The indices i and j range over the x, y, and z directions.

In this notation, it is understood that the partial derivatives ∂_i and ∂_j operate on everything to their right. If there is an unperturbed or background wave operator

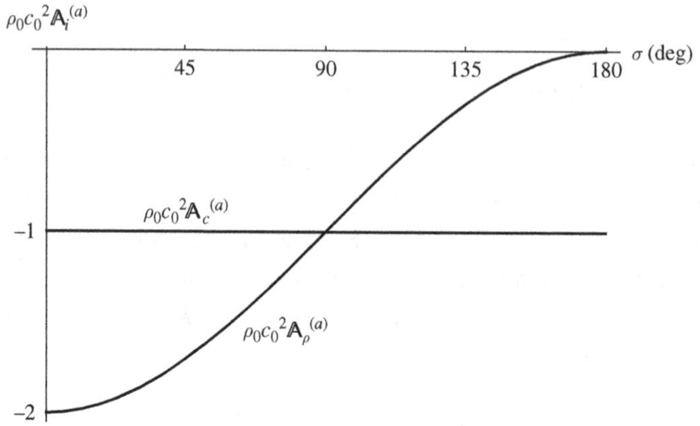

Figure 7.1 The components of the acoustic scattering potential $\mathbb{V}_a$, as a function of opening angle σ. The straight line is the coefficient $\rho_0 c_0{}^2 \mathbb{A}_c{}^{(a)}$ of the velocity perturbation a_c, and the sinusoidal line is the coefficient $\rho_0 c_0{}^2 \mathbb{A}_\rho{}^{(a)}$ of the density perturbation a_ρ. These curves are plotted for opening angles σ between 0° and 180°, corresponding to an incident angle γ between 0° and 90°. Realistically, data anywhere near $\gamma = 90°$ are very difficult to acquire.

$$\boldsymbol{\mathcal{L}}_{E0}(\boldsymbol{x}, \omega) = \begin{pmatrix} L_{0xx} & L_{0xy} & L_{0xz} \\ L_{0yx} & L_{0yy} & L_{0yz} \\ L_{0zx} & L_{0zy} & L_{0zz} \end{pmatrix}, \tag{7.25}$$

then we can form an elastic scattering potential

$$\boldsymbol{\mathcal{V}}_E(\boldsymbol{x}, \omega) = \boldsymbol{\mathcal{L}}_E(\boldsymbol{x}, \omega) - \boldsymbol{\mathcal{L}}_{E0}(\boldsymbol{x}, \omega), \quad = \begin{pmatrix} V_{xx} & V_{xy} & V_{xz} \\ V_{yx} & V_{yy} & V_{yz} \\ V_{zx} & V_{zy} & V_{zz} \end{pmatrix}, \tag{7.26}$$

with elements

$$V_{ii} = \rho_0 \left(\omega^2 a_\rho + \alpha_0{}^2 \partial_i a_\gamma \partial_i + \beta_0{}^2 \sum_{j \neq i} \partial_j a_\mu \partial_j \right), \quad i, j = x, y, z;$$
$$V_{ij} = \rho_0 \left(\alpha_0{}^2 \partial_i a_\gamma \partial_j - 2\beta_0{}^2 \partial_i a_\mu \partial_j + \beta_0{}^2 \partial_j a_\mu \partial_i \right), \; j \neq i. \tag{7.27}$$

In these expressions, the subscripts i and j have been used to indicate x, y, and z components. The quantity γ stands for the P-modulus $\lambda + 2\mu = \rho\alpha^2$, and is not to be confused with half-opening angle. The quantities ρ_0, α_0, and β_0 are the unperturbed or background values of density, P-velocity, and S-velocity, respectively. The parameters a_ρ, a_γ, and a_μ are the perturbations

$$a_\rho = \frac{\rho - \rho_0}{\rho_0} \simeq \frac{\Delta\rho}{\rho}, \tag{7.28}$$

$$a_\gamma = \frac{\gamma - \gamma_0}{\gamma_0} \simeq \frac{\Delta\gamma}{\gamma}, \tag{7.29}$$

$$a_\mu = \frac{\mu - \mu_0}{\mu_0} \simeq \frac{\Delta\mu}{\mu}. \tag{7.30}$$

While the perturbations are defined slightly differently than for the scalar and acoustic cases, to first order they are equivalent.

7.4.1 Diagonalization operators

Before substituting a wavenumber vector for $\boldsymbol{\nabla}$ in the elastic scattering potential (7.27), we must first separate out the P- and S-components. Then, P- and S-components may each be assumed locally planar with wavenumber magnitude ω/α_0 and ω/β_0, respectively.

The background wave operator $\mathcal{L}_{E0}$ can be diagonalized into P- and S-components with the transformation matrix $\boldsymbol{\Pi}\mathcal{L}_{E0}\boldsymbol{\Pi}^T\mathcal{J}^{-1}$ as defined in section 5.2.3.

Application of the same transformation to the scattering potential will separate it into compressional and shear components. The upper left diagonal term will correspond to an incoming and a reflected P-wave. Off-diagonal terms will correspond to converted waves of one sort or another.

We next apply the transformation (5.29) to$\mathcal{V}_E$, forming $\boldsymbol{\Pi}\mathcal{V}_E\boldsymbol{\Pi}^T\mathcal{J}^{-1}$. With P- and S-components effectively separated, we can, using the logic of the last section, replace derivatives with their corresponding wavenumbers. In the matrix $\boldsymbol{\Pi}$ on the left, we replace derivatives with i times a local reflected-wave wavenumber. For $\boldsymbol{\Pi}^T$ on the right, we replace derivatives with i times a local incident-wave wavenumber. The operator $\mathcal{J}^{-1}$ evaluates to the unit operator inside the Born approximation, and can be ignored. The Laplacian where it appears is replaced by minus local wavenumber magnitude squared, or minus frequency squared over velocity squared. We have to be a little careful in making these replacements, because P- and S-waves have different wavenumbers and velocities. The first row in $\boldsymbol{\Pi}$ and the first column in $\boldsymbol{\Pi}^T$ will have P-wave wavenumbers. All remaining elements will have S-wave wavenumbers. Denoting $\boldsymbol{k}_{Pr}$ and $\boldsymbol{k}_{Pi}$ to be the P-wave wavenumbers, and $\boldsymbol{k}_{Sr}$ and $\boldsymbol{k}_{Si}$ to be the S-wave wavenumbers, with respective magnitudes

$$k_{Pr} = k_{Pi} = \frac{\omega}{\alpha_0}, \tag{7.31}$$

and

$$k_{Sr} = k_{Si} = \frac{\omega}{\beta_0}, \tag{7.32}$$

then the matrices $\boldsymbol{\Pi}$ and $\boldsymbol{\Pi}^T$ on the left and right of $\mathcal{V}_E$ become

$$\boldsymbol{\Pi} \to \boldsymbol{\Pi}_r = \frac{-1}{\omega} \begin{pmatrix} \alpha_0 k_{Prx} & \alpha_0 k_{Pry} & \alpha_0 k_{Prz} \\ 0 & -\beta_0 k_{Srz} & \beta_0 k_{Sry} \\ \beta_0 k_{Srz} & 0 & -\beta_0 k_{Srx} \\ -\beta_0 k_{Sry} & \beta_0 k_{Srx} & 0 \end{pmatrix} = i \begin{pmatrix} \alpha_0 \boldsymbol{k}_{Pr} \cdot \\ \beta_0 \boldsymbol{k}_{Sr} \times \end{pmatrix}, \tag{7.33}$$

and

$$\begin{aligned} \boldsymbol{\Pi}^T \to \boldsymbol{\Pi}^T{}_i &= \frac{-1}{\omega} \begin{pmatrix} \alpha_0 k_{Pix} & 0 & \beta_0 k_{Siz} & -\beta_0 k_{Siy} \\ \alpha_0 k_{Piy} & -\beta_0 k_{Siz} & 0 & \beta_0 k_{Six} \\ \alpha_0 k_{Piz} & \beta_0 k_{Siy} & -\beta_0 k_{Six} & 0 \end{pmatrix} \\ &= \frac{-1}{\omega} \left(\alpha_0 \boldsymbol{k}_{Pi} \cdot^T \quad \beta_0 \left(\boldsymbol{k}_{Si} \times \right)^T \right). \end{aligned} \tag{7.34}$$

7.4.2 Rotation into SV and SH components

The product $\boldsymbol{\Pi}_r \mathcal{V}_E \boldsymbol{\Pi}^T{}_i$ is a 4×4 matrix, with elements corresponding to P-waves and the *x*-, *y*-, and *z*-components of S-waves. Since the shear waves are transverse, there are really only three independent dimensions. It would be convenient to explicitly reduce the number of dimensions to three, and deal with P-waves and the two independent S-wave components. We can do this by rotating the shear-wave components to a local coordinate system in which the third (longitudinal) S-wave component is zero.

At this point, we are ready to perform a further rotation of the shear-wave portion of these matrices. Incoming and outgoing wavenumber vectors define a reference plane at the reflection point $\boldsymbol{x}$. A shear-wave vector normal to this reference plane is referred to as an SH-wave. A shear-wave vector lying within the reference plane is called an SV-wave (see Figure 7.2). For an incoming wavenumber vector $\boldsymbol{k}_i$ and an outgoing wavenumber vector $\boldsymbol{k}_r$, the SH-direction is defined by a unit vector

$$\hat{\boldsymbol{e}}_{\mathrm{SH}} \equiv \frac{\boldsymbol{k}_r \times \boldsymbol{k}_i}{|\boldsymbol{k}_r \times \boldsymbol{k}_i|}. \tag{7.35}$$

Whether $\boldsymbol{k}_r$ and $\boldsymbol{k}_i$ are P- or S-wave vectors will depend on the incoming and outgoing wave type under consideration. For an incoming S-wave, the SV-direction is defined by a unit vector

$$\hat{\boldsymbol{e}}_{\mathrm{SV}i} \equiv \frac{\boldsymbol{k}_{Si} \times \hat{\boldsymbol{e}}_{\mathrm{SH}}}{\left|\boldsymbol{k}_{Si} \times \hat{\boldsymbol{e}}_{\mathrm{SH}}\right|} = \frac{\beta_0}{\omega} \boldsymbol{k}_{Si} \times \hat{\boldsymbol{e}}_{\mathrm{SH}}. \tag{7.36}$$

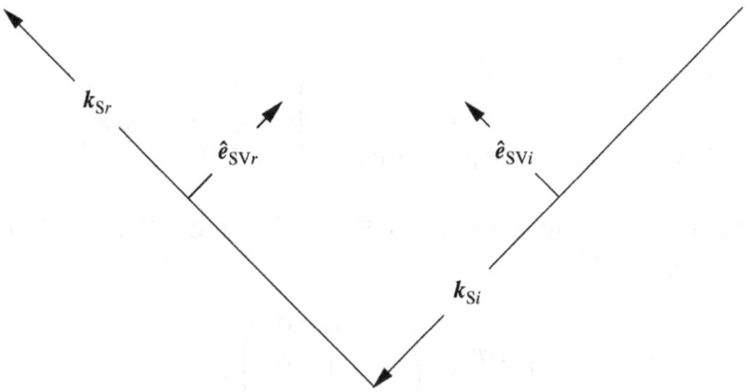

Figure 7.2 Incident and reflected SV rays. A depiction of the SV directions $\hat{\boldsymbol{e}}_{SVi}$ and $\hat{\boldsymbol{e}}_{SVr}$ for incident and reflected rays. For $\hat{\boldsymbol{e}}_{SH}$, $\hat{\boldsymbol{e}}_{SV}$, and $\hat{\boldsymbol{k}}_S$ to form a right-handed coordinate system, $\hat{\boldsymbol{e}}_{SH}$ must for both incident and reflected wave point directly outward towards the reader.

For completeness, a third longitudal direction vector can also be defined:

$$\hat{\boldsymbol{e}}_{SLi} \equiv \frac{\boldsymbol{k}_{Si}}{|\boldsymbol{k}_{Si}|} = \frac{\beta_0}{\omega}\boldsymbol{k}_{Si}. \tag{7.37}$$

For an outgoing S-wave, the SV-direction is defined by a unit vector

$$\hat{\boldsymbol{e}}_{SVr} \equiv \frac{\boldsymbol{k}_{Sr} \times \hat{\boldsymbol{e}}_{SH}}{|\boldsymbol{k}_{Sr} \times \hat{\boldsymbol{e}}_{SH}|} = \frac{\beta_0}{\omega}\boldsymbol{k}_{Sr} \times \hat{\boldsymbol{e}}_{SH}, \tag{7.38}$$

and the third direction vector is

$$\hat{\boldsymbol{e}}_{SLr} \equiv \frac{\boldsymbol{k}_{Sr}}{|\boldsymbol{k}_{Sr}|} = \frac{\beta_0}{\omega}\boldsymbol{k}_{Sr}. \tag{7.39}$$

The unit vectors $\hat{\boldsymbol{e}}_{SH}$, $\hat{\boldsymbol{e}}_{SVi}$, and $\hat{\boldsymbol{e}}_{SLi}$, taken in that order, form a right-handed coordinate system, as do $\hat{\boldsymbol{e}}_{SH}$, $\hat{\boldsymbol{e}}_{SVr}$, and $\hat{\boldsymbol{e}}_{SLr}$. Thus,

$$\hat{\boldsymbol{e}}_{SLi} \times \hat{\boldsymbol{e}}_{SH} = \hat{\boldsymbol{e}}_{SVi}, \quad \hat{\boldsymbol{e}}_{SVi} \times \hat{\boldsymbol{e}}_{SLi} = \hat{\boldsymbol{e}}_{SH}, \quad \boldsymbol{k}_{Si} \times \hat{\boldsymbol{e}}_{SLi} = 0, \tag{7.40}$$

$$\hat{\boldsymbol{e}}_{SLr} \times \hat{\boldsymbol{e}}_{SH} = \hat{\boldsymbol{e}}_{SVr}, \quad \hat{\boldsymbol{e}}_{SVr} \times \hat{\boldsymbol{e}}_{SLr} = \hat{\boldsymbol{e}}_{SH}, \quad \boldsymbol{k}_{Sr} \times \hat{\boldsymbol{e}}_{SLr} = 0. \tag{7.41}$$

Define 4 by 3 rotation matrices for the incoming and outgoing waves

$$\boldsymbol{E}_i \equiv \begin{pmatrix} 1 & 0 & 0 & 0 \\ 0 & e_{SVix} & e_{SViy} & e_{SViz} \\ 0 & -e_{SHx} & -e_{SHy} & -e_{SHz} \end{pmatrix} = \begin{pmatrix} 1 & \boldsymbol{0}^T \\ 0 & \hat{\boldsymbol{e}}_{SVi}{}^T \\ 0 & -\hat{\boldsymbol{e}}_{SH}{}^T \end{pmatrix} \tag{7.42}$$

and

$$\boldsymbol{E}_r \equiv \begin{pmatrix} 1 & 0 & 0 & 0 \\ 0 & e_{\text{SV}rx} & e_{\text{SV}ry} & e_{\text{SV}rz} \\ 0 & -e_{\text{SH}x} & -e_{\text{SH}y} & -e_{\text{SH}z} \end{pmatrix} = \begin{pmatrix} 1 & \boldsymbol{0}^T \\ 0 & \hat{\boldsymbol{e}}_{\text{SV}r}{}^T \\ 0 & -\hat{\boldsymbol{e}}_{\text{SH}}{}^T \end{pmatrix} \tag{7.43}$$

that transform a PS decomposed wave $\boldsymbol{\Psi} = \boldsymbol{\Pi u}$ into SV and SH components $\boldsymbol{\Phi} = \boldsymbol{E\Pi u}$. We have

$$\boldsymbol{E}\boldsymbol{E}^T \equiv \begin{pmatrix} 1 & 0 & 0 \\ 0 & 1 & 0 \\ 0 & 0 & 1 \end{pmatrix}. \tag{7.44}$$

The matrix $\boldsymbol{E}^T\boldsymbol{E}$ is 4 by 4 and not the unit matrix. However,

$$\boldsymbol{E}^T\boldsymbol{E}\boldsymbol{\Pi} = \boldsymbol{\Pi}, \tag{7.45}$$

so that effectively, the transpose of $\boldsymbol{E}$ is its inverse.

The combination of the transposes of $\boldsymbol{E}$ and $\boldsymbol{\Pi}$ is, for incident waves,

$$\begin{aligned} \boldsymbol{\Pi}^T{}_i\boldsymbol{E}_i{}^T &= \frac{-1}{\omega}\begin{pmatrix} \alpha_0 k_{Pix} & -\beta_0\left(\boldsymbol{k}_{Si}\times\hat{\boldsymbol{e}}_{\text{SV}i}\right)_x & \beta_0\left(\boldsymbol{k}_{Si}\times\hat{\boldsymbol{e}}_{\text{SH}}\right)_x \\ \alpha_0 k_{Piy} & -\beta_0\left(\boldsymbol{k}_{Si}\times\hat{\boldsymbol{e}}_{\text{SV}i}\right)_y & \beta_0\left(\boldsymbol{k}_{Si}\times\hat{\boldsymbol{e}}_{\text{SH}}\right)_y \\ \alpha_0 k_{Piz} & -\beta_0\left(\boldsymbol{k}_{Si}\times\hat{\boldsymbol{e}}_{\text{SV}i}\right)_z & \beta_0\left(\boldsymbol{k}_{Si}\times\hat{\boldsymbol{e}}_{\text{SH}}\right)_z \end{pmatrix} \\ &= \frac{-1}{\omega}\begin{pmatrix} \alpha_0\boldsymbol{k}_{Pi} & -\beta_0\left(\boldsymbol{k}_{Si}\times\hat{\boldsymbol{e}}_{\text{SV}i}\right) & +\beta_0\left(\boldsymbol{k}_{Si}\times\hat{\boldsymbol{e}}_{\text{SH}}\right) \end{pmatrix}. \end{aligned} \tag{7.46}$$

Invoking the orthogonality relations (7.40), the product matrix can be written

$$\begin{aligned} \boldsymbol{\Pi}^T{}_i\boldsymbol{E}_i{}^T &= -\begin{pmatrix} \frac{\alpha_0}{\omega}k_{Pix} & e_{\text{SH}x} & e_{\text{SV}ix} \\ \frac{\alpha_0}{\omega}k_{Piy} & e_{\text{SH}y} & e_{\text{SV}iy} \\ \frac{\alpha_0}{\omega}k_{Piz} & e_{\text{SH}z} & e_{\text{SV}iz} \end{pmatrix} \\ &= -\begin{pmatrix} \frac{\alpha_0}{\omega}\boldsymbol{k}_{Pi} & \hat{\boldsymbol{e}}_{\text{SH}} & \hat{\boldsymbol{e}}_{\text{SV}i} \end{pmatrix}. \end{aligned} \tag{7.47}$$

Similarly, the product of $\boldsymbol{E}$ and $\boldsymbol{\Pi}$ is, for reflected waves,

$$\begin{aligned} \boldsymbol{E}_r\boldsymbol{\Pi}_r &= \frac{-1}{\omega}\begin{pmatrix} \alpha_0 k_{Prx} & \alpha_0 k_{Pry} & \alpha_0 k_{Prz} \\ -\beta_0\left(\boldsymbol{k}_{Sr}\times\hat{\boldsymbol{e}}_{\text{SV}r}\right)_x & -\beta_0\left(\boldsymbol{k}_{Sr}\times\hat{\boldsymbol{e}}_{\text{SV}r}\right)_y & -\beta_0\left(\boldsymbol{k}_{Sr}\times\hat{\boldsymbol{e}}_{\text{SV}r}\right)_z \\ \beta_0\left(\boldsymbol{k}_{Sr}\times\hat{\boldsymbol{e}}_{\text{SH}}\right)_x & \beta_0\left(\boldsymbol{k}_{Sr}\times\hat{\boldsymbol{e}}_{\text{SH}}\right)_y & \beta_0\left(\boldsymbol{k}_{Sr}\times\hat{\boldsymbol{e}}_{\text{SH}}\right)_y \end{pmatrix} \\ &= \frac{-1}{\omega}\begin{pmatrix} \alpha_0\boldsymbol{k}_{Pr}{}^T \\ -\beta_0\left(\boldsymbol{k}_{Sr}\times\hat{\boldsymbol{e}}_{\text{SV}r}\right)^T \\ \beta_0\left(\boldsymbol{k}_{Sr}\times\hat{\boldsymbol{e}}_{\text{SH}}\right)^T \end{pmatrix}. \end{aligned} \tag{7.48}$$

Invoking the orthogonality relations (7.41), this product matrix can be written

$$
\begin{aligned}
\boldsymbol{E}_r\boldsymbol{\Pi}_r &= -\begin{pmatrix} \frac{\alpha_0}{\omega}k_{Prx} & \frac{\alpha_0}{\omega}k_{Pry} & \frac{\alpha_0}{\omega}k_{Prz} \\ e_{\mathrm{SH}x} & e_{\mathrm{SH}y} & e_{\mathrm{SH}z} \\ e_{\mathrm{SV}rx} & e_{\mathrm{SV}ry} & e_{\mathrm{SV}rz} \end{pmatrix} \\
&= -\begin{pmatrix} \frac{\alpha_0}{\omega}\boldsymbol{k}_{Pr}{}^T \\ \hat{\boldsymbol{e}}_{\mathrm{SH}}{}^T \\ \hat{\boldsymbol{e}}_{\mathrm{SV}r}{}^T \end{pmatrix}.
\end{aligned} \tag{7.49}
$$

Applied to a displacement vector, the top row of $\boldsymbol{E}_r\boldsymbol{\Pi}_r$ and first column of $\boldsymbol{\Pi}^T{}_i\boldsymbol{E}_i{}^T$ separate out the P-component. The second row of $\boldsymbol{E}_r\boldsymbol{\Pi}_r$ and second column of $\boldsymbol{\Pi}^T{}_i\boldsymbol{E}_i{}^T$ separate out the SH-component, and the third row of $\boldsymbol{E}_r\boldsymbol{\Pi}_r$ and third column of $\boldsymbol{\Pi}^T{}_i\boldsymbol{E}_i{}^T$ separate out the SV-component.

The operator we now want to evaluate is the scattering potential in the P–SH–SV coordinate system, or

$$
\boldsymbol{\mathcal{V}}_W \equiv \boldsymbol{E}_r\boldsymbol{\Pi}_r\boldsymbol{\mathcal{V}}_E\boldsymbol{\Pi}^T{}_i\boldsymbol{E}_i{}^T. \tag{7.50}
$$

The elements of this potential matrix can be identified with P- , SH- , and SV-wave components as follows:

$$
\begin{aligned}
\boldsymbol{\mathcal{V}}_W &= \begin{pmatrix} \mathcal{V}_{\mathrm{PP}} & \mathcal{V}_{\mathrm{PSH}} & \mathcal{V}_{\mathrm{PSV}} \\ \mathcal{V}_{\mathrm{SHP}} & \mathcal{V}_{\mathrm{SHSH}} & \mathcal{V}_{\mathrm{SHSV}} \\ \mathcal{V}_{\mathrm{SVP}} & \mathcal{V}_{\mathrm{SVSH}} & \mathcal{V}_{\mathrm{SVSV}} \end{pmatrix} \qquad (7.51) \\
&= \begin{pmatrix} \left(\alpha_0{}^2/\omega^2\right)\boldsymbol{k}_{Pr}{}^T\boldsymbol{\mathcal{V}}_E\boldsymbol{k}_{Pi} & \left(\alpha_0/\,\omega\right)\boldsymbol{k}_{Pr}{}^T\boldsymbol{\mathcal{V}}_E\hat{\boldsymbol{e}}_{\mathrm{SH}} & \left(\alpha_0/\,\omega\right)\boldsymbol{k}_{Pr}{}^T\boldsymbol{\mathcal{V}}_E\hat{\boldsymbol{e}}_{\mathrm{SV}i} \\ (\alpha_0/\omega)\,\hat{\boldsymbol{e}}_{\mathrm{SH}}{}^T\boldsymbol{\mathcal{V}}_E\boldsymbol{k}_{Pi} & \hat{\boldsymbol{e}}_{\mathrm{SH}}{}^T\boldsymbol{\mathcal{V}}_E\hat{\boldsymbol{e}}_{\mathrm{SH}} & \hat{\boldsymbol{e}}_{\mathrm{SH}}{}^T\boldsymbol{\mathcal{V}}_E\hat{\boldsymbol{e}}_{\mathrm{SV}i} \\ (\alpha_0/\omega)\,\hat{\boldsymbol{e}}_{\mathrm{SV}r}{}^T\boldsymbol{\mathcal{V}}_E\boldsymbol{k}_{Pi} & \hat{\boldsymbol{e}}_{\mathrm{SV}r}{}^T\boldsymbol{\mathcal{V}}_E\hat{\boldsymbol{e}}_{\mathrm{SH}} & \hat{\boldsymbol{e}}_{\mathrm{SV}r}{}^T\boldsymbol{\mathcal{V}}_E\boldsymbol{e}_{\mathrm{SV}i} \end{pmatrix}.
\end{aligned}
$$

The elements of $\boldsymbol{\mathcal{V}}_E$, given in equation (7.27), also contain derivatives. Consistent with our treatment of $\boldsymbol{\Pi}$, we replace the derivatives with wavenumbers as follows: in the first row of $\boldsymbol{\mathcal{V}}_W$, derivatives appearing to the left of the elastic parameters a_ρ, a_γ, or a_μ are replaced by $i\boldsymbol{k}_{Pr}$. In remaining rows, left-derivatives are replaced by $i\boldsymbol{k}_{Sr}$. In the first column of $\boldsymbol{\mathcal{V}}_W$, derivatives appearing on the right of the elastic parameters are replaced by $i\boldsymbol{k}_{Pi}$. In the remaining columns, right-derivatives are replaced by $i\boldsymbol{k}_{Si}$.

7.4.3 The P-to-P scattering potential

The upper left element of $\boldsymbol{\mathcal{V}}_W$, corresponding to input and output P-waves, we denote $\mathcal{V}_{\mathrm{PP}}$. We have

$$
\mathcal{V}_{\mathrm{PP}} = \left(\alpha_0{}^2/\omega^2\right)\boldsymbol{k}_{Pr}{}^T\boldsymbol{\mathcal{V}}_E\boldsymbol{k}_{Pi}. \tag{7.52}
$$

With (7.27) for $\mathcal{V}_E$, we replace left-derivatives with elements of $i\boldsymbol{k}_{Pr}$ and right-derivatives with elements of $i\boldsymbol{k}_{Pi}$. Then,

$$\mathcal{V}_{\text{PP}} = -\frac{\rho_0\alpha_0^2}{\omega^2}\left(a_\gamma \alpha_0^2 k_{Pr}{}^2 k_{Pi}{}^2 - a_\rho \boldsymbol{k}_{Pr} \cdot \boldsymbol{k}_{Pi}\omega^2 - 2a_\mu \beta_0^2 |\boldsymbol{k}_{Pr} \times \boldsymbol{k}_{Pi}|^2\right). \tag{7.53}$$

The magnitudes k_{Pr} and k_{Pi} are equal to ω/α_0. The dot and cross products of the incoming and outgoing P-wave wavenumber vectors can be expressed in terms of the opening angle σ between them:

$$\boldsymbol{k}_{Pr} \cdot \boldsymbol{k}_{Pi} = -\frac{\omega^2}{\alpha_0^2}\cos\sigma, \quad |\boldsymbol{k}_{Pr} \times \boldsymbol{k}_{Pi}|^2 = \frac{\omega^4}{\alpha_0^4}\sin^2\sigma. \tag{7.54}$$

With these expressions, within the Born integral, $\mathcal{V}_{\text{PP}}$ becomes the angle-dependent function

$$\mathcal{V}_{\text{PP}} \to \omega^2 \mathbb{V}_{\text{PP}}(\boldsymbol{x},\sigma) = -\frac{\omega^2}{\alpha_0^2}\mathcal{C}_P{}^2\left(a_\gamma + a_\rho\cos\sigma - 2\frac{\beta_0^2}{\alpha_0^2}a_\mu \sin^2\sigma\right), \tag{7.55}$$

with $\mathcal{C}_P$ the P-wave scaling factor $\left(\rho_0\alpha_0^2\right)^{1/2}$. As for the acoustic case, we have used a new symbol $\mathbb{V}$ to represent the potential in its angular-dependent frequency-independent form. Setting

$$a_\gamma = a_\rho + a_\alpha, \tag{7.56}$$

$$a_\mu = a_\rho + a_\beta, \tag{7.57}$$

we can express the PP scattering potential in terms of velocity and density perturbations:

$$\begin{aligned}\mathbb{V}_{\text{PP}}(\boldsymbol{x},\sigma) &\equiv \mathbb{A}_\alpha{}^{(\text{PP})}a_\alpha + \mathbb{A}_\beta{}^{(\text{PP})}a_\beta + \mathbb{A}_\rho{}^{(\text{PP})}a_\rho \\ &= -\frac{\mathcal{C}_P{}^2}{\alpha_0^2}\left(a_\alpha + a_\rho\left(1+\cos\sigma - 2\frac{\beta_0^2}{\alpha_0^2}\sin^2\sigma\right) - 2\frac{\beta_0^2}{\alpha_0^2}a_\beta\sin^2\sigma\right).\end{aligned} \tag{7.58}$$

The components of the PP potential are illustrated in Figure 7.3.

Relation to scalar and acoustic potentials

Where the shear velocity is zero, $\mathbb{V}_{\text{PP}}$ reduces, except for a factor of density, to the acoustic potential (7.24). Where density and shear velocity remain constant, $\mathbb{V}_{\text{PP}}$ reduces, except for a factor of $\mathcal{C}_P{}^2 = \rho_0\alpha_0^2$, to the scalar potential (7.16). Thus, the acoustic and scalar scattering potentials can be treated as special cases of the elastic scattering potential.

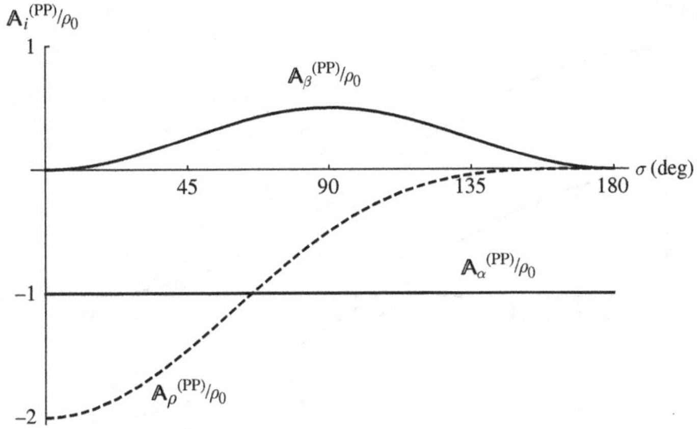

Figure 7.3 PP scattering potential. The components of the P to P frequency-reduced scattering potential $\mathbb{V}_{PP}$, as a function of opening angle σ. In this illustration, we have chosen the background ratio β_0/α_0 of S to P velocity to be 1/2. The upper curve is the coefficient $\rho_0{}^{-1}\mathbb{A}_\beta{}^{(PP)}$ of the S-velocity component a_β, the middle line is the coefficient $\rho_0{}^{-1}\mathbb{A}_\alpha{}^{(PP)}$ of the P-velocity perturbation a_α, and the sinusoidal dashed curve is the coefficient $\rho_0{}^{-1}\mathbb{A}_\rho{}^{(PP)}$ of the density perturbation a_ρ.

7.4.4 The SH to SH scattering potential

The SH to SH component of the scattering potential is the second diagonal element of $\boldsymbol{\mathcal{V}}_W$, or

$$\mathcal{V}_{\text{SHSH}} = \hat{\boldsymbol{e}}_{\text{SH}}{}^T \boldsymbol{\mathcal{V}}_E \hat{\boldsymbol{e}}_{\text{SH}}. \tag{7.59}$$

This evaluates to

$$\mathcal{V}_{\text{SHSH}} = \rho_0 \omega^2 \left(a_\rho \; - a_\mu \frac{(\boldsymbol{k}_{Sr} \cdot \boldsymbol{k}_{Si})}{\omega^2} \right). \tag{7.60}$$

Because the SH direction is the same for incoming and outgoing waves, this expression simplifies to

$$\mathbb{V}_{\text{SHSH}} = \frac{\mathcal{C}_S{}^2}{\beta_0{}^2} \left(a_\rho + a_\mu \cos\sigma \right), \tag{7.61}$$

or, in terms of density and shear velocity perturbations,

$$\mathbb{V}_{\text{SHSH}} \equiv \mathbb{A}_\beta{}^{(\text{SHSH})} a_\beta + \mathbb{A}_\rho{}^{(\text{SHSH})} a_\rho = \rho_0 \left(a_\beta \cos\sigma + a_\rho (1 + \cos\sigma) \right). \tag{7.62}$$

The two components of the frequency-reduced SHSH scattering potential are shown in Figure 7.4.

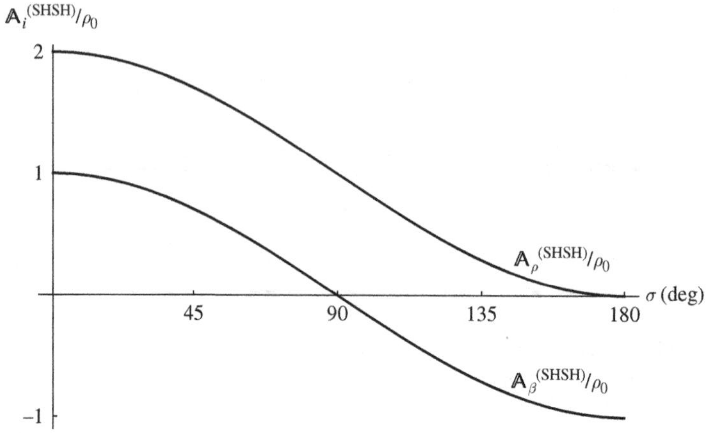

Figure 7.4 SHSH scattering potential. The components of the SH to SH frequency-reduced scattering potential $\mathbb{V}_{\mathrm{SHSH}}$, as a function of opening angle σ. The lower line is the coefficient $\rho_0{}^{-1}\mathbb{A}_\beta{}^{(\mathrm{SHSH})}$ of the shear velocity perturbation a_β, and the upper line is the coefficient $\rho_0{}^{-1}\mathbb{A}_\rho{}^{(\mathrm{SHSH})}$ of the density perturbation a_ρ.

7.4.5 The SV to SV scattering potential

The third diagonal element of $\boldsymbol{\mathcal{V}}_W$, corresponding to input and output SV-waves, we denote $\mathcal{V}_{\mathrm{SVSV}}$. We have

$$\mathcal{V}_{\mathrm{SVSV}} = \hat{\boldsymbol{e}}_{\mathrm{SV}r}{}^T \boldsymbol{\mathcal{V}}_E \hat{\boldsymbol{e}}_{\mathrm{SV}i}. \tag{7.63}$$

In this case, both the incoming and outgoing waves are shear waves. The wavenumber vectors that define the SH-direction in equation (7.35) are the shear-wave directions $\boldsymbol{k}_{Sr}$ and $\boldsymbol{k}_{Si}$. Expanding $\boldsymbol{\mathcal{V}}_E$ and substituting wavenumbers for derivative operators, this becomes

$$\mathcal{V}_{\mathrm{SVSV}} = \frac{\rho_0}{k_{Sr}k_{Si}}\left(a_\mu\left(k_{Sr}{}^2k_{Si}{}^2 - 2\left(\boldsymbol{k}_{Sr}\cdot\boldsymbol{k}_{Si}\right)^2\right)\beta_0{}^2 + a_\rho\,\omega^2\boldsymbol{k}_{Sr}\cdot\boldsymbol{k}_{Si}\right). \tag{7.64}$$

Now,

$$k_{Sr} = k_{Si} = \frac{\omega}{\beta_0}, \quad \boldsymbol{k}_{Sr}\cdot\boldsymbol{k}_{Si} = -\frac{\omega^2}{\beta_0{}^2}\cos\sigma, \tag{7.65}$$

whence

$$\mathcal{V}_{\mathrm{SVSV}} \to \omega^2\mathbb{V}_{\mathrm{SVSV}}(\boldsymbol{x},\sigma) = -\frac{\omega^2}{\beta_0{}^2}C_S{}^2\left(a_\rho\,\cos\sigma + a_\mu\cos 2\sigma\right), \tag{7.66}$$

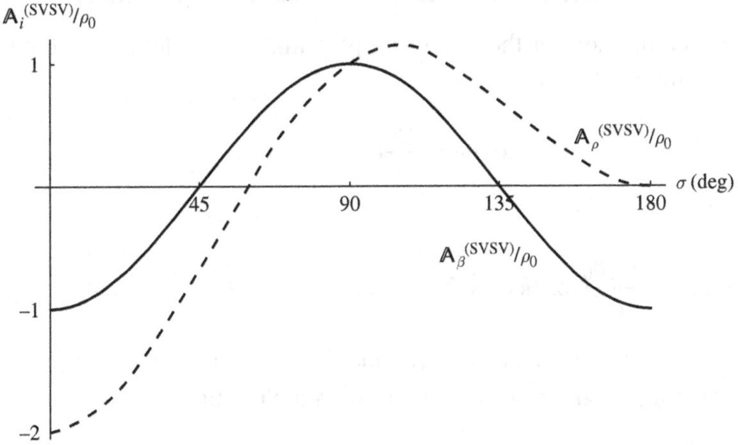

Figure 7.5 SVSV scattering potential. The components of the SV to SV frequency-reduced scattering potential $\mathbb{V}_{SVSV}$, as a function of opening angle σ. The solid curve is the coefficient $\rho_0{}^{-1}\mathbb{A}_\beta{}^{(SVSV)}$ of the shear velocity perturbation a_β, and the dashed curve is the coefficient $\rho_0{}^{-1}\mathbb{A}_\rho{}^{(SVSV)}$ of the density perturbation a_ρ.

with $C_S{}^2 = \rho_0\beta_0{}^2$. Invoking (7.57), this may be written in terms of the density and shear-velocity perturbations as

$$\begin{aligned}\mathbb{V}_{SVSV}(\boldsymbol{x}, \sigma) &\equiv \mathbb{A}_\beta{}^{(SVSV)} a_\beta + \mathbb{A}_\rho{}^{(SVSV)} a_\rho \\ &= -\rho_0 \left(a_\rho \left(\cos\sigma + \cos 2\sigma\right) + a_\beta \cos 2\sigma\right). \end{aligned} \tag{7.67}$$

The component coefficients of $\mathbb{V}_{SVSV}$ are plotted in Figure 7.5.

7.4.6 The SH to SV or P, and SV or P to SH, scattering potentials

The SH to SV scattering potential is

$$\mathcal{V}_{SVSH} = \hat{\boldsymbol{e}}_{SV_r}{}^T \boldsymbol{\mathcal{V}}_E \hat{\boldsymbol{e}}_{SH}, \tag{7.68}$$

which evaluates to

$$\mathcal{V}_{SVSH} = 0. \tag{7.69}$$

The SV to SH scattering potential is also zero:

$$\mathcal{V}_{SHSV} = 0, \tag{7.70}$$

as are the SH to P and P to SH potentials:

$$\mathcal{V}_{PSH} = \mathcal{V}_{SHP} = 0. \tag{7.71}$$

7.4.7 The P to SV and SV to P scattering potentials

The P to SV component of the scattering potential is the element in the third row and first column of $\mathcal{V}_W$, or

$$\mathcal{V}_{\mathrm{SVP}} = \frac{\alpha_0}{\omega} \hat{\boldsymbol{e}}_{\mathrm{SV}r}{}^T \mathcal{V}_E \boldsymbol{k}_{Pi}, \tag{7.72}$$

or

$$\mathcal{V}_{\mathrm{SVP}} = -\frac{\alpha_0 \beta_0}{\omega^2} \rho_0 \left|\boldsymbol{k}_{Sr} \times \boldsymbol{k}_{Pi}\right| \left(a_\rho\, \omega^2 - 2a_\mu \left(\boldsymbol{k}_{Sr} \cdot \boldsymbol{k}_{Pi}\right) \beta_0{}^2\right). \tag{7.73}$$

Since the incoming wave is a P-wave and the outgoing wave is a shear wave, we define the opening angle σ between them with the relation

$$\boldsymbol{k}_{Sr} \cdot \boldsymbol{k}_{Pi} = -\frac{\omega^2}{\alpha_0 \beta_0} \cos\sigma. \tag{7.74}$$

Likewise,

$$\left|\boldsymbol{k}_{Sr} \times \boldsymbol{k}_{Pi}\right| = \frac{\omega^2}{\alpha_0 \beta_0} |\sin\sigma|. \tag{7.75}$$

Thus,

$$\mathcal{V}_{\mathrm{SVP}} \to \omega^2 \mathbb{V}_{\mathrm{SVP}}(\boldsymbol{x}, \sigma) = -\mathcal{C}_P{}^2 \frac{\omega^2}{\alpha_0{}^2} |\sin\sigma| \left(a_\rho + 2a_\mu \frac{\beta_0}{\alpha_0} \cos\sigma\right). \tag{7.76}$$

In terms of density and velocity perturbations,

$$\begin{aligned} \mathbb{V}_{\mathrm{SVP}}(\boldsymbol{x}, \omega, \sigma) &\equiv \mathbb{A}_\beta{}^{(\mathrm{SVP})} a_\beta + \mathbb{A}_\rho{}^{(\mathrm{SVP})} a_\rho \\ &= -\rho_0 \left|\sin\sigma\right| \left(a_\rho \left(1 + 2\frac{\beta_0}{\alpha_0}\cos\sigma\right) + 2a_\beta \frac{\beta_0}{\alpha_0}\cos\sigma\right). \end{aligned} \tag{7.77}$$

The components of the frequency-reduced potential are illustrated in Figure 7.6. For the SV to P scattering potential, a similar analysis shows

$$\mathbb{V}_{\mathrm{PSV}}(\boldsymbol{x}, \omega, \sigma) = +\rho_0 \left|\sin\sigma\right| \left(a_\rho + 2a_\mu \frac{\beta_0}{\alpha_0}\cos\sigma\right), \tag{7.78}$$

or

$$\mathbb{V}_{\mathrm{PSV}}(\boldsymbol{x}, \omega, \sigma) = +\rho_0 \left|\sin\sigma\right| \left(a_\rho \left(1 + 2\frac{\beta_0}{\alpha_0}\cos\sigma\right) + 2a_\beta \frac{\beta_0}{\alpha_0}\cos\sigma\right). \tag{7.79}$$

It is apparent that $\mathbb{V}_{\mathrm{PSV}}$ is equal to the negative of $\mathbb{V}_{\mathrm{SVP}}$.

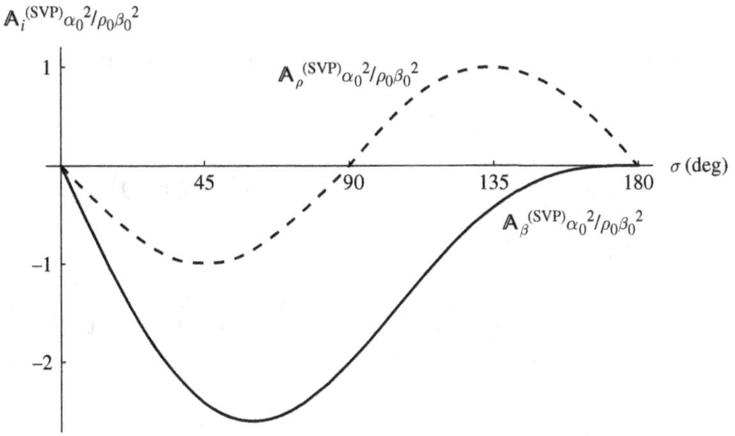

Figure 7.6 SVP scattering potential. The components of the P to SV frequency-reduced scattering potential $\mathbb{V}_{SVP}$, as a function of opening angle σ. The ratio α_0/β_0 has been set to two in this figure. The dashed curve is the coefficient $\mathbb{A}_\beta{}^{(SVP)}\alpha_0^2/\rho_0\beta_0^2$ of the shear velocity perturbation a_β, and the solid curve is the coefficient $\mathbb{A}_\rho{}^{(SVP)}\alpha_0^2/\rho_0\beta_0^2$ of the density perturbation a_ρ. The SV to P to potential $\mathbb{V}_{PSV}$ is proportional to the negative of $\mathbb{V}_{SVP}$.

7.5 Summary

The frequency-reduced scattering potentials discussed in this chapter have all had the form (7.10), or

$$\mathbb{V}(\boldsymbol{x},\sigma) = \sum_i a_i(\boldsymbol{x})\mathbb{A}_i(\boldsymbol{x},\sigma), \tag{7.80}$$

where $a_i(\boldsymbol{x})$ is a (rapidly changing with spatial location $\boldsymbol{x}$) perturbation in a velocity or density, and $\mathbb{A}_i(\boldsymbol{x},\sigma)$ is a (slowly varying in $\boldsymbol{x}$ and opening angle σ) coefficient. Specifically, the potentials for the various wave types are

$$\mathbb{V}_{\text{scalar}}(\boldsymbol{x}) = -\frac{a_c(\boldsymbol{x})}{c_0(\boldsymbol{x})^2} \tag{7.81}$$

for scalar waves,

$$\mathbb{V}_a(\boldsymbol{x},\sigma) = -\frac{a_\rho(\boldsymbol{x})(1+\cos\sigma) + a_c(\boldsymbol{x})}{\rho_0(\boldsymbol{x})c_0{}^2(\boldsymbol{x})} \tag{7.82}$$

for acoustic waves,

$$\mathbb{V}_{PP}(\boldsymbol{x},\sigma) = -\rho_0(\boldsymbol{x}) \tag{7.83}$$
$$\times\left(a_\alpha(\boldsymbol{x}) + a_\rho(\boldsymbol{x})\left(1+\cos\sigma - 2\frac{\beta_0{}^2(\boldsymbol{x})}{\alpha_0{}^2(\boldsymbol{x})}\sin^2\sigma\right) - a_\beta(\boldsymbol{x})2\frac{\beta_0{}^2(\boldsymbol{x})}{\alpha_0{}^2(\boldsymbol{x})}\sin^2\sigma\right)$$

for P to P waves,

$$\mathbb{V}_{\mathrm{SVSV}}(\boldsymbol{x},\sigma) = -\rho_0(\boldsymbol{x})\left(a_\rho(\boldsymbol{x})(\cos\sigma + \cos 2\sigma) + a_\beta(\boldsymbol{x})\cos 2\sigma\right) \tag{7.84}$$

for SV to SV waves,

$$\mathbb{V}_{\mathrm{SHSH}}(\boldsymbol{x}) = \rho_0(\boldsymbol{x})\left(a_\beta \cos\sigma + a_\rho(1+\cos\sigma)\right) \tag{7.85}$$

for SH to SH waves,

$$\mathbb{V}_{\mathrm{SVP}}(\boldsymbol{x},\sigma) = -\rho_0(\boldsymbol{x})\,|\sin\sigma|\left(a_\rho(\boldsymbol{x})\left(1+2\frac{\beta_0(\boldsymbol{x})}{\alpha_0(\boldsymbol{x})}\cos\sigma\right) + a_\beta(\boldsymbol{x})2\frac{\beta_0(\boldsymbol{x})}{\alpha_0(\boldsymbol{x})}\cos\sigma\right) \tag{7.86}$$

for P to SV waves, and

$$\mathbb{V}_{\mathrm{PSV}}\,(\boldsymbol{x},\sigma_0) = \rho_0(\boldsymbol{x})\,|\sin\sigma|\left(a_\rho(\boldsymbol{x})\left(1+2\frac{\beta_0(\boldsymbol{x})}{\alpha_0(\boldsymbol{x})}\cos\sigma\right) + 2a_\beta(\boldsymbol{x})\frac{\beta_0(\boldsymbol{x})}{\alpha_0(\boldsymbol{x})}\cos\sigma\right). \tag{7.87}$$

for SV to P waves. Other converted wave potentials are zero. In the above expressions, c_0, α_0, β_0, and ρ_0 are background (or unperturbed) scalar velocity, P-velocity, S-velocity, and density. The perturbation terms are (to first order)

$$a_c(\boldsymbol{x}) = 1 - \frac{{c_0}^2(\boldsymbol{x})}{c^2(\boldsymbol{x})} = \frac{\Delta c^2(\boldsymbol{x})}{c^2(\boldsymbol{x})} \simeq \frac{\Delta c^2(\boldsymbol{x})}{{c_0}^2(\boldsymbol{x})} \simeq \frac{2\Delta c(\boldsymbol{x})}{c_0(\boldsymbol{x})}, \tag{7.88}$$

$$a_\rho(\boldsymbol{x}) = 1 - \frac{\rho_0(\boldsymbol{x})}{\rho(\boldsymbol{x})} = \frac{\Delta\rho(\boldsymbol{x})}{\rho(\boldsymbol{x})} \simeq \frac{\Delta\rho(\boldsymbol{x})}{\rho_0(\boldsymbol{x})}, \tag{7.89}$$

$$a_\alpha(\boldsymbol{x}) = 1 - \frac{{\alpha_0}^2(\boldsymbol{x})}{\alpha^2(\boldsymbol{x})} = \frac{\Delta\alpha^2(\boldsymbol{x})}{\alpha^2(\boldsymbol{x})} \simeq \frac{\Delta\alpha^2(\boldsymbol{x})}{{\alpha_0}^2(\boldsymbol{x})} \simeq \frac{2\Delta\alpha(\boldsymbol{x})}{\alpha_0(\boldsymbol{x})}, \tag{7.90}$$

and

$$a_\beta(\boldsymbol{x}) = 1 - \frac{{\beta_0}^2(\boldsymbol{x})}{\beta^2(\boldsymbol{x})} = \frac{\Delta\beta^2(\boldsymbol{x})}{\beta^2(\boldsymbol{x})} \simeq \frac{\Delta\beta^2(\boldsymbol{x})}{{\beta_0}^2(\boldsymbol{x})} \simeq \frac{\Delta\beta(\boldsymbol{x})}{\beta_0(\boldsymbol{x})}. \tag{7.91}$$

Exercises

7.1 Suppose the PP scattering potential $\mathbb{V}_{\mathrm{PP}}$ has been determined at three opening angles $\sigma_1 = 0°$, $\sigma_2 = 45°$, and $\sigma_3 = 90°$, to have values $\mathbb{V}_1 = -0.3$, $\mathbb{V}_2 = -0.22071$, and $\mathbb{V}_3 = -0.1$, with background values of the elastic parameters $\rho_0 = 1$ and $\beta_0/\alpha_0 = 0.5$.

(a) From equation (7.58) for the P to P scattering potential, determine the elastic perturbations a_α, a_ρ, and a_β.
(b) What is the sensitivity of the perturbations to an error of 0.001 in any of the three potential values?
(c) What is the sensitivity of the perturbations to a 1 percent error in β_0/α_0?

8
Reflectivity

A complete seismic inversion would solve for the physical parameters in the scattering potential. Most imaging algorithms stop short of that, solving instead for a reflectivity function. To produce the physical model behind the reflectivity, further analysis is required. Similarly, to synthesize seismic data one often starts from a reflectivity rather than a scattering potential.

Inversion for or modeling from a reflectivity model is legitimate, if incomplete, though one does have to ask, just what is reflectivity? The reflection coefficient is a well defined concept for plane waves impinging upon an abrupt boundary of no curvature. Asymptotically, the concept extends to waves far from source or receiver impinging upon boundaries of reasonable curvature. Consequently, when we think about reflectivity we are usually thinking about discrete specular reflections produced at reflecting boundary surfaces. However, changes in real-Earth properties may or may not be abrupt, so confining reflectivity to a boundary surface is at best an idealization, at worst wrong. Moreover, imaging algorithms normally produce images of volumes, not surfaces. That is, they produce a value at every point within a volume representing the Earth's subsurface, not just along selected surfaces within the volume. Imaging algorithms solve for a *pointwise* reflectivity function, not a specular reflectivity function.

This is as it should be. Specular reflectivity is the integrated response of a boundary surface, not the response of a single point on that surface. In the ray-theoretical limit, a specular reflection may appear to come from a single point, but that is an illusion. If an imaging algorithm is solving for properties at a point, it is not producing a specular reflectivity.

8.1 Point reflectivity

In image space, a point reflector would appear as a delta function. That is, if $\boldsymbol{x}_0 = (x_0, y_0, z_0)$ is the reflection point, then

$$\mathcal{R}p\left(\boldsymbol{x}\,|\boldsymbol{x}_0\right) = R\delta\left(z - z_0\right)\delta\left(y - y_0\right)\delta\left(x - x_0\right) \tag{8.1}$$

is a point reflectivity at that point. The amplitude R of a reflecting point will in general depend upon the direction from which the point is approached, more about which later.

Any reflectivity function can be considered to be a superposition of point reflectivities. In particular, consider a specular reflectivity located on the surface $b(\boldsymbol{x}) = 0$. Its image will be

$$\mathcal{R}_s(\boldsymbol{x}) = R\delta\left(\frac{b(\boldsymbol{x})}{|\nabla b|}\right) = R|\nabla b|\delta(b(\boldsymbol{x})). \tag{8.2}$$

This image can be trivially decomposed into point reflectors as

$$\mathcal{R}_s(\boldsymbol{x}) = \int\int\int dx_0 dy_0 dz_0 \mathcal{R}_{\mathcal{P}}\left(\boldsymbol{x}\,|\boldsymbol{x}_0\right)|\nabla_0 b|\,\delta\left(b\left(\boldsymbol{x}_0\right)\right), \tag{8.3}$$

with $\nabla_0 b$ representing the gradient of $b\left(\boldsymbol{x}_0\right)$ with respect to $\boldsymbol{x}_0$.

Point reflectivity is less intuitive and more difficult to define than specular reflectivity. A specular reflection can be produced by a simple step function change in seismic properties. For example, the scalar reflection (5.13) is produced at an interface at depth z, when velocity above z is c and velocity beneath z is c'. There is no way, however, for a step function to hide behind a point reflector. Whatever produces a point reflection, it cannot be a step function.

One way to view point reflectivity is as a purely mathematical construct. A point reflectivity can be written as a superposition of specular reflectivity functions (see Appendix D). This implies that one can form a basis of specular reflectivity functions from which any possible image, including a point reflectivity function, can be constructed. From this point of view, specular reflectivity is the fundamental entity, and pointwise reflectivity a derived quantity.

One can argue that seismic data lend themselves more readily to the reflectivity representation of Earth properties than to the scattering potential. For conventionally processed seismic data, a reflection appears as a (bandlimited) delta function. At a reflecting boundary, the scattering potential, on the other hand, will appear as a (bandlimited) step function. If migration is to preserve the delta-function character of reflections, it must solve not for the potential $\mathcal{V}$, but for a reflectivity function $\mathcal{R}$.

A step function can be converted into a delta function by differentiation. This suggests that a scattering potential can be related to an associated pointwise reflectivity function by taking its normal derivative or gradient magnitude; i.e.

$$\mathcal{R} \propto \frac{\partial \mathcal{V}}{\partial m} = \hat{\boldsymbol{m}} \cdot \nabla \mathcal{V}, \tag{8.4}$$

with $\hat{\boldsymbol{m}}$ a unit vector pointing in the (downward) direction of the potential gradient. If this is the case, then within the context of the Born approximation, the gradient magnitude can be evaluated. Look at the Born-approximation integral (5.59), with $\mathcal{V}$ replaced by $\nabla\mathcal{V}$:

$$\mathcal{U} = \mathfrak{w}(\omega)\int\int\int d^3x G_0\left(\tilde{\boldsymbol{x}}_g\middle|\boldsymbol{x};\omega\right)\nabla\mathcal{V}(\boldsymbol{x},\omega)G_0\left(\boldsymbol{x}\left|\tilde{\boldsymbol{x}}_s;\omega\right.\right). \tag{8.5}$$

If the Green's functions are plane waves or local approximations to plane waves, so that (7.4) holds, then (8.5) can be integrated by parts to obtain

$$\mathcal{U} = -i\mathfrak{w}(\omega)\int\int\int d^3x G_0\left(\tilde{\boldsymbol{x}}_g\middle|\boldsymbol{x};\omega\right)\mathcal{V}(\boldsymbol{x},\ \omega)G_0\left(\boldsymbol{x}\left|\tilde{\boldsymbol{x}}_s;\omega\right.\right)\left(\boldsymbol{k}_i(\boldsymbol{x})-\boldsymbol{k}_r(\boldsymbol{x})\right), \tag{8.6}$$

where $\boldsymbol{k}_i$ and $\boldsymbol{k}_r$ are the local wavenumber vectors of the incident and reflected Green's functions. Thus, inside the integral,

$$\nabla\mathcal{V}(\boldsymbol{x}) \to -i\left(\boldsymbol{k}_i(\boldsymbol{x})-\boldsymbol{k}_r(\boldsymbol{x})\right)\mathcal{V}(\boldsymbol{x}). \tag{8.7}$$

The vector $\hat{\boldsymbol{m}}$ points in the direction $\boldsymbol{k}_i(\boldsymbol{x})-\boldsymbol{k}_r(\boldsymbol{x})$, whence

$$\frac{\partial\mathcal{V}}{\partial m} \to -i\left|\boldsymbol{k}_i(\boldsymbol{x})-\boldsymbol{k}_r(\boldsymbol{x})\right|\mathcal{V}. \tag{8.8}$$

Strictly, it is not true that the left and right sides of (8.8) are equal. However, inside the Born integral, the two sides are equivalent and interchangeable.

If the incoming and reflected waves are of the same type, then they travel at the same velocity c_0, and $k_i = k_r = \omega/c_0$. In this case, $|\boldsymbol{k}_i - \boldsymbol{k}_r| = \omega\sqrt{2(1+\cos\sigma)}/c_0$, where σ is the opening angle between the two waves. For converted waves, where one wave travels at velocity α_0 and one at β_0, $|\boldsymbol{k}_i - \boldsymbol{k}_r| = 2\omega c_e(\sigma)/\alpha_0\beta_0$, with $c_e(\sigma) = \sqrt{\alpha_0{}^2+\beta_0{}^2+2\alpha_0\beta_0\cos\sigma}/2$ an angle-dependent average velocity introduced in equation (5.38). In either case, if reflectivity is proportional to the normal derivative of the potential, it is proportional to the potential also. Summarizing, if $\gamma = \sigma/2$,

$$\begin{aligned}\frac{\partial\mathcal{V}}{\partial m} &\to -\frac{i\omega\sqrt{2(1+\cos\sigma)}}{c_0}\mathcal{V} = -\frac{2i\omega\cos\gamma}{c_0}\mathcal{V} \quad \text{(non-converted waves)},\\ &\to -\frac{2i\omega c_e(\sigma)}{\alpha_0\beta_0}\mathcal{V} \quad \text{(converted waves)}.\end{aligned} \tag{8.9}$$

8.2 A scalar reflectivity function

The scattering potential and reflectivity representations of reflections are not inconsistent (Stolt & Weglein, 1985; Stolt & Benson, 1986; Bleistein, 1987). Returning to the scalar wave example, suppose a reflecting boundary exists on the plane

$$\hat{\boldsymbol{m}} \cdot (\boldsymbol{x} - \boldsymbol{x}_r) = 0, \tag{8.10}$$

where $\hat{\boldsymbol{m}}$ is a unit vector in the direction pointing downward normal to the plane, and $\boldsymbol{x}_r$ is a point on the plane. Above the plane, $c = c_0$, and below the plane, $c = c_1$.

The reflectivity at the boundary depends on angle of incidence γ_0 and angle of refraction γ_1 (see equation (5.13)):

$$\mathcal{R}_{\text{scalar}}(\boldsymbol{x}, \sigma) = \frac{c_1 \cos\gamma_0 - c_0 \cos\gamma_1}{c_1 \cos\gamma_0 + c_0 \cos\gamma_1} \delta\left(\hat{\boldsymbol{m}} \cdot (\boldsymbol{x} - \boldsymbol{x}_r)\right), \tag{8.11}$$

where $\gamma_0 = \sigma/2$, and

$$\sin\gamma_1 = \frac{c_1}{c_0} \sin\gamma_0. \tag{8.12}$$

Following (7.15), define the velocity perturbation a_{c0} as

$$a_{c0} \equiv \left(1 - \frac{{c_0}^2}{{c_1}^2}\right) = \frac{\Delta c^2}{{c_1}^2} \simeq 2\frac{\Delta c}{c}. \tag{8.13}$$

Expanding $\mathcal{R}_{scalar}$ in powers of a_{c0}, we get

$$\mathcal{R}_{\text{scalar}}(\boldsymbol{x}, \sigma) = \left[\frac{a_{c0}}{4\cos^2\gamma_0} + \mathcal{O}\left({a_{c0}}^2\right)\right] \delta\left(\hat{\boldsymbol{m}} \cdot (\boldsymbol{x} - \boldsymbol{x}_r)\right). \tag{8.14}$$

This approximation is illustrated in Figure 8.1, along with the exact scalar reflection coefficient for a 5 percent velocity increase. The linear approximation breaks

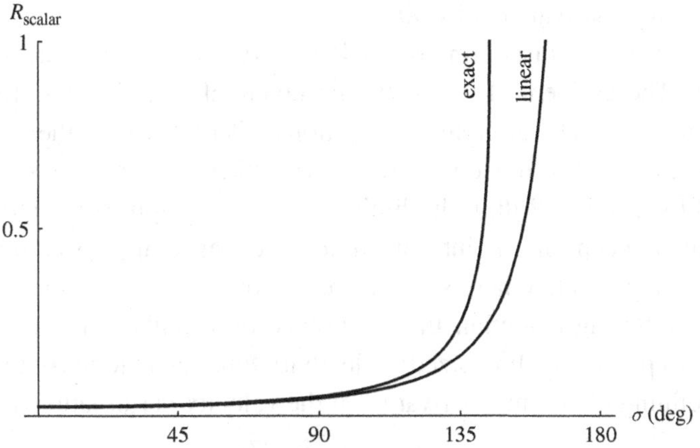

Figure 8.1 The scalar reflectivity function as a function of opening angle for a boundary with a 5 percent velocity increase. Both the exact and linearized reflectivity are depicted. The two curves begin to diverge near $\sigma = 90°$. The exact curve approaches 1 at critical angle (about 145° in this case) and is complex beyond it, while the linearized curve remains real and increasing for all $\sigma < 180°$.

down near critical angle (σ about 145° in this example), but is otherwise a good match.

Now look at the scalar scattering potential $\mathcal{V}_{\text{scalar}}$. It is zero above the plane. Beneath the plane (see equation (7.16)),

$$\mathcal{V}_{\text{scalar}} = \omega^2 \mathbb{V}_{\text{scalar}} = -\frac{\omega^2}{c_0{}^2} a_{c0}. \tag{8.15}$$

The potential is proportional to a step function at the boundary, whereas the reflectivity function is proportional to a delta function. The derivative of $\mathcal{V}_{\text{scalar}}$ with respect to the direction $\hat{\boldsymbol{m}}$ normal to the boundary is

$$\frac{\partial \mathcal{V}_{\text{scalar}}}{\partial m} \equiv \hat{\boldsymbol{m}} \cdot \nabla \mathcal{V}_{\text{scalar}} = -\frac{\omega^2}{c_0{}^2} a_{c0} \delta \left(\hat{\boldsymbol{m}} \cdot (\boldsymbol{x} - \boldsymbol{x}_r) \right), \tag{8.16}$$

which is closer to the form of the reflectivity. Comparing reflectivity to the normal derivative of scattering potential (8.16) leads to the relation

$$\mathcal{R}_{\text{scalar}}(\boldsymbol{x}, \sigma) = -\frac{c_0{}^2}{\omega^2} \frac{\partial \mathcal{V}_{\text{scalar}} / \partial m}{4 \cos^2 \gamma_0} + \mathcal{O}\left(a_{c0}{}^2 \right). \tag{8.17}$$

To first order in the perturbation a_{c0}, reflectivity is proportional to the normal derivative of the scattering potential. With the substitution (8.9), we relate reflectivity directly to the potential:

$$\mathcal{R}_{\text{scalar}}(\boldsymbol{x}, \sigma) \simeq \frac{i \omega c_0}{2 \cos \gamma_0} \mathbb{V}_{\text{scalar}}. \tag{8.18}$$

The linearized scalar reflectivity is proportional to the scalar scattering potential, rescaled and with phase altered by 90°.

Suppose, instead of a discontinuity at a boundary, that velocity varies rapidly but continuously. The gradient of the scattering potential loses its delta function and becomes continuous. The discrete perturbation a_0 is replaced by the log derivative of c^2, i.e. by $2\nabla\, c/c$. The reflectivity function is likewise continuous, and relation (8.17) should still hold. Within the limits of the linear approximation, scattering theory ties the concept of a point reflectivity to changes in physical properties in the vicinity of the point. Where specular reflections exist, equation (8.18) relates them to the scattering potential. In the absence of specular reflections, equation (8.18) defines a pointwise (linearized) reflectivity function in terms of the scattering potential. Defining the normal derivative of the velocity perturbation as

$$b_c \equiv \frac{d a_c}{d m} \simeq 2 \frac{\hat{\boldsymbol{m}} \cdot \nabla c}{c}, \tag{8.19}$$

the scalar reflectivity can be written to first order as

$$\mathcal{R}_{\text{scalar}}(\boldsymbol{x}, \sigma) \simeq \frac{b_c(\boldsymbol{x})}{4 \cos^2 \gamma} \simeq \frac{i \omega c_0}{2 \cos \gamma} \mathbb{V}_{\text{scalar}}(\boldsymbol{x}). \tag{8.20}$$

Using (8.18), the Born approximation integral (5.59) for scalar waves can be expressed in terms of reflectivity:

$$D_{\text{scalar}}\left(\tilde{\boldsymbol{x}}_g|\tilde{\boldsymbol{x}}_s;\omega\right) = -2i\omega\mathfrak{w}(\omega)\int d^3x G_{\text{scalar}}\left(\tilde{\boldsymbol{x}}_g|\boldsymbol{x};\omega\right)$$
$$\times\, \mathcal{R}_{\text{scalar}}(\boldsymbol{x},\sigma)\frac{\cos\gamma}{c_0}G_{\text{scalar}}\left(\boldsymbol{x}|\tilde{\boldsymbol{x}}_s;\omega\right). \qquad (8.21)$$

This formula is a straightforward prescription for generating scalar-wave reflection data from a reflectivity function.

8.3 An acoustic reflectivity function

The acoustic specular reflection coefficient (5.16) at a reflecting boundary is

$$R_a(\gamma_0) = \left(\frac{\rho_1 c_1\cos\gamma_0 - \rho_0 c_0\cos\gamma_1}{\rho_1 c_1\cos\gamma_0 + \rho_0 c_0\cos\gamma_1}\right), \qquad (8.22)$$

with $\gamma_0 = \sigma/2$ the incidence angle and γ_1 the refraction angle given by

$$\frac{\sin\gamma_1}{c_1} = \frac{\sin\gamma_0}{c_0}. \qquad (8.23)$$

Define velocity and density perturbations at the boundary as (see equations (7.20))

$$a_\rho(\boldsymbol{x}) \equiv 1 - \frac{\rho_0}{\rho_1} \simeq \frac{\Delta\rho}{\rho}, \quad a_c(\boldsymbol{x}) \equiv 1 - \frac{{c_0}^2}{{c_1}^2} \simeq 2\frac{\Delta c}{c}. \qquad (8.24)$$

The reflection coefficient can be written as a power series in a_ρ and a_c. Keeping only the lowest order term,

$$R_a \simeq \frac{a_\rho(1+\cos\sigma)+a_c}{2(1+\cos\sigma)} = \frac{1}{2}\left(a_\rho + \frac{a_c}{1+\cos\sigma}\right). \qquad (8.25)$$

Repeating the analysis of Section 8.2, if $\hat{\boldsymbol{m}}$ is a direction vector normal to the boundary, the linearized reflectivity function is

$$\mathcal{R}_a(\boldsymbol{x},\sigma) \simeq \frac{1}{2}\left(a_\rho + \frac{a_c}{1+\cos\sigma}\right)\delta\left(\hat{\boldsymbol{m}}\cdot(\boldsymbol{x}-\boldsymbol{x}_r)\right). \qquad (8.26)$$

The density and velocity components of this function are shown in Figure 8.2.

The normal derivative of the acoustic scattering potential (7.24) across the reflecting boundary is

$$\frac{\partial \mathbb{V}_a}{\partial m}(\boldsymbol{x},\sigma) = -\frac{\delta\left(\hat{\boldsymbol{m}}\cdot(\boldsymbol{x}-\boldsymbol{x}_r)\right)}{\rho_0 {c_0}^2}\left(a_\rho(1+\cos\sigma)+a_c\right), \qquad (8.27)$$

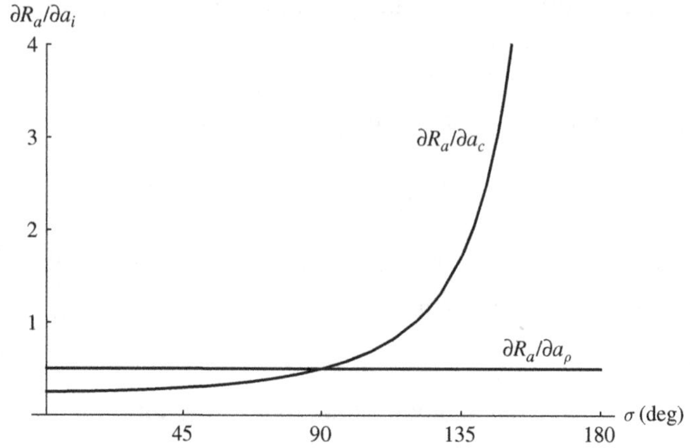

Figure 8.2 The components of the linearized acoustic reflectivity, as a function of opening angle σ. The line curving upward is the coefficient $\partial R_a / \partial a_c$ of the velocity perturbation a_c, and the flat line is the coefficient $\partial R_a / \partial a_\rho$ of the density perturbation a_ρ.

Comparing the equations for the reflectivity and the potential normal derivative, we find

$$\mathcal{R}_a(\boldsymbol{x}, \sigma) \simeq -\frac{c_0{}^2 \mathcal{C}_a{}^{-2}}{4\cos^2\gamma_0} \frac{\partial \mathbb{V}_a}{\partial m}(\boldsymbol{x}, \gamma_0), \tag{8.28}$$

where $\mathcal{C}_a = \rho_0{}^{-1/2}$. Using (8.9), we can replace the normal derivative of $\mathbb{V}_a$ with $-\frac{2i\omega\cos\gamma}{c_0}\mathbb{V}_a$, to obtain

$$\mathcal{R}_a(\boldsymbol{x}, \sigma) \simeq \frac{i\omega c_0 \mathcal{C}_a{}^{-2}}{2\cos\gamma_0}\mathbb{V}_a(\boldsymbol{x}, \sigma). \tag{8.29}$$

Except for the factor of density or $\mathcal{C}_a{}^{-2}$, the relation between reflectivity and scattering potential is the same as the scalar relation (8.17). The factor of density appears because we have normalized the acoustic wave equation differently than we have the scalar wave equation, so that the scalar potential has dimension of one over distance squared while the acoustic potential has dimensionality of distance over mass. Reflectivity, in either case, must have dimensionality of inverse distance.

Though (8.29) was derived for the case of a step-function discontinuity, we use it to define reflectivity in a general acoustic medium. It may be used to replace $\mathbb{V}_a$ in the Born integral with an acoustic reflectivity function.

Alternatively, we can write the reflectivity in terms of normal derivatives of the perturbations. Defining

$$b_\rho \equiv \frac{da_\rho}{dm} \simeq \frac{\hat{\boldsymbol{m}} \cdot \nabla \rho}{\rho}, \tag{8.30}$$

we have

$$\mathcal{R}_a(\boldsymbol{x}, \sigma) \simeq \frac{b_c(\boldsymbol{x}) + b_\rho(\boldsymbol{x})(1 + \cos\sigma)}{4\cos^2\gamma}. \tag{8.31}$$

8.4 Elastic reflectivity functions

8.4.1 P to P reflectivity

Across a boundary, the linearized PP reflection coefficient (see equation 5.44 of Aki & Richards, 1980), in the current notation is

$$R_{\mathrm{PP}}(\sigma) \simeq \left(\frac{1}{2} - \frac{{\beta_0}^2}{{\alpha_0}^2}(1 - \cos\sigma)\right) a_\rho + \frac{1}{2(1 + \cos\sigma)} a_\alpha - \frac{{\beta_0}^2}{{\alpha_0}^2}(1 - \cos\sigma) a_\beta. \tag{8.32}$$

The components of this function are presented in Figure 8.3.

Comparing $\mathbb{V}_{\mathrm{PP}}$ (7.58) to the linearized reflection coefficient (8.32), applying the same argument as for the scalar and acoustic cases, we see that the P to P reflectivity function in this case is

$$\mathcal{R}_{\mathrm{PP}}(\boldsymbol{x}, \sigma) \simeq -\frac{{\alpha_0}^2 \mathcal{C}_P{}^{-2}}{4\cos^2\gamma} \frac{\partial \mathbb{V}_{\mathrm{PP}}}{\partial m}(\boldsymbol{x}, \sigma) \simeq i \frac{\omega \alpha_0 \mathcal{C}_P{}^{-2}}{2\cos\gamma} \mathbb{V}_{\mathrm{PP}}(\boldsymbol{x}, \sigma), \tag{8.33}$$

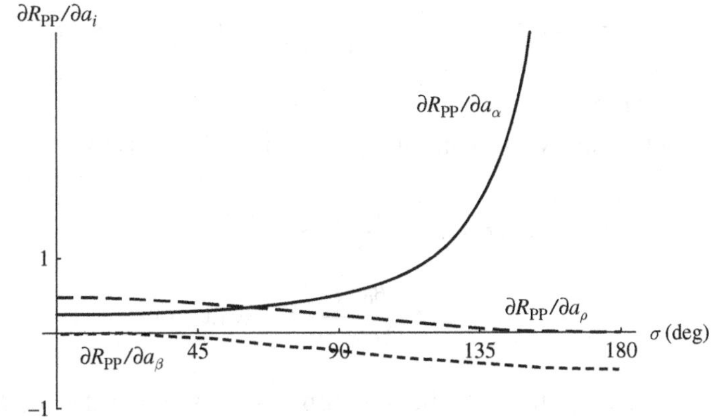

Figure 8.3 The components of the P to P reflectivity R_{PP}, as a function of opening angle σ. In this illustration, we have chosen the background ratio β_0/α_0 of S to P velocity to be 1/2. The finely dashed curve is the coefficient $\partial R_{\mathrm{PP}}/\partial a_\beta$ of the S-velocity component a_β, the solid line is the coefficient $\partial R_{\mathrm{PP}}/\partial a_\alpha$ of the P-velocity perturbation a_α, and the coarsely dashed curve is the coefficient $\partial R_{\mathrm{PP}}/\partial a_\rho$ of the density perturbation a_ρ.

which, except for a factor of $\mathcal{C}_P{}^2 = \rho_0\alpha_0{}^2$, is the same formula as (8.17) linking the scalar wave equation reflectivity and scattering potential, and except for the factor of density, is the same as the acoustic formula (8.28). We are at liberty to use equation (8.33) as a definition of P to P reflectivity.

We may also define normal derivatives of the perturbations. With

$$b_\alpha \equiv \frac{da_\alpha}{dm} \simeq \frac{\hat{\boldsymbol{m}} \cdot \nabla\alpha}{\alpha}, \tag{8.34}$$

$$b_\beta \equiv \frac{da_\beta}{dm} \simeq \frac{\hat{\boldsymbol{m}} \cdot \nabla\beta}{\beta}, \tag{8.35}$$

the linearized PP reflectivity function is

$$\begin{aligned}\mathcal{R}_{\mathrm{PP}}(\boldsymbol{x},\sigma) \simeq &\left(\frac{1}{2} - \frac{\beta_0{}^2(\boldsymbol{x})}{\alpha_0{}^2(\boldsymbol{x})}(1-\cos\sigma)\right) b_\rho(\boldsymbol{x}) \\ &+ \frac{b_\alpha(\boldsymbol{x})}{2(1+\cos\sigma)} - \frac{\beta_0{}^2(\boldsymbol{x})}{\alpha_0{}^2(\boldsymbol{x})}(1-\cos\sigma) b_\beta(\boldsymbol{x}).\end{aligned} \tag{8.36}$$

8.4.2 SV to SV reflectivity

The linearized SVSV reflection coefficient across a boundary is, according to equation 5.44, Aki and Richards (1980),

$$R_{\mathrm{SVSV}}(\sigma) \simeq \frac{a_\rho}{2}(2\cos\sigma - 1) + \frac{a_\beta}{2(1+\cos\sigma)}\left(2\cos^2\sigma - 1\right), \tag{8.37}$$

as depicted in Figure 8.4.

Comparison with the SVSV scattering potential (7.67) leads to

$$\begin{aligned}\mathcal{R}_{\mathrm{SVSV}}(\boldsymbol{x},\sigma) &\simeq -\frac{\beta_0{}^2\mathcal{C}_S{}^{-2}}{4\cos^2\gamma}\frac{\partial\mathbb{V}_{\mathrm{SVSV}}}{\partial m}(\boldsymbol{x},\sigma) \\ &\simeq \boldsymbol{i}\frac{\omega\beta_0\mathcal{C}_S{}^{-2}}{2\cos\gamma}\mathbb{V}_{\mathrm{SVSV}}(\boldsymbol{x},\sigma),\end{aligned} \tag{8.38}$$

with $\mathcal{C}_S{}^2 = \rho_0\beta_0{}^2$. One shouldn't be too surprised to see that the relation is the same as the PP relation (8.33), with S-wave velocities and incident angle instead of P-wave.

As a linearized function of the perturbation derivatives, the reflectivity is

$$\mathcal{R}_{\mathrm{SVSV}}(\boldsymbol{x},\sigma) \simeq \frac{b_\rho(\boldsymbol{x})}{2}(2\cos\sigma - 1) + \frac{b_\beta(\boldsymbol{x})}{2(1+\cos\sigma)}\left(2\cos^2\sigma - 1\right). \tag{8.39}$$

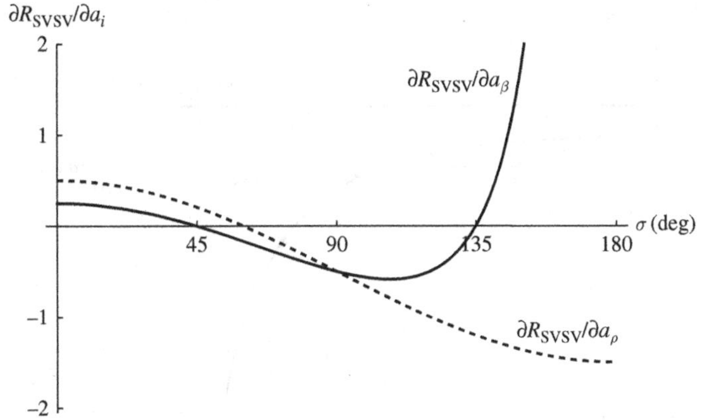

Figure 8.4 The components of the SV to SV reflectivity R_{SVSV}, as a function of opening angle σ. The solid line is the coefficient $\rho_0{}^{-1}\partial R_{\mathrm{SVSV}}/\partial a_\beta$ of the shear velocity perturbation a_β, and the dashed line is the coefficient $\rho_0{}^{-1}\partial R_{\mathrm{SVSV}}/\partial a_\rho$ of the density perturbation a_ρ.

8.4.3 SH to SH reflectivity

According to equation 5.32, Aki and Richards (1980), the plane-layer SH to SH reflection coefficient is

$$R_{\mathrm{SHSH}}(\gamma_0) = \left(\frac{\rho_1\beta_1\cos\gamma_1 - \rho_0\beta_0\cos\gamma_0}{\rho_1\beta_1\cos\gamma_1 + \rho_0\beta_0\cos\gamma_0}\right), \tag{8.40}$$

with γ_1 related to γ_0 by Snell's law. Expanding in powers of a_ρ and a_β, and retaining only the lowest-order terms,

$$R_{\mathrm{SHSH}}(\gamma_0) \simeq \frac{1}{2}\left(a_\rho + a_\beta\frac{\cos\sigma}{1+\cos\sigma}\right). \tag{8.41}$$

The components of the linearized reflectivity R_{SHSH} are shown in Figure 8.5.

The SH to SH scattering potential is given by (7.62). This gives the SH to SH reflectivity function a form identical to (8.38) :

$$\mathcal{R}_{\mathrm{SHSH}}(\boldsymbol{x},\sigma) \simeq -\frac{\beta_0{}^2\mathcal{C}_S{}^{-2}}{4\cos^2\gamma}\frac{\partial\mathbb{V}_{\mathrm{SHSH}}}{\partial m}(\boldsymbol{x}) \simeq \frac{i\omega\beta_0}{2\mathcal{C}_S{}^2\cos\gamma}\mathbb{V}_{\mathrm{SHSH}}(\boldsymbol{x}). \tag{8.42}$$

In terms of normal derivatives of the perturbations, the SH to SH reflectivity is

$$\mathcal{R}_{\mathrm{SHSH}}(\boldsymbol{x},\sigma) = \frac{b_\rho(1+\cos\sigma) + b_\beta\cos\sigma}{4\cos^2\gamma}. \tag{8.43}$$

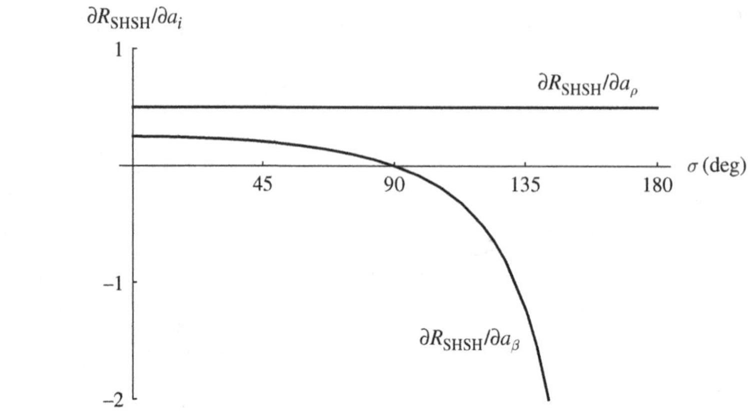

Figure 8.5 The components of the SH to SH reflectivity R_{SHSH}, as a function of opening angle σ. The lower downward curving line is the coefficient $\partial R_{\text{SHSH}}/\partial a_\beta$ of the shear velocity perturbation a_β, and the upper flat line is the coefficient $\partial R_{\text{SHSH}}/\partial a_\rho$ of the density perturbation a_ρ.

8.4.4 P to SV reflectivity

According to equation 5.44 of Aki and Richards (1980), the linearized reflection coefficient for a wave converting from P to SV at a planar boundary is

$$\mathcal{R}_{\text{SVP}}(\boldsymbol{x},\sigma) = \frac{|\sin\sigma|}{4\cos\gamma_S}\frac{\alpha_0}{c_e(\sigma)}\left(a_\rho\left(1+2\frac{\beta_0}{\alpha_0}\cos\sigma\right)+2a_\beta\frac{\beta_0}{\alpha_0}\cos\sigma\right) \times \delta\left(\hat{\boldsymbol{m}}\cdot(\boldsymbol{x}-\boldsymbol{x}_r)\right), \tag{8.44}$$

with σ the opening angle between the incident P wave and the reflected SV wave, $\gamma_S = \arccos(\frac{\alpha_0+\beta_0\cos\sigma}{2c_e(\sigma)})$ the angle between the surface normal $\hat{\boldsymbol{m}}$ and the reflected SV wave, and $c_e(\sigma)$ the angle-dependent average velocity $\sqrt{\alpha_0{}^2+\beta_0{}^2+2\alpha_0\beta_0\cos\sigma}\,/2$ (see equation (5.38)). The components of this function are illustrated in Figure 8.6. The corresponding scattering potential is

$$\mathbb{V}_{\text{SVP}}(\boldsymbol{x},\sigma) = -\rho_0|\sin\sigma|\left(a_\rho\left(1+2\frac{\beta_0}{\alpha_0}\cos\sigma\right)+2a_\beta\frac{\beta_0}{\alpha_0}\cos\sigma\right)\Theta\left(\hat{\boldsymbol{m}}\cdot(\boldsymbol{x}-\boldsymbol{x}_r)\right). \tag{8.45}$$

The normal derivative of the scattering potential is

$$\frac{\partial}{\partial m}\mathbb{V}_{\text{SVP}}(\boldsymbol{x},\sigma) = -\rho_0|\sin\sigma|\left(a_\rho\left(1+2\frac{\beta_0}{\alpha_0}\cos\sigma\right)+2a_\beta\frac{\beta_0}{\alpha_0}\cos\sigma\right) \times \delta\left(\hat{\boldsymbol{m}}\cdot(\boldsymbol{x}-\boldsymbol{x}_r)\right), \tag{8.46}$$

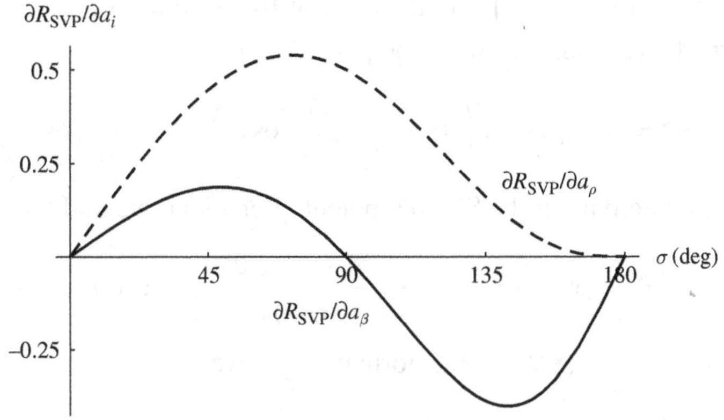

Figure 8.6 The components of the P to SH reflectivity R_{SVP}, as a function of opening angle σ. The lower (solid) line is the coefficient $\partial R_{\text{SVP}}/\partial a_\beta$ of the shear velocity perturbation a_β, and the upper (dashed) line is the coefficient $\partial R_{\text{SVP}}/\partial a_\rho$ of the density perturbation a_ρ.

so that

$$R_{\text{SVP}}(\boldsymbol{x},\sigma) \simeq -\frac{\alpha_0}{4c_e(\sigma)\rho_0 \cos\gamma_S} \frac{\partial}{\partial m} \mathbb{V}_{\text{SVP}}(\boldsymbol{x},\sigma). \tag{8.47}$$

We also have the relation (8.9), or

$$\frac{\partial}{\partial m}\mathbb{V}_{\text{SVP}}(\boldsymbol{x},\sigma) = -\frac{2i\omega c_e(\sigma)}{\alpha_0\beta_0}\mathbb{V}_{\text{SVP}}(\boldsymbol{x},\sigma), \tag{8.48}$$

from which it follows

$$R_{\text{SVP}}(\boldsymbol{x},\sigma) \simeq \frac{i\omega}{2\rho_0\beta_0\cos\gamma_S}\mathbb{V}_{\text{SVP}}(\boldsymbol{x},\sigma). \tag{8.49}$$

In terms of the perturbation derivatives, the P to SV reflectivity is

$$\mathcal{R}_{\text{SVP}}(\boldsymbol{x},\sigma) \simeq \frac{|\sin\sigma|}{4\cos\gamma_S}\frac{\alpha_0(\boldsymbol{x})}{c_e(\sigma)}\left(\left(1+2\frac{\beta_0(\boldsymbol{x})}{\alpha_0(\boldsymbol{x})}\cos\sigma\right)b_\rho(\boldsymbol{x}) + 2\frac{\beta_0(\boldsymbol{x})}{\alpha_0(\boldsymbol{x})}\cos\sigma\, b_\beta(\boldsymbol{x})\right). \tag{8.50}$$

8.4.5 SV to P reflectivity

The linearized reflection coefficient for a wave converting from SV to P at a planar reflector is

$$R_{\text{PSV}}(\boldsymbol{x},\sigma) = -\frac{|\sin\sigma|}{4\cos\gamma_P}\frac{\beta_0}{c_e(\sigma)}\left(a_\rho\left(1+2\frac{\beta_0}{\alpha_0}\cos\sigma\right)+2a_\beta\frac{\beta_0}{\alpha_0}\cos\sigma\right), \tag{8.51}$$

with $\gamma_P = \arccos\left(\frac{\beta_0+\alpha_0\cos\sigma}{2c_e(\sigma)}\right)$ the angle made by the reflected P wave with the reflector normal. We compare this to the potential (7.79), or

$$\mathbb{V}_{\text{PSV}}(\boldsymbol{x},\omega,\sigma) = +\rho_0|\sin\sigma|\left(a_\rho\left(1+2\frac{\beta_0}{\alpha_0}\cos\sigma\right)+2a_\beta\frac{\beta_0}{\alpha_0}\cos\sigma\right). \tag{8.52}$$

Comparing reflectivity to the SV to P potential gradient, we find

$$\mathcal{R}_{\text{PSV}}(\boldsymbol{x},\ \sigma) = -\frac{\beta_0}{4\rho_0 c_e(\sigma)\cos\gamma_{\text{P}}}\frac{\partial\mathbb{V}_{\text{PSV}}}{\partial m}(\boldsymbol{x},\sigma), \tag{8.53}$$

or, with the substitution (8.9) for the normal derivative,

$$\mathcal{R}_{\text{PSV}}(\boldsymbol{x},\ \sigma) = \frac{i\omega}{2\rho_0\alpha_0\cos\gamma_{\text{P}}}\mathbb{V}_{\text{PSV}}(\boldsymbol{x},\sigma). \tag{8.54}$$

The two converted-wave reflectivity formulas (8.44) and (8.54) represent slight generalizations of the formulas which preserve wave types. If one sets $\alpha_0 = \beta_0$ in (8.44) or (8.54), the expressions reduce, except for the $\mathcal{C}$-factor which differs case by case, to the same formula as the others.

In terms of the perturbation derivatives, the SV to P reflectivity is

$$\mathcal{R}_{\text{PSV}}(\boldsymbol{x},\ \sigma) \simeq -\frac{|\sin\sigma|\ \beta_0(\boldsymbol{x})}{4\cos\gamma_P\ c_e(\sigma)}\left(\left(1+2\frac{\beta_0(\boldsymbol{x})}{\alpha_0(\boldsymbol{x})}\cos\sigma\right)b_\rho(\boldsymbol{x})\right.$$
$$\left.+\ 2\frac{\beta_0(\boldsymbol{x})}{\alpha_0(\boldsymbol{x})}\cos\sigma\, b_\beta(\boldsymbol{x})\right). \tag{8.55}$$

8.5 A general formula relating scattering potential and reflectivity

For the wave equations studied here, reflectivity in the frequency domain has been found to be effectively proportional to the scattering potential. The general form for the relation has been

$$\mathcal{R}_{ri}(\boldsymbol{x},\ \sigma) \simeq \frac{ic_r\omega}{2\mathcal{C}_r{}^2\cos\gamma_r}\mathbb{V}_{ri}(\boldsymbol{x},\sigma). \tag{8.56}$$

where $\mathcal{C}$ is a factor that gives reflectivity the units of inverse depth. The quantity c_i is the velocity of the incoming wave, and $\mathcal{C}_i$ is the factor appropriate to the incoming wave. For the scalar wave equation,

$$\mathcal{C}_{\text{scalar}} = 1. \tag{8.57}$$

For the acoustic wave equation,

$$\mathcal{C}_a = \rho_0{}^{-1/2}. \tag{8.58}$$

For the P-wave equation,

$$\mathcal{C}_P = {\rho_0}^{1/2}\alpha_0, \tag{8.59}$$

while for shear waves

$$\mathcal{C}_S = {\rho_0}^{1/2}\beta_0. \tag{8.60}$$

For either type of elastic wave, we can write

$$\mathcal{C}_i = {\rho_0}^{1/2}c_i, \tag{8.61}$$

where c_i represents the velocity of the incoming wave.

For reflections that preserve wave type, the angle γ_r is half the opening angle σ. For converted-wave reflections, γ_r is the reflected-wave angle γ_P or γ_S, related to the opening angle by (5.42).

It would be possible to normalize the different wave operators, and consequently the scattering potentials and Green's functions, to have the same dimensionality. As long as one recognizes that such renormalization alters the measured wave attribute, no harm is done. However, we chose to allow each wave equation to find its own dimension.

Because the reflectivity and scattering potential are effectively proportional, the Born approximation (5.59) for any pair of wave types can be written in terms of reflectivity:

$$D_{ri}\left(\tilde{\boldsymbol{x}}_g|\tilde{\boldsymbol{x}}_s;\omega\right) = -i\omega\mathfrak{w}(\omega)\int\int\int d^3x \tag{8.62}$$

$$\times\, G_{0r}\left(\tilde{\boldsymbol{x}}_g|\boldsymbol{x};\omega\right)\mathcal{R}_{ri}(\boldsymbol{x},\sigma)\frac{2{\mathcal{C}_i}^2\cos\gamma_r}{c_i}G_{0i}\left(\boldsymbol{x}\,|\tilde{\boldsymbol{x}}_s;\omega\right).$$

8.6 Summary of linearized reflectivity functions

Similar to the scattering potential, the linearized reflectivity functions for the various wave types examined here can be expressed as a sum of rapidly varying perturbations $b_i(\boldsymbol{x})$ weighted by slowly varying coefficients $\mathbb{B}_i(\boldsymbol{x},\sigma)$.

$$\mathcal{R}(\boldsymbol{x},\sigma) \simeq \sum_i b_i(\boldsymbol{x})\mathbb{B}_i(\boldsymbol{x},\sigma), \tag{8.63}$$

The perturbations depend only on spatial location $\boldsymbol{x}$, whereas the coefficients depend also on opening angle σ. Specifically, the scalar reflectivity is (8.20)

$$\mathcal{R}_{\text{scalar}}(\boldsymbol{x},\sigma) \simeq \frac{b_c(\boldsymbol{x})}{4\cos^2\gamma}, \tag{8.64}$$

acoustic reflectivity is (8.31)

$$\mathcal{R}_a(\boldsymbol{x},\sigma) \simeq \frac{b_c(\boldsymbol{x}) + b_\rho(\boldsymbol{x})(1+cos\sigma)}{4\cos^2\gamma}, \tag{8.65}$$

P to P reflectivity is (8.36)

$$\mathcal{R}_{\mathrm{PP}}(\boldsymbol{x},\ \sigma) \simeq \left(\frac{1}{2} - \frac{{\beta_0}^2(\boldsymbol{x})}{{\alpha_0}^2(\boldsymbol{x})}(1-\cos\sigma)\right) b_\rho(\boldsymbol{x}) + \frac{b_\alpha(\boldsymbol{x})}{2(1+\cos\sigma)} - \frac{{\beta_0}^2(\boldsymbol{x})}{{\alpha_0}^2(\boldsymbol{x})}(1-\cos\sigma) b_\beta(\boldsymbol{x}), \tag{8.66}$$

SV to SV reflectivity is (8.39)

$$\mathcal{R}_{\mathrm{SVSV}}(\boldsymbol{x},\ \sigma) \simeq \frac{b_\rho(\boldsymbol{x})}{2}(2\cos\sigma - 1) + \frac{b_\beta(\boldsymbol{x})}{2(1+\cos\sigma)}\left(2\cos^2\sigma - 1\right), \tag{8.67}$$

SH to SH reflectivity is (8.43)

$$\mathcal{R}_{\mathrm{SHSH}}(\boldsymbol{x},\ \sigma) = \frac{b_\rho(1+\cos\sigma) + b_\beta\cos\sigma}{4\cos^2\gamma}, \tag{8.68}$$

P to SV reflectivity is (8.50)

$$\mathcal{R}_{\mathrm{SVP}}(\boldsymbol{x},\sigma) \simeq \frac{|\sin\sigma|}{4\cos\gamma_S}\frac{\alpha_0}{c_e(\sigma)}\left(\left(1 + 2\frac{\beta_0(\boldsymbol{x})}{\alpha_0(\boldsymbol{x})}\cos\sigma\right) b_\rho(\boldsymbol{x}) + 2\frac{\beta_0(\boldsymbol{x})}{\alpha_0(\boldsymbol{x})}\cos\sigma\, b_\beta(\boldsymbol{x})\right), \tag{8.69}$$

and SV to P reflectivity is (8.55)

$$\mathcal{R}_{\mathrm{PSV}}(\boldsymbol{x},\sigma) \simeq -\frac{|\sin\sigma|}{4\cos\gamma_P}\frac{\beta_0(\boldsymbol{x})}{c_e(\sigma)}\left(\left(1 + 2\frac{\beta_0(\boldsymbol{x})}{\alpha_0(\boldsymbol{x})}\cos\sigma\right) b_\rho(\boldsymbol{x}) + 2\frac{\beta_0(\boldsymbol{x})}{\alpha_0(\boldsymbol{x})}\cos\sigma\, b_\beta(\boldsymbol{x})\right). \tag{8.70}$$

Appearing in the last two expressions are γ_P and γ_S, which obey $\gamma_P + \gamma_S = \sigma$, and $\sin\gamma_P/\alpha_0 = \sin\gamma_S/\beta_0$. The perturbation terms are first-order expressions

$$b_c = 2\frac{\hat{\boldsymbol{m}}\cdot\nabla c}{c}, \tag{8.71}$$

$$b_\rho = \frac{\hat{\boldsymbol{m}}\cdot\nabla\rho}{\rho}, \tag{8.72}$$

$$b_\alpha = \frac{\hat{\boldsymbol{m}}\cdot\nabla\alpha}{\alpha}, \tag{8.73}$$

and

$$b_\beta = \frac{\hat{\boldsymbol{m}}\cdot\nabla\beta}{\beta}. \tag{8.74}$$

Exercises

8.1 Starting with the reflection coefficient for a scalar wave at a flat interface, expand in powers of the velocity perturbation $a_c = 1 - c_0^2/c_1^2$, confirm that the term linear in a_c is (8.14), and derive the terms proportional to the second and third powers of a_c.

8.2 Starting with the reflection coefficient for an acoustic wave at a flat interface, expand in powers of the velocity and density perturbations $a_c = 1 - c_0^2/c_1^2$ and $a_\rho = 1 - \rho_0/\rho_1$, confirm that the term linear in a_c and a_ρ is (8.25), and derive the the second and third order terms.

8.3 Equation 5.44 of Aki and Richards (1980) presents the reflection coefficients in terms of the angles of incidence γ_P and reflection γ_S (using our notation). Demonstrate the equivalence of these formulae to (8.44) and (8.51).

9

Synthesizing reflection data

9.1 The Born model for seismic reflections

As discussed in Section 5.4, we choose to model seismic reflections with the Born approximation. Abstractly, we use the scattering theory equation (5.57), or

$$D \simeq \mathfrak{w}\, G_0 \mathcal{V} G_0. \tag{9.1}$$

Here, D is data, $\mathfrak{w}$ a source wavelet, $\mathbf{G}_0$ a "background" or approximate one-way Green's function, and $\mathcal{V}$ the scattering potential. Written explicitly as an integral equation, we have equation (5.59), or

$$D\left(\boldsymbol{x}_g|\boldsymbol{x}_s;\omega\right) = \mathfrak{w}(\omega)\int\int\int d^3x G_0\left(\boldsymbol{x}_g\middle|\boldsymbol{x};\omega\right)\mathcal{V}(\boldsymbol{x},\omega)G_0\left(\boldsymbol{x}\,|\boldsymbol{x}_s;\;\omega\right). \tag{9.2}$$

9.2 A constant background

9.2.1 The Born approximation in f–k space

Where the background medium is unchanging, reflection response can be determined relatively simply. For a constant background, Green's functions are invariant under translation and source–receiver reciprocity (see Section 6.1), so that the upgoing Green's function is

$$\begin{aligned} G_r\left(\boldsymbol{x}_g\middle|\boldsymbol{x};\omega\right) &= \mathcal{G}_r\left(\boldsymbol{x}_g-\boldsymbol{x};\omega\right) \\ &= \mathcal{G}_r\left(\boldsymbol{x}-\boldsymbol{x}_g;\omega\right), \end{aligned} \tag{9.3}$$

and the downgoing Green's function is

$$\begin{aligned} G_i\left(\boldsymbol{x}\,|\boldsymbol{x}_s\,;\omega\right) &= \mathcal{G}_i\left(\boldsymbol{x}-\boldsymbol{x}_s;\omega\right) \\ &= \mathcal{G}_i\left(\boldsymbol{x}_s-\boldsymbol{x};\omega\right). \end{aligned} \tag{9.4}$$

For surface data, the source and receiver positions $\boldsymbol{x}_s$ and $\boldsymbol{x}_g$ have no depth component, whereas the internal coordinate $\boldsymbol{x} = (x, y, z)$ is a 3-vector. In terms of the 2-vectors $\tilde{\boldsymbol{x}} = (x, y)$, $\tilde{\boldsymbol{x}}_s = (x_s, y_s)$ and $\tilde{\boldsymbol{x}}_g = \left(x_g, y_g\right)$,

$$G_r\left(\tilde{\boldsymbol{x}}_g, 0\,|\tilde{\boldsymbol{x}}\,, z; \omega\right) = \mathcal{G}_r\left(\tilde{\boldsymbol{x}}_g - \tilde{\boldsymbol{x}}, -z; \omega\right) = \mathcal{G}_r\left(\tilde{\boldsymbol{x}} - \tilde{\boldsymbol{x}}_g, z; \omega\right), \tag{9.5}$$

and

$$G_i\left(\tilde{\boldsymbol{x}}, z\,|\tilde{\boldsymbol{x}}_s\,, 0; \omega\right) = \mathcal{G}_i\left(\tilde{\boldsymbol{x}} - \tilde{\boldsymbol{x}}_s, z; \omega\right) = \mathcal{G}_i\left(\tilde{\boldsymbol{x}}_s - \tilde{\boldsymbol{x}},\ -z; \omega\right). \tag{9.6}$$

With Fourier transforms over $\tilde{\boldsymbol{x}}_g$ and $\tilde{\boldsymbol{x}}_s$, the reflection-data expression (5.59) becomes

$$D_{ri}\left(\tilde{\boldsymbol{k}}_g|\tilde{\boldsymbol{k}}_g; \omega\right) = \mathfrak{w}(\omega) \int\int\int d^3x \mathcal{G}_r\left(\tilde{\boldsymbol{k}}_g, z, \omega\right) \times e^{-i\tilde{\boldsymbol{k}}_g\cdot\tilde{\boldsymbol{x}}} \mathcal{V}_{ri}(x, \omega) e^{+i\tilde{\boldsymbol{k}}_s\cdot\tilde{\boldsymbol{x}}} \mathcal{G}_i\left(\tilde{\boldsymbol{k}}_s,\ z, \omega\right), \tag{9.7}$$

with $\tilde{\boldsymbol{k}}_g$ and $\tilde{\boldsymbol{k}}_s$ 2-vector wavenumbers conjugate to the coordinates $\tilde{\boldsymbol{x}}_g$ and $\tilde{\boldsymbol{x}}_s$. The potential $\mathcal{V}_{ri}(\boldsymbol{x}, \omega)$ as written might be the perturbation of any type of wavefunction, and is taken to be non-zero only for positive depths. For specificity, let us consider the medium to be elastic, with data D_{ri} and potential $\mathcal{V}_{ri}$ one of the possible elastic components (D_{PP}, D_{SVSV}, D_{SHSH}, D_{SVP}, or D_{PSV}, and similarly for $\mathcal{V}_{ri}$). The Green's functions $\mathcal{G}_r$ and $\mathcal{G}_i$ are then the appropriate S- or P-wave background-velocity impulse responses. In the $\left(k_x, k_y, z, \omega\right)$ domain, they reduce to the scalar form (6.14), reweighted by a factor of $\rho_0 c_r^2$ or $\rho_0 c_i^2$, with c_r and c_i the appropriate background velocity (α_0 for P-waves, and β_0 for S-waves):

$$\mathcal{G}_r(\tilde{\boldsymbol{k}}_g, z\,, \omega) = \frac{e^{-ik_{gz}|z|}}{2ik_{gz}\rho_0 c_r^2}, \quad \mathcal{G}_i(\tilde{\boldsymbol{k}}_s, z, \omega) \ = -\frac{e^{ik_{sz}|z|}}{2ik_{sz}\rho_0 c_i^2}, \tag{9.8}$$

with

$$k_{gz} = -\frac{\omega}{c_r}\sqrt{1 - \frac{\left({k_{gx}}^2 + {k_{gy}}^2\right)c_r^2}{\omega^2}}, \quad k_{sz} = +\frac{\omega}{c_i}\sqrt{1 - \frac{\left({k_{sx}}^2 + {k_{sy}}^2\right)c_i^2}{\omega^2}}. \tag{9.9}$$

In these formulas, we have added subscripts r and i to various quantities, to indicate whether they are associated with the reflected or the incoming waves. We have defined the receiver vertical wavenumber k_{gz} to have the sign opposite to frequency, and the source vertical wavenumber k_{sz} to have the same sign as frequency.

The complete expression of $\mathcal{G}_r$ and $\mathcal{G}_i$ includes evanescent waves, where k_{gz} and/or k_{sz} become imaginary. In the interest of simplicity, we confine attention to propagating waves only, where the vertical wavenumbers remain real.

The scattering potential $\mathcal{V}_{ri}(x\ ,\ y\ , z, \omega)$ is assumed to be non-zero only for $z > 0$. Consequently, inside the integral (9.7), the absolute values of z appearing in the exponents of the Green's functions (9.8) can be replaced by z, and (9.7) written as

$$D_{ri}\left(\tilde{\boldsymbol{k}}_g|\tilde{\boldsymbol{k}}_s; \omega\right) = \frac{\mathfrak{w}(\omega)}{4k_{gz}k_{sz}{\rho_0}^2{c_r}^2{c_i}^2} \int\int\int d^3x e^{-i\boldsymbol{k}_g\cdot\boldsymbol{x}} \mathcal{V}_{ri}(\boldsymbol{x}, \omega) e^{+i\boldsymbol{k}_s\cdot\boldsymbol{x}}, \tag{9.10}$$

with $\boldsymbol{k}_g$ and $\boldsymbol{k}_s$ the 3-vectors

$$\boldsymbol{k}_s = \left(\tilde{\boldsymbol{k}}_s, k_{sz}\right) = \left(k_{sx}, k_{sy}, k_{sz}\right), \tag{9.11}$$

$$\boldsymbol{k}_g = \left(\tilde{\boldsymbol{k}}_g, k_{gz}\right) = \left(k_{gx}, k_{gy}, k_{gz}\right). \tag{9.12}$$

Conversion of the scattering potential $\mathcal{V}_{ri}(\boldsymbol{x}, \omega)$ from differential to angular-dependent form is exact in this case, because the Fourier transforms have decomposed the incoming and outgoing Green's functions into plane-wave components. The potential $\mathcal{V}_{ri}$ is a sum of perturbation terms. Each perturbation is surrounded by coefficients of differential operators, with some derivatives appearing left of the perturbation, and some to the right. In (9.10), derivatives to the right of the perturbation are applied to the exponential $e^{+i\boldsymbol{k}_s\cdot\boldsymbol{x}}$, the operator ∂_i producing ik_{si}. Derivatives to the left of the perturbation can be applied via integration by parts to the exponential $e^{-i\boldsymbol{k}_g\cdot\boldsymbol{x}}$, the operator ∂_i producing $+ik_{gi}$. Given that the magnitude of $\boldsymbol{k}_g$ is ω / c_r and $\boldsymbol{k}_s$ is ω / c_i, and that the angle between $\boldsymbol{k}_g$ and $\boldsymbol{k}_s$ is the opening angle σ, the potential can be expressed in angle-dependent form $\omega^2 \mathbb{V}_{ri}(\boldsymbol{x}, \sigma)$ as given by equations (7.83)–(7.87).

Since $\mathbb{V}_{ri}(\boldsymbol{x}, \sigma)$ is not a differential operator, we are free to rearrange the terms in (9.10). Collecting terms,

$$D_{ri}\left(\tilde{\boldsymbol{k}}_g | \tilde{\boldsymbol{k}}_s; \omega\right) = \frac{\omega^2 \mathfrak{w}(\omega)}{4k_{gz}k_{sz}\rho_0^2 c_r^2 c_i^2} \int\int\int d^3x e^{-i(\boldsymbol{k}_g - \boldsymbol{k}_s)\cdot\boldsymbol{x}} \mathbb{V}_{ri}(\boldsymbol{x}, \sigma). \tag{9.13}$$

The integral over x_m is recognizable as a 3-D Fourier transform of $\mathbb{V}_{ri}$, evaluated at wavenumbers

$$\boldsymbol{k}_m \equiv \boldsymbol{k}_g - \boldsymbol{k}_s. \tag{9.14}$$

Thus, since the only spatial dependence is in the perturbations, we find that the Fourier transform of the reflection response is proportional to the Fourier transform of the scattering potential:

$$D_{ri}\left(\tilde{\boldsymbol{k}}_g | \tilde{\boldsymbol{k}}_s; \omega\right) = \mathfrak{w}(\omega) \frac{\omega^2 \mathbb{V}_{ri}\left(\boldsymbol{k}_m, \sigma\right)}{4k_{gz}k_{sz}\rho_0^2 c_r^2 c_i^2}. \tag{9.15}$$

To form an expression in terms of a reflectivity function, invoke its definition (8.56), with $\mathcal{C}_r{}^2 = \rho_0 c_r{}^2$; i.e.

$$\mathbb{V}_{ri}(\boldsymbol{x}, \sigma) = \frac{-2i\rho_0 c_r \cos\gamma_r}{\omega} \mathcal{R}_{ri}(\boldsymbol{x}, \sigma). \tag{9.16}$$

Then,

$$D_{ri}\left(\tilde{\boldsymbol{k}}_g | \tilde{\boldsymbol{k}}_s; \omega\right) = -i\omega\, \mathfrak{w}(\omega) \frac{\mathcal{R}_{ri}\left(\boldsymbol{k}_m, \sigma\right) \cos\gamma_r}{2k_{gz}k_{sz}\rho_0 c_r c_i{}^2}. \tag{9.17}$$

When synthesizing seismic data, the starting point is $\mathbb{V}_{ri}(\boldsymbol{x},\sigma)$ or $\mathcal{R}_{ri}(\boldsymbol{x},\sigma)$, which is transformed into $\mathbb{V}_{ri}(\boldsymbol{k}_m,\sigma)$ or $\mathcal{R}_{ri}(\boldsymbol{k}_m,\sigma)$ with a 3-D Fourier transform. Given k_m and σ, we can deduce the other parameters. We have

$$\begin{aligned} k_m^2 \equiv |\boldsymbol{k}_m|^2 &= |\boldsymbol{k}_g|^2 + |\boldsymbol{k}_s|^2 - 2\boldsymbol{k}_g\cdot\boldsymbol{k}_s \\ &= \frac{\omega^2}{c_r^2 c_i^2}\left(c_r^2 + c_i^2 + 2c_r c_i \cos\sigma\right) \\ &= \frac{4\omega^2 c_e(\sigma)^2}{c_r^2 c_i^2}, \end{aligned} \tag{9.18}$$

with $c_e(\sigma)$ an effective velocity defined in Appendix H as

$$c_e(\sigma) = \frac{\sqrt{c_r^2 + c_i^2 + 2c_r c_i \cos\sigma}}{2}. \tag{9.19}$$

For wave-type-preserving reflections, where $c_r = c_i = c_0$ and $\gamma_r = \gamma_i = \gamma = \sigma/2$,

$$c_e(\sigma) = \frac{\sqrt{2(1+\cos\sigma)}}{2} = c_0 \cos\gamma \,(\text{non-converted waves}). \tag{9.20}$$

In any event, the magnitude of the image wavenumber $\boldsymbol{k}_m$ is constrained to be less than or equal to $\omega\left(1/c_r + 1/c_i\right)$.

Solving for ω,

$$\omega = \frac{c_r c_i}{2c_e(\sigma)} k_m, \tag{9.21}$$

with $c_e(\sigma)$ the effective, opening-angle-dependent velocity defined in (9.19).

The unit vector $\hat{\boldsymbol{k}}_m$ pointing in the direction of $\boldsymbol{k}_m$ is

$$\hat{\boldsymbol{k}}_m = \frac{\boldsymbol{k}_g - \boldsymbol{k}_s}{k_m} = \frac{\hat{\boldsymbol{k}}_g c_i - \hat{\boldsymbol{k}}_s c_r}{2c_e(\sigma)}. \tag{9.22}$$

The angle of reflection γ_r and the angle of incidence γ_i are found in Appendix H to obey (equations (H.56)–(H.59))

$$\cos\gamma_r = \frac{(c_i + c_r\cos\sigma)}{2c_e(\sigma)}, \tag{9.23}$$

$$\sin\gamma_i = \frac{c_i \sin\sigma}{2c_e(\sigma)}, \tag{9.24}$$

$$\cos\gamma_i = \frac{(c_r + c_i\cos\sigma)}{2c_e(\sigma)}, \tag{9.25}$$

$$\sin\gamma_r = \frac{c_r \sin\sigma}{2c_e(\sigma)}. \tag{9.26}$$

The wavenumbers $\boldsymbol{k}_g$ and $\boldsymbol{k}_s$ are not completely determined by $\boldsymbol{k}_m$ and σ. Defining offset wavenumber

$$\boldsymbol{k}_h \equiv \boldsymbol{k}_s + \boldsymbol{k}_g, \tag{9.27}$$

then the magnitude of $\boldsymbol{k}_h$ is also fixed:

$$\begin{aligned} k_h^2 &= |\boldsymbol{k}_g|^2 + |\boldsymbol{k}_s|^2 + 2\boldsymbol{k}_g\cdot\boldsymbol{k}_s \\ &= \frac{\omega^2}{c_r^2 c_i^2}\left(c_r^2 + c_i^2 - 2c_r c_i \cos\sigma\right). \end{aligned} \tag{9.28}$$

Thus,

$$k_h = k_m\sqrt{\frac{c_r^2 + c_i^2 - 2c_r c_i \cos\sigma}{c_r^2 + c_i^2 + 2c_r c_i \cos\sigma}}. \tag{9.29}$$

The direction of $\boldsymbol{k}_h$ is not fixed by $\boldsymbol{k}_m$ and σ; however, the angle ψ between $\boldsymbol{k}_h$ and $\boldsymbol{k}_m$ is given by

$$\cos\psi = \frac{\boldsymbol{k}_h\cdot\boldsymbol{k}_m}{k_h k_m} = \frac{\left(\boldsymbol{k}_g - \boldsymbol{k}_s\right)\cdot\left(\boldsymbol{k}_g + \boldsymbol{k}_s\right)}{k_h k_m} = \frac{\left(c_i^2 - c_r^2\right)}{\sqrt{\left(c_r^2 + c_i^2\right)^2 - 4c_r^2 c_i^2 \cos^2\sigma}}. \tag{9.30}$$

Thus, for a given $\boldsymbol{k}_m$ and σ, the set of all possible $\boldsymbol{k}_h$ traces a cone of vectors of length k_h at angle ψ to $\boldsymbol{k}_m$.

If we begin with a point $\left(\tilde{\boldsymbol{k}}_g|\tilde{\boldsymbol{k}}_s;\ \omega\right)$ in data *f–k* space, then a point in image $(\boldsymbol{k}_m, \sigma)$-space is uniquely determined. The vertical wavenumbers k_{gz} and k_{sz} are determined from (9.9). The wavenumber $\boldsymbol{k}_m$ is then determined from (9.14). The opening angle σ is determined by (H.37) and (H.48), or

$$\cos\sigma = -\frac{\boldsymbol{k}_s\cdot\boldsymbol{k}_g c_i c_r}{\omega^2}, \tag{9.31}$$

$$\sin\sigma = \frac{\left|\boldsymbol{k}_g \times \boldsymbol{k}_s\right| c_i c_r}{\omega^2} S\left(\left(\boldsymbol{k}_g \times \boldsymbol{k}_s\right)_y\right). \tag{9.32}$$

To construct a constant-background data set, one first Fourier transforms the image (reflectivity or potential), then performs the transformation of coordinates from image coordinates to data coordinates, and finally inverse Fourier transforms the data back to space and time.

The elastic potential, being isotropic, does not change with the direction of $\boldsymbol{k}_h$, and consequently, neither does the reflectivity or the reflected data. We can, however, imagine a potential which does vary with azimuth, in which case the mapping between potential or reflectivity and data becomes one to one, as described in Appendix H.

Change the variables in the above expressions from the 2-vectors $\tilde{\boldsymbol{k}}_g$ and $\tilde{\boldsymbol{k}}_s$ to the two-vectors $\tilde{\boldsymbol{k}}_m = \tilde{\boldsymbol{k}}_g - \tilde{\boldsymbol{k}}_s$ and $\tilde{\boldsymbol{k}}_h = \tilde{\boldsymbol{k}}_s + \tilde{\boldsymbol{k}}_g$. We can then write the relations

$$k_{gz} = -\sqrt{\frac{\omega^2}{c_r^2} - \frac{|\tilde{\boldsymbol{k}}_m + \tilde{\boldsymbol{k}}_h|^2}{4}}, \qquad k_{sz} = \sqrt{\frac{\omega^2}{c_i^2} - \frac{|\tilde{\boldsymbol{k}}_m - \tilde{\boldsymbol{k}}_h|^2}{4}}, \tag{9.33}$$

$$\begin{aligned} k_{mz} &= k_{gz} - k_{sz} \\ &= -\sqrt{\frac{\omega^2}{c_r^2} - \frac{|\tilde{\boldsymbol{k}}_m + \tilde{\boldsymbol{k}}_h|^2}{4}} - \sqrt{\frac{\omega^2}{c_i^2} - \frac{|\tilde{\boldsymbol{k}}_m - \tilde{\boldsymbol{k}}_h|^2}{4}}, \end{aligned} \tag{9.34}$$

and, expressing ω as a function of $\tilde{\boldsymbol{k}}_m$, $\tilde{\boldsymbol{k}}_h$, and k_{mz},

$$\omega = \frac{c_r c_i}{\left|c_r^2 - c_i^2\right|} \sqrt{\tilde{\boldsymbol{k}}_m \cdot \tilde{\boldsymbol{k}}_h \left(c_i^2 - c_r^2\right) + k_{mz}^2 f_{ri}}, \tag{9.35}$$

where

$$f_{ri} = c_r^2 + c_i^2 \mp 2c_r c_i \sqrt{1 + \frac{\left(c_i^4 - c_r^4\right)}{2c_r^2 c_i^2} \frac{\tilde{\boldsymbol{k}}_m \cdot \tilde{\boldsymbol{k}}_h}{k_{mz}^2} - \frac{\left(c_r^2 - c_i^2\right)^2}{4c_r^2 c_i^2} \frac{\left(\tilde{k}_m^2 + \tilde{k}_h^2\right)}{k_{mz}^2}}. \tag{9.36}$$

For reflections that preserve wave type, some of the above relations simplify. In this case, $c_i = c_r = c_0$, and there is only one finite solution for ω. We have $\cos\gamma_r \to \cos(\sigma/2) = k\, c_0 \big/ 2\omega$, and $\omega \to c_0 k/2 \cos(\sigma/2)$. Moreover, $k_h = 2\omega \sin(\sigma/2) \big/ c_0$, and $\boldsymbol{k}_h \cdot \boldsymbol{k}_m = 0$. The expression for frequency reduces to

$$\omega = -\frac{k_{mz} c_0}{2} \sqrt{1 + \frac{\tilde{k}_m^2}{k_{mz}^2} + \frac{\tilde{k}_h^2}{k_{mz}^2} + \frac{\left(\tilde{\boldsymbol{k}}_m \cdot \tilde{\boldsymbol{k}}_h\right)^2}{k_{mz}^4}} \text{(non-converted waves)}. \tag{9.37}$$

For type-preserving reflections, there are also simple expressions for k_{gz} and k_{sz}.

$$\begin{aligned} k_{gz} &= \frac{k_{mz}}{2} \left(1 + \frac{\tilde{\boldsymbol{k}}_m \cdot \tilde{\boldsymbol{k}}_h}{k_{mz}^2}\right), \\ k_{sz} &= -\frac{k_{mz}}{2} \left(1 - \frac{\tilde{\boldsymbol{k}}_m \cdot \tilde{\boldsymbol{k}}_h}{k_{mz}^2}\right). \end{aligned} \tag{9.38}$$

9.2.2 Point scatterers and reflectors

For a point scatterer, the scattering potential $\mathbb{V}_{ri}$ is proportional to a delta function:

$$\mathbb{V}_{ri}(\boldsymbol{x}, \sigma) = V_{ri}(\sigma) \delta\left(\boldsymbol{x} - \boldsymbol{x}_0\right). \tag{9.39}$$

The response to the point scatterer is found from Fourier transforming (9.39) and substituting into (9.15):

$$D_{ri}\left(\tilde{\boldsymbol{k}}_g|\tilde{\boldsymbol{k}}_s;\ \omega\right) = \mathfrak{w}(\omega)\frac{\omega^2 V_{ri}(\sigma)e^{-i\boldsymbol{k}_m\cdot\boldsymbol{x}_0}}{4k_{gz}k_{sz}\rho_0^2 c_r^2 c_i^2}. \tag{9.40}$$

For a point reflector, it is the reflectivity function that becomes a delta function:

$$\mathcal{R}_{ri}(\boldsymbol{x},\sigma) = R_{ri}(\sigma)\delta\left(\boldsymbol{x}-\boldsymbol{x}_0\right), \tag{9.41}$$

leading to the response formula

$$D_{ri}\left(\tilde{\boldsymbol{k}}_g|\tilde{\boldsymbol{k}}_s;\omega\right) = -i\omega\,\mathfrak{w}(\omega)\frac{R_{ri}(\sigma)e^{-i\boldsymbol{k}_m\cdot\boldsymbol{x}_0}\cos\gamma_r}{2k_{gz}k_{sz}\rho_0 c_r c_i{}^2}. \tag{9.42}$$

The responses to the point scatterer and the point reflector have very similar expressions, and they are clearly related. However, they are not identical. They have different phases and different angular dependence. The physical meaning of a point scatterer is reasonably clear – it is just a localized perturbation in the scattering potential. The physical meaning of a point reflector is less clear – one can only say that it represents a point on a reflectivity image. Nevertheless, it is a useful concept, perhaps even necessary if one wishes to treat reflectivity as if it were a physical attribute.

9.2.3 Planar scatterers and reflectors

Suppose the scattering potential is non-zero only on the plane $z = z_0 + q_x x + q_y y$, so that

$$\mathbb{V}_{ri}(\boldsymbol{x},\sigma) = V_{ri}(\sigma)\sqrt{1+q_x{}^2+q_y{}^2}\delta\left(z-z_0-q_x x-q_y y\right), \tag{9.43}$$

with corresponding response to the planar scatterer of

$$D_{ri}\left(\tilde{\boldsymbol{k}}_g|\tilde{\boldsymbol{k}}_s;\ \omega\right) = (2\pi)^2\omega^2\mathfrak{w}(\omega)\sqrt{1+q_x{}^2+q_y{}^2}$$
$$\times\frac{V_{ri}(\sigma)e^{-ik_z z_0}\delta\left(k_x+q_x k_z\right)\delta\left(k_y+q_y k_z\right)}{4k_{gz}k_{sz}\rho_0^2 c_r^2 c_i^2}. \tag{9.44}$$

For a planar reflector,

$$\mathcal{R}_{ri}(\boldsymbol{x},\sigma) = R_{ri}(\sigma)\sqrt{1+q_x{}^2+q_y{}^2}\delta\left(z-z_0-q_x x-q_y y\right). \tag{9.45}$$

The response to a planar reflector is

$$D_{ri}\left(\tilde{\boldsymbol{k}}_g|\tilde{\boldsymbol{k}}_s;\omega\right) = -(2\pi)^2 i\omega\,\mathfrak{w}(\omega)\sqrt{1+q_x{}^2+q_y{}^2}$$
$$\times\frac{R_{ri}(\sigma)\cos\gamma_r e^{-ik_z z_0}\delta\left(k_x+q_x k_z\right)\delta\left(k_y+q_y k_z\right)}{2k_{gz}k_{sz}\rho_0 c_r c_i{}^2}. \tag{9.46}$$

The physical meanings of both the planar scatterer and the planar reflector, though distinct, are clear. The planar scatterer is a perturbation of the scattering potential along a plane. The planar reflector is a specular reflecting plane. The reflecting plane, if related back to scattering theory, corresponds to a step change in the scattering potential at the boundary.

9.2.4 Space–frequency

The relation (9.15) between data and scattering potential can be expressed in space–frequency as

$$D_{ri}\left(\tilde{\boldsymbol{x}}_g|\tilde{\boldsymbol{x}}_s;\omega\right) = \frac{1}{(2\pi)^4}\frac{\mathfrak{w}(\omega)}{4\rho_0^2c_r^2c_i^2}\int\!\!\int d^2\tilde{k}_g e^{i\tilde{\boldsymbol{k}}_g\cdot\tilde{\boldsymbol{x}}_g}\int\!\!\int d^2\tilde{k}_s e^{-i\tilde{\boldsymbol{k}}_s\cdot\tilde{\boldsymbol{x}}_s} \times\frac{\omega^2}{k_{gz}k_{sz}}\int\!\!\int\!\!\int d^3x e^{-i\boldsymbol{k}_m\cdot\boldsymbol{x}}\mathbb{V}_{ri}(\boldsymbol{x},\sigma). \tag{9.47}$$

In this expression, the vertical wavenumbers k_{sz} and k_{gz} are given by (9.9) or (9.33), with k_{mz} the difference of k_{gz} and k_{sz}, and opening angle σ given by (9.31).

We can change the integration variables in this expression from $\tilde{\boldsymbol{k}}_g$ and $\tilde{\boldsymbol{k}}_s$ to $\tilde{\boldsymbol{k}}_m = \tilde{\boldsymbol{k}}_g - \tilde{\boldsymbol{k}}_s$ and $\tilde{\boldsymbol{k}}_h = \tilde{\boldsymbol{k}}_s + \tilde{\boldsymbol{k}}_g$. Then, setting midpoint $\tilde{\boldsymbol{x}}_m = \left(\tilde{\boldsymbol{x}}_s + \tilde{\boldsymbol{x}}_g\right)/2$ and half-offset $\tilde{\boldsymbol{x}}_h = \left(\tilde{\boldsymbol{x}}_s - \tilde{\boldsymbol{x}}_g\right)/2$, (9.47) becomes

$$\begin{aligned} D_{ri}\left(\tilde{\boldsymbol{x}}_m,\tilde{\boldsymbol{x}}_h;\omega\right) &\equiv D_{ri}\left(\tilde{\boldsymbol{x}}_g|\tilde{\boldsymbol{x}}_s;\omega\right) \\ &= \frac{1}{(2\pi)^4}\frac{\mathfrak{w}(\omega)}{16\rho_0^2c_r^2c_i^2}\int\!\!\int d^2\tilde{k}_m e^{i\tilde{\boldsymbol{k}}_m\cdot\tilde{\boldsymbol{x}}_m}\int\!\!\int d^2\tilde{k}_h e^{i\tilde{\boldsymbol{k}}_h\cdot\tilde{\boldsymbol{x}}_h} \\ &\quad\times\frac{\omega^2}{k_{gz}k_{sz}}\int\!\!\int\!\!\int d^3x e^{-i\boldsymbol{k}_m\cdot\boldsymbol{x}}\mathbb{V}_{ri}(\boldsymbol{x},\sigma). \end{aligned} \tag{9.48}$$

The relation (9.17) between data and reflectivity can also be expressed in space-frequency. The result is

$$D_{ri}\left(\tilde{\boldsymbol{x}}_m,\tilde{\boldsymbol{x}}_h;\omega\right) = \frac{-1}{(2\pi)^4}\frac{i\omega\mathfrak{w}(\omega)}{8\rho_0c_rc_i{}^2}\int\!\!\int d^2\tilde{k}_m e^{i\tilde{\boldsymbol{k}}_m\cdot\tilde{\boldsymbol{x}}_m}\int\!\!\int d^2\tilde{k}_h e^{i\tilde{\boldsymbol{k}}_h\cdot\tilde{\boldsymbol{x}}_h} \times\frac{\cos\gamma_r}{k_{gz}k_{sz}}\int\!\!\int\!\!\int d^3x e^{-i\boldsymbol{k}_m\cdot\boldsymbol{x}}\,\mathcal{R}_{ri}(\boldsymbol{x},\sigma), \tag{9.49}$$

with angle of reflection γ_r given by (10.22).

9.3 Restricted data sets

One typically wants or will settle for less than a full 5-D data set. One can restrict the range of the output variables $\tilde{\boldsymbol{x}}_m$ and $\tilde{\boldsymbol{x}}_h$, but the Fourier-transform formulas (9.48) and (9.49) still expect all five inverse transform integrals. There may be fewer output points to compute, but if one is using a fast Fourier transform, there may be little if any computational savings. Under some circumstances, however, it may be possible to eliminate a transform or reduce it to a simple integral.

9.3.1 3-D zero offset

For a zero–offset data set, the space-frequency formula (9.49) becomes

$$D_{ri}^{(3DZO)}(\tilde{\boldsymbol{x}}_m;\omega) = D_{ri}\left(\tilde{\boldsymbol{x}}_m, \tilde{\boldsymbol{0}};\omega\right) \tag{9.50}$$
$$= \frac{-1}{(2\pi)^4}\frac{i\omega\mathfrak{w}(\omega)}{8\rho_0 c_r c_i{}^2}\int\!\!\int d^2\tilde{k}_m e^{i\tilde{\boldsymbol{k}}_m\cdot\tilde{\boldsymbol{x}}_m}\int\!\!\int d^2\tilde{k}_h\frac{\cos\gamma_r}{k_{gz}k_{sz}}\int\!\!\int\!\!\int d^3x e^{-i\boldsymbol{k}_m\cdot\boldsymbol{x}}\,\mathcal{R}_{ri}(\boldsymbol{x},\sigma).$$

Because the formula outputs only a single offset ($\tilde{\boldsymbol{0}}$), the two offset-wavenumber transforms reduce to simple integrals. This formula may be implemented as is, or one might eliminate the offset-wavenumber integrals entirely with a stationary phase approximation. To do the latter, we rearrange the integrals as follows:

$$D_{ri}^{(3DZO)}(\tilde{\boldsymbol{x}}_m;\omega) = \frac{-1}{(2\pi)^4}\frac{i\omega\mathfrak{w}(\omega)}{8\rho_0 c_r c_i{}^2}\int\!\!\int d^2\tilde{k}_m e^{i\tilde{\boldsymbol{k}}_m\cdot\tilde{\boldsymbol{x}}_m}\int\!\!\int\!\!\int d^3x e^{-i\tilde{\boldsymbol{k}}_m\cdot\tilde{\boldsymbol{x}}}$$
$$\times\int\!\!\int d^2\tilde{k}_h e^{-ik_{mz}z}\frac{\cos\gamma_r}{k_{gz}k_{sz}}\,\mathcal{R}_{ri}(\boldsymbol{x},\sigma). \tag{9.51}$$

Functions of opening angle σ and wavenumbers k_{gz}, k_{sz}, and k_{mz} all depend on offset wavenumber, hence remain inside the offset-wavenumber integrals. The rapid variation with $\tilde{\boldsymbol{k}}_h$ is confined to the exponential term $e^{-ik_{mz}z}$, and we can expect the significant contribution to the offset-wavenumber integrals will be from the stationary point of this exponential, i.e. the point $\tilde{\boldsymbol{k}}_h^{ZO}$ at which $\partial k_z/\partial k_{hx} = \partial k_z/\partial k_{hy} = 0$. The 2-dimensional stationary-phase approximation to the offset-wavenumber double integral is

$$\int\!\!\int d^2\tilde{k}_h e^{-ik_{mz}z}\frac{\cos\gamma_r}{k_{gz}k_{sz}}\mathcal{R}_{ri}(\boldsymbol{x},\sigma) \simeq 2\pi\frac{\mathcal{R}_{ri}\left(\boldsymbol{x},\sigma^{ZO}\right)}{z\sqrt{|\det[K^{ZO}]|}}\frac{\cos\gamma_r^{ZO}}{k_{gz}^{ZO}k_{sz}^{ZO}}e^{-i\left(k_{mz}^{ZO}z+\text{sig}[K^{ZO}]\pi/4\right)}. \tag{9.52}$$

In this expression, quantities superscripted with a "ZO" are meant to be evaluated at the stationary point $\tilde{\boldsymbol{k}}_h^{ZO}$. $\boldsymbol{K}^{ZO}$ is the matrix of second derivatives of k_{mz} with respect to k_{hx} and k_{hy}, also evaluated at the stationary point. The quantity $\det\left[\boldsymbol{K}^{ZO}\right]$ is the determinant of $\boldsymbol{K}^{ZO}$, and $\text{sig}\left[\boldsymbol{K}^{ZO}\right]$ is its signature, or sum of the signs of its eigenvalues. The vertical wavenumbers k_{gz}, k_{sz}, and k_{mz} depend on $\tilde{\boldsymbol{k}}_h$, hence also have stationary values. The two-vector $\tilde{\boldsymbol{k}}_m$, however, is determined independently of $\tilde{\boldsymbol{k}}_h$, hence has no stationary value.

The first derivatives of k_{mz} are

$$\frac{\partial k_{mz}}{\partial k_{hx}} = \frac{k_{sx}}{2k_{sz}} - \frac{k_{gx}}{2k_{gz}}, \quad \frac{\partial k_{mz}}{\partial k_{hy}} = \frac{k_{sy}}{2k_{sz}} - \frac{k_{gy}}{2k_{gz}}. \tag{9.53}$$

These derivatives are both zero at the stationary point, from which it follows

$$\tilde{\boldsymbol{k}}_h^{\mathrm{ZO}} = \frac{c_i - c_r}{c_r + c_i}\tilde{\boldsymbol{k}}_m, \tag{9.54}$$

and

$$\begin{aligned} \tilde{\boldsymbol{k}}_g^{\mathrm{ZO}} &= +\frac{c_i}{c_r + c_i}\tilde{\boldsymbol{k}}_m, \\ \tilde{\boldsymbol{k}}_s^{\mathrm{ZO}} &= -\frac{c_r}{c_r + c_i}\tilde{\boldsymbol{k}}_m, \end{aligned} \tag{9.55}$$

$$\begin{aligned} k_{gz}^{\mathrm{ZO}} &= -\frac{c_i}{c_r + c_i}\sqrt{\frac{\omega^2}{\underline{c}^2} - \tilde{k}_m^2} = +\frac{c_i}{c_r + c_i}k_{mz}^{\mathrm{ZO}}, \\ k_{sz}^{\mathrm{ZO}} &= +\frac{c_r}{c_r + c_i}\sqrt{\frac{\omega^2}{\underline{c}^2} - \tilde{k}_m^2} = -\frac{c_r}{c_r + c_i}k_{mz}^{\mathrm{ZO}}, \end{aligned} \tag{9.56}$$

where

$$k_{mz}^{\mathrm{ZO}} = -\sqrt{\frac{\omega^2}{\underline{c}^2} - \tilde{k}_m^2}, \tag{9.57}$$

with $\underline{c}^{-1}$ the summed slowness, or

$$\frac{1}{\underline{c}} = \frac{1}{c_r} + \frac{1}{c_i} = \left(\frac{c_r + c_i}{c_r c_i}\right). \tag{9.58}$$

We note that for non-converted waves, where the two velocities are equal, $\underline{c} = c_0 / 2$. In any case, c can be thought of as an effective one-way velocity for zero-offset waves, in the sense that vertical traveltime to depth z and back again is $\tau = z/c_i + z/c_r = z/\underline{c}$.

The second derivatives of k_{mz} are

$$\begin{aligned} \frac{\partial^2 k_{mz}}{\partial k_{hx}^2} &= \frac{1}{4}\left(-\frac{k_{gx}^2 + k_{gz}^2}{k_{gz}^3} + \frac{k_{sx}^2 + k_{sz}^2}{k_{sz}^3}\right), \\ \frac{\partial^2 k_{mz}}{\partial k_{hy}^2} &= \frac{1}{4}\left(-\frac{k_{gy}^2 + k_{gz}^2}{k_{gz}^3} + \frac{k_{sy}^2 + k_{sz}^2}{k_{sz}^3}\right), \\ \frac{\partial^2 k_{mz}}{\partial k_{hx}\partial k_{hy}} &= \frac{1}{4}\left(-\frac{k_{gx}k_{gy}}{k_{gz}^3} + \frac{k_{sx}k_{sy}}{k_{sz}^3}\right), \end{aligned} \tag{9.59}$$

becoming, at the stationary point,

$$\frac{\partial^2 k_{mz}}{\partial k_{hx}^2} \to -\frac{c_i c_r}{4\left(k_{mz}^{\mathrm{ZO}}\right)^3 \underline{c}^2}\left(\frac{\omega^2}{\underline{c}^2} - k_{my}^2\right),$$
$$\frac{\partial^2 k_{mz}}{\partial k_{hy}^2} \to -\frac{c_i c_r}{4\left(k_{mz}^{\mathrm{ZO}}\right)^3 \underline{c}^2}\left(\frac{\omega^2}{\underline{c}^2} - k_{mx}^2\right),$$
$$\frac{\partial^2 k_{mz}}{\partial k_{hx}\partial k_{hy}} \to -\frac{k_{mx}k_{my}}{4\left(k_{mz}^{\mathrm{ZO}}\right)^3}\frac{c_i c_r}{\underline{c}^2}. \tag{9.60}$$

This leads to

$$\mathrm{sig}\left[\boldsymbol{K}^{\mathrm{ZO}}\right] = 2, \tag{9.61}$$

and

$$\mathrm{Det}\left[\boldsymbol{K}^{\mathrm{ZO}}\right] = \frac{\omega^2}{16\left(k_{mz}^{\mathrm{ZO}}\right)^4}\frac{c_i^2 c_r^2}{\underline{c}^6} \tag{9.62}$$

From expression (9.57) for k_{mz}^{ZO}, it follows that the magnitude of $\boldsymbol{k}_m$ at the stationary point is (taking frequency to be positive)

$$k_m^{\mathrm{ZO}} = \sqrt{\left(k_{mz}^{\mathrm{ZO}}\right)^2 + \tilde{k}_m^2} = \frac{\omega}{\underline{c}}, \tag{9.63}$$

and, from (9.31), cosine of opening angle at the stationary point is one, implying $\sigma^{\mathrm{ZO}} = 0$. Likewise, from equation (9.23) for $\cos\gamma_r$, the stationary value of γ_r is zero.

The stationary phase approximation thus becomes

$$\int\int d^2\tilde{k}_h \frac{\cos\gamma_r}{k_{gz}k_{sz}} e^{-ik_{mz}z}\,\mathcal{R}_{ri}(\boldsymbol{x},\sigma) \simeq \frac{8\pi i\underline{c}}{z\omega}\mathcal{R}_{ri}(\boldsymbol{x},0)e^{-ik_{mz}^{\mathrm{ZO}}z}, \tag{9.64}$$

and the data integral (9.51)

$$D_{ri}^{(\mathrm{3DZO})}(\tilde{\boldsymbol{x}}_m;\omega) \simeq \frac{1}{(2\pi)^3}\frac{\mathfrak{w}(\omega)\underline{c}}{2\rho_0 c_r c_i{}^2} \tag{9.65}$$
$$\times \int\int d^2\tilde{k}_m e^{i\tilde{\boldsymbol{k}}_m\cdot\tilde{\boldsymbol{x}}_m}\int\int\int d^3x e^{-i\left(\tilde{\boldsymbol{k}}_m\cdot\tilde{\boldsymbol{x}}+k_{mz}^{\mathrm{ZO}}z\right)}\left(\frac{\mathcal{R}_{ri}}{z}\right)(\boldsymbol{x},0).$$

The expression $\left(\frac{\mathcal{R}_{ri}}{z}\right)(\boldsymbol{x},0)$ represents the reflectivity function for zero opening angle at point $\boldsymbol{x} = (x, y, z)$, rescaled by the factor $1/z$.

The stationary-phase expression for zero-offset data is, like the expression for the full data set, a combination of forward and inverse Fourier transforms.

The three-fold Fourier transform $\left(\tilde{\boldsymbol{x}}_m, t \to \tilde{\boldsymbol{k}}_m, \omega\right)$ of the zero-offset data is proportional to the three-fold Fourier transform $\left(\boldsymbol{x} \to \tilde{\boldsymbol{k}}_m, k_{mz}^{\mathrm{ZO}}\right)$ of the rescaled reflectivity function $\left(\frac{\mathcal{R}_{ri}}{z}\right)$. That is,

$$D_{ri}^{(\mathrm{3DZO})}\left(\tilde{\boldsymbol{k}}_m; \omega\right) \simeq \frac{1}{(2\pi)} \frac{\mathfrak{w}(\omega)\underline{c}}{2\rho_0 c_r c_i{}^2} \left(\frac{\mathcal{R}_{ri}}{z}\right)\left(\tilde{\boldsymbol{k}}_m, k_{mz}{}^{\mathrm{ZO}}, 0\right). \tag{9.66}$$

While (9.51) is exact, (9.65) or (9.66) represent a far-field approximation.

Time-rescaled data

One could, in lieu of rescaling reflectivity, write an equally valid expression in terms of time-rescaled data (tD_{ri}). An easy way to make the transition is to differentiate both sides of the above equation by k_{mz}^{ZO}. Assuming that $\mathfrak{w}$ varies slowly with ω and hence with k_{mz}^{ZO},

$$\frac{d}{dk_{\mathrm{mz}}}^{\mathrm{ZO}} D_{ri}^{(\mathrm{3DZO})}\left(\tilde{\boldsymbol{k}}_m; \omega\right) \simeq \frac{1}{(2\pi)} \frac{\underline{c}\,\mathfrak{w}(\omega)}{2\rho_0 c_r c_i{}^2} \frac{d}{dk_{mz}^{\mathrm{ZO}}} \left(\frac{\mathcal{R}_{ri}}{z}\right)\left(\tilde{\boldsymbol{k}}_m, k_{\mathrm{mz}}^{\mathrm{ZO}}, 0\right). \tag{9.67}$$

The derivative on the right-hand side of this equation can be written

$$\begin{aligned}
\frac{d}{dk_{mz}^{\mathrm{ZO}}} \left(\frac{\mathcal{R}_{ri}}{z}\right)\left(\tilde{\boldsymbol{k}}_m, k_{mz}^{\mathrm{ZO}}, 0\right) &= \frac{d}{dk_{\mathrm{mz}}^{\mathrm{ZO}}} \int dz \left(\frac{\mathcal{R}_{ri}}{z}\right)\left(\tilde{\boldsymbol{k}}_m, z, 0\right) e^{-ik_{\mathrm{mz}}^{\mathrm{ZO}} z} \\
&= -i \int dz \mathcal{R}_{ri}\left(\tilde{\boldsymbol{k}}_m, z, 0\right) e^{-ik_{\mathrm{mz}}^{\mathrm{ZO}} z} \\
&= -i\mathcal{R}_{ri}\left(\tilde{\boldsymbol{k}}_m, k_{\mathrm{mz}}^{\mathrm{ZO}}, 0\right).
\end{aligned} \tag{9.68}$$

On the left-hand side, the derivative can be written

$$\begin{aligned}
\frac{d}{dk_{\mathrm{mz}}^{\mathrm{ZO}}} D_{ri}^{(\mathrm{3DZO})}\left(\tilde{\boldsymbol{k}}_m; \omega\right) &= \frac{d\omega}{dk_{\mathrm{mz}}^{\mathrm{ZO}}} \frac{d}{d\omega} D_{ri}^{(\mathrm{3DZO})}\left(\tilde{\boldsymbol{k}}_m; \omega\right) \\
&= \frac{d\omega}{dk_{\mathrm{mz}}^{\mathrm{ZO}}} \frac{d}{d\omega} \int dt e^{i\omega t} D_{ri}^{(\mathrm{3DZO})}\left(\tilde{\boldsymbol{k}}_m; t\right) \\
&= i \frac{d\omega}{dk_{\mathrm{mz}}^{\mathrm{ZO}}} \int dt e^{i\omega t} t D_{ri}^{(\mathrm{3DZO})}\left(\tilde{\boldsymbol{k}}_m; t\right) \\
&= i \frac{d\omega}{dk_{\mathrm{mz}}^{\mathrm{ZO}}} \left(t D_{ri}^{(\mathrm{3DZO})}\right)\left(\tilde{\boldsymbol{k}}_m; \omega\right).
\end{aligned} \tag{9.69}$$

From (9.57), the derivative of frequency with respect to vertical wavenumber is

$$\frac{d\omega}{dk_{\mathrm{mz}}^{\mathrm{ZO}}} = -\underline{c}\sqrt{1 - \tilde{k}_m^2 \underline{c}^2/\omega^2}. \tag{9.70}$$

Putting these expressions together, equation (9.67) becomes

$$\left(tD_{ri}^{(3DZO)}\right)\left(\tilde{\boldsymbol{k}}_m;\omega\right) \simeq \frac{1}{(2\pi)}\frac{\mathfrak{w}(\omega)}{2\rho_0 c_r c_i{}^2}\frac{\mathcal{R}_{ri}\left(\tilde{\boldsymbol{k}}_m, k_{\text{mz}}^{\text{ZO}}, 0\right)}{\sqrt{1-\tilde{k}_m^2\underline{c}^2/\omega^2}}. \tag{9.71}$$

Thus, the zero-offset formula involves either a depth-rescaled reflectivity function or time-rescaled data. Both formulas are correct only asymptotically.

9.3.2 Zero azimuth data

Consider a data set for which the y-component of offset is zero. The data is given by

$$\begin{aligned} D_{ri}^{(ZA)}\left(\tilde{\boldsymbol{x}}_m, x_h; \omega\right) &= D_{ri}\left(\tilde{\boldsymbol{x}}_m, x_h, 0; \omega\right) \\ &= -\frac{1}{(2\pi)^4}\frac{i\omega\mathfrak{w}(\omega)}{8\rho_0 c_r c_i{}^2}\int\!\!\int d^2\tilde{k}_m e^{i\tilde{\boldsymbol{k}}_m\cdot\tilde{\boldsymbol{x}}_m}\int dk_{hx}e^{i\,k_{hx}x_h} \\ &\cdot\int\!\!\int\!\!\int d^3x e^{-i\,\tilde{\boldsymbol{k}}_m\cdot\tilde{\boldsymbol{x}}}\int dk_{hy}\,e^{-ik_{mz}z}\frac{\cos\gamma_r}{k_{gz}k_{sz}}\mathcal{R}_{ri}(\boldsymbol{x},\sigma). \end{aligned} \tag{9.72}$$

This may be evaluated in the far field with a single stationary phase approximation. The stationary point k_{hy}^{ZA} of the k_{hy} integral is the value of k_{hy} at which the derivative of the phase term is zero, i.e. where $dk_{mz}/dk_{hy} = 0$. The stationary-phase approximation to this integral is

$$\int dk_{hy}\,e^{-ik_{mz}z}\frac{\cos\gamma_r}{k_{gz}k_{sz}}\mathcal{R}_{ri}(\boldsymbol{x},\sigma) \simeq \sqrt{\frac{-2\pi i}{zk_{mz}^{\text{ZA}\prime\prime}}}\frac{\cos\overset{\wedge}{\gamma_r}}{k_{gz}^{\text{ZA}}k_{sz}^{\text{ZA}}}\mathcal{R}_{ri}\left(\boldsymbol{x},\underset{\wedge}{\sigma}\right)e^{-ik_{mz}^{\text{ZA}}z}, \tag{9.73}$$

where the superscript "ZA" indicates that the variable is to be evaluated at its stationary point, where the derivative of k_{mz} with respect to k_{hy} is zero. Now, from (9.34), this derivative is

$$\frac{dk_{mz}}{dk_{hy}} = -\frac{k_{hy}+k_{my}}{4k_{gz}} + \frac{k_{hy}-k_{my}}{4k_{sz}}. \tag{9.74}$$

At the stationary point, this derivative is zero, yielding the relations

$$k_{gy}^{\text{ZA}} = k_{sy}^{\text{ZA}}\frac{k_{gz}^{\text{ZA}}}{k_{sz}^{\text{ZA}}},\; k_{hy}^{\text{ZA}} = k_{my}\frac{k_{hz}^{\text{ZA}}}{k_{mz}^{\text{ZA}}}. \tag{9.75}$$

In Appendix H, azimuthal angle χ is defined in terms of ray parameters as (H.39), or, in terms of $\boldsymbol{k}_g$ and $\boldsymbol{k}_s$,

$$\tan\chi = \frac{k_{sy}k_{gz}-k_{gy}k_{sz}}{k_{sx}k_{gz}-k_{gx}k_{sz}}. \tag{9.76}$$

At $\chi = 0$, this equation reduces to

$$k_{sy}k_{gz} = k_{gy}k_{sz}, \tag{9.77}$$

consistent with (9.75), indicating that asymptotically, azimuth may be equivalently defined either in terms of wavenumbers or source and receiver location.

The relations (9.75) define an angle ν:

$$\frac{k_{gy}^{ZA}}{k_{gz}^{ZA}} = \frac{k_{sy}^{ZA}}{k_{sz}^{ZA}} = \frac{k_{hy}^{ZA}}{k_{hz}^{ZA}} = \frac{k_{my}}{k_{mz}^{ZA}} \equiv \tan\nu. \tag{9.78}$$

Define the 2-D vertical wavenumbers

$$k_{gz0} = -\sqrt{\frac{\omega^2}{{c_r}^2} - \frac{(k_{mx}+k_{hx})^2}{4}} = -\frac{\omega}{c_r}\cos\phi_r,$$

$$k_{sz0} = +\sqrt{\frac{\omega^2}{c_i^2} - \frac{(k_{mx}-k_{hx})^2}{4}} = +\frac{\omega}{c_i}\cos\phi_i, \tag{9.79}$$

and

$$k_{mz0} = k_{gz0} - k_{sz0},$$

$$k_{hz0} = k_{gz0} + k_{sz0}. \tag{9.80}$$

The 2-D wavenumbers are related to their 3-D counterparts:

$${k_{gz0}}^2 = {k_{gz}}^2 + {k_{gy}}^2,$$

$${k_{sz0}}^2 = {k_{sz}}^2 + {k_{sy}}^2. \tag{9.81}$$

At the stationary point (9.75), the above equations become

$${k_{gz0}}^2 = (k_{gz}^{ZA})^2\left((k_{sz}^{ZA})^2 + (k_{sy}^{ZA})^2\right) / (k_{sz}^{ZA})^2,$$

$${k_{sz0}}^2 = (k_{sz}^{ZA})^2 + (k_{sy}^{ZA})^2. \tag{9.82}$$

Combining,

$$\frac{k_{sz}^{ZA}}{k_{sz0}} = \frac{k_{gz}^{ZA}}{k_{gz0}} = \frac{k_{hz}^{ZA}}{k_{\mathrm{hz}0}} = \frac{k_{\mathrm{mz}}^{\ ZA}}{k_{\mathrm{mz}0}} \equiv \cos\nu. \tag{9.83}$$

With these relations and the stationarity equations, we also have

$$\frac{k_{sy}^{ZA}}{k_{sz0}} = \frac{k_{gy}^{ZA}}{k_{gz0}} = \frac{k_{hy}^{ZA}}{k_{hz0}} = \frac{k_{my}}{k_{mz0}} \equiv \sin\nu. \tag{9.84}$$

From the last of these relations, we explicitly determine the stationary value of k_{hy} to be

$$k_{hy}^{ZA} = k_{hz0}\sin\nu = k_{hz0}\frac{k_{my}}{k_{mz0}}. \tag{9.85}$$

Likewise, the stationary value of k_{mz} is

$$k_{mz}^{\mathrm{ZA}} = k_{mz0} \cos \nu = k_{mz0} \sqrt{1 - \frac{k_{my}^2}{k_{mz0}^2}}. \tag{9.86}$$

The second derivative of k_{mz} is

$$\frac{d^2 k_{mz}}{dk_{hy}^2} = -\frac{{k_{gz0}}^2}{4k_{gz}^3} + \frac{{k_{sz0}}^2}{4k_{sz}^3}, \tag{9.87}$$

which becomes at the stationary point

$$k_{mz}^{\mathrm{ZA}\prime\prime} = \frac{k_{mz0}}{4\cos^3 \nu} \left(\frac{1}{k_{gz0} k_{sz0}} \right). \tag{9.88}$$

The opening angle σ^{ZA} at the stationary point is given by

$$\begin{aligned} \cos\left(\sigma^{\mathrm{ZA}}\right) &= -\frac{c_i c_r}{\omega^2} \boldsymbol{k}_g{}^{\mathrm{ZA}} \cdot \boldsymbol{k}_s{}^{\mathrm{ZA}} \\ &= -\frac{c_i c_r}{\omega^2} \left(k_{gx} k_{sx} + k_{gz0} k_{sz0} \right). \end{aligned} \tag{9.89}$$

In the stationary phase approximation, the relation between data and reflectivity thus becomes

$$D_{ri}^{(\mathrm{ZA})}\left(\tilde{\boldsymbol{x}}_m, x_h; \omega\right) = \frac{1}{(2\pi)^{7/2}} \frac{\sqrt{i\omega}\mathfrak{w}(\omega)}{4\rho_0 c_r {c_i}^2} \int\!\!\int d^2\tilde{k}_m e^{i\tilde{\boldsymbol{k}}_m \cdot \tilde{\boldsymbol{x}}_m} \times \int dk_{hx} e^{i\, k_{hx} x_h} \frac{\cos \gamma_r^{\mathrm{ZA}}}{\sqrt{k_{mz}^{\mathrm{ZA}} k_{gz0} k_{sz0}}} \int\!\!\int\!\!\int d^3x e^{-i\left(\tilde{\boldsymbol{k}}_m \cdot \tilde{\boldsymbol{x}} + k_{mz}^{\mathrm{ZA}} \mathbf{z}\right)} \frac{\mathcal{R}_{ri}\left(\boldsymbol{x}, \sigma^{\mathrm{ZA}}\right)}{\sqrt{z}}, \tag{9.90}$$

with k_{gz0} and k_{sz0} given by (9.79), k_{mz}^{ZA} by (9.86), and σ^{ZA} by (9.89).

While the full data function D_{ri} depends on five variables, the zero-azimuth data function $D_{ri}^{(\mathrm{ZA})}$ depends on only four. The six integrals in (9.90) are Fourier transforms, from $(\tilde{\boldsymbol{x}}, z)$ to $\left(\tilde{\boldsymbol{k}}_m, k_{mz}^{\mathrm{ZA}}\right)$ and from $\left(\tilde{\boldsymbol{k}}_m, k_{hx}\right)$ to $(\tilde{\boldsymbol{x}}_m, x_h)$. Consequently, we can express the equation in wavenumber–frequency:

$$D_{ri}^{(\mathrm{ZA})}\left(\tilde{\boldsymbol{k}}_m, k_{hx}; \omega\right) = \frac{1}{(2\pi)^{1/2}} \frac{\sqrt{i\omega}\mathfrak{w}(\omega)}{4\rho_0 c_r {c_i}^2} \frac{\cos \gamma_r^{\mathrm{ZA}}}{\sqrt{k_{mz}^{\mathrm{ZA}} k_{gz0} k_{sz0}}} \left(\frac{\mathcal{R}_{ri}}{\sqrt{z}}\right) \left(\tilde{\boldsymbol{k}}_m, k_{mz}^{\mathrm{ZA}}, \sigma^{\mathrm{ZA}}\right), \tag{9.91}$$

with $\mathcal{R}_{ri}/\sqrt{z}$ a depth-rescaled reflectivity function, expressed in the wavenumber–angle domain. Since y-offset is not a variable in $D_{ri}^{(\mathrm{ZA})}$, the y-offset wavenumber k_{hy} does not appear in its argument list. Rather, the stationary value $k_{hy}^{\mathrm{ZA}} = k_{hz0} \frac{k_{my}}{k_{mz0}}$ is fixed by the other wavenumbers. This value then determines the relations between the data and reflectivity parameters.

9.3.3 2.5-D

Another opportunity to simplify occurs where reflectivity and data depend on x and z only. In this case the y-integral in the zero-azimuth formula (9.90) produces a delta function $2\pi\ \delta\left(k_{my}\right)$ which in turn eliminates the k_{my} integral, leaving

$$D_{ri}^{(2.5\mathrm{D})}\left(x_m,\ x_h;\ \omega\right) = \frac{\sqrt{i}}{(2\pi)^{5/2}} \frac{\omega\mathfrak{w}(\omega)}{4\rho_0 c_r {c_i}^2} \int dk_{mx} e^{ik_{mx}x_m} \int dk_{hx} e^{ik_{hx}x_h} \frac{\cos\ \gamma_r^{\mathrm{ZA}}}{\sqrt{{k_{mz}}^{\mathrm{ZA}} k_{gz0} k_{sz0}}}$$
$$\times \int dx \int dz e^{-i\left(k_{mx}x + {k_{mz}}^{\mathrm{ZA}} z\right)} \left(\frac{\mathcal{R}_{ri}}{\sqrt{z}}\right)\left(x,\ z, \sigma^{\mathrm{ZA}}\right). \tag{9.92}$$

The stationary point (9.75) reduces to $k_{hy} = 0$, from which it follows that

$$k_{gz} \to k_{gz0} = -\sqrt{\frac{\omega^2}{{c_r}^2} - \frac{\left(k_x - k_{hx}\right)^2}{4}},$$
$$k_{sz} \to k_{sz0} = \sqrt{\frac{\omega^2}{c_i^2} - \frac{\left(k_x + k_{hx}\right)^2}{4}}, \tag{9.93}$$

$$k_{mz} \to k_{mz0} = k_{gz0} - k_{sz0}, \tag{9.94}$$

$$k_{mz}{''} \to \frac{1}{4} \frac{k_{mz0}}{k_{gz0} k_{sz0}}, \tag{9.95}$$

and

$$D_{ri}^{(2.5\mathrm{D})}\left(x_m,\ x_h;\ \omega\right) = \frac{\sqrt{i}}{(2\pi)^{5/2}} \frac{\omega\mathfrak{w}(\omega)}{4\rho_0 c_r {c_i}^2} \int dk_{mx} e^{ik_{mx}x_m} \int dk_{hx} e^{ik_{hx}x_h}$$
$$\frac{\cos\gamma_{r0}}{\sqrt{k_{mz0} k_{gz0} k_{sz0}}} \int dx \int dz e^{-i\left(k_{mx}x + k_{mz0} z\right)} \left(\frac{\mathcal{R}_{ri}}{\sqrt{z}}\right)\left(x, z, \sigma_0\right). \tag{9.96}$$

The data formula (9.96) is sometimes referred to as a 2.5-dimensional formula because it describes point sources and receivers in an otherwise 2-D world.

Since the integrals in (9.96) are forward and inverse Fourier transforms, the equation can be expressed in the Fourier domain as

$$D_{ri}^{(2.5\mathrm{D})}\left(k_{mx},\ k_{hx};\ \omega\right) = \frac{\sqrt{i}}{(2\pi)^{1/2}} \frac{\omega\mathfrak{w}(\omega)}{4\rho_0 c_r {c_i}^2} \frac{\cos\ \gamma_{r0}}{\sqrt{k_{mz0} k_{gz0} k_{sz0}}} \left(\frac{\mathcal{R}_{ri}}{\sqrt{z}}\right)\left(k_{mx},\ k_{mz0}, \sigma_0\right). \tag{9.97}$$

9.3.4 2.5-D zero offset

For zero offset, the 2.5-D formula becomes even simpler. We have

$$D_{ri}^{(2.5DZO)}(x_m;\omega) = \frac{\sqrt{i}}{(2\pi)^{5/2}} \frac{\omega \mathfrak{w}(\omega)}{4\rho_0 c_r c_i{}^2} \int dk_{mx} e^{ik_{mx}x_m} \int dx e^{-ik_{mx}x} \int dz \int dk_{hx} \frac{\cos\,\gamma_{r0}}{\sqrt{k_{mz0}k_{gz0}k_{sz0}}} e^{-i\,k_{mz0}z} \frac{\mathcal{R}_{ri}(\boldsymbol{x},\sigma_0)}{\sqrt{z}}. \quad (9.98)$$

At the stationary point for the k_{hx} integral,

$$k_{gx} \to -k_{sx}c_i/c_r, \quad (9.99)$$

which, when combined with the requirement that $k_{gx} - k_{sx} = k_{mx}$, yields

$$\begin{aligned} k_{gx} &\to k_{mx}c_i/(c_r+c_i) &= k_{mx}\underline{c}/c_r\,, \\ k_{sx} &\to -k_{mx}c_r/(c_r+c_i) &= -k_{mx}\underline{c}/c_i\,, \end{aligned} \quad (9.100)$$

$$\begin{aligned} k_{gz} &\to -\frac{\omega}{c_r}\sqrt{1-\frac{k_{mx}^2\underline{c}^2}{\omega^2}}, \\ k_{sz} &\to +\frac{\omega}{c_i}\sqrt{1-\frac{k_{mx}^2\underline{c}^2}{\omega^2}}, \\ k_{mz} = k_{gz} - k_{sz} &\to -\frac{\omega}{\underline{c}}\sqrt{1-\frac{k_{mx}^2\underline{c}^2}{\omega^2}}, \end{aligned} \quad (9.101)$$

or

$$\omega \to \underline{c}\sqrt{k_{mz}{}^2 + k_{mx}^2}, \quad (9.102)$$

with $\underline{c}$ the inverse of the sum of inverse velocities $\left(\underline{c}^{-1} = c_r^{-1} + c_i^{-1}\right)$.

The second derivative of k_{mz} with respect to k_{hx} at the stationary point is

$$k_{mz}{}'' \to \frac{c_i c_r}{4\omega\left(1 - k_{mx}^2\underline{c}^2/\omega^2\right)^{3/2}\underline{c}}. \quad (9.103)$$

The opening angle at the stationary point is zero. Thus, the stationary phase approximation for the k_{hx} integral is

$$\int dk_{hx} \frac{\cos\,\overset{\wedge}{\gamma_r}}{\sqrt{k_{mz}k_{gz0}k_{sz0}}} e^{-i\,k_{mz}z} \frac{\mathcal{R}_{ri}(\boldsymbol{x},\sigma)}{\sqrt{z}} = \sqrt{-2\pi i}\frac{2\underline{c}}{\omega} e^{-i\,k_{mz}z} \frac{\mathcal{R}_{ri}(\boldsymbol{x},0)}{z}, \quad (9.104)$$

and the expression for 2.5-D zero-offset data becomes

$$D_{ri}^{(2.5\mathrm{DZO})}(x_m;\omega) = \frac{1}{(2\pi)^2}\frac{\mathfrak{w}(\omega)\underline{c}}{2\rho_0 c_r {c_i}^2}\cdot\int dk_{mx}e^{ik_{mx}x_m}$$
$$\int dx\int dz e^{-ik_{mx}x}e^{-i\,k_{mz}z}\left(\frac{\mathcal{R}_{ri}}{z}\right)(x,\,z,0). \tag{9.105}$$

This expression could also be deduced directly from the 3-D zero-offset formula (9.65). Algorithmically, (9.105) consists of (1) rescaling reflectivity by $1/z$, (2) a 2-D forward Fourier transform of reflectivity to wavenumber coordinates (k_{mx}, k_{mz}), (3) a coordinate transform from k_{mz} to ω according to (9.102), (4) an inverse Fourier transform from k_{mx} to x_m, and (5) multiplication by the factor $\mathfrak{w}(\omega)\underline{c}\,/\left(16\pi\rho_0 c_r^2 c_i\right)$. To put the data in the time domain requires a final inverse Fourier transform from ω to t.

The Fourier-domain expression for (9.105) is

$$D_{ri}^{(2.5\mathrm{DZO})}(k_{mx};\omega) = \frac{\mathfrak{w}(\omega)\underline{c}}{4\pi\rho_0 c_r {c_i}^2}\left(\frac{\mathcal{R}_{ri}}{z}\right)(k_{mx},\,k_{mz},0)\,. \tag{9.106}$$

9.4 A depth-variable background

Suppose the background values of the elastic parameters ρ_0, α_0, and β_0 vary slowly with depth.

For a constant background, we were able to write an expression for reflection data in terms of scattering potential or reflectivity, making no far-field or high-frequency approximations. For a variable background, including a depth-variable background, we content ourselves with a WKBJ solution to the background wave equation.

9.4.1 Three-dimensional form

For a background that varies only with depth, one can still perform 2-D Fourier transforms over the surface coordinates $\tilde{\boldsymbol{x}}_s$ and $\tilde{\boldsymbol{x}}_g$. The resulting elastic-wave background Green's functions are

$$\mathcal{G}_i(\tilde{\boldsymbol{k}}_s, z, \omega) = -\frac{e^{i\int_0^z dz' k_{sz}(z')}}{2ic_i(0)c_i(z)\sqrt{\rho_0(0)k_{sz}(0)\rho_0(z)k_{sz}(z)}}, \tag{9.107}$$

and

$$\mathcal{G}_r(\tilde{\boldsymbol{k}}_g, z, \omega) = -\frac{e^{-i\int_0^z dz' k_{gz}(z')}}{2ic_r(0)c_r(z)\sqrt{\rho_0(0)k_{gz}(0)\rho_0(z)k_{gz}(z)}}, \tag{9.108}$$

with the vertical wavenumbers given by

$$k_{gz}(z) = -\frac{\omega}{c_r(z)}\sqrt{1 - \frac{c_r^2(z)}{\omega^2}\left({k_{gx}}^2 + {k_{gy}}^2\right)},$$

$$k_{sz}(z) = +\frac{\omega}{c_i(z)}\sqrt{1 - \frac{c_i^2(z)}{\omega^2}\left({k_{sx}}^2 + {k_{sy}}^2\right)}. \tag{9.109}$$

The reflection response for any elastic component is

$$D_{ri}\left(\tilde{\boldsymbol{k}}_g|\tilde{\boldsymbol{k}}_s;\omega\right) = \frac{-\mathfrak{w}(\omega)}{4\sqrt{\left|k_{gz}(0)k_{sz}(0)\right|\rho_0(0)c_r(0)c_i(0)}}$$

$$\int\int\int d^3x \cdot e^{-i\left(\tilde{\boldsymbol{k}}_g\cdot\tilde{\boldsymbol{x}}+\int_0^z dz' k_{gz}(z')\right)} \frac{\mathcal{V}_{ri}(\boldsymbol{x},\omega)}{\sqrt{\left|k_{gz}(z)k_{sz}(z)\right|\rho_0(z)c_r(z)c_i(z)}} e^{i\left(\tilde{\boldsymbol{k}}_s\cdot\tilde{\mathbf{x}}+\int_0^z dz' k_{sz}(z')\right)}. \tag{9.110}$$

Conversion of the scattering potential from a differential operator to a function of angle, exact for a constant background, is for a depth-variable background a high-frequency approximation. Within this approximation, we can write

$$D_{ri}\left(\tilde{\boldsymbol{k}}_g|\tilde{\boldsymbol{k}}_s;\omega\right) \simeq -\frac{\omega^2\mathfrak{w}(\omega)}{4\sqrt{\left|k_{gz}(0)k_{sz}(0)\right|\rho_0(0)c_r(0)c_i(0)}}\int dx\int dy\int dz$$

$$e^{-i\left(\tilde{\boldsymbol{k}}_m\cdot\tilde{\boldsymbol{x}}+\int_0^z dz' k_{mz}(z')\right)} \frac{\mathbb{V}_{ri}(\boldsymbol{x},\sigma(z))}{\sqrt{\left|k_{gz}(z)k_{sz}(z)\right|\rho_0(z)c_r(z)c_i(z)}}, \tag{9.111}$$

with

$$\tilde{\boldsymbol{k}}_m(z) = \tilde{\boldsymbol{k}}_g(z) - \tilde{\boldsymbol{k}}_s(z), \tag{9.112}$$

and

$$k_{mz}(z) = k_{gz}(z) - k_{sz}(z). \tag{9.113}$$

Note that for positive ω, k_{gz} and k_{mz} are negative, while k_{sz} is positive.

As for the constant-background case, the magnitude $k_m(z)$ of the 3-vector $\boldsymbol{k}_m(z) = \left(\tilde{\boldsymbol{k}}_m, k_{mz}(z)\right)$ is related to the opening angle σ:

$$k_m(z) = -\omega\frac{\sqrt{2c_r(z)c_i(z)\cos\sigma(z) + c_r(z)^2 + c_i(z)^2}}{c_r(z)c_i(z)}, \tag{9.114}$$

or

$$\cos\sigma(z) = \frac{k_m(z)^2 c_r(z)c_i(z)}{2\omega^2} - \frac{1}{2}\left(\frac{c_r(z)}{c_i(z)} + \frac{c_i(z)}{c_r(z)}\right). \tag{9.115}$$

The x- and y-integrals in (9.111) are Fourier transforms, leaving us with

$$D_{ri}\left(\tilde{\boldsymbol{k}}_g|\tilde{\boldsymbol{k}}_s;\omega\right) = \frac{-\omega^2\mathfrak{w}(\omega)}{4\sqrt{\left|k_{gz}(0)k_{sz}(0)\right|\rho_0(0)c_r(0)c_i(0)}}$$
$$\times\int dz e^{-i\int_0^z dz' k_{mz}(z')}\frac{\mathbb{V}_{ri}\left(\tilde{\boldsymbol{k}}_m, z, \sigma(z)\right)}{\sqrt{\left|k_{gz}(z)k_{sz}(z)\right|\rho_0(z)c_r(z)c_i(z)}}. \tag{9.116}$$

The remaining integral over depth is a phase-shift operation. Thus, given a scattering potential and a depth-dependent background, one can synthesize reflection data by first Fourier transforming the x- and y-coordinates of the potential to k_{mx} and k_{my}, then rescaling and performing the phase-shift integration as per (9.116), The resulting data function is inverse transformed back to space and time.

If desired, one can couch (9.116) in terms of reflectivity, invoking the relation (8.56). Whereas for a constant background this relation was exact, here it is approximate in that we are neglecting z-derivatives of background velocities, densities, and incident angle. We have

$$D_{ri}\left(\tilde{\boldsymbol{k}}_g|\tilde{\boldsymbol{k}}_s;\omega\right) = \frac{i\omega\mathfrak{w}(\omega)}{2\sqrt{\left|k_{gz}(0)k_{sz}(0)\right|\rho_0(0)c_r(0)c_i(0)}}$$
$$\times\int dz e^{-i\int_0^z dz' k_{mz}(z')}\frac{\mathcal{R}_{ri}\left(\tilde{\boldsymbol{k}}_m, z, \sigma\right)\cos\left(\gamma_r(z)\right)}{\sqrt{\left|k_{gz}(z)k_{sz}(z)\right|c_i(z)}}. \tag{9.117}$$

9.4.2 Radial parameter form

Radial parameters provide an alternative to the phase-shift formulations (9.116) and (9.117). Define

$$\begin{aligned}
\tilde{\boldsymbol{p}}_g &\equiv \tilde{\boldsymbol{k}}_g/\omega,\\
\tilde{\boldsymbol{p}}_s &\equiv \tilde{\boldsymbol{k}}_s/\omega,\\
\tilde{\boldsymbol{p}}_m &\equiv \tilde{\boldsymbol{k}}_m/\omega = \tilde{\boldsymbol{p}}_g - \tilde{\boldsymbol{p}}_s,\\
p_{gz} &\equiv k_{gz}/\,\omega,\\
p_{sz} &\equiv k_{sz}/\,\omega,\\
p_{mz} &\equiv p_{gz} - p_{sz},\\
\boldsymbol{p}_m &= \left(\tilde{\boldsymbol{p}}_m, p_{mz}\right).
\end{aligned} \tag{9.118}$$

We can express the data in terms of these parameters:

$$D_{ri}\left(\tilde{\boldsymbol{p}}_g|\tilde{\boldsymbol{p}}_s;\omega\right) \equiv D_{ri}\left(\tilde{\boldsymbol{k}}_g|\tilde{\boldsymbol{k}}_s;\omega\right). \tag{9.119}$$

$D_{ri}\left(\tilde{p}_g|\tilde{p}_s;\omega\right)$ can represent a change of variable from wavenumbers $\tilde{\boldsymbol{k}}_g$ and $\tilde{\boldsymbol{k}}_s$ to $\tilde{\boldsymbol{p}}_g$ and $\tilde{\boldsymbol{p}}_s$. Alternatively, starting from data in space–time, it may be realized by a four-fold Radon transform of the source and receiver coordinates and a Fourier transform from time to frequency. Either way, the formula (9.117) becomes

$$D_{ri}\left(\tilde{\boldsymbol{p}}_g|\tilde{\boldsymbol{p}}_s;\omega\right) = \frac{i\,\mathfrak{w}(\omega)}{2\omega\sqrt{p_{gz}(0)p_{sz}(0)}\rho_0(0)c_r(0)c_i(0)}$$
$$\times \int dz e^{-i\omega\int_0^z dz' p_{mz}(z')} \frac{\mathcal{R}_{ri}\left(\omega\tilde{\boldsymbol{p}}_m, z, \sigma(z)\right)\cos\left(\gamma_r(z)\right)}{\sqrt{p_{gz}(z)p_{sz}(z)}c_i(z)}. \tag{9.120}$$

Defining a time-parameter τ as

$$\tau\left(\tilde{\boldsymbol{p}}_g|\tilde{\boldsymbol{p}}_s;z\right) = \int_0^z dz' p_{mz}(z'), \tag{9.121}$$

we can change integration variable from z to τ, so that

$$D_{ri}\left(\tilde{\boldsymbol{p}}_g|\tilde{\boldsymbol{p}}_s;\omega\right) = \frac{i\,\mathfrak{w}(\omega)}{2\omega\sqrt{p_{gz}(0)p_{sz}(0)}\rho_0(0)c_r(0)c_i(0)}$$
$$\times \int d\tau e^{-i\omega\tau} \frac{\mathcal{R}_{ri}\left(\omega\tilde{\boldsymbol{p}}_m, z, \sigma(z)\right)\cos\left(\gamma_r(z)\right)}{p_{mz}(z)\sqrt{p_{gz}(z)p_{sz}(z)}c_i(z)}. \tag{9.122}$$

In this expression, z must be considered a function of τ, $\tilde{\boldsymbol{p}}_g$ and $\tilde{\boldsymbol{p}}_s$ as determined by (9.121). The integral in (9.122) is a Fourier transform; however, the dependence of $\mathcal{R}_{ri}$ upon lateral wavenumber $\tilde{\boldsymbol{k}}_m$, which in turn depends upon frequency ω, does complicate matters somewhat.

9.4.3 2.5-dimensional form

Suppose the scattering potential (or equivalently, reflectivity) depends on x and z but not y. Then, in equation (9.111), the y-integral evaluates to a delta function:

$$D_{ri}^{(2.5\mathrm{D})}\left(\tilde{\boldsymbol{k}}_g|\tilde{\boldsymbol{k}}_s;\omega\right) = \frac{-2\pi\omega^2\mathfrak{w}(\omega)\delta\left(k_{my}\right)}{4\sqrt{\left|k_{gz}(0)k_{sz}(0)\right|}\rho_0(0)c_r(0)c_i(0)}$$
$$\times \int dx \int dz e^{-i\left(k_{mx}x+\int_0^z dz' k_{mz}(z')\right)} \frac{\mathbb{V}_{ri}^{(2.5\mathrm{D})}(x,z,\sigma(z))}{\sqrt{\left|k_{gz}(z)k_{sz}(z)\right|}\rho_0(z)c_r(z)c_i(z)}. \tag{9.123}$$

The x-integral is still a Fourier transform, so we may write

$$D_{ri}^{(2.5\mathrm{D})}\left(\tilde{\boldsymbol{k}}_g|\tilde{\boldsymbol{k}}_s;\omega\right) = -\frac{2\pi\omega^2\mathfrak{w}(\omega)\delta\left(k_{my}\right)}{4\sqrt{\left|k_{gz}(0)k_{sz}(0)\right|\rho_0(0)c_r(0)c_i(0)}}$$
$$\times\int dz e^{-i\int_0^z dz' k_{mz}(z')}\frac{\mathbb{V}_{ri}^{(2.5\mathrm{D})}\left(k_{mx},z,\sigma(z)\right)}{\sqrt{\left|k_{gz}(z)k_{sz}(z)\right|\rho_0(z)c_r(z)c_i(z)}}. \tag{9.124}$$

In the space domain, this expression becomes

$$D_{ri}^{(2.5\mathrm{D})}\left(\tilde{\boldsymbol{x}}_m,\tilde{\boldsymbol{x}}_h;\omega\right) \equiv D_{ri}^{(2.5\mathrm{D})}\left(\tilde{\boldsymbol{x}}_g|\tilde{\boldsymbol{x}}_s;\omega\right) = -\frac{1}{(2\pi)^3}\frac{\omega^2\mathfrak{w}(\omega)}{16\rho_0(0)c_r(0)c_i(0)}\int\!\!\int d^2\tilde{k}_h$$
$$\times e^{i\tilde{\boldsymbol{k}}_h\cdot\tilde{\boldsymbol{x}}_h}\int dk_{mx}\frac{e^{ik_{mx}x_m}}{\sqrt{\left|k_{gz}(0)k_{sz}(0)\right|}}\int dz e^{-i\int_0^z dz' k_{mz}(z')}$$
$$\times\frac{\mathbb{V}_{ri}^{(2.5\mathrm{D})}\left(k_{mx},z,\sigma(z)\right)}{\sqrt{\left|k_{gz}(z)k_{sz}(z)\right|\rho_0(z)c_r(z)c_i(z)}}. \tag{9.125}$$

As one might expect, the data does not depend on y_m. For 2.5-D, the data is restricted to $y_h = 0$. That is,

$$D_{ri}^{(2.5\mathrm{D})}\left(x_m,x_h;\omega\right) \equiv D_{ri}^{(2.5\mathrm{D})}\left(\tilde{\boldsymbol{x}}_m,\tilde{\boldsymbol{x}}_h=(x_h,0);\omega\right) = -\frac{1}{(2\pi)^3}\frac{\omega^2\mathfrak{w}(\omega)}{16\rho_0(0)c_r(0)c_i(0)}$$
$$\times\int dk_{hx}e^{ik_{hx}x_h}\int dk_{mx}e^{ik_{mx}x_m}\int dz\int dk_{hy}$$
$$\times\frac{e^{-i\int_0^z dz' k_{mz}(z')}\mathbb{V}_{ri}^{(2.5\mathrm{D})}\left(k_{mx},\ z,\sigma(z)\right)}{\sqrt{k_{gz}(0)k_{sz}(0)k_{gz}(z)k_{sz}(z)\rho_0(z)c_r(z)c_i(z)}}. \tag{9.126}$$

We can perform a stationary-phase approximation to the k_{hy} integral. Rapid dependence on k_{hy} is confined to the term in the exponential, which we write as

$$\lambda(z) = -\int_0^z dz' k_{mz}(z'). \tag{9.127}$$

The stationary point for k_{hy} is where $\partial\lambda/\partial k_{hy} = 0$, which is at $k_{hy} = 0$. At the stationary point, the second derivative of λ is

$$\lambda'' \equiv \left[\frac{\partial^2\lambda(z)}{\partial {k_{hy}}^2}\right]_{k_{hy}=0} = -\frac{1}{4}\int_0^z dz'\frac{\underset{\wedge}{k}_{mz}(z')}{\underset{\wedge}{k}_{gz}(z')\underset{\wedge}{k}_{sz}(z')}, \tag{9.128}$$

with

$$\begin{aligned} \underset{\wedge}{k}_{mz}(z) &= \underset{\wedge}{k}_{gz}(z) - \underset{\wedge}{k}_{sz}(z) \\ &= -\frac{\omega}{c_r(z)}\sqrt{1 - \frac{k_{gx}{}^2 c_r^2(z)}{\omega^2}} - \frac{\omega}{c_i(z)}\sqrt{1 - \frac{k_{sx}{}^2 c_i^2(z)}{\omega^2}}. \end{aligned} \tag{9.129}$$

Invoking the stationary phase approximation for the k_{hy} integral, we obtain

$$\begin{aligned} D_{ri}^{(2.5\,\mathrm{D})}(x_m, x_h; \omega) &= -\frac{1}{(2\pi)^{5/2}} \frac{\omega^2 \mathfrak{w}(\omega)}{8\rho_0(0)c_r(0)c_i(0)} \int dk_{hx} e^{ik_{hx}x_h} \int dk_{mx} e^{ik_x x_m} \\ &\times \int dz \frac{e^{-i\int_0^z dz' \underset{\wedge}{k}_{mz}(z')}}{\sqrt{\int_0^z dz' \frac{\underset{\wedge}{k}_{mz}(z')}{\underset{\wedge}{k}_{gz}(z')\underset{\wedge}{k}_{sz}(z')}}} \frac{e^{-i\pi/4}\mathbb{V}_{ri}^{(2.5D)}(k_{mx}, z, \sigma(z))}{\sqrt{\underset{\wedge}{k}_{gz}(0)\underset{\wedge}{k}_{sz}(0)\underset{\wedge}{k}_{gz}(z)\underset{\wedge}{k}_{sz}(z)\rho_0(z)c_r(z)c_i(z)}}. \end{aligned} \tag{9.130}$$

In the wavenumber domain, this can be written

$$\begin{aligned} D_{ri}^{(2.5\mathrm{D})}(k_{mx}, k_{hx}; \omega) &= -\frac{1}{(2\pi)^{1/2}} \frac{\omega^2 \mathfrak{w}(\omega)}{8\rho_0(0)c_r(0)c_i(0)} \int dz \frac{e^{-i\int_0^z dz' \underset{\wedge}{k}_{mz}(z')}}{\rho_0(z)c_r(z)c_i(z)} \\ &\times \frac{e^{-\mathrm{i}\pi/4}\mathbb{V}_{ri}^{(2.5\mathrm{D})}(k_{mx}, z, \sigma(z))}{\sqrt{\underset{\wedge}{k}_{gz}(0)\underset{\wedge}{k}_{sz}(0)\underset{\wedge}{k}_{gz}(z)\underset{\wedge}{k}_{sz}(z) \int_0^z dz' \frac{\underset{\wedge}{k}_{mz}(z')}{\underset{\wedge}{k}_{gz}(z')\underset{\wedge}{k}_{sz}(z')}}}. \end{aligned} \tag{9.131}$$

Substituting reflectivity for the scattering potential, this expression becomes

$$\begin{aligned} D_{ri}^{(2.5\mathrm{D})}(k_{mx}, k_{hx}; \omega) &= \frac{\sqrt{i}}{(2\pi)^{1/2}} \frac{\omega \mathfrak{w}(\omega)}{4\rho_0(0)c_r(0)c_i(0)} \int dz \frac{e^{-i\int_0^z dz' \underset{\wedge}{k}_{mz}(z')}}{c_i(z)} \\ &\times \frac{\cos\gamma_r \mathcal{R}_{ri}^{(2.5\mathrm{D})}(k_{mx}, z, \sigma(z))}{\sqrt{\underset{\wedge}{k}_{gz}(0)\underset{\wedge}{k}_{sz}(0)\underset{\wedge}{k}_{gz}(z)\underset{\wedge}{k}_{sz}(z) \int_0^z dz' \frac{\underset{\wedge}{k}_{mz}(z')}{\underset{\wedge}{k}_{gz}(z')\underset{\wedge}{k}_{sz}(z')}}}. \end{aligned} \tag{9.132}$$

9.5 A generally variable background

For a generally variable background, the Green's functions for the incoming and outgoing waves should have the asymptotic form (6.24)

$$G_i(\boldsymbol{x}\,|\boldsymbol{x}_s;\ \omega) = g_i(\boldsymbol{x}\,|\boldsymbol{x}_s)\, e^{i\omega\tau_i(\boldsymbol{x}|\boldsymbol{x}_s)}, \tag{9.133}$$

$$G_r\left(\boldsymbol{x}_g\middle|\,\boldsymbol{x}; \omega\right) = g_r(\boldsymbol{x}\,|\boldsymbol{x}_s)\, e^{i\omega\tau_r(\boldsymbol{x}_g|\boldsymbol{x})}, \tag{9.134}$$

with $\tau_i(\boldsymbol{x}|\boldsymbol{x}_s)$ the traveltime from $\boldsymbol{x}_s$ to $\boldsymbol{x}$, $\tau_r(\boldsymbol{x}_g|\boldsymbol{x})$ the traveltime from $\boldsymbol{x}$ to $\boldsymbol{x}_g$, and $g_i(\boldsymbol{x}|\boldsymbol{x}_s)$ and $g_r(\boldsymbol{x}_g|\boldsymbol{x})$ the asymptotic amplitude functions. That rays exist between $\boldsymbol{x}_s$ and $\boldsymbol{x}$, and between $\boldsymbol{x}$ and $\boldsymbol{x}_g$, must be assumed at this point. We further assume for now that only one ray links each of the two pairs of points, so that the traveltime function is single-valued. Extension to multiple raypaths is straightforward – one need only sum the single-raypath formula over all possible raypaths. The number of raypaths in the sum depends, of course, on the starting and end points of the Green's functions.

In terms of the Green's functions (9.133) and (9.134), surface reflection data can be written as the Born integral

$$D_{ri}(\tilde{\boldsymbol{x}}_g|\tilde{\boldsymbol{x}}_s;\omega) = \mathfrak{w}(\omega)\int\int\int d^3x g_r(\boldsymbol{x}_g|\boldsymbol{x})\, e^{i\omega\tau_r(\boldsymbol{x}_g|\boldsymbol{x})}\mathcal{V}_{ri}(\boldsymbol{x},\omega) g_i(\boldsymbol{x}|\boldsymbol{x}_s)\, e^{i\omega\tau_i(\boldsymbol{x}|\boldsymbol{x}_s)}. \tag{9.135}$$

Equation (9.135) gives the total reflection response as a weighted, phase-delayed volume integral of the scattering potential.

The asymptotic Green's functions are assumed locally planar, in that

$$\nabla_x\left(g_i(\boldsymbol{x}|\boldsymbol{x}_s)\, e^{-i\omega\tau_i(x|\boldsymbol{x}_s)}\right) \simeq \mathrm{i}\omega\nabla_x\tau_i(\boldsymbol{x}|\boldsymbol{x}_s)\, g_i(x|\boldsymbol{x}_s)\, e^{i\omega\tau_i(\boldsymbol{x}|\boldsymbol{x}_s)}, \tag{9.136}$$

and

$$\nabla_x\left(g_r(\boldsymbol{x}_g|\boldsymbol{x})\, e^{-i\omega\tau_r(\boldsymbol{x}_g|\boldsymbol{x})}\right) \simeq \mathrm{i}\omega\nabla_x\tau_r(\boldsymbol{x}_g|\boldsymbol{x})\, g_r(\boldsymbol{x}_g|\boldsymbol{x})\, e^{i\omega\tau_r(\boldsymbol{x}_g|\boldsymbol{x})}. \tag{9.137}$$

If c_i and c_r are the propagation velocities of the incoming and outgoing Green's functions, then, according to the Eikonal equation,

$$\nabla_x\tau_i(\boldsymbol{x}|\boldsymbol{x}_s) = \frac{\hat{n}_i(\boldsymbol{x})}{c_i(\boldsymbol{x})}, \tag{9.138}$$

and

$$\nabla_x\tau_r(\boldsymbol{x}_g|\boldsymbol{x}) = -\frac{\hat{n}_r(\boldsymbol{x})}{c_r(\boldsymbol{x})}, \tag{9.139}$$

with $\hat{\boldsymbol{n}}_i(\boldsymbol{x})$ the direction at $\boldsymbol{x}$ of the ray from $\boldsymbol{x}_s$ to $\boldsymbol{x}$, and $\hat{\boldsymbol{n}}_r(\boldsymbol{x})$ the direction at $\boldsymbol{x}$ of the ray from $\boldsymbol{x}$ to $\boldsymbol{x}_g$. Thus the gradients of the Green's functions can be written as

$$\begin{aligned}\nabla_x G_i(\boldsymbol{x}|\boldsymbol{x}_s;\omega) &\simeq +i\frac{\omega}{c_i}\hat{n}_i(\boldsymbol{x})G_i(\boldsymbol{x}|\boldsymbol{x}_s;\omega),\\ \nabla_x G_r(\boldsymbol{x}_g|\boldsymbol{x};\omega) &\simeq -i\frac{\omega}{c_r}\hat{n}_r(\boldsymbol{x})G_r(\boldsymbol{x}_g|\boldsymbol{x};\omega).\end{aligned} \tag{9.140}$$

These relations are all we need to express reflection data in terms of scattering potential as per Chapter 7, or in terms of reflectivity as per Chapter 8. Thus we can write

$$D_{ri}\left(\tilde{\boldsymbol{x}}_g|\tilde{\boldsymbol{x}}_s;\omega\right)=\omega^2\mathfrak{w}(\omega)\int\int\int d^3x\mathbb{V}_{ri}(\boldsymbol{x},\sigma)g_i\left(\boldsymbol{x}\,|\tilde{\boldsymbol{x}}_s\right)g_r\left(\tilde{\boldsymbol{x}}_g\middle|\,\boldsymbol{x}\right)e^{i\omega\tau\left(\tilde{\boldsymbol{x}}_g|\boldsymbol{x}|\tilde{\boldsymbol{x}}_s\right)},\tag{9.141}$$

where τ is the total traveltime

$$\tau\left(\tilde{\boldsymbol{x}}_g|\boldsymbol{x}|\tilde{\boldsymbol{x}}_s\right)\equiv\tau_i\left(\boldsymbol{x}\,|\tilde{\boldsymbol{x}}_s\right)+\tau_r\left(\tilde{\boldsymbol{x}}_g\middle|\,\boldsymbol{x}\right),\tag{9.142}$$

Alternatively, exploiting equation (8.56),

$$D_{ri}\left(\tilde{\boldsymbol{x}}_g|\tilde{\boldsymbol{x}}_s;\omega\right)=-2i\omega\mathfrak{w}(\omega)\int\int\int d^3x\,\mathcal{R}_{ri}(\boldsymbol{x},\sigma)\frac{\mathcal{C}_r{}^2\cos\gamma_r}{c_r}$$
$$\times g_i\left(\boldsymbol{x}\,|\tilde{\boldsymbol{x}}_s\right)g_r\left(\tilde{\boldsymbol{x}}_g\middle|\,\boldsymbol{x}\right)e^{i\omega\tau\left(\tilde{\boldsymbol{x}}_g|\boldsymbol{x}|\tilde{\boldsymbol{x}}_s\right)},\tag{9.143}$$

with $\mathcal{C}_i$ and γ_r the wave-type-dependent factor and reflection angle as defined in Section 8.5.

When multipathing is present, equation (9.143) would represent the contribution to D_{ri} from one of the possible sets of raypaths. The total data set would be the sum of contributions from all possible raypaths.

In the absence of caustics, the amplitude factors g_i and g_r are real. If a ray passes through caustics (where the area of an elemental ray tube shrinks to zero in one or both dimensions) its amplitude factor picks up a phase of $\pi/2$ times the number of caustics traversed.

9.5.1 Reflection at a boundary

At a reflecting boundary, the reflectivity function takes the form (8.2). If the equation for the reflecting boundary is $b(\boldsymbol{x}\,)=z-\varsigma(x,y)=0$, then

$$\mathcal{R}(\boldsymbol{x},\sigma)=R(\boldsymbol{x},\sigma)\sqrt{1+(\partial_x\varsigma)^2+\left(\partial_y\varsigma\right)^2}\delta(z-\varsigma(x,y)).\tag{9.144}$$

The expression (9.143) reduces to a surface integral

$$D_{ri}\left(\tilde{\boldsymbol{x}}_g|\tilde{\boldsymbol{x}}_s;\omega\right)=-2i\omega\mathfrak{w}(\omega)\int\int_S d^2x\mathcal{R}_{ri}(x,y,\varsigma,\sigma)\frac{\mathcal{C}_r{}^2\cos\gamma_r}{c_r\cos\alpha}$$
$$\times g_i\left(x\,,\,y,\,\varsigma\,|\tilde{\boldsymbol{x}}_s\right)g_r\left(\tilde{\boldsymbol{x}}_g\middle|\,x,\,y,\,\varsigma\right)e^{i\omega\tau\left(\tilde{\boldsymbol{x}}_g|x,y,\varsigma|\tilde{\boldsymbol{x}}_s\right)},\tag{9.145}$$

where

$$\cos\alpha=1\bigg/\sqrt{1+(\partial_x\varsigma)^2+\left(\partial_y\varsigma\right)^2}.\tag{9.146}$$

Alternatively, if we define a local cartesian coordinate system (s_1, s_2, s_3) on the boundary, with s_1 and s_2 parallel to the boundary and s_3 normal to it, then

$$D_{ri}\left(\tilde{\boldsymbol{x}}_g|\tilde{\boldsymbol{x}}_s;\omega\right) = -2i\omega\mathfrak{w}(\omega)\int\int_S ds_1 ds_2\, \mathcal{R}_{ri}(x, y, \varsigma, \sigma)\frac{\mathcal{C}_r{}^2\cos\gamma_r}{c_r}$$
$$\times g_i\left(x, y, \varsigma\,|\tilde{\boldsymbol{x}}_s\right) g_r\left(\tilde{\boldsymbol{x}}_g\middle| x, y, \varsigma\right) e^{i\omega\tau\left(\tilde{x}_g|x, y, \varsigma|\tilde{x}_s\right)}. \qquad (9.147)$$

In the time domain, the exponential term is replaced by the time derivative of the wavelet $d\mathfrak{w}/dt = \dot{\mathfrak{w}}$:

$$D_{ri}\left(\tilde{\boldsymbol{x}}_g|\tilde{\boldsymbol{x}}_s;t\right) = \frac{1}{\pi}\int\int_S ds_1 ds_2 \mathcal{R}_{ri}(x, y, \varsigma, \sigma)\frac{\mathcal{C}_r{}^2\cos\gamma_r}{c_r}$$
$$\times g_i\left(x, y, \varsigma\,|\tilde{\boldsymbol{x}}_s\right) g_r\left(\tilde{\boldsymbol{x}}_g\middle| x, y, \varsigma\right)\dot{\mathfrak{w}}\left(t - \tau\left(\tilde{\mathbf{x}}_g\middle| x, y, \varsigma\,|\tilde{\boldsymbol{x}}_s\right)\right). \qquad (9.148)$$

The surface integral in this expression can often be further simplified with tools developed in Appendix G. Most of the contributions to the integral interfere destructively, leaving apparent contributions (e.g. specular reflections) from a few points (in the simplest case, one point) on the surface.

Points that contribute to the integral include minima, maxima, saddle points, and inflection points in the total traveltime function. If the boundary or its slope has discontinuities, the discontinuities may also contribute to the integral. As the surface point moves to infinity in any direction traveltime becomes infinite, implying that somewhere on the surface τ goes through a minimum value. In the simplest case, traveltime will undergo a single minimum τ_e at some point $\boldsymbol{x}_e = (x_e, y_e, \varsigma(x_e y_e))$ on the reflector. At a traveltime minimum, $\partial\tau/\partial s_1 = \partial\tau/\partial s_2 = 0$. If

$$\operatorname{Det}\tau_e{}'' = \frac{\partial^2\tau}{\partial s_1^2}\frac{\partial^2\tau}{\partial s_2^2} - \left(\frac{\partial^2\tau}{\partial s_1\partial s_2}\right)^2 \qquad (9.149)$$

is the determinant of the matrix of second derivatives of τ with respect to the surface variables s_1 and s_2 at $\boldsymbol{x}_e$, then according to equation (G.75) (note that a minimum in τ corresponds to a maximum in γ), the resulting event is

$$D_e\left(\tilde{\boldsymbol{x}}_g|\tilde{\boldsymbol{x}}_s;t\right) = \frac{2}{c_r}\mathcal{C}_r{}^2\cos\gamma_r R_{ri}\left(\boldsymbol{x}_e, \sigma_e\right)\times\left(\frac{g_i\left(\boldsymbol{x}_e|\tilde{\boldsymbol{x}}_s\right) g_r\left(\tilde{\boldsymbol{x}}_g|\boldsymbol{x}_e\right)}{\sqrt{|\operatorname{Det}\tau_e{}''|}}\right)\mathfrak{w}$$
$$\times\left(t - \tau_e\left(\tilde{\boldsymbol{x}}_g|\boldsymbol{x}_e|\tilde{\boldsymbol{x}}_s\right)\right). \qquad (9.150)$$

The event in this case is a specular reflection which appears to come from the point $\boldsymbol{x}_e$ on the reflecting boundary. The amplitude depends upon the curvature of the traveltime function at $\boldsymbol{x}_e$. A narrow, sharp traveltime minimum will yield a small-amplitude reflection, whereas a broad, shallow minimum will have a large amplitude.

If, in addition to the minimum, traveltime goes through a local maximum at some point, the resulting event has the same form as (9.150), with opposite sign (see equation (G.75)). More common would be a local saddle point, where traveltime goes through a minimum in one direction and a maximum in the other. In this case the resulting event is a Hilbert transform of the original wavelet:

$$D_e\left(\tilde{\boldsymbol{x}}_g|\tilde{\boldsymbol{x}}_s;t\right)=\frac{2}{c_r}C_r{}^2\cos\gamma_r R_{ri}(\boldsymbol{x}_e,\sigma_e)\frac{g_i\left(\boldsymbol{x}_e|\tilde{\boldsymbol{x}}_s\right)g_r\left(\tilde{\boldsymbol{x}}_g|\boldsymbol{x}_e\right)}{\sqrt{|\mathrm{Det}\,\tau_e''|}}[H*\mathfrak{w}]$$
$$\times\left(t-\tau_e\left(\tilde{\boldsymbol{x}}_g|\boldsymbol{x}_e|\tilde{\boldsymbol{x}}_s\right)\right). \tag{9.151}$$

Though this exhausts the list of events predicted by Snell's law, others are possible. Traveltime may go through an inflection point, at which the gradient of τ goes through a minimum close to zero. The resulting event looks like (see equation (G.89))

$$D_e\left(\tilde{\boldsymbol{x}}_g|\tilde{\boldsymbol{x}}_s;t\right)=\frac{2}{3c_r\sqrt{\pi}}\sqrt{\frac{\tau_1}{\tau_{111}\,|\tau_{22}|}}C_r{}^2\cos\gamma_r R_{ri}\left(\boldsymbol{x}_e,\sigma_e\right)$$
$$\times g_i\left(\boldsymbol{x}_e|\tilde{\boldsymbol{x}}_s\right)g_r\left(\tilde{\boldsymbol{x}}_g|\boldsymbol{x}_e\right)\left[w_{hS_2}\left(\tau_d\right)*\mathfrak{w}\right]\left(t-\tau_e\left(\tilde{\boldsymbol{x}}_g|\boldsymbol{x}_e|\tilde{\boldsymbol{x}}_s\right)\right). \tag{9.152}$$

The operator $w_{h\pm}\left(\tau_d\right)$ appearing in this expression is the half-differentiated inflection filter defined in equation (E.72) and plotted in Figure E.17. The inflection filter is not scalable, instead expands or contracts as per the dilation factor τ_d which, invoking (G.29), is given by

$$\tau_d=\frac{2}{3}\sqrt{\frac{2\tau_1^3}{\tau_{111}}}, \tag{9.153}$$

where τ_1 and τ_{111} are the first and third derivatives of τ in the direction of $\nabla\tau$ at $\boldsymbol{x}_e$. Also appearing in (9.152) is τ_{22}, which is the second derivative of τ in the direction normal to the gradient. The convolution operation appearing in (9.152) is as defined for dilated operators in equation (E.11).

Appendix G also contains formulas for other event types. For a smooth reflector, we have covered the most common.

9.5.2 An example

As a simple example, consider the reflector shown in Figure 9.1, from which we construct a synthetic zero-offset data set along the x-axis.

Assuming a scalar medium of constant velocity c_i above the reflector and c_t below, traveltime is a minimum with respect to y at $y=0$. The diffraction integral simplifies to

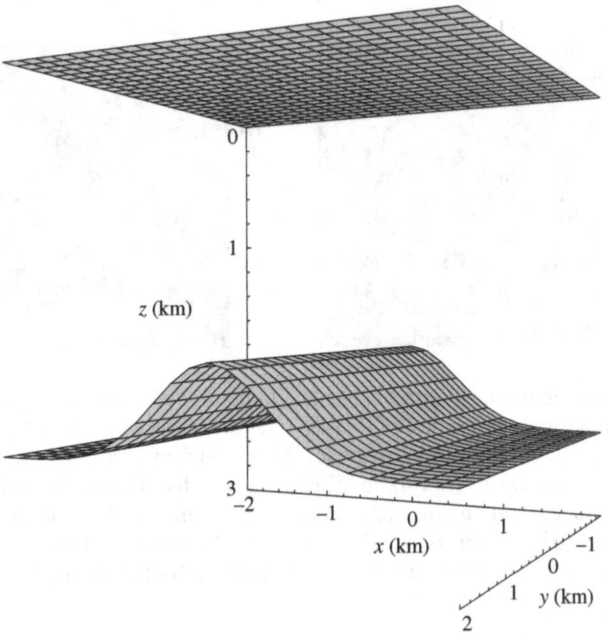

Figure 9.1 A reflecting surface, invariant in the y-direction, at depth z beneath the Earth's surface z= 0.

$$D_{ri}\left(\tilde{\boldsymbol{x}}_g|\tilde{\boldsymbol{x}}_s;t\right)=\frac{1}{8\pi^2\sqrt{2\pi c_i}}\int ds_1\frac{c_t\cos\gamma_i-c_i\cos\gamma_t}{c_t\cos\gamma_i+c_i\cos\gamma_t}\frac{\cos\gamma_i}{\sqrt{r_g r_s r}}$$
$$\times\left[d_{h+}*\mathfrak{w}\right]\left(t-\tau\left(\tilde{x}_g\middle|\,x,\,0,\,\varsigma\,|\tilde{\boldsymbol{x}}_s\right)\right), \qquad (9.154)$$

with

$$\cos\gamma_i=\frac{1}{\sqrt{2}}\sqrt{1+\frac{\left(x-x_g\right)\left(x-x_s\right)+\varsigma^2}{r_g r_s}}, \qquad (9.155)$$

and

$$\cos\gamma_t=\sqrt{1-\frac{{c_t}^2}{{c_i}^2}\left(1-\cos^2\gamma_i\right)}. \qquad (9.156)$$

The left panel of Figure 9.2 shows a synthetic zero-offset data set constructed using the diffraction integral equation (9.154).

Near the center of the section we can see a single event, presumably a traveltime minimum. As we move outward from the center, we pick up a second event, and shortly after a third event branching off from the second. Moving further outward, the third event merges with the first, the combined event gradually fading away as we approach the outer edge of the data set. Where three events are present, it

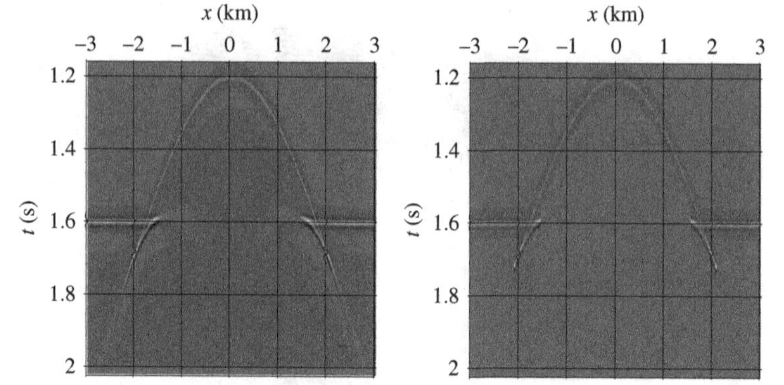

Figure 9.2 Dome model synthetics. Left: Dome model zero-offset synthetic generated by numerical integration of (9.148) over the surface of Figure 9.1, assuming the source wavelet to be a Ricker wavelet. This modeling technique admits diffractive effects, allowing events to fade away gradually. Right: A synthetic from the same surface, made using the Snell event formulas (9.150) and (9.151). The min–max–min triplication is clearly visible on the flanks of the structure. Since diffractions are not included, events terminate abruptly in this approximation.

is safe to assume that they represent a min–max–min triad of traveltimes (more precisely, a min–saddle–min). Where two events are present, they likely represent a minimum and an inflection point in traveltime.

In lieu of numerical integration, one can evaluate (9.154) through event decomposition. At a traveltime minimum, (9.154) reduces to

$$D_{\min}\left(\tilde{\boldsymbol{x}}_g|\tilde{\boldsymbol{x}}_s;t\right)=\frac{1}{8\pi^2\sqrt{\tau_{\min}{}''c_i}}\frac{c_t\cos\gamma_i-c_i\cos\gamma_t}{c_t\cos\gamma_i+c_i\cos\gamma_t}\frac{\cos\gamma_i}{\sqrt{r_g r_s r}}\mathfrak{w}\left(t-\tau_{\min}\right). \tag{9.157}$$

At a maximum,

$$D_{\max}\left(\tilde{\boldsymbol{x}}_g|\tilde{\boldsymbol{x}}_s;t\right)=\frac{1}{8\pi^2\sqrt{-\tau_{\max}{}''c_i}}\frac{c_t\cos\gamma_i-c_i\cos\gamma_t}{c_t\cos\gamma_i+c_i\cos\gamma_t}\frac{\cos\gamma_i}{\sqrt{r_g r_s r}}\left[H*\mathfrak{w}\right]\left(t-\tau_{\max}\right). \tag{9.158}$$

At an inflection,

$$D_{\max}\left(\tilde{\boldsymbol{x}}_g|\tilde{\boldsymbol{x}}_s;t\right)=\frac{1}{24\pi^2\sqrt{\pi c_i}}\times\sqrt{\frac{\tau_{infl}{}'}{\tau_{infl}{}'''}}\frac{c_t\cos\gamma_i-c_i\cos\gamma_t}{c_t\cos\gamma_i+c_i\cos\gamma_t}\frac{\cos\gamma_i}{\sqrt{r_g r_s r}}\left[w_{hS_2}\left(\tau_d\right)*\mathfrak{w}\right]\left(t-\tau_{infl}\right). \tag{9.159}$$

The right panel of Figure 9.2 shows the extremal events generated by equations (9.157) and (9.158); i.e.the events predicted by Snell's law. This section closely matches the previous one in the one- and three-event regions, cutting off abruptly in the two-event regions. We conclude that Snell's law, even for simple geometries, may not give a complete representation of a seismic reflection.

9.6 Summary

3-D space–frequency Born approximation

$$D\left(\boldsymbol{x}_g|\boldsymbol{x}_s;\omega\right) = \mathfrak{w}(\omega)\int\int\int d^3x G_0\left(\boldsymbol{x}_g\middle|\boldsymbol{x};\omega\right)\mathcal{V}(\boldsymbol{x},\omega)G_0\left(\boldsymbol{x}\,|\boldsymbol{x}_s\,;\omega\right). \tag{9.160}$$

3-D constant velocity wavenumber–frequency Born approximation

Green's functions

$$\mathcal{G}_r(\tilde{\boldsymbol{k}}_g, z, \omega) = \frac{e^{-ik_{gz}|z|}}{2ik_{gz}\rho_0 c_r^2}, \quad \mathcal{G}_i(\tilde{\boldsymbol{k}}_s, z, \omega) = -\frac{e^{ik_{sz}|z|}}{2ik_{sz}\rho_0 c_i^2}. \tag{9.161}$$

Vertical wavenumbers

$$k_{gz} = -\frac{\omega}{c_r}\sqrt{1-\frac{\left(k_{gx}{}^2+k_{gy}{}^2\right)c_r^2}{\omega^2}},$$

$$k_{sz} = +\frac{\omega}{c_i}\sqrt{1-\frac{\left(k_{sx}{}^2+k_{sy}{}^2\right)c_i^2}{\omega^2}}. \tag{9.162}$$

Midpoint and offset wavenumbers

$$\boldsymbol{k}_m \equiv \boldsymbol{k}_g - \boldsymbol{k}_s, \tag{9.163}$$

$$\boldsymbol{k}_h \equiv \boldsymbol{k}_s + \boldsymbol{k}_g. \tag{9.164}$$

Born approximation

$$D_{ri}\left(\tilde{\boldsymbol{k}}_g|\tilde{\boldsymbol{k}}_s;\omega\right) = \mathfrak{w}(\omega)\int\int\int d^3x \times \mathcal{G}_r\left(\tilde{\boldsymbol{k}}_g, z, \omega\right)e^{-i\tilde{\boldsymbol{k}}_g\cdot\tilde{\boldsymbol{x}}}\mathcal{V}_{ri}(\boldsymbol{x},\omega)e^{+i\tilde{\boldsymbol{k}}_s\cdot\tilde{\boldsymbol{x}}}\mathcal{G}_i\left(\tilde{\boldsymbol{k}}_s, z, \omega\right), \tag{9.165}$$

$$D_{ri}\left(\tilde{\boldsymbol{k}}_g|\tilde{\boldsymbol{k}}_s;\omega\right) = \mathfrak{w}(\omega)\frac{\omega^2\mathbb{V}_{ri}\left(\boldsymbol{k}_m,\sigma\right)}{4k_{gz}k_{sz}\rho_0^2c_r^2c_i^2}, \tag{9.166}$$

$$D_{ri}\left(\tilde{\boldsymbol{k}}_g|\tilde{\boldsymbol{k}}_s;\omega\right) = -i\omega\,\mathfrak{w}(\omega)\frac{\mathcal{R}_{ri}\left(\boldsymbol{k}_m,\sigma\right)\cos\gamma_r}{2k_{gz}k_{sz}\rho_0c_rc_i{}^2}. \tag{9.167}$$

3-D constant-velocity space–frequency Born approximation

$$D_{ri}\left(\tilde{\boldsymbol{x}}_m, \tilde{\boldsymbol{x}}_h; \omega\right) \equiv D_{ri}\left(\tilde{\boldsymbol{x}}_g|\tilde{\boldsymbol{x}}_s; \omega\right) = \frac{1}{(2\pi)^4}\frac{\mathfrak{w}(\omega)}{16\rho_0^2 c_r^2 c_i^2}\int\int d^2\tilde{k}_m e^{i\tilde{\boldsymbol{k}}_m\cdot\tilde{\boldsymbol{x}}_m}$$
$$\times\int\int d^2\tilde{k}_h e^{i\tilde{\boldsymbol{k}}_h\cdot\tilde{\boldsymbol{x}}_h}\frac{\omega^2}{k_{gz}k_{sz}}\int\int\int d^3x e^{-i\boldsymbol{k}_m\cdot\boldsymbol{x}}\mathbb{V}_{ri}(\boldsymbol{x},\sigma). \quad (9.168)$$

$$D_{ri}\left(\tilde{\boldsymbol{x}}_m, \tilde{\boldsymbol{x}}_h; \omega\right) = \frac{-1}{(2\pi)^4}\frac{i\omega\mathfrak{w}(\omega)}{8\rho_0 c_r {c_i}^2}\int\int d^2\tilde{k}_m e^{i\tilde{\boldsymbol{k}}_m\cdot\tilde{\boldsymbol{x}}_m}\int\int d^2\tilde{k}_h e^{i\tilde{\boldsymbol{k}}_h\cdot\tilde{\boldsymbol{x}}_h}\frac{\cos\gamma_r}{k_{gz}k_{sz}}$$
$$\times\int\int\int d^3x e^{-i\boldsymbol{k}_m\cdot\boldsymbol{x}}\,\mathcal{R}_{ri}(\boldsymbol{x},\sigma). \quad (9.169)$$

3-D constant-velocity zero-offset wavenumber–frequency Born approximation

$$D_{ri}^{(\mathrm{3DZO})}\left(\tilde{\boldsymbol{k}}_m;\omega\right) \simeq \frac{1}{(2\pi)}\frac{\mathfrak{w}(\omega)\underline{c}}{2\rho_0 c_r {c_i}^2}\left(\frac{\mathcal{R}_{ri}}{z}\right)\left(\tilde{\boldsymbol{k}}_m, {k_{mz}}^{\mathrm{ZO}}, 0\right). \quad (9.170)$$

$$\left(tD_{ri}^{(\mathrm{3DZO})}\right)\left(\tilde{\boldsymbol{k}}_m;\omega\right) \simeq \frac{1}{(2\pi)}\frac{\mathfrak{w}(\omega)}{2\rho_0 c_r {c_i}^2}\frac{\mathcal{R}_{ri}\left(\tilde{\boldsymbol{k}}_m, k_{\mathrm{mz}}^{\mathrm{ZO}}, 0\right)}{\sqrt{1-\tilde{k}_m^2\underline{c}^2/\omega^2}}. \quad (9.171)$$

3-D constant-velocity zero-azimuth wavenumber–frequency Born approximation

$$D_{ri}^{(\mathrm{ZA})}\left(\tilde{\boldsymbol{k}}_m, k_{hx};\omega\right) = \frac{1}{(2\pi)^{1/2}}\frac{\sqrt{i}\,\omega\mathfrak{w}(\omega)}{4\rho_0 c_r {c_i}^2}\frac{\cos\gamma_r^{\mathrm{ZA}}}{\sqrt{k_{mz}^{\mathrm{ZA}}k_{\mathrm{gz0}}k_{\mathrm{sz0}}}}\left(\frac{\mathcal{R}_{ri}}{\sqrt{z}}\right)\left(\tilde{\boldsymbol{k}}_m, k_{mz}^{\mathrm{ZA}}, \sigma^{\mathrm{ZA}}\right). \quad (9.172)$$

2.5-D constant-velocity wavenumber–frequency Born approximation

$$\mathrm{D}_{ri}^{(2.5\mathrm{D})}\left(k_{mx}, k_{hx};\omega\right) = \frac{\sqrt{i}}{(2\pi)^{1/2}}\frac{\omega\mathfrak{w}(\omega)}{4\rho_0 c_r {c_i}^2}\frac{\cos\,\gamma_{r0}}{\sqrt{k_{mz0}k_{gz0}k_{sz0}}}\left(\frac{\mathcal{R}_{ri}}{\sqrt{z}}\right)\left(k_{mx}, k_{mz0}, \sigma_0\right). \quad (9.173)$$

2.5-D constant-velocity zero-offset wavenumber–frequency Born approximation

$$D_{ri}^{(2.5\mathrm{DZO})}\left(k_{mx};\omega\right) = \frac{\mathfrak{w}(\omega)\underline{c}}{4\pi\rho_0 c_r {c_i}^2}\left(\frac{\mathcal{R}_{ri}}{z}\right)\left(k_{mx}, k_{mz}, 0\right), \quad (9.174)$$

with

$$k_{mz} = -\frac{\omega}{\underline{c}}\sqrt{1-\frac{k_{mx}^2\underline{c}^2}{\omega^2}}, \quad (9.175)$$

3-D depth-variable velocity wavenumber–frequency Born approximation

$$D_{ri}\left(\tilde{\boldsymbol{k}}_g|\tilde{\boldsymbol{k}}_s;\omega\right) = \frac{-\omega^2\mathfrak{w}(\omega)}{4\sqrt{\left|k_{gz}(0)k_{sz}(0)\right|}\rho_0(0)c_r(0)c_i(0)} \times \int dz e^{-i\int_0^z dz' k_{mz}(z')} \frac{\mathbb{V}_{ri}\left(\tilde{\boldsymbol{k}}_m, z, \sigma(z)\right)}{\sqrt{\left|k_{gz}(z)k_{sz}(z)\right|}\rho_0(z)c_r(z)c_i(z)}. \quad (9.176)$$

$$D_{ri}\left(\tilde{\boldsymbol{k}}_g|\tilde{\boldsymbol{k}}_s;\omega\right) = \frac{i\omega\mathfrak{w}(\omega)}{2\sqrt{\left|k_{gz}(0)k_{sz}(0)\right|}\rho_0(0)c_r(0)c_i(0)} \times \int dz e^{-i\int_0^z dz' k_{mz}(z')} \frac{\mathcal{R}_{ri}\left(\tilde{\boldsymbol{k}}_m, z, \sigma\right)\cos\left(\gamma_r(z)\right)}{\sqrt{\left|k_{gz}(z)k_{sz}(z)\right|}c_i(z)}. \quad (9.177)$$

2.5-D depth-variable velocity wavenumber–frequency Born approximation

$$D_{ri}^{(2.5\mathrm{D})}(k_{mx}, k_{hx};\omega) = -\frac{1}{(2\pi)^{1/2}}\frac{\omega^2\mathfrak{w}(\omega)}{8\rho_0(0)c_r(0)c_i(0)}\int dz \frac{e^{-i\int_0^z dz' \underset{\wedge}{k}_{mz}(z')}}{\rho_0(z)c_r(z)c_i(z)} \times \left(\frac{e^{-i\pi/4}\mathbb{V}_{ri}\left(k_{mx}, z, \sigma(z)\right)}{\sqrt{\underset{\wedge}{k}_{gz}(0)\underset{\wedge}{k}_{sz}(0)\underset{\wedge}{k}_{gz}(z)\underset{\wedge}{k}_{sz}(z)\int_0^z dz' \frac{\underset{\wedge}{k}_{mz}(z')}{\underset{\wedge}{k}_{gz}(z')\underset{\wedge}{k}_{sz}(z')}}}\right). \quad (9.178)$$

$$D_{ri}^{(2.5\mathrm{D})}(k_{mx}, k_{hx};\omega) = \frac{\sqrt{i}}{(2\pi)^{1/2}}\frac{\omega\mathfrak{w}(\omega)}{4\rho_0(0)c_r(0)c_i(0)}\int dz \frac{e^{-i\int_0^z dz' \underset{\wedge}{k}_{mz}(z')}}{c_i(z)} \times \left(\frac{\cos\gamma_r\mathcal{R}_{ri}(\boldsymbol{x}, \sigma)}{\sqrt{\underset{\wedge}{k}_{gz}(0)\underset{\wedge}{k}_{sz}(0)\underset{\wedge}{k}_{gz}(z)\underset{\wedge}{k}_{sz}(z)\int_0^z dz' \frac{\underset{\wedge}{k}_{mz}(z')}{\underset{\wedge}{k}_{gz}(z')\underset{\wedge}{k}_{sz}(z')}}}\right). \quad (9.179)$$

3-D generally variable velocity Born approximation

$$D_{ri}\left(\tilde{\boldsymbol{x}}_g|\tilde{\boldsymbol{x}}_s;\omega\right) = \omega^2\mathfrak{w}(\omega)\int\int\int d^3x\mathbb{V}_{ri}(\boldsymbol{x}, \sigma)g_i\left(\boldsymbol{x}\,|\tilde{\boldsymbol{x}}_s\right)g_r\left(\tilde{\boldsymbol{x}}_g\middle|\,\boldsymbol{x}\right)e^{i\omega\tau\left(\tilde{\boldsymbol{x}}_g|\boldsymbol{x}|\tilde{\boldsymbol{x}}_s\right)}, \quad (9.180)$$

$$D_{ri}\left(\tilde{\boldsymbol{x}}_g|\tilde{\boldsymbol{x}}_s;\omega\right) = -2i\omega\mathfrak{w}(\omega)\int\int\int d^3x\mathcal{R}_{ri}(\boldsymbol{x},\sigma)\frac{C_r{}^2\cos\gamma_r}{c_r}$$
$$\times g_i\left(\boldsymbol{x}\,|\tilde{\boldsymbol{x}}_s\right)g_r\left(\tilde{\boldsymbol{x}}_g\middle|\,\boldsymbol{x}\right)e^{i\omega\tau\left(\tilde{x}_g|x|\tilde{x}_s\right)}. \tag{9.181}$$

Exercises

9.1 Consider a surface reflection experiment with a depth-dependent scalar potential in a constant velocity background. (*a*) Write the first four terms of the Born expansion in the $(\tilde{\boldsymbol{k}}, z, \omega)$ domain. (*b*) Evaluate the first four terms for a step potential function, and with the identification $\cos\gamma_0 = k_z\, c_0/\omega$, compare this expansion with the expansion of the scalar reflection coefficient in Exercise 8.1. (*c*) Evaluate the same terms for a box car potential function. What do these terms become as $z_1 \to z_0$ and as $z_1 \to \infty$? If the potential amplitude a_0 increases inversely as $z_1 - z_0$ as $z_1 \to z_0$, what do the terms become?

9.2 For a constant-velocity background, extend the development of the first Born approximation in *f*–*k* space to the second Born approximation for a scalar potential.

9.3 Derive equations (9.53)–(9.57) from the stationarity conditions.

10
Frequency–wavenumber migration

10.1 Full constant-velocity migration

We consider seismic migration to be the process of recovering a reflectivity function from reflection data. Breaking down the reflectivity function into its constituent perturbations can be incorporated into the migration process, but for present purposes we will consider that to be a separate exercise. For *f–k* migration, most of the work is already done, in that a simple reciprocal relation exists between the forward modeling formulas of Chapter 9, and the desired migration formulas.

10.1.1 Prestack migration of a complete 3-D data set

For a complete, 5-D data set, equation (9.17) can be inverted for components of the elastic reflectivity function $\mathcal{R}_{ri}$,

$$\mathcal{R}_{ri}\left(\boldsymbol{k}_m, \sigma\right) = \frac{2ik_{gz}k_{sz}\rho_0 c_r c_i^{\,2}}{\omega \cos \gamma_r} D_{ri}\left(\tilde{\boldsymbol{k}}_m, \tilde{\boldsymbol{k}}_h; \omega\right), \tag{10.1}$$

or, from (9.15), for components of the scattering potential $\mathbb{V}_{ri}$,

$$\mathbb{V}_{ri}\left(\boldsymbol{k}_m, \sigma\right) = 4\frac{k_{gz}k_{sz}}{\omega^2}\rho_0 c_r^{\,2} c_i^{\,2} D_{ri}\left(\tilde{\boldsymbol{k}}_m, \tilde{\boldsymbol{k}}_h; \omega\right), \tag{10.2}$$

In this expression, c_i and c_r are the velocities of the incoming and reflected waves, respectively. The 3-vector $\boldsymbol{k}_m$ is the wavenumber of $\mathcal{R}_{ri}$ or $\mathbb{V}_{ri}$, and σ is the opening angle between an incident and reflected plane wave. Of course, $\mathcal{R}_{ri}$ or $\mathbb{V}_{ri}$ can be determined only within the bandwidth of the source wavelet, which for simplicity we have assumed to be infinite. In expressions (10.1) and (10.2), vertical receiver and source wavenumbers k_{gz} and k_{sz} can be expressed in terms of data frequency ω, outgoing wavenumber 2-vector $\tilde{\boldsymbol{k}}_g$, and incoming wavenumber 2-vector $\tilde{\boldsymbol{k}}_s$ by equations (9.9). The outgoing and incoming wavenumber 2-vectors are in turn related to midpoint and offset wavenumber 2-vectors as $\tilde{\boldsymbol{k}}_m = \tilde{\boldsymbol{k}}_g - \tilde{\boldsymbol{k}}_s$

and $\tilde{\boldsymbol{k}}_h = \tilde{\boldsymbol{k}}_g + \tilde{\boldsymbol{k}}_s$. The opening angle σ can be expressed as a function of ω, k_m, c_r, and c_i by (9.31). Frequency ω, midpoint wavenumber 2-vector $\tilde{\boldsymbol{k}}_m$, and vertical wavenumber k_{mz} are uniquely determined by $\boldsymbol{k}_m$ and σ:

$$\boldsymbol{k}_m = \left(\tilde{\boldsymbol{k}}_m, k_{mz}\right), \tag{10.3}$$

and (equation (9.21))

$$\omega = \frac{c_r c_i}{2c_e(\sigma)} k_m, \tag{10.4}$$

with $c_e(\sigma)$ the effective velocity introduced in Appendix H:

$$c_e(\sigma) = \frac{\sqrt{{c_r}^2 + {c_i}^2 + 2c_r c_i \cos\sigma}}{2}. \tag{10.5}$$

Offset wavenumber $\tilde{\boldsymbol{k}}_h$ is not uniquely determined by $\boldsymbol{k}_m$ and σ. If

$$\boldsymbol{k}_h = \left(\tilde{\boldsymbol{k}}_h, k_{gz} + k_{sz}\right), \tag{10.6}$$

then the magnitude k_h of $\boldsymbol{k}_h$ is given by (9.29), and the angle $\psi(\sigma)$ between $\boldsymbol{k}_h$ and $\boldsymbol{k}_m$ is given by equation (9.30). For a given $\boldsymbol{k}_m$ and σ, the set of compatible $\boldsymbol{k}_h$ form a cone of angle $\psi(\sigma)$ with axis $\boldsymbol{k}_m$ and length k_h. (For non-converted reflections, where $c_i = c_r$, the angle ψ is $\pi/2$.) Since the relation (10.1) overdetermines $\mathcal{R}_{ri}$, the right-hand side of (10.1) can be replaced by any normalized linear combination of D_{ri} at compatible values of $\boldsymbol{k}_h$.

Linearized response: When solving for the individual perturbations within the linearized reflectivity function or the scattering potential, the data appear even more redundant. For example, the elastic PP-scattering potential is given by (7.58), or, for constant background velocity,

$$\mathbb{V}_{\mathrm{PP}}(\boldsymbol{x}, \sigma) = -\rho_0 \left(a_\alpha + a_\rho \left(1 + \cos\sigma - 2\frac{{\beta_0}^2}{{\alpha_0}^2} \sin^2\sigma \right) - 2\frac{{\beta_0}^2}{{\alpha_0}^2} a_\beta \sin^2\sigma \right), \tag{10.7}$$

with a_α, a_β, and a_ρ the dimensionless perturbations in P-velocity, S-velocity, and density, respectively. Thus, the equation for the perturbations in terms of data is

$$\begin{aligned} a_\alpha + a_\rho \left(1 + \cos\sigma - 2\frac{{\beta_0}^2}{{\alpha_0}^2} \sin^2\sigma \right) - 2\frac{{\beta_0}^2}{{\alpha_0}^2} a_\beta \sin^2\sigma \\ = -4\frac{k_{gz}k_{sz}}{\omega^2} {c_r}^2 {c_i}^2 D_{ri}\left(\tilde{\boldsymbol{k}}, \tilde{\boldsymbol{k}}_h; \omega\right). \end{aligned} \tag{10.8}$$

In principle, to solve for the three perturbation terms a_α, a_β, and a_ρ, data from three points of distinct σ are required. Alternatively, if data from other data types are available, they could be combined with the PP data to determine the individual perturbations. In practice, statistical methods would be applied to all available data. Resolution of all three elastic parameters from PP data alone is difficult, and often two-parameter simplifications are employed (Shuey, 1985).

10.1.2 3-D zero-offset migration

The far-field zero-offset *f*–*k* space migration formula for reflectivity is, from (9.66),

$$\left(\frac{\mathcal{R}_{ri}}{z}\right)\left(\tilde{\boldsymbol{k}}_m, {k_{mz}}^{\mathrm{ZO}}, 0\right) = \frac{4\pi\rho_0 c_r {c_i}^2}{\underline{c}} D_{ri}^{(\mathrm{3DZO})}\left(\tilde{\boldsymbol{k}}_m; \omega\right), \tag{10.9}$$

where (see Figure 10.1)

$$k_{mz}^{\mathrm{ZO}} = -\frac{\omega}{\underline{c}}\sqrt{1 - \frac{\tilde{k}_m{}^2 \underline{c}^2}{\omega^2}}, \tag{10.10}$$

$$\underline{c} = \frac{c_r c_i}{c_r + c_i}, \tag{10.11}$$

and

$$\left(\frac{\mathcal{R}_{ri}}{z}\right)(\boldsymbol{x}, \sigma) \equiv \frac{\mathcal{R}_{ri}(\boldsymbol{x}, \sigma)}{z}. \tag{10.12}$$

In (10.9), opening angle σ is zero. Consequently, the relation (10.4) between frequency and wavenumber becomes, in this case,

$$\omega = k_m \underline{c}. \tag{10.13}$$

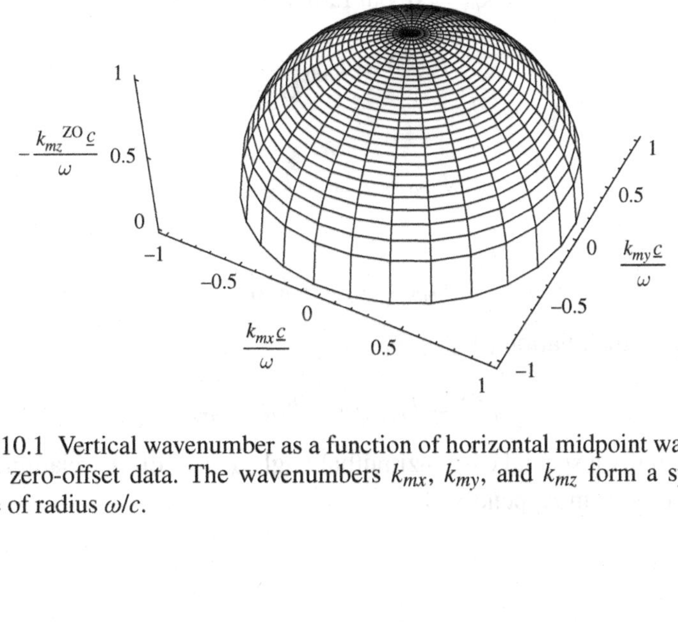

Figure 10.1 Vertical wavenumber as a function of horizontal midpoint wavenumber for zero-offset data. The wavenumbers k_{mx}, k_{my}, and k_{mz} form a spherical surface of radius ω/c.

Only the zero-opening angle component of reflectivity is recoverable from zero-offset data. With this limitation, the mapping (10.9) is one-to-one.

10.1.3 Prestack migration of a 3-D zero-azimuth data set

Suppose we have a zero-azimuth data set $D^{(ri)}{}_{\mathrm{ZA}}\,(\boldsymbol{x}_m, x_h; \omega) = D_{ri}\,(\boldsymbol{x}_m, x_h, y_h = 0; \omega)$.

Defining

$$\left(\frac{\mathcal{R}_{ri}}{\sqrt{z}}\right)(\boldsymbol{x}, \sigma) \equiv \frac{\mathcal{R}_{ri}(\boldsymbol{x}, \sigma)}{\sqrt{z}}, \tag{10.14}$$

the far-field approximate equation (9.91) for modeling of zero-azimuth data in *f–k* space (again assuming the source wavelet to be a delta function) can be expressed as a migration equation

$$\left(\frac{\mathcal{R}_{ri}}{\sqrt{z}}\right)\left(\tilde{\boldsymbol{k}}_m, k_{mz}{}^{\mathrm{ZA}}, \sigma^{\mathrm{ZA}}\right) = \sqrt{-2\pi i k_{mz}{}^{\mathrm{ZA}} k_{gz0} k_{sz0}}\,\frac{4\rho_0 c_r c_i{}^2}{\omega \cos\,\gamma_r{}^{ZA}}\,D^{(\mathrm{ZA})}{}_{ri}\left(\tilde{\boldsymbol{k}}_m, k_{hx}; \omega\right) \tag{10.15}$$

with (treating the data coordinates k_{mx}, k_{my}, k_{hx}, and ω as the independent variables)

$$k_{gz0} = -\sqrt{\frac{\omega^2}{c_r{}^2} - \frac{(k_{mx} - k_{hx})^2}{4}},$$

$$k_{sz0} = \sqrt{\frac{\omega^2}{c_i{}^2} - \frac{(k_{mx} + k_{hx})^2}{4}}. \tag{10.16}$$

As discussed in Chapter 9, the stationary value of k_{hy} corresponding to the zero-azimuth point $y_h = 0$ is not $k_{hy} = 0$, but rather (9.75), or

$$k_{hy}{}^{\mathrm{ZA}} = k_{hz0}\frac{k_{my}}{k_{mz0}}, \tag{10.17}$$

with

$$k_{hz0} = k_{gz0} + k_{sz0}, \tag{10.18}$$

and

$$k_{mz0} = k_{gz0} - k_{sz0}. \tag{10.19}$$

This leads to the relation (9.86),

$$k_{mz}{}^{\mathrm{ZA}} = k_{mz0}\sqrt{1 - k_{my}{}^2/k_{mz0}{}^2}. \tag{10.20}$$

This is actually a special case (azimuthal angle $\chi = 0$) of the relation (H.39) for azimuth developed in Appendix H.

The zero-azimuth expression for cosine of opening angle is (9.89),

$$\cos\sigma^{\mathrm{ZA}} = -\frac{c_i c_r}{\omega^2}\left(k_{gx}k_{sx} + k_{gz0}k_{sz0}\right). \tag{10.21}$$

Reflection angle γ_r is related to opening angle as (9.23), or

$$\cos\gamma_r{}^{\mathrm{ZA}} = \frac{(c_i + c_r\cos\sigma)}{2c_e(\sigma)}. \tag{10.22}$$

In this case, the mapping between reflectivity and data is one to one. If one wishes to solve for the individual perturbations within the reflectivity function, then the data contains redundancy, and statistical methods can be employed.

10.1.4 2.5-D migration

Zero-azimuth migration of a 2.5-D data set follows from the formula (9.97), expressible in *f–k* space as

$$\left(\frac{\mathcal{R}_{ri}}{\sqrt{z}}\right)(k_{mx}, k_{mz0}, \sigma_0) = \frac{4\rho_0 c_r c_i{}^2\sqrt{-2\pi i k_{mz0}k_{gz0}k_{sz0}}}{\omega\ \cos\gamma_{r0}} D^{ri}_{(2.5\mathrm{D})}(k_{mx}, k_{hx}; \omega). \tag{10.23}$$

The reflectivity variables can be expressed in terms of the data variables:

$$k_{gz0} = -\sqrt{\frac{\omega^2}{c_r{}^2} - \frac{(k_{mx} - k_{hx})^2}{4}},$$
$$k_{sz0} = +\sqrt{\frac{\omega^2}{c_i{}^2} - \frac{(k_{mx} + k_{hx})^2}{4}}, \tag{10.24}$$

$$k_{mz0} = k_{gz0} - k_{sz0} \tag{10.25}$$
$$= -\frac{\omega}{\underline{c}}\left(\frac{c_i}{c_r + c_i}\sqrt{1 - \frac{(k_{mx} - k_{hx})^2 c_r{}^2}{4\omega^2}} + \frac{c_r}{c_r + c_i}\sqrt{1 - \frac{(k_{mx} - k_{hx})^2 c_i{}^2}{4\omega^2}}\right),$$

and

$$\cos\sigma_0 = -\frac{c_r c_i}{\omega^2}\left(\frac{(k_{hx}{}^2 - k_{mx}{}^2)}{4} + k_{gz0}k_{sz0}\right), \tag{10.26}$$

with the relation between σ_0 and γ_{r0} still (10.22). The mapping between data and reflectivity is again one-to-one. Figure 10.2 shows k_{mz0} versus k_{mx} and k_{hx} for the case where $c_r = c_i = c$.

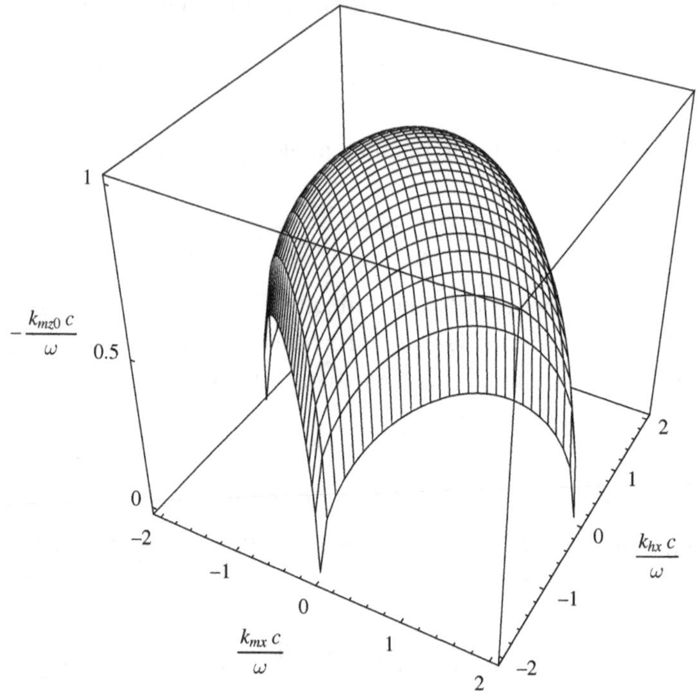

Figure 10.2 Vertical wavenumber k_{mz0} as a function of k_{mx} and k_{hx} for 2.5-D data.

10.1.5 2.5-D zero-offset migration

The 2.5-D zero-offset *f–k* migration formula is from (9.106)

$$\left(\frac{\mathcal{R}_{ri}}{z}\right)(k_{mx}, k_{mz}, \sigma = 0) \simeq 4\pi\rho_0 \frac{c_r c_i{}^2}{\underline{c}} D^{ri}_{(2.5\mathrm{DZO})}(k_{mx}; \omega)\,. \tag{10.27}$$

In this expression, the relation between k_{mz} and ω is

$$\omega = -\underline{c}k_{mz}\sqrt{1 + k_{mx}{}^2/k_{mz}{}^2}. \tag{10.28}$$

10.1.6 Output in vertical traveltime

The above formulas yield elastic reflectivity as a function of horizontal and vertical wavenumbers. When inverse-Fourier transformed, the result is reflectivity as a function of x and z or x, y, and z. To obtain reflectivity as a function of vertical traveltime τ instead of depth, we perform the coordinate transform $z \to \tau = z\underline{c}^{-1}$, with $\underline{c}^{-1}$ the summed slowness as defined in equation (9.58). In the wavenumber domain, this is equivalent to a change of output variable from vertical wavenumber

k_z to a migrated frequency $\omega_{\mathrm{Mig}} = -k_z \underline{c}$. Where incident and reflected velocities are equal, $c_i = c_r = c$, the transformation reduces to $\tau = 2z/c$, as described in Chapter 2.

10.2 Partial *f–k* migration

On prestack data, a full migration actually performs several operations at once. Since an event occurs at different times for different source–receiver offsets, part of the migration is a time normalization or moveout operation. Since all offsets contribute to the same image, migration also includes a summing or stacking operation. Thus, a prestack migration operator may be considered a product of multiple operations (Claerbout, 1976; Stolt & Benson, 1986; Black *et al.*, 1993; Tygel *et al.*, 1998; Bleistein *et al.*, 1999; Sava, 2003; Schleicher *et al.*, 2007). For example, we can write

$$\begin{aligned} \mathcal{R} &= \mathcal{M}\mathcal{D} \\ &= \mathcal{M}_{\mathrm{ZOM}}\mathcal{M}_{\mathrm{MZO}}\mathcal{D}. \end{aligned} \tag{10.29}$$

The operator $\mathcal{M}_{\mathrm{MZO}}$ performs the moveout and stacking operation, kinematically reducing the data to zero offset. The other operator $\mathcal{M}_{\mathrm{ZOM}}$ completes the full migration with a "zero-offset" migration operation.

Part of the appeal of the full migration operator is its ability to combine all tasks into a single operation. However, in practice, there may be advantages to a decomposition into several operations. Approximations and imperfections in the background parameter model may mean that the moveout operation responds best to a different velocity than the migration operation. Multiple operations allow a peek at intermediate results, and tweaking of parameters as necessary.

10.2.1 MZO plus ZOM

Let us try a two-operator decomposition of the 2.5-D migration operator (10.23), the first operation (MZO) reducing the data to zero offset, and the second (ZOM) performing a zero-offset migration. If we express the final result in vertical traveltime, then the full operator consists of a shift in frequency from ω to $\omega_{\mathrm{Mig}} = -k_z \underline{c}$, plus a change in amplitude and phase. We demand the individual operations also to take the form of a frequency shift plus a phase and amplitude change:

$$D_{\mathrm{MZO}}\left(k_{mx}, k_{hx}, 0; \omega_{\mathrm{MZO}}\right) = \mathcal{A}_{\mathrm{MZO}} D_{ri}\left(k_{mx}, k_{hx}, 0; \omega\right), \tag{10.30}$$

and

$$\left(\frac{\mathcal{R}_{ri}}{\sqrt{z}}\right)\left(k_{mx}, k_{hx}, 0; \omega_{\mathrm{ZOM}}\right) = \mathcal{A}_{\mathrm{ZOM}} D_{\mathrm{MZO}}\left(k_{mx}, k_{hx}, 0; \omega_{\mathrm{MZO}}\right). \tag{10.31}$$

The frequency after each step is determined by the nature of the operation. Since the final output frequency is the full-migration output frequency, we must have $\omega_{\text{ZOM}} = \omega_{\text{Mig}}$. There is some discretion in the choice of amplitude-phase terms, subject to the constraint that the product of the terms equals the phase-amplitude term in the 2.5-D migration operator (10.23):

$$\mathcal{A}_{\text{ZOM}}\mathcal{A}_{\text{MZO}} = \frac{4\rho_0 c_r {c_i}^2 \sqrt{-2\pi i k_{mz0} k_{gz0} k_{sz0}}}{\omega \; \cos\gamma_{r0}}. \tag{10.32}$$

The frequency shift for the second operation $\mathcal{M}_{\text{ZOM}}$ is that of a zero-offset migration (10.10):

$$\omega_{\text{Mig}} = \omega_{\text{ZOM}} = \sqrt{{\omega_{\text{MZO}}}^2 - {k_{mx}}^2 \underline{c}^2}. \tag{10.33}$$

The frequency-shift for the first operation $\boldsymbol{\mathcal{M}}_{\text{MZO}}$ is just what is needed to make $\omega_{\text{ZOM}} = \omega_{\text{Mig}}$. Start with the relation between ω and ω_{Mig}, which can be written as

$$\begin{aligned} \omega_{\text{Mig}} &= -\underline{c}k_{mz0} \\ &= \omega\left(\frac{c_i}{c_r + c_i}\sqrt{1 - \frac{(k_{hx} - k_{mx})^2 {c_r}^2}{4\omega^2}} + \frac{c_r}{c_r + c_i}\sqrt{1 + \frac{(k_{hx} + k_{mx})^2 {c_i}^2}{4\omega^2}}\right), \end{aligned} \tag{10.34}$$

and use (10.33) to calculate ω_{MZO} (see Figure 10.3):

$$\omega_{\text{MZO}} = \sqrt{{\omega_{\text{Mig}}}^2 + {k_{mx}}^2 \underline{c}^2}. \tag{10.35}$$

To fix the phase-amplitude terms, choose the last term to be that of the zero-offset migration operator (10.27):

$$\mathcal{A}_{\text{Mig}} = 4\pi\rho_0 c_r {c_i}^2 \big/ \underline{c}. \tag{10.36}$$

The phase-amplitude term for the first operator should be the remainder of the total phase-amplitude term, to wit:

$$\mathcal{A}_{\text{MZO}} = \frac{2\underline{c}\sqrt{-i k_{mz0} k_{gz0} k_{sz0}}}{\sqrt{2\pi}\omega \; \cos\gamma_r}. \tag{10.37}$$

The sequence of operations $\boldsymbol{\mathcal{M}}_{\text{ZOM}}\boldsymbol{\mathcal{M}}_{\text{MZO}}$ preserves the angle-of-incidence information in the data, so if desired an amplitude-preserving migration can be performed in this manner. The angle of incidence is given by equation (9.89), or

$$\cos\sigma = -\frac{\boldsymbol{k}_g \cdot \boldsymbol{k}_s}{k_g k_s} = -\frac{c_i c_r}{\omega^2}\left(k_{gx}k_{sx} + k_{gz0}k_{sz0}\right). \tag{10.38}$$

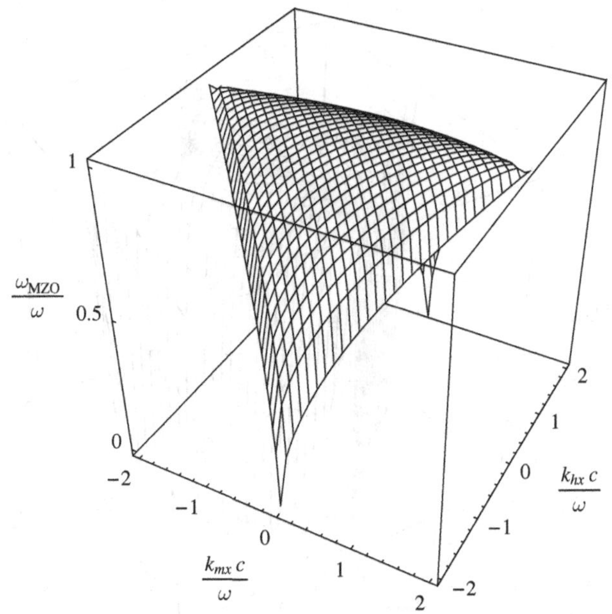

Figure 10.3 Frequency ω_{MZO} after migration to zero offset as a function of frequency and horizontal wavenumbers, for non-converted waves. Frequency ω_{MZO} is seen to be relatively flat in midpoint wavenumber.

10.2.2 MAD, NMO-stack, and ZOM

The MZO operator can be further divided. One can, for instance, start with an operator that adjusts moveout for reflector dip (MAD), followed by an operator that does a dip-independent moveout and reduction to zero offset (NMO):

$$\mathcal{M}_{MZO} = \mathcal{M}_{NMO}\mathcal{M}_{MAD}, \tag{10.39}$$

or

$$D_{MZO}\left(k_{mx}, k_{hx}, 0; \omega_{MZO}\right) = \mathcal{A}_{NMO}D_{MAD}\left(k_{mx}, k_{hx}, 0; \omega_{MAD}\right), \tag{10.40}$$

and

$$D_{MAD}\left(k_{mx}, k_{hx}, 0; \omega_{MAD}\right) = \mathcal{A}_{MAD}D_{PP}\left(k_{mx}, k_{hx}, 0; \omega\right). \tag{10.41}$$

The frequency shift for $\mathcal{M}_{NMO}$ should perform normal moveout at some chosen velocity c_0. That is,

$$\omega_{NMO} = \omega_{MAD}\sqrt{1 - {k_{hx}}^2\, {c_0}^2 / 4{\omega_{MAD}}^2}. \tag{10.42}$$

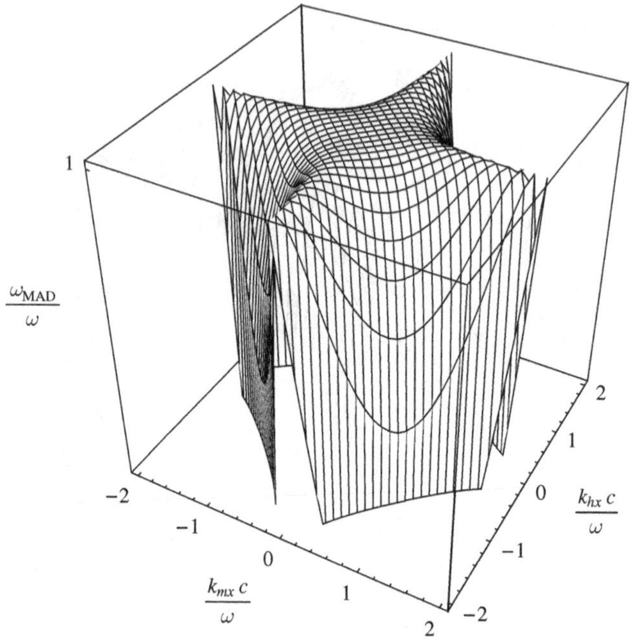

Figure 10.4 Frequency ω_{MAD} after moveout adjustment for dip as a function of frequency and horizontal wavenumbers, for non-converted waves. Frequency ω_{MAD} is pretty flat in both midpoint and offset wavenumber.

Since the output frequency ω_{NMO} must equal ω_{MZO}, the frequency shift from ω to ω_{MAD} is easily computed:

$$\omega_{\text{MAD}} = \sqrt{\underline{c}^2 \left(k_{mz0}{}^2 + k_{mx}{}^2\right) + k_{hx}{}^2\, c_0{}^2 / 4}, \tag{10.43}$$

with k_{mz0} as given by (10.34). This operator is shown in Figure 10.4 for the case where $c_r = c_i = c_0$.

The phase and amplitude of the operator $\mathcal{M}_{\text{NMO}}$ should not involve dip. We can choose

$$\mathcal{A}_{\text{NMO}} = \sqrt{\frac{-i\,\omega_{\text{NMO}}}{\pi\, c_0}}. \tag{10.44}$$

This leaves for the moveout-adjustment amplitude

$$\mathcal{A}_{\text{MZO}} = \frac{2\underline{c}\sqrt{-i\,k_{mz0}k_{gz0}k_{sz0}}}{\sqrt{2\pi}\,\omega\ \cos\gamma_r\,\mathcal{A}_{\text{NMO}}} = \frac{\underline{c}}{\omega\cos\gamma_r}\sqrt{\frac{2k_{mz0}k_{gz0}k_{sz0}c_0}{\omega_{\text{NMO}}}}. \tag{10.45}$$

The above has decomposed the full prestack migration operator into three separate operations. The first adjusts moveout to compensate for reflector dip (and, for converted waves, dual velocities), the second performs a dip-independent normal moveout correction and reduction to zero offset, and the third does a zero-offset migration. Amplitudes and incidence angles are preserved in these operations.

10.2.3 Other decompositions

One is obviously free to partition the full migration operator any number of ways. While the above operators have been expressed in the *f–k* domain, some of them could be implemented as equivalent space–time or space–frequency operations. Asymptotic space-domain formulas for residual migration are discussed in Chapter 12.

10.3 Residual *f–k* migration

Migration with the wrong or an approximate velocity can be considered a partial migration operation (Rothman *et al.*, 1985; Stolt & Benson, 1986; Sava, 2003; Fomel, 2003). If one applies an initial approximate migration operator $\mathcal{M}_A$, the result is an approximate image function

$$\mathcal{R}_A = \mathcal{M}_A \mathcal{D}. \tag{10.46}$$

To achieve the correct result

$$\mathcal{R} = \mathcal{M}\mathcal{D}, \tag{10.47}$$

we must apply a residual migration operator $\mathcal{M}_R$ to the approximate image function $\mathcal{R}_A$

$$\mathcal{R}_R = \mathcal{M}_R \mathcal{R}_A, \tag{10.48}$$

with

$$\mathcal{M} = \mathcal{M}_R \mathcal{M}_A. \tag{10.49}$$

As for partial migration, each operation involves a frequency shift and a change in amplitude and possibly phase.

10.3.1 2.5-D zero-offset residual migration

The full 2.5-D zero-offset migration operator is given by (10.27), which we express in terms of vertical traveltime $\tau = z/\underline{c}$ as

$$\left(\frac{\mathcal{R}_{ri}}{\tau}\right)\left(k_{mx}, \omega_{\text{Mig}}, \sigma_0 = 0\right) = 4\pi \rho_0 c_r {c_i}^2 D_{ri}\left(k_{mx}, x_h = 0; \omega\right), \tag{10.50}$$

with

$$\omega_{\text{Mig}} = \sqrt{\omega^2 - k_{mx}{}^2\underline{c}^2}. \tag{10.51}$$

If we migrate initially with incorrect velocities c_{iA}, c_{rA}, and $\underline{c}_A = c_{iA}c_{rA}/(c_{iA} + c_{rA})$, the result is

$$\left(\frac{\mathcal{R}_A}{\tau}\right)(k_x, \omega_A, \sigma_0 = 0) = 4\pi\rho_0 c_{iA}{}^2 c_{rA} D_{ri}(k_{mx}, x_h = 0; \omega), \tag{10.52}$$

with

$$\omega_A = \sqrt{\omega^2 - k_{mx}{}^2\underline{c}_A{}^2}. \tag{10.53}$$

The residual migration operator $\mathcal{M}_R$ is given by

$$\left(\frac{\mathcal{R}_{ri}}{\tau}\right)\left(k_x, \omega_{\text{Mig}}, \sigma_0 = 0\right) = \frac{c_r c_i{}^2}{c_{rA}c_{iA}{}^2}\left(\frac{\mathcal{R}_A}{\tau}\right)(k_x, \omega_A, \sigma_0 = 0). \tag{10.54}$$

Combining (10.51) and (10.53), we discover

$$\omega_{\text{Mig}} = \sqrt{\omega_A{}^2 - k_{mx}{}^2\left(\underline{c}^2 - \underline{c}_A{}^2\right)}. \tag{10.55}$$

Provided the incorrect velocity is less than the correct velocity, the residual migration operator frequency shift is that for migration at an effective velocity

$$\underline{c}_e = \sqrt{\underline{c}^2 - \underline{c}_A{}^2}. \tag{10.56}$$

If the incorrect velocity is greater than the correct velocity, the residual migration operator becomes an inverse migration or modeling operator.

10.3.2 2.5-D prestack residual migration

Prestack, the residual operator takes a slightly more complex form. We have for the final migrated frequency,

$$\omega_{\text{Mig}} = \omega\underline{c}\left(\sqrt{c_r{}^{-2} - \frac{k_{gx}{}^2}{\omega^2}} + \sqrt{c_i{}^{-2} - \frac{k_{sx}{}^2}{\omega^2}}\right), \tag{10.57}$$

and for the initial, approximate migration,

$$\omega_A = \omega\underline{c}_A\left(\sqrt{c_{rA}{}^{-2} - \frac{k_{gx}{}^2}{\omega^2}} + \sqrt{c_{iA}{}^{-2} - \frac{k_{sx}{}^2}{\omega^2}}\right). \tag{10.58}$$

It is convenient to define radial parameters

$$p_r \equiv \frac{k_{gx}}{\omega}, \qquad p_i \equiv \frac{k_{sx}}{\omega}, \tag{10.59}$$

in terms of which

$$\omega_{\text{Mig}} = \omega \underline{c} \left(\sqrt{c_r{}^{-2} - p_r{}^2} + \sqrt{c_i{}^{-2} - p_i{}^2} \right), \tag{10.60}$$

and

$$\omega_A = \omega \underline{c}_A \left(\sqrt{c_{rA}{}^{-2} - p_r{}^2} + \sqrt{c_{iA}{}^{-2} - p_i{}^2} \right). \tag{10.61}$$

The data could have been expressed directly in terms of the radial parameters p_r and p_i by effecting Radon transforms from the spatial variables x_g and x_s. Here, we started with Fourier transforms, so to put the data in terms of p_r and p_s requires the change of variable (10.59).

The relation between ω_{Mig} and ω_A is (see Figure 10.5)

$$\omega_{\text{Mig}} = \omega_A \frac{\underline{c} \left(\sqrt{c_r{}^{-2} - p_r{}^2} + \sqrt{c_i{}^{-2} - p_i{}^2} \right)}{\underline{c}_A \left(\sqrt{c_{rA}{}^{-2} - p_r{}^2} + \sqrt{c_{iA}{}^{-2} - p_i{}^2} \right)}, \tag{10.62}$$

The prestack residual migration frequency shift (10.62) is more complex than a migration at an effective velocity, as was the case for the zero-offset operation. Before returning the migrated data to the space-domain, one must change variables back to k_{gx} and k_{sx}, or equivalently, k_{mx} and k_{hx}. We have

$$k_{gx} = p_r \omega, \qquad k_{sx} = p_i \omega, \tag{10.63}$$

or, in terms of migrated frequency,

$$\begin{aligned} k_{gx} &= \frac{p_r \omega_{\text{Mig}}}{\underline{c} \left(\sqrt{c_r{}^{-2} - p_r{}^2} + \sqrt{c_i{}^{-2} - p_i{}^2} \right)}, \\ k_{sx} &= \frac{p_i \omega_{\text{Mig}}}{\underline{c} \left(\sqrt{c_r{}^{-2} - p_r{}^2} + \sqrt{c_i{}^{-2} - p_i{}^2} \right)}. \end{aligned} \tag{10.64}$$

The opening angle is determined from the final velocities:

$$\cos \sigma = p_i c_i p_r c_r + \sqrt{\left(1 - p_r{}^2 c_r{}^2\right)\left(1 - p_i{}^2 c_i{}^2\right)}. \tag{10.65}$$

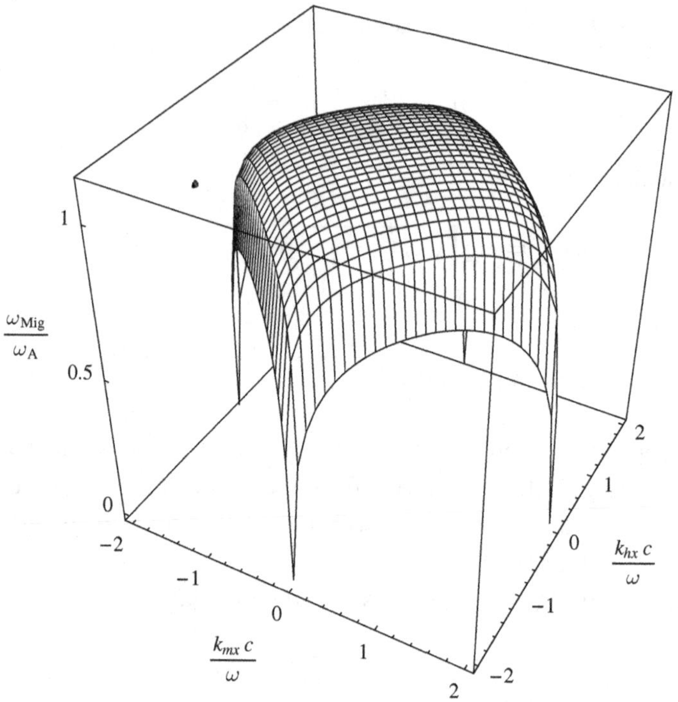

Figure 10.5 An *f–k* residual migration operator for non-converted waves. In this example, the initial migration velocity was 90% of the correct value.

10.4 Migration in a depth-variable medium

10.4.1 Elastic reflection response in a depth-variable medium

In a depth-variable medium, the elastic response D_{ri} in *f–k* space to a scattering potential $\mathbb{V}_{ri}$ is given by (9.111), or

$$D_{ri}\left(\tilde{\boldsymbol{k}}_g|\tilde{\boldsymbol{k}}_s;\omega\right) = -\frac{\omega^2}{4\sqrt{k_{gz}(0)k_{sz}(0)}\rho_0(0)c_r(0)c_i(0)} \times \int dz e^{-i\int_0^z dz' k_{mz}(z')} \frac{\mathbb{V}_{ri}\left(\tilde{\boldsymbol{k}}_m, z, \sigma(z)\right)}{\sqrt{k_{gz}(z)k_{sz}(z)}\rho_0(z)c_r(z)c_i(z)}. \quad (10.66)$$

Here,

$$\tilde{\boldsymbol{k}}_m = \tilde{\boldsymbol{k}}_g - \tilde{\boldsymbol{k}}_s, \quad (10.67)$$

$$k_{mz}(z) = k_{gz}(z) - k_{sz}(z), \quad (10.68)$$

$$k_{gz}(z) = -\frac{\omega}{c_r(z)}\sqrt{1 - \frac{\left(k_{gx}{}^2 + k_{gy}{}^2\right) c_r{}^2(z)}{\omega^2}},$$

$$k_{sz}(z) = +\frac{\omega}{c_i(z)}\sqrt{1 - \frac{\left(k_{sx}{}^2 + k_{sy}{}^2\right) c_i{}^2(z)}{\omega^2}}, \tag{10.69}$$

$$\boldsymbol{k}_g = \left(\tilde{\boldsymbol{k}}_g, k_{gz}\right), \tag{10.70}$$

$$\boldsymbol{k}_s = \left(\tilde{\boldsymbol{k}}_s, k_{sz}\right), \tag{10.71}$$

$$\boldsymbol{k}_m = \left(\tilde{\boldsymbol{k}}_m, k_{mz}\right), \tag{10.72}$$

and (see Appendix H)

$$\cos\sigma(z) = -\frac{\boldsymbol{k}_g \cdot \boldsymbol{k}_s}{\omega^2} c_r(z) c_i(z), \tag{10.73}$$

$$\sin\sigma(z) = \frac{\left|\boldsymbol{k}_g \times \boldsymbol{k}_s\right| \operatorname{Sign}\left(\left(\boldsymbol{k}_g \times \boldsymbol{k}_s\right)_y\right)}{\omega^2} c_r(z) c_i(z). \tag{10.74}$$

Opening angle σ, for a given set of input wavenumbers, will change with depth, which is to say, the angle between the incident and reflected waves is different at the surface $z = 0$ than at the reflector. Potential or reflectivity amplitude depends upon opening angle at the reflector, as has been indicated in equation (10.66).

10.4.2 3-D prestack phase-shift migration

Equation (10.66) is invertible for the potential $\mathbb{V}_{ri}$, or equivalently, a reflectivity function $\mathcal{R}_{ri}$. However, the data field is a function of five variables, while the potential is a function of four. We can handle this mathematically by introducing source–receiver azimuthal angle as a fifth variable in the potential or reflectivity. According to the elastic model we have employed to this point, the potential and reflectivity functions should be the same for all azimuths, but that need not stop us from adding it to the list of variables. Moreover, we concede that real-world potentials and reflectivities are likely to exhibit azimuthal variation, no matter what our models have to say about the matter. As in Appendix H, we define azimuthal angle in terms of wavenumbers as

$$\tan\chi(z) = \frac{k_{sy}k_{gz}(z) - k_{gy}k_{sz}(z)}{k_{sx}k_{gz}(z) - k_{gx}k_{sz}(z)}. \tag{10.75}$$

Because the vertical wavenumbers vary with depth, for constant horizontal wavenumbers k_{sx}, k_{sy}, k_{gx}, and k_{gy}, so does χ. If our physical model predicted variation of the potential or reflectivity function with χ, we would probably want to add $\chi(z)$ to the argument list of $\mathbb{V}$ or $\mathcal{R}$, rewriting (10.66) as

$$D_{ri}\left(\tilde{\boldsymbol{k}}_g|\tilde{\boldsymbol{k}}_s;\omega\right) = -\int dz e^{-i\int_0^z dz' k_{mz}(z')} E_{ri}\left(\tilde{\boldsymbol{k}}_g, \tilde{\boldsymbol{k}}_s;\omega, z\right) \mathbb{V}_{ri}\left(\tilde{\boldsymbol{k}}_m, z, \sigma(z), \chi(z)\right), \tag{10.76}$$

with the phase-amplitude term E_{ri} given by

$$E_{ri}\left(\tilde{\boldsymbol{k}}_g, \tilde{\boldsymbol{k}}_s;\omega, z\right) = \frac{\omega^2}{4\sqrt{k_{gz}(0)k_{sz}(0)}\rho_0(0)c_r(0)c_i(0)} \frac{1}{\sqrt{k_{gz}(z)k_{sz}(z)}\rho_0(z)c_r(z)c_i(z)}. \tag{10.77}$$

We expect a migration operator of the form

$$\mathbb{V}_{ri}\left(\tilde{\boldsymbol{k}}_m, z, \sigma, \chi\right) = \int d\omega F_{ri}(\omega, z, \ldots) D_{ri}\left(\tilde{\boldsymbol{k}}_g|\tilde{\boldsymbol{k}}_s;\omega\right) e^{i\int_0^z dz' k_{mz}(z')}, \tag{10.78}$$

with F_{ri} yet to be determined. In this expression, $\tilde{\boldsymbol{k}}_m$, z, opening angle σ, and azimuthal angle χ are independent variables. Since ω is an integration variable, it, too, should be considered independent. The wavenumbers $\tilde{\boldsymbol{k}}_g$ and $\tilde{\boldsymbol{k}}_s$ can be considered to be dependent on the four independent variables, though not all the dependence has been explicitly noted. Likewise, the complete argument list of F_{ri} would include all independent variables.

We find F_{ri} by substituting the modeling equation (10.76) into the migration equation (10.78):

$$\begin{aligned} \mathbb{V}_{ri}\left(\tilde{\boldsymbol{k}}_m, z, \sigma, \chi\right) = &-\int dz' \int d\omega F_{ri}(\omega, z, \ldots) E_{ri}(\omega, z', \ldots) \\ &\times \mathbb{V}_{ri}\left(\tilde{\boldsymbol{k}}_m, z', \sigma'(z'), \chi'(z')\right) e^{i\int_{z'}^z dz'' k_{mz}(z'')}. \end{aligned} \tag{10.79}$$

The frequency dependence of the exponential term in this equation is rapid and oscillatory. When integrated over frequency, it will sum to nearly zero for most intervals. However, near $z = z'$, the exponential slows down, and the integral can produce a finite result. In this neighborhood, $\int_{z'}^z dz'' k_{mz}(z'') \simeq k_{mz}(z)(z - z')$.

Except in the scattering potential, the dependence on z' is slow. Slowly varying functions of z' within (10.79) can be replaced with their values at z. Likewise, $\sigma'(z') \simeq \sigma$ and $\chi'(z') \simeq \chi$. Rearranging the integrals, we have

$$\begin{aligned} \mathbb{V}_{ri}\left(\tilde{\boldsymbol{k}}_m, z, \sigma, \chi\right) \simeq &-\int d\omega F_{ri}(\omega, z, \ldots) E_{ri}(\omega, z, \ldots) \\ &\times e^{ik_{mz}(z)z} \int dz' \mathbb{V}_{ri}\left(\tilde{\boldsymbol{k}}_m, z', \sigma_0, \chi_0\right) e^{-ik_{mz}(z)z'}, \end{aligned} \tag{10.80}$$

The z' integral in (10.80) is seen to be a Fourier transform from z' to $k_{mz}(z)$:

$$\mathbb{V}_{ri}\left(\tilde{\boldsymbol{k}}_m, z, \sigma, \chi\right) \simeq -\int d\omega F_{ri}(\omega, z, \ldots) E_{ri}(\omega, z, \ldots) e^{ik_{mz}(z)z} \times \mathbb{V}_{ri}\left(\tilde{\boldsymbol{k}}_m, k_{mz}(z), \sigma, \chi\right). \quad (10.81)$$

That k_{mz} changes with depth does not affect our ability to change integration variable from ω to k_{mz}, in effect making k_{mz} an independent variable and ω a dependent variable. From Appendix H.3,

$$\omega\left(k_{mz}, \tilde{\boldsymbol{k}}_m, z, \sigma\right) = -\text{Sign}\left(k_{mz}\right) \frac{\sqrt{k_{mz}{}^2 + \tilde{k}_m{}^2 c_i(z) c_r(z)}}{2c_e(\sigma, z)}, \quad (10.82)$$

with $c_e(\sigma, z)$ defined as in equation (H.51), or

$$c_e(\sigma, z) = \frac{1}{2}\sqrt{c_i{}^2(z) + c_r{}^2(z) + 2c_i(z)c_r(z)\cos\sigma}. \quad (10.83)$$

Frequency is seen to depend upon $\tilde{\boldsymbol{k}}_m$, z, and σ, but not χ. It has derivative with respect to k_{mz}

$$\left[\frac{\partial \omega}{\partial k_{mz}}\right]_{\tilde{\boldsymbol{k}}_m, \sigma_0, z} = -\frac{1}{\sqrt{1 + \tilde{k}_m{}^2/k_{mz}{}^2}} \frac{c_i(z)c_r(z)}{2c_e(\sigma, z)}. \quad (10.84)$$

With k_{mz} as integration variable, equation (10.81) becomes

$$\mathbb{V}_{ri}\left(\tilde{\boldsymbol{k}}_m, z, \sigma, \chi\right) \simeq \frac{c_i(z)c_r(z)}{2c_e(\sigma, z)} \int dk_{mz} \frac{F_{ri}(\omega, z, \ldots) E_{ri}(\omega, z, \ldots)}{\sqrt{1 + \tilde{k}_m{}^2/k_{mz}{}^2}} \cdot e^{ik_{mz}z} \mathbb{V}_{ri}\left(\tilde{\boldsymbol{k}}_m, k_{mz}, \sigma, \chi\right). \quad (10.85)$$

Choosing

$$F_{ri}(\omega, z, \ldots) = \frac{2c_e(\sigma, z)}{2\pi E_{ri}(\omega, z, \ldots)} \sqrt{1 + \tilde{k}_m{}^2/k_{mz}{}^2}, \quad (10.86)$$

we are left with the identity

$$\mathbb{V}_{ri}\left(\tilde{\boldsymbol{k}}_m, z, \sigma, \chi\right) \simeq \mathbb{V}_{ri}\left(\tilde{\boldsymbol{k}}_m, z, \sigma, \chi\right), \quad (10.87)$$

which validates (10.78) as the migration operator.

To implement (10.78), one needs to be able to determine $\tilde{\boldsymbol{k}}_g$ and $\tilde{\boldsymbol{k}}_s$ as well as ω from $\boldsymbol{k}_m$, z, σ, and χ. The prescription is given in Appendix H.3. Defining unit vectors $\hat{\boldsymbol{k}}_g$, $\hat{\boldsymbol{k}}_s$, and $\hat{\boldsymbol{k}}_m$ in the directions of $\boldsymbol{k}_g$, $\boldsymbol{k}_s$, and $\boldsymbol{k}_m$, respectively,

$$\hat{\boldsymbol{k}}_s = -\hat{\boldsymbol{k}}_m \cos\gamma_i - \hat{\boldsymbol{e}}_{\parallel} \sin\gamma_i. \quad (10.88)$$

and

$$\hat{\boldsymbol{k}}_g = \hat{\boldsymbol{k}}_m \cos\gamma_r - \hat{\boldsymbol{e}}_\parallel \sin\gamma_r . \tag{10.89}$$

where $\hat{\boldsymbol{e}}_\parallel$ is the unit vector

$$\hat{\boldsymbol{e}}_\parallel = \frac{1}{\sqrt{\hat{k}_{mz}{}^2 + (\hat{k}_{my}\cos\chi - \hat{k}_{mx}\sin\chi)^2}} \begin{pmatrix} \left(\hat{k}_{my}{}^2 + \hat{k}_{mz}{}^2\right)\cos\chi - \hat{k}_{mx}\hat{k}_{my}\sin\chi \\ \left(\hat{k}_{mx}{}^2 + \hat{k}_{mz}{}^2\right)\sin\chi - \hat{k}_{mx}\hat{k}_{my}\cos\chi \\ -\hat{k}_{mz}\left(\hat{k}_{mx}\cos\chi + \hat{k}_{my}\sin\chi\right) \end{pmatrix}, \tag{10.90}$$

γ_i is the angle of incidence, given by

$$\cos\gamma_i = \frac{(c_r(z) + c_i(z)\cos\sigma)}{2c_e(\sigma, z)} \tag{10.91}$$

$$\sin\gamma_i = \frac{c_i(z)\sin\sigma}{2c_e(\sigma, z)}, \tag{10.92}$$

and γ_r the angle of reflection, given by

$$\cos\gamma_r = \frac{(c_i(z) + c_r(z)\cos\sigma)}{2c_e(\sigma, z)}, \tag{10.93}$$

$$\sin\gamma_r = \frac{c_r(z)\sin\sigma}{2c_e(\sigma, z)}. \tag{10.94}$$

10.4.3 2.5-D prestack phase-shift migration

From (9.131), in 2.5 dimensions, the elastic-component reflected impulse response reduces to

$$\mathrm{D}_{ri}^{(2.5\mathrm{D})}(k_{mx}, k_{hx}; \omega) = -\int dz e^{-i\int_0^z dz' k_{mz}(z')} E_{ri}(k_{mx}, k_{hx}; \omega, z) \times \mathbb{V}_{(2.5\mathrm{D})}^{ri}(k_{mx}, z, \sigma(z)) . \tag{10.95}$$

with the phase-amplitude term E_{ri} given by

$$E_{ri}(k_{mx}, k_{hx}; \omega, z) = \frac{1}{8\rho_0(0)c_r(0)c_i(0)\rho_0(z)c_r(z)c_i(z)} \times \frac{\omega^2}{\sqrt{2\pi i k_{gz}(0)k_{sz}(0)k_{gz}(z)k_{sz}(z)\int_0^z dz' \frac{k_{mz}(z')}{k_{gz}(z')k_{sz}(z')}}}. \tag{10.96}$$

The phase-amplitude term is expected to change slowly relative to the potential.

Data are expressed in terms of the wavenumber–frequency variables k_{mx}, k_{hx}, and ω, while the scattering potential is expressed in terms of horizontal wavenumber k_{mx}, depth z, and opening angle σ. Treating the data variables as the independent variables, we have

$$k_{sx} = (k_{hx} - k_{mx})/2, \tag{10.97}$$

$$k_{gx} = (k_{hx} + k_{mx})/2, \tag{10.98}$$

$$\begin{aligned} k_{gz}(z) &= -\omega\sqrt{\frac{1}{c_r(z)^2} - \frac{k_{gx}{}^2}{\omega^2}}, \\ k_{sz}(z) &= +\omega\sqrt{\frac{1}{c_i(z)^2} - \frac{k_{sx}{}^2}{\omega^2}}, \end{aligned} \tag{10.99}$$

$$\begin{aligned} \boldsymbol{k}_g(z) &= \left(k_{gx}, k_{gz}(z)\right), \\ \boldsymbol{k}_s(z) &= \left(k_{sx}, k_{sz}(z)\right), \end{aligned} \tag{10.100}$$

$$\begin{aligned} k_{mz}(z) &= k_{gz}(z) - k_{sz}(z) \\ &= -\omega\left(\sqrt{\frac{1}{c_r(z)^2} - \frac{k_{gx}{}^2}{\omega^2}} + \sqrt{\frac{1}{c_i(z)^2} - \frac{k_{sx}{}^2}{\omega^2}}\right), \end{aligned} \tag{10.101}$$

$$\boldsymbol{k}_m(z) = \left(k_{mx}, k_{mz}(z)\right), \tag{10.102}$$

and

$$\cos\sigma(z) = -\frac{\boldsymbol{k}_g(z)\cdot\boldsymbol{k}_s(z)}{k_g(z)k_s(z)} = -\frac{c_r(z)c_i(z)}{\omega^2}\left(k_{gx}k_{sx} + k_{gz}(z)k_{sz}(z)\right). \tag{10.103}$$

Note that a complete argument list of these variables would include k_{mx} and k_{hx}, or equivalently, k_{sx} and k_{gx}.

Following the argument of Section 10.2, we find the migration operator corresponding to (10.78) to be

$$\mathbb{V}^{ri}_{(2.5\mathrm{D})}(k_{mx}, z, \sigma) = \int d\omega F_{ri}(\omega, z, \ldots) D^{ri}_{(2.5\mathrm{D})}(k_{mx}, k_{hx}; \omega)\, e^{i\int_0^z dz' k_{mz}(z')}, \tag{10.104}$$

with F_{ri} a weighted inverse of the modeling amplitude-phase term E_{ri}:

$$F_{ri}(\omega, z, \ldots) = -\left(2\pi E_{ri}(\omega, z, \ldots)\left[\frac{\partial\omega}{\partial k_{mz}}\right]_{k_{mx},\sigma,z}\right)^{-1}, \tag{10.105}$$

It remains to find the value of $\partial\omega/\partial k_{mz}$. The derivative is to be carried out keeping k_{mx} and σ constant. Under these conditions, the wavenumbers k_{gx} and k_{sx} are not fixed, so the formula (10.101) is not directly usable. However, we can write the relation

$$\begin{aligned}
{k_{mx}}^2 + {k_{mz}}^2\,(k_{mx}, z, \sigma, \omega) &\equiv {k_m}^2(z, \sigma, \omega) \\
&= \left|\boldsymbol{k}_g - \boldsymbol{k}_s\right|^2 \\
&= {k_g}^2 + {k_s}^2 - 2\boldsymbol{k}_g \cdot \boldsymbol{k}_s \\
&= \omega^2\left(\frac{1}{{c_r}^2(z)} + \frac{1}{{c_i}^2(z)} + \frac{2\cos\sigma}{c_r(z)c_i(z)}\right) \\
&= \frac{4\omega^2{c_e}^2(\sigma, z)}{{c_r}^2(z){c_i}^2(z)},
\end{aligned} \tag{10.106}$$

with c_e defined as in equation (H.51). Thus,

$$k_m(z, \sigma, \omega) = \frac{2\omega c_e(\sigma, z)}{c_r(z)c_i(z)}, \tag{10.107}$$

and

$$k_{mz}\,(k_{mx}, z, \sigma, \omega) = -\sqrt{\frac{4\omega^2{c_e}^2(\sigma, z)}{{c_r}^2(z){c_i}^2(z)} - {k_{mx}}^2}. \tag{10.108}$$

Solving for ω, we have frequency as a function of k_{mz}, k_{mx}, and σ:

$$\omega = \frac{c_r(z)c_i(z)}{2c_e(\sigma, z)}\sqrt{{k_{mx}}^2 + {k_{mz}}^2}. \tag{10.109}$$

Differentiating ω with respect to k_{mz},

$$\left[\left|\frac{\partial\omega}{\partial k_{mz}(z)}\right|\right]_{k_{mx},\sigma} = \frac{c_r(z)c_i(z)}{2c_e(\sigma, z)}\frac{1}{\sqrt{1 + {k_{mx}}^2/{k_{mz}}^2(z)}}. \tag{10.110}$$

Thus, from (10.105), (10.77), and (10.110), the amplitude-phase term for the migration operator is

$$\begin{aligned}
F_{ri} = &-\frac{16\rho_0(0)\rho_0(z)c_r(0)c_i(0)c_e(\sigma, z)}{\omega^2} \\
&\times\sqrt{\frac{{k_m}^2(z)}{{k_{mz}}^2(z)}\frac{k_{gz}(0)k_{sz}(0)k_{gz}(z)k_{sz}(z)}{-2\pi i}\int_0^z dz'\frac{k_{mz}(z')}{k_{gz}(z')k_{sz}(z')}}.
\end{aligned} \tag{10.111}$$

To express $k_{gz}(z)$ and $k_{sz}(z)$ in terms of k_{mx}, $k_{mz}(z)$, and σ, note the following. The incident and reflected angles relative to the direction of $\boldsymbol{k}_m$ are given by

$$\cos\gamma_i = -\frac{\boldsymbol{k}_s\cdot\boldsymbol{k}_m}{k_s k_m} = \frac{k_s{}^2 - \boldsymbol{k}_s\cdot\boldsymbol{k}_g}{k_s k_m} = \frac{c_r(z) + c_i(z)\cos\sigma}{2c_e(z,\sigma)}, \tag{10.112}$$

$$\cos\gamma_r = \frac{\boldsymbol{k}_g\cdot\boldsymbol{k}_m}{k_g k_m} = \frac{k_g{}^2 - \boldsymbol{k}_s\cdot\boldsymbol{k}_g}{k_g k_m} = \frac{c_i(z) + c_r(z)\cos\sigma}{2c_e(z,\sigma)}, \tag{10.113}$$

from which it follows that

$$\frac{\sin\gamma_i}{c_i(z)} = \frac{\sin\gamma_r}{c_r(z)} = \frac{\sin\sigma}{2c_e(z,\sigma)}. \tag{10.114}$$

This relation is an expression of Snell's law. Where $c_i = c_r$, the relation reduces to $\gamma_i = \gamma_r = \sigma/2$.

The individual wavenumbers $\boldsymbol{k}_g$ and $\boldsymbol{k}_s$ can also be expressed in terms of $\boldsymbol{k}_m$ and σ, as per Figure 10.6. We have the following angles:

$$\boldsymbol{k}_m = \frac{2\omega c_e(\sigma,z)}{c_r(z)c_i(z)}\left(\sin\theta_m, -\cos\theta_m\right), \tag{10.115}$$

$$\boldsymbol{k}_s = \frac{\omega}{c_i(z)}\left(-\sin\phi_i, \cos\phi_i\right), \tag{10.116}$$

$$\boldsymbol{k}_g = \frac{\omega}{c_r(z)}\left(-\sin\phi_r, -\cos\phi_r\right), \tag{10.117}$$

with

$$\phi_i = \gamma_i + \theta_m, \tag{10.118}$$

$$\phi_r = \gamma_r - \theta_m. \tag{10.119}$$

The magnitudes of the wavenumber vectors are

$$k_m = \frac{2\omega c_e(\sigma,z)}{c_r(z)c_i(z)}, \tag{10.120}$$

$$k_s = \frac{\omega}{c_i(z)}, \tag{10.121}$$

and

$$k_g = \frac{\omega}{c_r(z)}. \tag{10.122}$$

The source and receiver wavenumbers are expressible as (suppressing the arguments of the velocities)

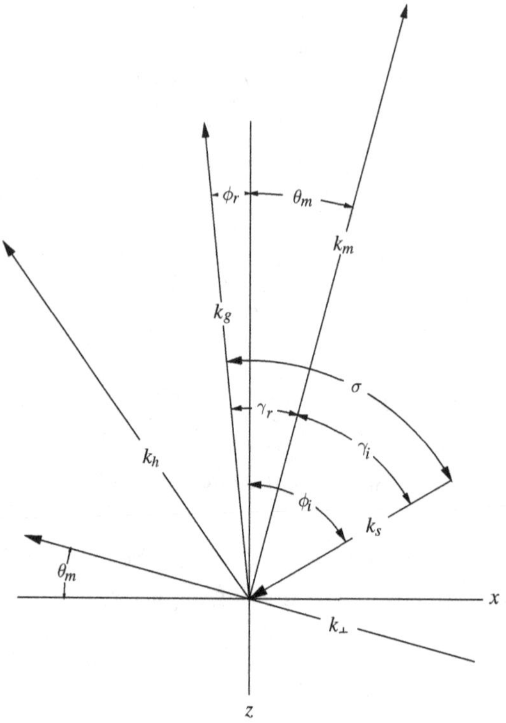

Figure 10.6 Illustration of a planar reflection. An incoming wave has wavenumber $\boldsymbol{k}_s$ of magnitude ω/c_i, the reflected wave $\boldsymbol{k}_g$ with magnitude ω/c_r. The wavenumber $\boldsymbol{k}_s$ makes an angle $\phi_i = -\arctan(k_{sx}/k_{sz})$ with respect to the vertical, and $\boldsymbol{k}_g$ the angle $\phi_r = +\arctan(k_{gx}/k_{gz})$. The difference of these two wavenumbers is $\boldsymbol{k}_m$, which is normal to the Snell reflecting surface $\boldsymbol{k}_\perp$, which slopes at angle $\theta_m = -\arctan(k_{mx}/k_{mz})$. With respect to the Snell surface normal, $\boldsymbol{k}_s$ makes the incident angle $\gamma_i = \phi_i - \theta_m$, and $\boldsymbol{k}_g$ the reflection angle $\gamma_r = \phi_r + \theta_m$. In this figure, $c_i = 2c_r$. When $c_i = c_r$, the incident and reflected angles are equal. With the sign conventions adopted here, all the angles in this figure are positive.

$$\begin{aligned} k_{sx} &= -\frac{\omega}{c_i}\sin\phi_i \\ &= -\frac{\omega}{c_i}(\sin\gamma_i\cos\theta_m + \cos\gamma_i\ \sin\theta_m) \\ &= \frac{c_r c_i}{4{c_e}^2}\left(k_{mz}\sin\sigma - k_{mx}\left(\frac{c_r}{c_i} + \cos\sigma\right)\right), \end{aligned} \tag{10.123}$$

$$\begin{aligned} k_{sz} &= \frac{\omega}{c_i}\cos\phi_i \\ &= \frac{\omega}{c_i}(\cos\gamma_i\cos\theta_m - \sin\gamma_i\ \sin\theta_m) \\ &= -\frac{c_r c_i}{4{c_e}^2}\left(k_{mz}\left(\frac{c_r}{c_i} + \cos\sigma\right) + k_{mx}\sin\sigma\right), \end{aligned} \tag{10.124}$$

$$k_{gx} = -\frac{\omega}{c_r}\sin\phi_r$$
$$= -\frac{\omega}{c_r}(\sin\gamma_r\cos\theta_m - \cos\gamma_r\ \sin\theta_m)$$
$$= \frac{c_r c_i}{4{c_e}^2}\left(k_{mz}\sin\sigma + k_{mx}\left(\frac{c_i}{c_r} + \cos\sigma\right)\right), \tag{10.125}$$

$$k_{gz} = -\frac{\omega}{c_r}\cos\phi_r$$
$$= -\frac{\omega}{c_r}(\cos\gamma_r\cos\theta_m + \sin\gamma_r\ \sin\theta_m)$$
$$= \frac{c_r c_i}{4{c_e}^2}\left(k_{mz}\left(\frac{c_i}{c_r} + \cos\sigma\right) - k_{mx}\sin\sigma\right). \tag{10.126}$$

In matrix form,

$$\boldsymbol{k}_s = \frac{c_r c_i}{4{c_e}^2}\begin{pmatrix} -\left(\frac{c_r}{c_i} + \cos\sigma\right) & \sin\sigma \\ -\sin\sigma & -\left(\frac{c_r}{c_i} + \cos\sigma\right)\end{pmatrix}\boldsymbol{k}_m, \tag{10.127}$$

and

$$\boldsymbol{k}_g = \frac{c_r c_i}{4{c_e}^2}\begin{pmatrix} \left(\frac{c_i}{c_r} + \cos\sigma\right) & \sin\sigma \\ -\sin\sigma & \left(\frac{c_i}{c_r} + \cos\sigma\right)\end{pmatrix}\boldsymbol{k}_m. \tag{10.128}$$

With the above relations, the terms in the amplitude-phase coefficient F_{ri} can be computed from $\boldsymbol{k}_m$ and σ at z. The integral in F_{ri} requires values at all depths shallower than z. Practically, this just means that the integral at z is computed from the integral at $z - dz$ by adding the quantity $\frac{k_{mz}(z)dz}{k_{gz}(z)k_{sz}(z)}$. Given F_{ri}, equation (10.78) can be evaluated directly as a phase-shift and sum.

10.4.4 2.5-D prestack Radon-transform migration

An alternative to *f–k* phase-shift migration evaluates the migration integral in the Radon transform domain (Levin, 1980; Ottolini and Claerbout, 1984; Beylkin, 1985; Bisset and Durrani, 1990).

We introduce radial parameters

$$p_{mx} = k_{mx}/\,\omega = p_m\sin\theta_m, \tag{10.129}$$

$$p_{hx} = k_{hx}/\,\omega = -\sin\sigma\,/c_e\,, \tag{10.130}$$

$$p_{sx} = k_{sx}/\,\omega = -\sin\phi_i/c_i = -\sin\sigma/2c_e, \tag{10.131}$$

$$p_{gx} = k_{gx}/\,\omega = -\sin\phi_r/c_r = -\sin\sigma/2c_e, \tag{10.132}$$

$$p_{gz}(z) = \frac{k_{gz}(z)}{\omega} = -\sqrt{\frac{1}{c_r(z)^2} - p_{gx}{}^2} = -\cos\phi_r/c_r,$$
$$p_{sz}(z) = \frac{k_{sz}(z)}{\omega} = +\sqrt{\frac{1}{c_i(z)^2} - p_{sx}{}^2} = +\cos\phi_i/c_i, \tag{10.133}$$

$$p_{mz}(z) = \frac{k_{mz}(z)}{\omega} = -\left(\sqrt{\frac{1}{c_r(z)^2} - p_{gx}{}^2} + \sqrt{\frac{1}{c_i(z)^2} - p_{sx}{}^2}\right) = -p_m\cos\theta_m, \tag{10.134}$$

with

$$p_m(z) = \frac{k_m(z)}{\omega}. \tag{10.135}$$

In terms of these parameters, opening angle becomes

$$\cos\sigma(z) = c_r(z)c_i(z)\left(p_{gx}p_{sx} + p_{gz}(z)p_{sz}(z)\right). \tag{10.136}$$

We can define a traveltime τ as

$$\tau(z) = \frac{1}{\omega}\int_0^z dz' k_{mz}(z') = \int_0^z dz' p_{mz}(z'). \tag{10.137}$$

We also define a divergence factor

$$U(z) = \int_0^z dz' \frac{p_{mz}(z')}{p_{gz}(z')p_{sz}(z')}, \tag{10.138}$$

in terms of which the coefficient E_{ri} (equation (10.77)) becomes

$$E_{ri}(\omega, z) = \sqrt{\frac{-i\omega}{2\pi}\frac{1}{8\rho_0(0)c_r(0)c_i(0)\rho_0(z)c_r(z)c_i(z)}} \times \frac{1}{\sqrt{p_{gz}(0)p_{sz}(0)p_{gz}(z)p_{sz}(z)U(z)}}. \tag{10.139}$$

For simplicity, the functional dependence of σ, τ, U, and E_{ri} on the radial parameters has been omitted from their list of arguments.

In x–z, the migration operator (10.78) is

$$\mathbb{V}^{ri}_{(2.5\mathrm{D})}(x, z, \sigma) = \frac{1}{2\pi}\int dk_{mx}\int d\omega F_{ri}(\omega, z) D^{ri}_{(2.5\mathrm{D})}(k_{mx}, k_{hx}; \omega)\, e^{i(\omega\tau(z) + k_{mx}x)}. \tag{10.140}$$

Change variables from wavenumbers to radial parameters, as per equations (10.129)–(10.134), and set

$$D^{ri}_{(2.5\mathrm{D})}\left(p_{mx}, p_{hx}; \omega\right) \equiv D^{ri}_{(2.5\mathrm{D})}(k_{mx}, k_{hx}; \omega). \tag{10.141}$$

Data in radial parameter coordinates can be produced from space–time data by a triple Fourier transform from (x_m, x_h, t) to (k_{mx}, k_{hx}, ω), followed by a change of variables from (k_{mx}, k_{hx}) to (p_{mx}, p_{hx}). Alternatively, one can do a double Radon transform of the data from (x_m, x_h) to (p_{mx}, p_{hx}), followed by a Fourier transform from time to frequency.

In terms of the radial parameters,

$$\mathbb{V}_{ri}(x, z, \sigma) = \frac{1}{2\pi} \int dp_{mx} \int d\omega\, |\omega| F_{ri}(\omega, z) D^{ri}_{(2.5D)}\left(p_{mx}, p_{hx}; \omega\right) e^{i\omega(\tau(z)+p_{mx}x)}. \tag{10.142}$$

In terms of the radial parameters, the phase-amplitude term F_{ri} (equation (10.111)) is

$$F_{ri} = -16\rho_0(0)\rho_0(z)c_r(0)c_i(0)c_e(\sigma, z) \times\sqrt{\frac{1+{p_{mx}}^2/{p_{mz}}^2(z)}{-2\pi i\omega} p_{g_z}(0)p_{s_z}(0)p_{g_z}(z)p_{s_z}(z)U(z)}. \tag{10.143}$$

The frequency dependence of F_{ri} is confined to the term $\sqrt{-2\pi i\omega}$ in the denominator. Removing the frequency dependence,

$$F_{0ri} \equiv F_{ri}\sqrt{-i\omega}, \tag{10.144}$$

we have

$$\mathbb{V}_{ri}(x, z, \sigma) = -\frac{1}{(2\pi)^{3/2}} \int dp_{mx} F_{0ri}(p_{mx}, z) \times \int d\omega\, \sqrt{i\omega} D^{ri}_{(2.5D)}\left(p_{mx}, p_{hx}; \omega\right) e^{i\omega(\tau(z)+p_{mx}x)}. \tag{10.145}$$

The frequency integral amounts to an inverse Fourier transform from frequency to time of the half-derivative of the transformed data, evaluated at time $\tau(z) + p_{mx}x$.

Considering z, σ, p_x, and ω to be the independent parameters, we can solve for remaining variables in terms of them as follows. From (10.108), we obtain a relation for p_{mz}:

$$p_{mz}(p_{mx}, z, \sigma) = \sqrt{\frac{4{c_e}^2(\sigma, z)}{{c_r}^2(z){c_i}^2(z)} - {p_{mx}}^2}. \tag{10.146}$$

From (10.123)–(10.126), we obtain relations for the source and receiver radial parameters:

$$p_{sx} = \frac{c_r c_i}{4{c_e}^2}\left(p_{mz}\sin\sigma - p_{mx}\left(\frac{c_r}{c_i} + \cos\sigma\right)\right), \tag{10.147}$$

$$p_{sz} = -\frac{c_r c_i}{4{c_e}^2}\left(p_{mz}\left(\frac{c_r}{c_i} + \cos\sigma\right) + p_{mx}\sin\sigma\right), \tag{10.148}$$

$$p_{gx} = \frac{c_r c_i}{4{c_e}^2}\left(p_{mz}\sin\sigma + p_{mx}\left(\frac{c_i}{c_r} + \cos\sigma\right)\right), \tag{10.149}$$

$$p_{gz} = \frac{c_r c_i}{4{c_e}^2}\left(p_{mz}\left(\frac{c_i}{c_r} + \cos\sigma\right) - p_{mx}\sin\sigma\right). \tag{10.150}$$

The offset ray parameter p_{hx} is seen to be

$$p_{hx} = p_{gx} + p_{sx} = \frac{c_r c_i}{4{c_e}^2}\left(2p_{mz}\sin\sigma + p_{mx}\left(\frac{c_i}{c_r} - \frac{c_r}{c_i}\right)\right). \tag{10.151}$$

Given the radial parameters, F_{0ri} and $\tau(z)$ can then be computed. $\mathbb{V}_{ri}^{(2.5D)}$ is obtained at each x by performing the p_{mx} integral (which looks very much like an inverse Radon transform).

Finally, we note that there is nothing (except computational complexity) in the above formula to prevent c_r and c_i varying with x as well as z. We would expect that moderate horizontal velocity could be accommodated by this formula with reasonable accuracy.

Exercises

10.1 In the absence of wave-type conversion, compare the zero-offset migration formula (10.27) to the simple exploding-reflector formula (2.99), and explain the similarities and differences.

10.2 Verify that the migration formula (10.104), with F_{ri} given by (10.105), is the asymptotic inverse of (10.95).

11

Asymptotic modeling and migration

11.1 Migration and modeling as mapping

Migration and modeling are herein defined as linear operations mapping a data function $\mathcal{D}$ onto an image function $\mathcal{R}$ and vice versa. Abstractly, modeling is defined as an operation to obtain the data function from the image function, as per equation (2.1), whereas migration, as per equation (2.2), is the operation that recovers the image function from the data function.

Since the modeling operation proceeds in the direction that nature intended, it can be referred to as a *forward* operation. As written, the migration operation is an inverse to the forward modeling operation. Migration can be treated as a statistical inverse to be applied to a complete, presumably redundant data set. In this chapter, we wish to emphasize the symmetry between the forward and inverse operations, and hence pose migration as a deterministic operation to be applied to a subset, presumably not redundant, of a complete data set.

The data and image are considered as functions of two or three variables, depending on whether we are working in two dimensions or in three. In an image function $\mathcal{R}(\tilde{x}, z)$, the coordinates represent a spatial location. For a "depth" image, all variables in the image function have the dimensions of length. For a "time" image, the vertical coordinate z has dimensions of time and represents a vertical traveltime to the location.

The data function $D(\tilde{x}, t)$ depends on one or two spatial coordinates $\tilde{x}$ and one time coordinate t. The spatial coordinates specify a location on some line or surface, often the Earth's surface. The data function clearly does not represent a complete seismic experiment, which would specify both a source and a receiver location for at least one component of seismic data. For the subset of the complete experiment spanned by the data function, source and receiver locations are related in some predefined fashion (for example, as a constant-offset or constant-angle data set). In general, the data function does not represent a single experiment, in the sense of recording the response to a single source at a set of receiver locations. More often, it combines many sources and many receivers in some prearranged configuration. Though some authors worry about such composite data sets, in fact whether the set corresponds to a single experiment or many is irrelevant. The only question that matters is whether the appropriate mappings can be defined.

That the mappings can be defined is guaranteed, ultimately, by the wave equation, which provides nature's mapping of the image function onto the complete seismic data set. Having chosen the data function as a subset of the complete set, the mapping to the subset is a subset of the complete mapping. From the forward mapping, one has only to find an inverse to define the inverse mapping, or migration. Generally, this is possible, though there are some limitations. Any image function locations not sampled by the forward mapping will not be recovered by the inverse mapping. For example, suppose the data function is a common-midpoint gather. Every offset in this case samples essentially the same subsurface locations, hence insufficient information resides in the data function to construct a multidimensional image. On the other hand, a common-offset data set provides relatively complete subsurface coverage, and hence, in the absence of "shadow zones", makes for an invertible mapping.

11.2 Traveltime and depth functions

At the heart of a Kirchhoff-like mapping operation (Schneider, 1978; Schleicher *et al.*, 1993; Bleistein *et al.*, 2001; Lambar *et al.*, 2003; Schleicher *et al.*, 2007) is a traveltime function $\tau(\tilde{\boldsymbol{x}}_i, \tilde{\boldsymbol{x}}, z)$ relating two points $(\tilde{\boldsymbol{x}}_i, z_i(\tilde{\boldsymbol{x}}_i))$ and $\boldsymbol{x} = (\tilde{\boldsymbol{x}}, z)$. Except for zero-offset mappings, additional coordinates are also involved, but we are restricting consideration to a subset of a complete data set in which the independent coordinates are those mentioned above. Thus, in general, the traveltime function does not belong to a single ray between $\boldsymbol{x}_i$ and $\boldsymbol{x}$, but to something more complicated determined by $\boldsymbol{x}_i$, $\boldsymbol{x}$, and additional constraining information. In 2-D mappings, $\tilde{\boldsymbol{x}}_i$ and $\tilde{\boldsymbol{x}}$ are scalars; in three-dimensional mappings, they are 2-vectors. The coordinate z is the vertical coordinate of the image function $\mathcal{R}$. It has been written as depth, though it might (for time migrations) represent vertical traveltimes. In the simplest case, the traveltime function between any two points is single-valued, representing the time required to traverse the minimum-time path between the two points. As the distance between the two points increases in any direction, traveltime will in the simplest case increase monotonically.

In nature, the simplest case rarely obtains. To see how easily these assumptions can be violated, look at a medium with velocity $c = 1$ for $z < 1$, and $c = 2$ for $z > 1$. Figure 11.1 shows a traveltime contour plot for traveltime from $(0, 0)$ to (x, z) in such a medium.

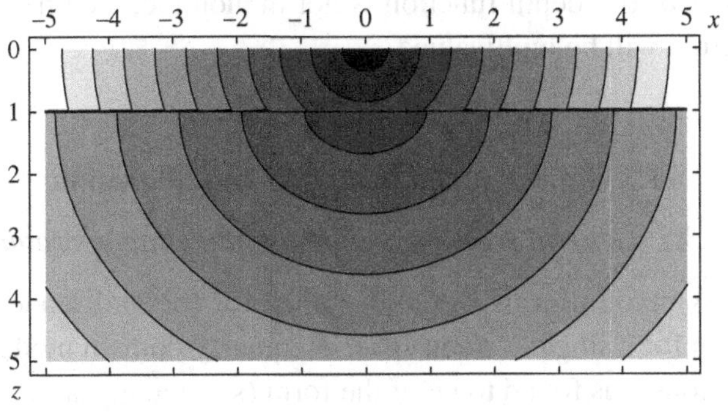

Figure 11.1 Contour plot of traveltime as a function of x and z for a two-layer velocity model. Velocity doubles at the boundary $z = 1$.This function is neither monotonic nor (if refractions are considered) single-valued.

Within each layer, traveltime is single-valued and increases monotonically with depth. However, at the layer boundary, past critical angle in the first layer, there is a discontinuity at which traveltime decreases abruptly. For the medium as a whole, traveltime is not monotonic. Nor, if refractions from the boundary are considered, is traveltime single-valued. Single-valued migration and modeling operations of necessity ignore these complications. Remarkably, it is often possible to construct single-valued operations that perform reasonably well even in complicated situations.

Corresponding to any traveltime function $\tau\left(\tilde{\boldsymbol{x}}_i, \tilde{\boldsymbol{x}}, z\right)$ is a depth function $\zeta\left(\tilde{\boldsymbol{x}}, \tilde{\boldsymbol{x}}_i, t\right)$, defined by the relation

$$t = \tau\left(\tilde{\boldsymbol{x}}_i, \tilde{\boldsymbol{x}}, \zeta\left(\tilde{\boldsymbol{x}}, \boldsymbol{x}_i, t\right)\right), \tag{11.1}$$

or equivalently,

$$z = \zeta\left(\tilde{\boldsymbol{x}}, \tilde{\boldsymbol{x}}_i, \tau\left(\tilde{\boldsymbol{x}}_i, \tilde{\boldsymbol{x}}, z\right)\right). \tag{11.2}$$

From either of these two formulas, it follows that the derivatives of the two variables are reciprocals:

$$\partial_t \zeta\left(\tilde{\boldsymbol{x}}, \tilde{\boldsymbol{x}}_i, \tau\right) \cdot \partial_z \tau\left(\tilde{\boldsymbol{x}}_i, \tilde{\boldsymbol{x}}, \zeta\right) = 1. \tag{11.3}$$

Given a starting point and the horizontal coordinates of the ending point, the depth function yields the end-point depth as a function of traveltime between the two points. From the defining equations (11.1) and (11.2), it is clear that if the traveltime function is monotonic in depth and single-valued, then the corresponding depth function is also single-valued and monotonic in time. However, if a traveltime function is not monotonic, then the corresponding depth function will be multivalued. If a depth function is not monotonic, then the corresponding traveltime function will be multivalued.

11.3 Single-valued modeling and migration

11.3.1 General frequency-domain modeling operator

In Chapter 9, asymptotic forms were developed for forward modeling operators. The general form for a single–valued space-frequency domain modeling operation in $n+1$ dimensions was found to be of the form (see e.g. equation (9.143))

$$D(\tilde{\boldsymbol{x}}_i, \omega) = f(\omega) \int d^n x \int dz \mathcal{R}(x, z) B^{fd}(\tilde{\boldsymbol{x}}_i, \tilde{x}, z)\, e^{i\omega\tau(\tilde{\boldsymbol{x}}_i, \tilde{x}, z)}. \tag{11.4}$$

In this expression, $\tilde{\boldsymbol{x}}_i$ is an n-vector representing a data surface position (e.g. source–receiver midpoint), ω is angular frequency, D is frequency-domain data at $\tilde{\boldsymbol{x}}_i$ and ω, f is some frequency filter, $\tilde{\boldsymbol{x}}$ is an n-vector that, together with the vertical coordinate z, defines a subsurface location, $\mathcal{R}$ is an image function at that subsurface location, B^{fd} is a slowly-varying amplitude function (the superscript fd indicates that the formula has been written in the frequency domain), and τ is a traveltime function associated with the surface point $\tilde{\boldsymbol{x}}_i$ and the subsurface point $(\tilde{\boldsymbol{x}}, z)$. Additional variables (e.g. offset, opening angle) appearing in specific formulas (see, for example, equation (9.143)) are superfluous to the current discussion, hence have been suppressed.

The filter $f(\omega)$ depends on the details of the operation. In two dimensions, for a single-valued modeling operation using a minimum-time raypath, the filter is the half-derivative filter d_{h+} defined in Appendix E. In three dimensions, the corresponding filter is the derivative operator. Since two applications of the half-derivative filter yield a differentiation, one could write the filter in either the 2- or 3-D case as $f(\omega) \to d_{h+}{}^n(\omega)$.

11.3.2 General time-domain modeling operator

The modeling equation (11.4) can be expressed in the time domain as

$$D(\tilde{\boldsymbol{x}}_i, t_i) = \frac{1}{2\pi} \int d\omega d_{h+}{}^n(\omega) \int d^n x \int dz \mathcal{R}(\tilde{\boldsymbol{x}}, z) B^{fd}(\tilde{\boldsymbol{x}}_i, \tilde{\boldsymbol{x}}, z)\, e^{-i\omega(t_i - \tau(\tilde{\boldsymbol{x}}_i, \tilde{\boldsymbol{x}}, z))}. \tag{11.5}$$

The frequency integral is just the inverse Fourier transform of the filter $d_{h+}{}^n$:

$$D(\tilde{\boldsymbol{x}}_i, t_i) = \int d^n x \int dz \mathcal{R}(\tilde{\boldsymbol{x}}, z) B^{fd}(\tilde{\boldsymbol{x}}_i, \tilde{\boldsymbol{x}}, z)\, d_{h+}{}^n\left(t_i - \tau(\tilde{\boldsymbol{x}}_i, \tilde{\boldsymbol{x}}, z)\right). \tag{11.6}$$

11.3.3 Post- and pre-modeling convolutional forms

The z-integral in equation (11.6) is almost a convolution with the filter ${d_{h+}}^n$, the problem being that the integration variable is an image coordinate whereas the filter argument is a data coordinate. The integral is easily transformed into a convolution, in either of two ways. One way is to change the integration variable from z to τ. τ is then the independent variable, and depth $\zeta(\tilde{\boldsymbol{x}}, \tilde{\boldsymbol{x}}_i, \tau)$ a function of τ, as per equation (11.2). We then have

$$D(\tilde{\boldsymbol{x}}_i, t_i) = \int d\tau {d_{h+}}^n (t_i - \tau) \times \int d^n x\, \mathcal{R}(\tilde{\boldsymbol{x}}, \zeta(\tilde{\boldsymbol{x}}, \tilde{\boldsymbol{x}}_i, \tau))\, \partial_\tau \zeta(\tilde{\boldsymbol{x}}, \tilde{\boldsymbol{x}}_i, \tau)\, B^{fd}(\tilde{\boldsymbol{x}}_i, \tilde{\boldsymbol{x}}, \zeta(\tilde{\boldsymbol{x}}, \tilde{\boldsymbol{x}}_i, \tau)). \tag{11.7}$$

Without the filter, the data set would be

$$D_0(\tilde{\boldsymbol{x}}_i, t_i) = \int d^n x\, \mathcal{R}(\tilde{\boldsymbol{x}}, \zeta(\tilde{\boldsymbol{x}}, \tilde{\boldsymbol{x}}_i, t_i)) B^{td}(\tilde{\boldsymbol{x}}_i, \tilde{\boldsymbol{x}}, t_i), \tag{11.8}$$

with

$$B^{td}(\tilde{\boldsymbol{x}}_i, \tilde{\boldsymbol{x}}, t_i) = \partial_\tau \zeta(\tilde{\boldsymbol{x}}, \tilde{\boldsymbol{x}}_i, t_i)\, B^{fd}(\tilde{\boldsymbol{x}}_i, \tilde{\boldsymbol{x}}, \zeta(\tilde{\boldsymbol{x}}, \tilde{\boldsymbol{x}}_i, t_i)). \tag{11.9}$$

Equation (11.7) is seen to be a convolution of the filter ${d_{h+}}^n$ with the pre-filter data D_0:

$$D(\tilde{\boldsymbol{x}}_i, t_i) = \left[{d_{h+}}^n * D_0\right](\tilde{\boldsymbol{x}}_i, t_i), \tag{11.10}$$

or, with the understanding that the filter acts on the output time coordinate,

$$D(\tilde{\boldsymbol{x}}_i, t_i) = {d_{h+}}^n * \int d^n x\, \mathcal{R}(\tilde{\boldsymbol{x}}, \zeta(\tilde{\boldsymbol{x}}, \tilde{\boldsymbol{x}}_i, t_i)) B^{td}(\tilde{\boldsymbol{x}}_i, \tilde{x}, t_i). \tag{11.11}$$

Another way to deal with the z-integral in (11.6) is to make it a convolutional filter acting on the image function z-coordinate. Since the filter ${d_{h+}}^n$ is scalable, and obeys the power law

$${d_{h+}}^n(at) = a^{-(n/2+1)} {d_{h+}}^n(t), \tag{11.12}$$

it is simple to make the z-integral a convolution. Note from (11.1) that at the point $t_i = \tau(\tilde{\boldsymbol{x}}_i, \tilde{\boldsymbol{x}}, z)$, that $z = \zeta(\tilde{\boldsymbol{x}}, \tilde{\boldsymbol{x}}_i, t_i)$. Near this point, we can write the argument of ${d_{h+}}^n$ as

$$t_i - \tau(\tilde{\boldsymbol{x}}_i, \tilde{\boldsymbol{x}}, z) \simeq (z - \zeta(\tilde{\boldsymbol{x}}, \tilde{\boldsymbol{x}}_i, t_i))\, \partial_z \tau. \tag{11.13}$$

Using the power law formula (11.12), we can remove $\partial_z \tau$ from the argument of ${d_{h+}}^n$:

$${d_{h+}}^n (t_i - \tau(\tilde{\boldsymbol{x}}_i, \tilde{\boldsymbol{x}}, z)) = (\partial_z \tau)^{-(n/2+1)} {d_{h+}}^n (z - \zeta(\tilde{\boldsymbol{x}}, \tilde{\boldsymbol{x}}_i, t_i)), \tag{11.14}$$

whence (11.6) becomes

$$D(\tilde{\boldsymbol{x}}_i, t_i) = \int d^n x \int dz \mathcal{R}(\tilde{\boldsymbol{x}}, z) B^{fd}(\tilde{\boldsymbol{x}}_i, \tilde{\boldsymbol{x}}, z) (\partial_z \tau)^{-(n/2+1)} d_{h+}{}^n (z - \zeta(\tilde{\boldsymbol{x}}, \tilde{\boldsymbol{x}}_i, t_i)). \tag{11.15}$$

Under the assumption that B^{fd} and $\partial_z \tau$ change slowly with z, they can be considered functions of ζ rather than z and removed from the z-integral:

$$D(\tilde{\boldsymbol{x}}_i, t_i) = \int d^n x B^{fd}(\tilde{\boldsymbol{x}}_i, \tilde{\boldsymbol{x}}, \zeta(\tilde{\boldsymbol{x}}, \tilde{\boldsymbol{x}}_i, t_i)) (\partial_z \tau)^{-(n/2+1)} \times \int dz \mathcal{R}(\tilde{\boldsymbol{x}}, z) d_{h+}{}^n (z - \zeta(\tilde{\boldsymbol{x}}, \tilde{\boldsymbol{x}}_i, t_i)). \tag{11.16}$$

The z-integral is now a convolutional filter acting upon the image function $\mathcal{R}$. We can write (11.16) as

$$D(\tilde{\boldsymbol{x}}_i, t_i) \simeq \int d^n x B(\tilde{\boldsymbol{x}}, \tilde{\boldsymbol{x}}_i, t_i) \left[d_{h+}{}^n * \mathcal{R} \right] (\tilde{\boldsymbol{x}}, \zeta(\tilde{\boldsymbol{x}}, \tilde{\boldsymbol{x}}_i, t_i)), \tag{11.17}$$

with amplitude function

$$B(\tilde{\boldsymbol{x}}, \tilde{\boldsymbol{x}}_i, t_i) = B^{fd}(\tilde{\boldsymbol{x}}_i, \tilde{\boldsymbol{x}}, \zeta(\tilde{\boldsymbol{x}}, \tilde{\boldsymbol{x}}_i, t_i)) (\partial_z \tau)^{-(n/2+1)}. \tag{11.18}$$

Consistent notation would put a superscript zd on the amplitude coefficient in (11.17) to indicate that the filter is performed in the z-domain. We have written the amplitude coefficient in this case simply as B without a modifying superscript, because we will employ this form most often.

The two time-domain modeling equations (11.11) and (11.17) are asymptotically equivalent. They differ in whether the convolutional filter is to be applied before or after the modeling operation. They also differ in amplitude, with

$$B(\tilde{\boldsymbol{x}}, \tilde{\boldsymbol{x}}_i, t_i) = B^{td}(\tilde{\boldsymbol{x}}_i, \tilde{\boldsymbol{x}}, \zeta(\tilde{\boldsymbol{x}}, \tilde{\boldsymbol{x}}_i, t_i)) (\partial_z \tau)^{-n/2}. \tag{11.19}$$

11.3.4 Time-domain migration operators

Corresponding to the single-valued modeling operation (11.17) or (11.7), we expect a single-valued migration operation to be of the form

$$\mathcal{R}(\tilde{\boldsymbol{x}}, z) \simeq \int d\tilde{\boldsymbol{x}}_i A(\tilde{\boldsymbol{x}}_i, \tilde{\boldsymbol{x}}, z) \left[d_{h-}{}^n * D \right] (\tilde{\boldsymbol{x}}_i, \tau(\tilde{\boldsymbol{x}}_i, \tilde{\boldsymbol{x}}, z)), \tag{11.20}$$

or equivalently, applying the filter to the output depth coordinate,

$$\mathcal{R}(\tilde{\boldsymbol{x}}, z) \simeq d_{h-}{}^n * \int d\tilde{\boldsymbol{x}}_i A^{zd}(\tilde{\boldsymbol{x}}_i, \tilde{\boldsymbol{x}}, z) D(\tilde{\boldsymbol{x}}_i, \tau(\tilde{\boldsymbol{x}}_i, \tilde{\boldsymbol{x}}, z)), \tag{11.21}$$

with

$$A^{zd}(\tilde{\boldsymbol{x}}_i, \tilde{\boldsymbol{x}}, z) = \frac{A(\tilde{\boldsymbol{x}}_i, \tilde{\boldsymbol{x}}, z)}{\left[\partial_z \tau(\tilde{\boldsymbol{x}}_i, \tilde{\boldsymbol{x}}, z)\right]^{n/2}}. \tag{11.22}$$

The amplitude function $A(\tilde{\boldsymbol{x}}_i, \tilde{\boldsymbol{x}}, z)$, like its modeling counterpart, is supposed to be slowly and regularly varying with respect to the integration variable. The filter, $d_{h-}{}^n$ in this case, is the half-derivative adjoint in two dimensions, or the negative of a derivative in three. In either two or three dimensions, the filter can be applied as in the first form to the time coordinate of the data function $D(\tilde{\boldsymbol{x}}_i, t)$ or, as in the second form, to the z-coordinate after migration.

11.4 Relating the forward and inverse mappings

The amplitude functions A and B that appear in the forward and inverse operators are closely related. To determine the relation, one can substitute the forward equation (11.17) into the inverse equation (11.20), and evaluate the result.

$$\begin{aligned}\mathcal{R}(\tilde{\boldsymbol{x}}, z) \simeq &\int d\tilde{\boldsymbol{x}}_i A(\tilde{\boldsymbol{x}}_i, \tilde{\boldsymbol{x}}, z)\, d_{h-}{}^n * \int d\tilde{\boldsymbol{x}}' B\left(\tilde{\boldsymbol{x}}', \tilde{\boldsymbol{x}}_i, \tau(\tilde{\boldsymbol{x}}_i, \tilde{\boldsymbol{x}}, z)\right) \\ &\times d_{h+}{}^n * \mathcal{R}\left(\tilde{\boldsymbol{x}}', \zeta\left(\tilde{\boldsymbol{x}}', \tilde{\boldsymbol{x}}_i, \tau(\tilde{\boldsymbol{x}}_i, \tilde{\boldsymbol{x}}, z)\right)\right).\end{aligned} \tag{11.23}$$

In this expression, the leftmost filter $d_{h-}{}^n$ is applied to the time coordinate τ, while the rightmost filter $d_{h+}{}^n$ is applied to the z-component of $\mathcal{R}$. Applying the formula (E.26), both filters can be made to act on the z-component of $\mathcal{R}$, yielding

$$\begin{aligned}\mathcal{R}(\tilde{\boldsymbol{x}}, z) \simeq &\int d\tilde{\boldsymbol{x}}_i \int d\tilde{\boldsymbol{x}}' A(\tilde{\boldsymbol{x}}_i, \tilde{\boldsymbol{x}}, z)\, B\left(\tilde{\boldsymbol{x}}', \tilde{\boldsymbol{x}}_i, \tau(\tilde{\boldsymbol{x}}_i, \tilde{\boldsymbol{x}}, z)\right) \\ &\times \left[\partial_t \zeta\left(\tilde{\boldsymbol{x}}', \tilde{\boldsymbol{x}}_i, \tau(\tilde{\boldsymbol{x}}_i, \tilde{\boldsymbol{x}}, z)\right)\right]^{n/2} \left[d_{h-}{}^n * d_{h+}{}^n * \mathcal{R}\right] \\ &\times \left(\tilde{\boldsymbol{x}}', \zeta\left(\tilde{\boldsymbol{x}}', \tilde{\boldsymbol{x}}_i, \tau(\tilde{\boldsymbol{x}}_i, \tilde{\boldsymbol{x}}, z)\right)\right).\end{aligned} \tag{11.24}$$

Now, the combined effect of the half-derivative filter and its adjoint is the "rho" filter ρ. Thus,

$$\begin{aligned}\mathcal{R}(\tilde{\boldsymbol{x}}, z) \simeq &\int d\tilde{\boldsymbol{x}}_i \int d\tilde{\boldsymbol{x}}' A(\tilde{\boldsymbol{x}}_i, \tilde{\boldsymbol{x}}, z)\, B\left(\tilde{\boldsymbol{x}}', \tilde{\boldsymbol{x}}_i, \tau(\tilde{\boldsymbol{x}}_i, \tilde{\boldsymbol{x}}, z)\right) \\ &\times \left[\partial_t \zeta\left(\tilde{\boldsymbol{x}}', \tilde{\boldsymbol{x}}_i, \tau(\tilde{\boldsymbol{x}}_i, \tilde{\boldsymbol{x}}, z)\right)\right]^{n/2} \left[\rho^n * \mathcal{R}\right]\left(\tilde{\boldsymbol{x}}', \zeta\left(\tilde{\boldsymbol{x}}', \tilde{\boldsymbol{x}}_i, \tau(\tilde{\boldsymbol{x}}_i, \tilde{\boldsymbol{x}}, z)\right)\right).\end{aligned} \tag{11.25}$$

By assumption, $\mathcal{R}$ is a high-frequency function, hence oscillates rapidly with respect to its depth variable. An integral of $\mathcal{R}$ over several wavelengths of depth will be nearly zero, even if $\mathcal{R}$ is multiplied by a coefficient that varies little over a wavelength of depth. The factor of $\mathcal{R}$ within the integrals of (11.25) is a function of ζ, which in turn depends on all the horizontal coordinates as well as output depth z. $\mathcal{R}$ depends on the integration variable $\tilde{\boldsymbol{x}}_i$ only through ζ. Considered as a function of $\tilde{\boldsymbol{x}}_i$, for the most part ζ should vary rapidly enough with $\tilde{\boldsymbol{x}}_i$ that $\mathcal{R}$ should average to zero. However, there should be regions of $(\tilde{\boldsymbol{x}}', \tilde{\boldsymbol{x}}_i)$-space where the variation slows enough to allow the integral over $\tilde{\boldsymbol{x}}_i$ to accumulate a value. Consider, in particular, the point $\tilde{\boldsymbol{x}}' = \tilde{\boldsymbol{x}}$. Here, according to the reciprocal relation (11.2) between ζ and τ,

$$\zeta\left(\tilde{\boldsymbol{x}}, \tilde{\boldsymbol{x}}_i, \tau\left(\tilde{\boldsymbol{x}}_i, \tilde{\boldsymbol{x}}, z\right)\right) = z. \tag{11.26}$$

That is, the depth coordinate ζ in $\mathcal{R}$, and hence $\mathcal{R}$ also , becomes completely independent of $\tilde{\boldsymbol{x}}_i$ at $\tilde{\boldsymbol{x}}' = \tilde{\boldsymbol{x}}$. Hence, we can expect a contribution from the $\tilde{\boldsymbol{x}}_i$ integral in the region of $\tilde{\boldsymbol{x}}'$ near $\tilde{\boldsymbol{x}}$.

For $\tilde{\boldsymbol{x}}'$ near enough to $\tilde{\boldsymbol{x}}$, we can approximate ζ with the linear expansion

$$\zeta\left(\tilde{\boldsymbol{x}}', \tilde{\boldsymbol{x}}_i, \tau\left(\tilde{\boldsymbol{x}}_i, \tilde{\boldsymbol{x}}, z\right)\right) \simeq z + \left(\tilde{\boldsymbol{x}}' - \tilde{\boldsymbol{x}}\right) \cdot \boldsymbol{\lambda}\left(\tilde{\boldsymbol{x}}_i, \tilde{\boldsymbol{x}}, z\right), \tag{11.27}$$

where $\boldsymbol{\lambda}$ is the n-dimensional gradient of ζ with respect to $\boldsymbol{x}'$, evaluated at $x' = x$:

$$\boldsymbol{\lambda}\left(\tilde{\boldsymbol{x}}_i, \tilde{\boldsymbol{x}}, z\right) = \boldsymbol{\nabla}_{\tilde{x}'} \zeta\left(\tilde{\boldsymbol{x}}', \tilde{\boldsymbol{x}}_i, \tau\left(\tilde{\boldsymbol{x}}_i, \tilde{\boldsymbol{x}}, z\right)\right)|_{\tilde{x}' = \tilde{x}}. \tag{11.28}$$

The linear approximation (11.27) can be substituted into (11.25). Moreover, since the factors multiplying $\mathcal{R}$ vary slowly with $\tilde{\boldsymbol{x}}'$, $\tilde{\boldsymbol{x}}'$ can be replaced by $\tilde{\boldsymbol{x}}$ in the arguments of these factors, and the factors brought outside the $\tilde{\boldsymbol{x}}'$ integral:

$$\begin{aligned} \mathcal{R}\left(\tilde{\boldsymbol{x}}, z\right) = & \int dx_i A\left(\tilde{\boldsymbol{x}}_i, \tilde{\boldsymbol{x}}, z\right) B\left(\tilde{\boldsymbol{x}}, \tilde{\boldsymbol{x}}_i, \tau\left(\tilde{\boldsymbol{x}}_i, \tilde{\boldsymbol{x}}, z\right)\right) \left[\partial_t \zeta\left(\tilde{\boldsymbol{x}}, \tilde{\boldsymbol{x}}_i, \tau\left(\tilde{\boldsymbol{x}}_i, \tilde{\boldsymbol{x}}, z\right)\right)\right]^{n/2} \\ & \times \int d\tilde{\boldsymbol{x}}' \rho^n * \mathcal{R}\left(\tilde{\boldsymbol{x}}', z + \left(\tilde{\boldsymbol{x}}' - \tilde{\boldsymbol{x}}\right) \cdot \boldsymbol{\lambda}\left(\tilde{\boldsymbol{x}}_i, \tilde{\boldsymbol{x}}, z\right)\right). \end{aligned} \tag{11.29}$$

Now, the n-dimensional inverse Radon transform of $\mathcal{R}$ has the form (C.47), or (interchanging the roles of $\boldsymbol{p}$ and $\tilde{\boldsymbol{x}}$)

$$\mathcal{R}(\boldsymbol{p}, z) \equiv \frac{1}{(2\pi)^n} \int d\tilde{\boldsymbol{x}}' \; \rho^n * \mathcal{R}\left(\tilde{\boldsymbol{x}}', z + \boldsymbol{p} \cdot \tilde{\boldsymbol{x}}'\right), \tag{11.30}$$

with ray parameter $\boldsymbol{p}$, like $\tilde{\boldsymbol{x}}$, an n-vector. The integral over $\tilde{\boldsymbol{x}}'$ in (11.29) is seen to be a double inverse Radon transform of $\mathcal{R}$, evaluated at ray parameter $\boldsymbol{p} = \boldsymbol{\lambda}$ and at depth $z - \boldsymbol{\lambda} \cdot \tilde{\boldsymbol{x}}$. Consequently, equation (11.29) can be written

$$\begin{aligned} \mathcal{R}\left(\tilde{\boldsymbol{x}}, z\right) = & (2\pi)^n \int d\tilde{\boldsymbol{x}}_i A\left(\tilde{\boldsymbol{x}}_i, \tilde{\boldsymbol{x}}, z\right) B\left(\tilde{\boldsymbol{x}}, \tilde{\boldsymbol{x}}_i, \tau\left(\tilde{\boldsymbol{x}}_i, \tilde{\boldsymbol{x}}, z\right)\right) \\ & \times \left[\partial_t \zeta\left(\tilde{\boldsymbol{x}}, \tilde{\boldsymbol{x}}_i, \tau\left(\tilde{\boldsymbol{x}}_i, \tilde{\boldsymbol{x}}, z\right)\right)\right]^{n/2} \mathcal{R}\left(\boldsymbol{\lambda}, z - \boldsymbol{\lambda} \cdot \tilde{\boldsymbol{x}}\right). \end{aligned} \tag{11.31}$$

The next step is to change integration variable from $\tilde{\boldsymbol{x}}_i$ to $\boldsymbol{\lambda}$. This is straightforward, provided $\boldsymbol{\lambda}$ is monotonic in $\tilde{\boldsymbol{x}}_i$. If not, then one is dealing with a multivalued mapping, which has been assumed not to be the case. Continuing with this assumption, $\tilde{\boldsymbol{x}}_i$ becomes a function of $\boldsymbol{\lambda}$, $\tilde{\boldsymbol{x}}$, and z, and (11.31) can be written

$$\mathcal{R}(\tilde{\boldsymbol{x}}, z) = (2\pi)^n \int\int d\boldsymbol{\lambda} A(\tilde{\boldsymbol{x}}_i, \tilde{\boldsymbol{x}}, z)\, B(\tilde{\boldsymbol{x}}, \tilde{\boldsymbol{x}}_i, \tau(\tilde{\boldsymbol{x}}_i, \tilde{\boldsymbol{x}}, z)) \times \frac{\left[\partial_t \zeta(\tilde{\boldsymbol{x}}, \tilde{\boldsymbol{x}}_i, \tau(\tilde{\boldsymbol{x}}_i, \tilde{\boldsymbol{x}}, z))\right]^{n/2}}{|J_\lambda(\tilde{\boldsymbol{x}}_i, \tilde{\boldsymbol{x}}, z)|} \mathcal{R}(\boldsymbol{\lambda}, z - \boldsymbol{\lambda}\cdot\tilde{\boldsymbol{x}}), \tag{11.32}$$

where J_λ is the Jacobian of the coordinate change from $\tilde{\boldsymbol{x}}_i$ to $\boldsymbol{\lambda}$. For $n = 1$, J_λ is the derivative of λ with respect to x_i. For $n = 2$, J_λ is the determinant of the matrix of derivatives of the components of $\boldsymbol{\lambda}$ with respect to the components of $\tilde{\boldsymbol{x}}_i$:

$$J_\lambda(\tilde{\boldsymbol{x}}_i, \tilde{\boldsymbol{x}}, z) = \mathrm{Det}\begin{pmatrix} \frac{\partial\lambda_x}{\partial x_i} & \frac{\partial\lambda_x}{\partial y_i} \\ \frac{\partial\lambda_y}{\partial x_i} & \frac{\partial\lambda_y}{\partial y_i} \end{pmatrix}. \tag{11.33}$$

Since $\mathcal{R}(\boldsymbol{\lambda}, z)$ is the inverse Radon transform of $\mathcal{R}(\tilde{\boldsymbol{x}}, z)$, it follows that $\mathcal{R}(\tilde{\boldsymbol{x}}, z)$ is the forward Radon transform of $\mathcal{R}(\boldsymbol{\lambda}, z)$. Equation (11.32) confirms this to be the case provided the product of all the coefficients in (11.32) is one. That is to say, the factor A must satisfy

$$A(\tilde{\boldsymbol{x}}_i, \tilde{\boldsymbol{x}}, z) = \frac{(2\pi)^{-n}\,|J_\lambda(\tilde{\boldsymbol{x}}_i, \tilde{\boldsymbol{x}}, z)|}{B(\tilde{\boldsymbol{x}}, \tilde{\boldsymbol{x}}_i, \tau(\tilde{\boldsymbol{x}}_i, \tilde{\boldsymbol{x}}, z))\,\left[\partial_t \zeta(\tilde{\boldsymbol{x}}, \tilde{\boldsymbol{x}}_i, \tau(\tilde{\boldsymbol{x}}_i, \tilde{\boldsymbol{x}}, z))\right]^{n/2}}. \tag{11.34}$$

With B related to A as per (11.34), this confirms the reciprocal relation between the forward and inverse mappings (3.17) and (3.20), under the assumptions outlined above.

Instead of substituting the inverse transformation into the forward transformation, we could just as easily have substituted the forward transformation into the inverse transformation. The resulting relation, equivalent to equation (11.34), is

$$A(\tilde{\boldsymbol{x}}_i, \tilde{\boldsymbol{x}}, \zeta(\tilde{\boldsymbol{x}}, \tilde{\boldsymbol{x}}_i, t)) = \frac{(2\pi)^{-n}\,|J_\kappa(\tilde{\boldsymbol{x}}, \tilde{\boldsymbol{x}}_i, t)|}{B(\tilde{\boldsymbol{x}}, \tilde{\boldsymbol{x}}_i, t)\left[\partial_z \tau(\tilde{\boldsymbol{x}}_i, \tilde{\boldsymbol{x}}, \zeta(\tilde{\boldsymbol{x}}, \tilde{\boldsymbol{x}}_i, t))\right]^{n/2}}, \tag{11.35}$$

where, for $n = 1$, J_κ is the derivative of κ with respect to x, with κ the derivative of τ with respect to x_i. For $n = 2$,

$$J_\kappa(\tilde{\boldsymbol{x}}, \tilde{\boldsymbol{x}}_i, t) = \mathrm{Det}\begin{pmatrix} \frac{\partial\kappa_x}{\partial x} & \frac{\partial\kappa_x}{\partial y} \\ \frac{\partial\kappa_y}{\partial x} & \frac{\partial\kappa_y}{\partial y} \end{pmatrix}, \tag{11.36}$$

and

$$\kappa(\tilde{\boldsymbol{x}}, \tilde{\boldsymbol{x}}_i, t) = \boldsymbol{\nabla}_{\tilde{x}_i'} \tau\left(\tilde{\boldsymbol{x}}_i', \tilde{\boldsymbol{x}}, \zeta(\tilde{\boldsymbol{x}}, \tilde{\boldsymbol{x}}_i, t)\right)|_{\tilde{x}_i' = \tilde{x}_i}. \tag{11.37}$$

11.5 Constant-angle migration

While we have developed above a general framework for asymptotic migration, there is one formulation that deserves some special attention. Constant-angle migration is conceptually a very simple way to image, with special powers in that multiple raypaths such as may appear in a constant-offset, constant-source, or constant-receiver migration generally do not appear in a constant-angle formulation. This is because once an initial slowness vector is specified, the Eikonal equation produces a single ray, whereas if we attempt to specify a ray from its end points, more than one ray may result.

We develop in this section 2-D and 3-D formulae for constant-angle migration, general enough to accommodate scalar, acoustic, and elastic waves, and perhaps a little bit beyond.

11.5.1 The forward problem

For an angle-dependent reflectivity, the general form, as developed above, for a forward modeling operation is

$$D\left(\tilde{\boldsymbol{x}}_g|\tilde{\boldsymbol{x}}_s;\omega\right) = f_n(\omega)\int d^n x \int dz \mathcal{R}(\tilde{\boldsymbol{x}}, z, \sigma) B_n\left(\tilde{\boldsymbol{x}}_g|\tilde{\boldsymbol{x}}, z|\tilde{\boldsymbol{x}}_s\right) e^{i\omega\tau\left(\tilde{\boldsymbol{x}}_g|\tilde{\boldsymbol{x}}, z|\tilde{\boldsymbol{x}}_s\right)}. \tag{11.38}$$

The traveltime function τ is found by computing raypaths from source position $\tilde{\boldsymbol{x}}_s$ to reflector location $\boldsymbol{x} = (\tilde{\boldsymbol{x}}, z)$, and from $\boldsymbol{x}$ to receiver position $\tilde{\boldsymbol{x}}_g$. Then

$$\tau\left(\tilde{\boldsymbol{x}}_g|\boldsymbol{x}|\tilde{\boldsymbol{x}}_s\right) = \tau_g\left(\tilde{\boldsymbol{x}}_g|\boldsymbol{x}\right) + \tau_s\left(\boldsymbol{x}|\tilde{\boldsymbol{x}}_s\right). \tag{11.39}$$

Opening angle is the angle between the incident and reflected rays at the reflection point. This can be calculated by defining local wavenumbers $\boldsymbol{k}_i$ and $\boldsymbol{k}_r$ or slowness vectors $\boldsymbol{p}_i$ and $\boldsymbol{p}_r$ at the reflection point. The slowness vectors are related to the gradients of the traveltimes:

$$\boldsymbol{p}_r(\boldsymbol{x}) = -\nabla_x \tau_g\left(\tilde{\boldsymbol{x}}_g|x\right) = \frac{\boldsymbol{k}_r}{\omega}, \quad \boldsymbol{p}_i(\boldsymbol{x}) = +\nabla_x \tau_s\left(\boldsymbol{x}|\tilde{\boldsymbol{x}}_s\right) = \frac{\boldsymbol{k}_i}{\omega}. \tag{11.40}$$

Assuming these quantities obey a local wave equation, their magnitudes are related to the velocities of the incident and reflected waves at the reflector:

$$\begin{aligned}
p_r(\boldsymbol{x}) &= \frac{1}{c_r(\boldsymbol{x})}, \\
p_i(\boldsymbol{x}) &= \frac{1}{c_i(\boldsymbol{x})}, \\
k_r(\boldsymbol{x}) &= \frac{\omega}{c_r(\boldsymbol{x})}, \\
k_i(\boldsymbol{x}) &= \frac{\omega}{c_i(\boldsymbol{x})}.
\end{aligned} \tag{11.41}$$

In terms of these quantities, opening angle is

$$\cos\sigma = -\boldsymbol{p}_r(x)\cdot\boldsymbol{p}_i(x)c_r(x)c_i(x). \tag{11.42}$$

A minus sign appears in this relation because the incident slowness vector is downgoing, while the reflected vector is upgoing.

In equation (11.38), the parameter n is one for 2-D Earths and two for 3-D Earths. For a 2-D Earth model, both the data set D and the reflectivity function R are three-dimensional. For a 3-D Earth, D is five-dimensional whereas R, as written, is four-dimensional. There is, however, a neglected dimension, in that elastic reflectivity is not supposed to depend upon source–receiver orientation or azimuth angle. If we were to allow for a more general reflectivity $R(\boldsymbol{x}, \sigma, \chi)$ which changes with source–receiver azimuth χ as well as location $\boldsymbol{x}$ and opening angle σ, then both reflectivity and data would be five-dimensional.

11.5.2 The 2-D inverse problem

For a 2-D Earth, the forward integral reduces to

$$D\left(x_g|x_s;\omega\right) = f_1(\omega)\int dx\int dz\mathcal{R}(x,\ z,\sigma)B_1\left(x_g|x,z|x_s\right)e^{i\omega\tau\left(x_g|x,z|x_s\right)}. \tag{11.43}$$

Though not mentioned explicitly, σ depends on x, z, x_g, and x_s.

If σ is fixed, we expect a 2-D inverse formula for $\mathcal{R}$ to be an integral over the 2-D subset of D corresponding asymptotically to σ. The integration variables could be taken to be frequency ω and source–receiver midpoint x_m. We choose instead a variable more closely related to the ray parameters and to σ, namely the local wavenumber, or

$$\boldsymbol{k}_m(\boldsymbol{x}) \equiv \boldsymbol{k}_r(\boldsymbol{x}) - \boldsymbol{k}_i(\boldsymbol{x}). \tag{11.44}$$

The magnitude of $\boldsymbol{k}_m$ is proportional to frequency scaled by the local velocities:

$$\boldsymbol{k}_m(\boldsymbol{x}) = \omega\sqrt{c_r^{-2} + c_i^{-2} + 2\cos\sigma\ /(c_r c_i)} \equiv 2\omega\frac{c_e(\boldsymbol{x},\sigma)}{c_r(\boldsymbol{x})c_i(\boldsymbol{x})}. \tag{11.45}$$

In the absence of wave-type conversion, where $c_r = c_i = c$, the magnitude of $\boldsymbol{k}_m$ reduces to $(2\omega/c)\cos\sigma/2$.

Given ω, x_s, $\boldsymbol{x}$, and x_g, the two rays are specified, hence $\boldsymbol{k}_m$ and σ are specified. Given $\boldsymbol{k}_m$ and σ, the ray parameters ω, $\boldsymbol{p}_i$ and $\boldsymbol{p}_r$ are specified, from which x_s and x_g can be determined. The fact that $\boldsymbol{k}_m$ depends on reflector location x does not prevent us using it as an integration variable. It only means that the mapping from $\boldsymbol{k}_m$ back to ω, $\boldsymbol{p}_i$ and $\boldsymbol{p}_r$ depends on $\boldsymbol{x}$. Hence we can write the asymptotic form

$$\mathcal{R}(\boldsymbol{x},\sigma) \simeq \frac{1}{(2\pi)^2}\int dk_m D\left(x_g|x_s;\omega\right)g_1(\omega)A_1\left(x_g|\boldsymbol{x}|x_s\right)e^{-i\omega\tau\left(x_g|\boldsymbol{x}|x_s\right)}. \tag{11.46}$$

To find the filter g_1 and amplitude function A_1, substitute (11.38) for D in this equation:

$$\mathcal{R}(\boldsymbol{x}, \sigma) \simeq \frac{1}{(2\pi)^2} \int dk_m f_1(\omega) \int dx' \int dz' \mathcal{R}(x', z', \sigma') B_1\left(x_g|x', z'|x_s\right) \times e^{i\omega\tau\left(x_g|x', z'|x_s\right)} g_1(\omega) A_1\left(x_g|\boldsymbol{x}|x_s\right) e^{-i\omega\tau\left(x_g|\boldsymbol{x}|x_s\right)}. \tag{11.47}$$

Following the argument of previous sections, we expect contributions to the integral to come from $\boldsymbol{x}$ near $\boldsymbol{x}'$, where the phase

$$\omega\tau\left(x_g|x', z'|x_s\right) - \omega\tau\left(x_g|x, z|x_s\right) \simeq \omega\nabla_x\tau\left(x_g|\boldsymbol{x}|x_s\right)\cdot(x' - x) = -\boldsymbol{k}_m\cdot(\boldsymbol{x}' - \boldsymbol{x}). \tag{11.48}$$

Thus

$$\mathcal{R}(\boldsymbol{x}, \sigma) \simeq \frac{1}{(2\pi)^2} \int dx' \int dz' \mathcal{R}(x', z', \sigma) \int d\boldsymbol{k}_m \times f_1(\omega) B_1\left(x_g|\boldsymbol{x}|x_s\right) g_1(\omega) A_1\left(x_g|\boldsymbol{x}|x_s\right) e^{-ik\cdot(\boldsymbol{x}'-\boldsymbol{x})}. \tag{11.49}$$

With the choice of $\boldsymbol{k}_m$ as integration variable, the integral has become an inverse Fourier transform from $\boldsymbol{k}_m$ to $\boldsymbol{x}' - \boldsymbol{x}$. If $g_1 = 1/f_1$ and $A_1 = 1/B_1$, the integral over $\boldsymbol{k}_m$ evaluates to $\delta(\boldsymbol{x}' - \boldsymbol{x})$ and (11.49) reduces to the identity

$$\mathcal{R}(\boldsymbol{x}, \sigma) \simeq \mathcal{R}(\boldsymbol{x}', \sigma). \tag{11.50}$$

Apparently, we are justified in writing the 2-D constant-σ imaging equation

$$\mathcal{R}(\boldsymbol{x}, \sigma) \simeq \frac{1}{(2\pi)^2} \int dk_m \frac{D\left(x_g|\tilde{x}_s; \omega\right) e^{-i\omega\tau\left(\tilde{x}_g|\boldsymbol{x}|x_s\right)}}{f_1(\omega) B_1\left(x_g|\boldsymbol{x}|x_s\right)}. \tag{11.51}$$

If we express $\boldsymbol{k}_m$ in cylindrical coordinates, then magnitude k can be converted to ω:

$$\boldsymbol{k}_m = -\omega\frac{2c_e}{c_r c_i}(\sin\alpha, \cos\alpha), \tag{11.52}$$

$$\frac{\partial\left(k_{mx}, k_{mz}\right)}{\partial(\omega, \alpha)} = -\frac{2c_e}{c_r c_i}\begin{pmatrix} \sin\alpha & \cos\alpha \\ \omega\cos\alpha & -\omega\sin\alpha \end{pmatrix}, \tag{11.53}$$

$$\left|\det\left[\frac{\partial\left(k_{mx}, k_{mz}\right)}{\partial(\omega, \alpha)}\right]\right| = \frac{4{c_e}^2\omega}{{c_r}^2{c_i}^2}, \tag{11.54}$$

and

$$\mathcal{R}(\boldsymbol{x}, \sigma) \simeq \frac{1}{(2\pi)^2}\frac{4{c_e}^2}{{c_r}^2{c_i}^2} \int \frac{d\alpha}{B_1\left(x_g|\boldsymbol{x}|x_s\right)} \int d\omega \frac{\omega}{f_1(\omega)} D\left(x_g|x_s; \omega\right) e^{-i\omega\tau\left(x_g|\boldsymbol{x}|x_s\right)}. \tag{11.55}$$

The frequency integral is then an inverse Fourier transform from ω to τ of $\omega f_1{}^{-1} * D$, leaving a single integral over wavenumber angle α:

$$\mathcal{R}(\boldsymbol{x}, \sigma) \simeq \frac{1}{2\pi} \frac{4c_e{}^2}{c_r{}^2 c_i{}^2} \int d\alpha \frac{\left[\omega f_1{}^{-1} * D\left(x_g|x_s\right)\right]\left(\tau\left(x_g|\boldsymbol{x}|x_s\right)\right)}{B_1\left(x_g|\boldsymbol{x}|x_s\right)}. \tag{11.56}$$

Algorithmically, to calculate $\mathcal{R}$ at $\boldsymbol{x}$ and σ, we apply the filter $\omega f_1{}^{-1}$ to D and store it in the time domain. We form a summation over realizable wavenumber angles α. For each α, we calculate the ray parameters $\boldsymbol{p}_i$ and $\boldsymbol{p}_r$, and follow the rays from $\boldsymbol{x}$ to their surface points x_s and x_g. We then weight D at this point by the inverse of B_1 and add it to the sum.

11.5.3 The 3-D inverse problem

For a 3-D Earth, the forward integral is

$$D\left(\tilde{\boldsymbol{x}}_g|\tilde{\boldsymbol{x}}_s; \omega\right) = f_2(\omega) \int d^3x\, \mathcal{R}(\boldsymbol{x}, \sigma, \chi_{ir}) B_2\left(\tilde{\boldsymbol{x}}_g|\boldsymbol{x}|\tilde{\boldsymbol{x}}_s\right) e^{i\omega\tau\left(\tilde{\boldsymbol{x}}_g|\boldsymbol{x}|\tilde{\boldsymbol{x}}_s\right)}. \tag{11.57}$$

We have added an azimuthal parameter χ_{ir} to the argument list of reflectivity, even though the scalar, acoustic, and elastic Earth models considered above all expect $\mathcal{R}$ to be indifferent to source–receiver azimuth. One may consider this precocious anticipation of a more general Earth model, or just simple ignorance. Either way, both data and reflectivity are treated as functions of five variables in this formula. The angles σ and χ_{ir} are considered to depend (slowly) upon source, receiver, and reflector location.

We have seen how to relate σ to ray parameters $\boldsymbol{p}_i$ and $\boldsymbol{p}_r$, and hence through ray tracing back to source and receiver positions $\tilde{\boldsymbol{x}}_s$ and $\tilde{\boldsymbol{x}}_g$. We define azimuthal angle χ_{ir} at the reflection point as follows: the two slowness vectors $\boldsymbol{p}_i(\boldsymbol{x})$ and $\boldsymbol{p}_r(\boldsymbol{x})$, together with their linear combinations, define a plane which we call the *plane of incidence and reflection or i–r plane* at $\boldsymbol{x}$, as depicted in Figure H.1. All horizontal lines in this plane make the same angle with respect to the x-axis. We define this angle to be azimuthal angle χ_{ir}. If velocity is constant between $\boldsymbol{x}$ and the surface, then the rays are straight and χ_{ir} is also the surface azimuthal angle χ_{gs} between source and receiver. In general, however, rays are not straight, and the azimuthal angle measured at the reflector may be very different than the angle measured at the source and receiver. Even so, the relation between the two angles can be established through ray tracing.

An easy way to obtain the azimuthal angle at the reflection point $\boldsymbol{x}$ is by taking the cross product of $\boldsymbol{p}_i(\boldsymbol{x})$ and $\boldsymbol{p}_r(\boldsymbol{x})$:

$$\boldsymbol{q}_\perp = \boldsymbol{p}_r \times \boldsymbol{p}_i. \tag{11.58}$$

Here $\boldsymbol{q}_\perp$ is a vector normal to the *i–r* plane. In particular, it is perpendicular to horizontal lines in that plane. Representing such a line by

$$\boldsymbol{\ell} = \ell \left(\cos \chi_{ir}, \sin \chi_{ir}\right), \tag{11.59}$$

we must have

$$\boldsymbol{q}_\perp \cdot \boldsymbol{\ell} = 0 = q_{\perp x} \cos \chi_{ir} + q_{\perp y} \sin \chi_{ir}, \tag{11.60}$$

from which it follows

$$\chi_{ir} = \arctan \left(-\frac{q_{\perp x}}{q_{\perp y}} \right). \tag{11.61}$$

Also in the *i–r* plane is the local wavenumber $\boldsymbol{k}_m = \boldsymbol{k}_r - \boldsymbol{k}_i$. As discussed in Appendix H, the angles σ and χ_{ir}, together with the wavenumber $\boldsymbol{k}_m$, comprise a 5-D set of reflection-centered coordinates equivalent to the five ray coordinates p_{ix}, p_{iy}, p_{rx}, p_{ry}, and ω.

To asymptotically invert (11.57), analogous to the 2-D case we form an integral over local wavenumber $\boldsymbol{k}_m$:

$$\mathcal{R}(\boldsymbol{x}, \sigma, \chi_{ir}) \simeq \frac{1}{(2\pi)^3} \int d^3k_m \frac{D\left(\tilde{\boldsymbol{x}}_g|\tilde{\boldsymbol{x}}_s; \omega\right) e^{-i\omega\tau\left(\tilde{\boldsymbol{x}}_g|\boldsymbol{x}|\tilde{\boldsymbol{x}}_s\right)}}{f_2(\omega) B_2\left(\tilde{\boldsymbol{x}}_g|\boldsymbol{x}|\tilde{\boldsymbol{x}}_s\right)}. \tag{11.62}$$

To confirm that this is the proper expression, substitute (11.57) for D in this equation.

$$\mathcal{R}(\boldsymbol{x}, \sigma, \chi_{ir}) \simeq \frac{1}{(2\pi)^3} \int d^3k_m \int d^3x' \frac{\mathcal{R}(\boldsymbol{x}', \sigma', \chi'_{ir}) B_2\left(\tilde{\boldsymbol{x}}_g|\boldsymbol{x}'|\tilde{\boldsymbol{x}}_s\right)}{B_2\left(\tilde{\boldsymbol{x}}_g|\boldsymbol{x}|\tilde{\boldsymbol{x}}_s\right)} \times e^{i\omega\left(\tau\left(\tilde{\boldsymbol{x}}_g|\boldsymbol{x}'|\tilde{\boldsymbol{x}}_s\right) - \tau\left(\tilde{\boldsymbol{x}}_g|\boldsymbol{x}|\tilde{\boldsymbol{x}}_s\right)\right)}. \tag{11.63}$$

Making the familiar argument that only $\boldsymbol{x}'$ near $\boldsymbol{x}$ contributes to the integral, and that slowly varying functions of $\boldsymbol{x}'$ may be replaced by their values at $\boldsymbol{x}$, we rewrite the expression as

$$\begin{aligned}\mathcal{R}(\boldsymbol{x}, \sigma, \chi_{ir}) &\simeq \frac{1}{(2\pi)^3} \int d^3k_m \int d^3x' \mathcal{R}(\boldsymbol{x}', \sigma, \chi_{ir}) e^{-i\omega\nabla_x\tau\left(\tilde{\boldsymbol{x}}_g|\boldsymbol{x}|\tilde{\boldsymbol{x}}_s\right)\cdot(\boldsymbol{x}'-\boldsymbol{x})} \\ &= \frac{1}{(2\pi)^3} \int d^3x' \mathcal{R}(\boldsymbol{x}', \sigma, \chi_{ir}) \int d^3k_m e^{-i\omega \boldsymbol{k}_m\cdot(\boldsymbol{x}'-\boldsymbol{x})} \\ &= \int d^3x' \mathcal{R}(\boldsymbol{x}', \sigma, \chi_{ir}) \delta(\boldsymbol{x} - \boldsymbol{x}') \\ &= \mathcal{R}(\boldsymbol{x}, \sigma, \chi_{ir}).\end{aligned} \tag{11.64}$$

It would appear that (11.62) is an asymptotic inverse of (11.57).

As for the 2-D case, one of the integrals in (11.62) can be reduced to an inverse Fourier transform, provided we transform magnitude of $\boldsymbol{k}_m$ to frequency. Express $\boldsymbol{k}_m$ in spherical coordinates as

$$\boldsymbol{k}_m = -\omega \frac{2c_e}{c_r c_i} \left(\cos\phi_k \sin\alpha, \sin\phi_k \sin\alpha, \cos\alpha \right), \tag{11.65}$$

In terms of frequency and the two angles,

$$\mathcal{R}(\boldsymbol{x}, \sigma, \chi_{ir}) \simeq \frac{1}{(2\pi)^3} \frac{8{c_e}^3}{{c_r}^3 {c_i}^3} \int d\alpha \sin\alpha \int \frac{d\phi_k}{B_2\left(\tilde{\boldsymbol{x}}_g|\boldsymbol{x}|\tilde{\boldsymbol{x}}_s\right)} \times \int d\omega \frac{\omega^2}{f_2(\omega)} D\left(\tilde{\boldsymbol{x}}_g|\tilde{\boldsymbol{x}}_s;\omega\right) e^{-i\omega\tau\left(\tilde{\boldsymbol{x}}_g|\boldsymbol{x}|\tilde{\boldsymbol{x}}_s\right)}. \tag{11.66}$$

The frequency integral is an inverse Fourier transform of the filter $\omega^2 {f_2}^{-1} D$, evaluated at time $t = \tau\left(\tilde{\boldsymbol{x}}_g|\boldsymbol{x}|\tilde{\boldsymbol{x}}_s\right)$.

Equation (11.66) is a reflection-centered algorithm. It begins at a reflection point $\boldsymbol{x}$, opening angle σ, and a reflection-point azimuth χ_{ir}. Summing over all possible wavenumber angles, it computes the reflection-point slowness vectors $\boldsymbol{p}_r$ and $\boldsymbol{p}_i$ (see Appendix H) , retrieves the corresponding raypaths, snatches the data at the raypath end points $\tilde{\boldsymbol{x}}_g$ and $\tilde{\boldsymbol{x}}_s$, weights the filtered data with the inverse of B_2, and adds it to the sum. Though for a fixed σ and χ_{ir}, only a subset of the full 5-D data is needed, it would be difficult a priori to separate out only the part that contributes. Thus, implicit is the assumption that a full, 5-D data set is available to pick from.

Practical data sets are not likely to sample the full 5-D data space. In particular, source–receiver azimuth is often severely restricted. This poses a problem for the algorithm in that, since the ray tracing begins at the reflector, even though reflection-point azimuth is known, source–receiver azimuth is unknown until the rays to the surface have been found. On the positive side, derivatives of source–receiver azimuth with respect to reflection point azimuth can be computed, and as long as variations of traveltime with reflection point, source, and receiver are well-behaved, there is no reason to believe that a limited-azimuth data set cannot be accommodated.

11.6 Special cases

11.6.1 Constant-velocity poststack 3-D migration

The constant-velocity, exploding-reflector model provides a simple illustration and confirmation of the modeling and migration formulas developed above.

In *f–k* space, according to the exploding-reflector model (see Chapter 2),

$$\mathcal{R}\left(k_x, k_y, k_z\right) = D\left(k_x, k_y, \omega\right)\left|\frac{d\omega}{dk_z}\right|, \tag{11.67}$$

where

$$k_z = -S(\omega)\sqrt{\frac{4\omega^2}{c^2} - {k_x}^2 - {k_y}^2}, \tag{11.68}$$

$$\frac{d\omega}{dk_z} = \frac{c^2 k_z}{4\omega}, \tag{11.69}$$

and $S(\omega)$ is the sign of frequency ω. Other, more realistic models of propagation and reflection have been shown above to yield slightly different expressions, generally of the form

$$\mathcal{R}\left(k_x, k_y, k_z\right) = D\left(k_x, k_y, \omega\right)\Omega\left(k_x/k_z, k_y/k_z\right), \tag{11.70}$$

with Ω a smooth and reasonably well-behaved function of the dip tangents k_x/k_z and k_y/k_z.

Returning this image to *x–y–z* space,

$$\mathcal{R}(x, y, z) = \frac{1}{(2\pi)^3}\int\int\int dk_x dk_y dk_z \Omega\left(k_x/k_z, k_y/k_z\right) D\left(k_x, k_y, \omega\right) \times e^{i\left(k_x x + k_y y + k_z z\right)}. \tag{11.71}$$

In this expression, decompose the data D into space–frequency components:

$$\mathcal{R}(x, y, z) = \frac{1}{(2\pi)^3}\int\int dx_i dy_i \int d\omega D\left(x_i, y_i, \omega\right) \times \int\int dk_x dk_y \Omega\left(k_x/k_z, k_y/k_z\right)\left|\frac{dk_z}{d\omega}\right| e^{i\left(k_x(x - x_i) + k_y(y - y_i) + k_z z\right)}. \tag{11.72}$$

Next, evaluate the k_x and k_y integrals using the 2-D stationary-phase approximation (see Appendix F.3.2), obtaining

$$\mathcal{R}(x, y, z) \simeq \frac{-1}{(2\pi)^2}\int\int dx_i dy_i \frac{4\Omega\left((x - x_i)/z, (y - y_i)/z\right)}{c^2 r} \times \int d\omega(-i\omega) D\left(x_i, y_i, \omega\right) e^{-i\omega\tau}, \tag{11.73}$$

where

$$\tau = \tau\left(\tilde{\boldsymbol{x}}_i, \tilde{\boldsymbol{x}}, z\right) = \frac{2}{c}\sqrt{|\tilde{\boldsymbol{x}} - \tilde{\boldsymbol{x}}_i|^2 + z^2} = \frac{2}{c}r. \tag{11.74}$$

The factor of $(-i\omega)$ that appears in (11.73) can be replaced by a derivative with respect to τ, and moved outside the frequency integral. The frequency integral is then just the inverse temporal Fourier transform of the data. Thus

$$\mathcal{R}(x, y, z) \simeq \frac{-4}{(2\pi)c^2} \int\int dx_i dy_i \frac{\Omega\big((x - x_i)/z, (y - y_i)/z\big)}{r} \dot{D}(x_i, y_i, \tau), \tag{11.75}$$

with the "dot" over the data function D indicating a derivative with respect to its last argument. Since τ implicitly depends on z, one can convert the time derivative into a depth-derivative, and (noting that $\partial\tau/\partial z = 2z/cr$, and neglecting the derivative of the amplitude term) move it outside the integral:

$$\mathcal{R}(x, y, z) \simeq \frac{-2}{(2\pi)c} \frac{\partial}{\partial z} \left[\int\int dx_i dy_i \frac{\Omega\big((x - x_i)/z, (y - y_i)/z\big)}{z} D(x_i, y_i, \tau) \right]. \tag{11.76}$$

Equations (11.75) and (11.76) are specific instances of (11.20) and (11.21), with traveltime given by (11.74) and amplitude given by

$$A(x_i, x, z) = \frac{2}{\pi c^2} \frac{\Omega\big((x - x_i)/z, (y - y_i)/z\big)}{r}. \tag{11.77}$$

If we employ the exploding reflector model, then $\Omega \rightarrow cz/2r$, and

$$A(x_i, x, z) \rightarrow \frac{1}{\pi c} \frac{z}{r^2}. \tag{11.78}$$

The 2-D analog of this formula is found in Section 2.6.5.

11.6.2 Constant-velocity zero-offset 3-D modeling

Starting with (11.67), we have, returning the data to space–time,

$$D(x_i, y_i, t) = \frac{1}{(2\pi)^3} \int\int\int dk_x dk_y d\omega \frac{\mathcal{R}(k_x, k_y, k_z)}{\Omega(k_x/k_z, k_y/k_z)} e^{i(k_x x_i + k_y y_i - \omega t)}. \tag{11.79}$$

Now return the image function to x–y:

$$D(x_i, y_i, t) = \frac{1}{(2\pi)^3} \int\int\int dx dy dk_z \mathcal{R}(x, y, k_z) \times \int\int dk_x dk_y \left| \frac{d\omega}{dk_z} \right| \frac{e^{i(k_x(x_i - x) + k_y(y_i - y) - \omega t)}}{\Omega(k_x/k_z, k_y/k_z)}. \tag{11.80}$$

Evaluating the k_x and k_y integrals via stationary phase,

$$D(x_i, y_i, t) \simeq \frac{2}{(2\pi)^2 c} \int\int \frac{dxdy}{\zeta\Omega\left((x_i - x)/\zeta,\, (y_i - y)/\zeta\right)} \times \int dk_z\, (ik_z)\, \mathcal{R}(x, y, k_z)\, e^{ik_z\zeta}, \tag{11.81}$$

where

$$\zeta = \zeta(\tilde{\boldsymbol{x}}, \tilde{\boldsymbol{x}}_i, t) = \sqrt{c^2\, t^2/4 - |\tilde{\boldsymbol{x}} - \tilde{\boldsymbol{x}}_i|^2}. \tag{11.82}$$

The filter (ik_z) can be replaced by a derivative with respect to ζ, and moved outside the k_z integral. Then, with the "dot" over the $\mathcal{R}$ indicating a derivative with respect to its last argument ζ,

$$D(x_i, y_i, t) \simeq \frac{2}{2\pi c} \int\int dxdy \frac{\dot{\mathcal{R}}(x, y, \zeta)}{\zeta\Omega\left((x_i - x)/\zeta,\, (y_i - y)/\zeta\right)}. \tag{11.83}$$

Neglecting the depth-dependence of the amplitude term, and noting that $\partial\zeta/\partial t = cr/2\zeta$, the derivative can optionally be converted to a time derivative and moved outside the x and y integrals. Then,

$$D(x_i, y_i, t) = \frac{4}{2\pi c^2 t} \frac{\partial}{\partial t} \left[\int\int dxdy \frac{\mathcal{R}(x, y, \zeta)}{\Omega\left((x_i - x)/\zeta,\, (y_i - y)/\zeta\right)} \right]. \tag{11.84}$$

Thus, with a slight modification of the amplitude factor, one can apply the differentiation either before or after the integrations. Equations (11.83) and (11.84) are instances of the modeling equations (11.17) and (11.7), with $\zeta(\tilde{\boldsymbol{x}}, \tilde{\boldsymbol{x}}_i, t)$ given by equation (11.82) and the amplitude factor $B(\tilde{\boldsymbol{x}}, \tilde{\boldsymbol{x}}_i, t)$ given by

$$B(\tilde{\boldsymbol{x}}, \tilde{\boldsymbol{x}}_i, t) = \frac{c}{4\pi\zeta\Omega\left((x_i - x)/\zeta,\, (y_i - y)/\zeta\right)}. \tag{11.85}$$

For the exploding reflector model, $\Omega \to \zeta/t$, and

$$B(\tilde{\boldsymbol{x}}, \tilde{\boldsymbol{x}}_i, t) \to \frac{ct}{4\pi\zeta^2}. \tag{11.86}$$

11.6.3 Constant-velocity elastic prestack modeling

To model prestack data, begin with the elastic reflectivity model (9.17), which we simplify for non-converted waves and an angle-independent reflectivity function to be

$$D\left(\tilde{k}_g, \tilde{k}_s, \omega\right) = -\frac{i\omega\cos\gamma}{2k_{gz}k_{sz}\rho c^3} \mathcal{R}(k_m), \tag{11.87}$$

with $\tilde{\boldsymbol{k}}_s$ and $\tilde{\boldsymbol{k}}_g$ 2-D source and receiver wavenumbers, ω frequency, and $\boldsymbol{k}_m$ the 3-vector

$$\boldsymbol{k}_m = \left(\tilde{\boldsymbol{k}}_m, k_z\right) = \left(\tilde{\boldsymbol{k}}_g - \tilde{\boldsymbol{k}}_s, k_{gz} - k_{sz}\right), \tag{11.88}$$

$$k_{gz} = -S(\omega)\sqrt{\frac{\omega^2}{c^2} - \tilde{k}_g{}^2},$$
$$k_{sz} = +S(\omega)\sqrt{\frac{\omega^2}{c^2} - \tilde{k}_s{}^2}. \tag{11.89}$$

Angle of incidence γ is given by the 3-vector dot product:

$$\cos 2\gamma = -\frac{c^2}{\omega^2} k_g \cdot k_s. \tag{11.90}$$

The wavenumber magnitude k_m has the value

$$k_m = \frac{2|\omega| \cos\gamma}{c}, \tag{11.91}$$

Equation (11.87) can be placed in the space–time domain as (with $\boldsymbol{x} = (\tilde{\boldsymbol{x}}, z)$)

$$D\left(\tilde{\boldsymbol{x}}_g, \tilde{\boldsymbol{x}}_s, t\right) = -\frac{1}{2(2\pi)^5 \rho c^3} \int d^2x \int dz \mathcal{R}(x) \int d\omega e^{-i\omega t} \int d^2k_g e^{i\left(\tilde{\boldsymbol{k}}_g \cdot (\tilde{\boldsymbol{x}}_g - \tilde{\boldsymbol{x}}) - k_{gz} z\right)}$$
$$\times \int d^2k_s e^{-i\left(\tilde{\boldsymbol{k}}_s \cdot (\tilde{\boldsymbol{x}}_s - \tilde{\boldsymbol{x}}) - k_{sz} z\right)} \frac{i\omega \cos\gamma}{k_{gz} k_{sz}}. \tag{11.92}$$

The integrals over source and receiver wavenumber can be evaluated using the stationary phase approximation, in which, if $\boldsymbol{r}_s$ is the vector pointing from $\tilde{\boldsymbol{x}}_s$ to $(\tilde{\boldsymbol{x}}, z)$ and $\boldsymbol{r}_g$ is the vector from $(\tilde{\boldsymbol{x}}, z)$ to $\tilde{\boldsymbol{x}}_g$,

$$k_s = \left(\tilde{\boldsymbol{k}}_s, k_{sz}\right) \to \frac{\omega}{c} \frac{\boldsymbol{r}_s}{r_s}, \tag{11.93}$$

$$k_g = \left(\tilde{\boldsymbol{k}}_g, k_{gz}\right) \to \frac{\omega}{c} \frac{\boldsymbol{r}_g}{r_g}, \tag{11.94}$$

and

$$D\left(\tilde{\boldsymbol{x}}_g, \tilde{\boldsymbol{x}}_s, t\right) = \frac{1}{2(2\pi)^3 \rho c^3} \int d^2x \int dz \mathcal{R}(x) \int d\omega e^{-i\omega(t - r/c)} \frac{-i\omega \cos\gamma}{r_g r_s}. \tag{11.95}$$

$$D\left(\tilde{\boldsymbol{x}}_g, \tilde{\boldsymbol{x}}_s, t\right) = D\left(\tilde{\boldsymbol{x}}_m, \tilde{\boldsymbol{x}}_h, t\right) \tag{11.96}$$
$$\simeq \frac{1}{(2\pi)^3 \rho c^3} \int d^2x \int dz \mathcal{R}\left(\tilde{x}, z\right) \frac{\cos\gamma}{r_g r_s} \int d\omega e^{-i\omega(t - r/c)}(-i\omega),$$

or, evaluating the frequency integral,

$$D\left(\tilde{\boldsymbol{x}}_m, \tilde{\boldsymbol{x}}_h, t\right) = \frac{1}{2(2\pi)^2 \rho c^3} \int d^2x \int dz \mathcal{R}(x) \frac{\cos\gamma}{r_g r_s} \dot{\delta}(t - r/c), \tag{11.97}$$

with half-offset $\tilde{\boldsymbol{x}}_h = \left(\tilde{\boldsymbol{x}}_s - \tilde{\boldsymbol{x}}_g\right)/2$ and midpoint $\tilde{\boldsymbol{x}}_m = \left(\tilde{\boldsymbol{x}}_s + \tilde{\boldsymbol{x}}_g\right)/2$. The angle of incidence is determined at the stationary point by the relation

$$\cos^2\gamma = \frac{1}{2}\left(1 - \frac{\boldsymbol{r}_g \cdot \boldsymbol{r}_s}{r_g r_s}\right) = \frac{1 - 4\left|\tilde{\boldsymbol{x}}_h\right|^2 / r^2}{1 - 16\left(\left(\tilde{\boldsymbol{x}} - \tilde{\boldsymbol{x}}_m\right) \cdot \tilde{\boldsymbol{x}}_h\right)^2 / r^4}. \tag{11.98}$$

The delta function in (11.97) can be used to reduce the volume integral to a surface integral, so that

$$D\left(\tilde{\boldsymbol{x}}_m, \tilde{\boldsymbol{x}}_h, t\right) = \frac{1}{2\pi\rho c^2} \frac{\partial}{\partial t} \int d^2x \frac{\mathcal{R}\left(\tilde{\boldsymbol{x}}, \zeta\right) \cos^2\gamma}{r\zeta}, \tag{11.99}$$

with ζ the depth function

$$\zeta\left(\tilde{\boldsymbol{x}}, \tilde{\boldsymbol{x}}_m, \tilde{\boldsymbol{x}}_h, t\right) = \frac{ct}{2}\sqrt{1 - \frac{4\left(\left|\tilde{\boldsymbol{x}} - \tilde{\boldsymbol{x}}_m\right|^2 + \tilde{\boldsymbol{x}}_h{}^2\right)}{c^2t^2} + \left(\frac{4\left(\tilde{\boldsymbol{x}} - \tilde{\boldsymbol{x}}_m\right) \cdot \tilde{\boldsymbol{x}}_h}{c^2t^2}\right)^2}. \tag{11.100}$$

The time derivative can be brought inside the integral and converted to a depth derivative of $\mathcal{R}$, under the assumption that $\mathcal{R}$ varies more rapidly with depth than the other factors. Then

$$D\left(\tilde{\boldsymbol{x}}_m, \tilde{\boldsymbol{x}}_h, t\right) \simeq \frac{1}{2\pi\rho c^2} \int d^2x \frac{\dot{\mathcal{R}}\left(\tilde{\boldsymbol{x}}, \zeta\right) \cos^2\gamma\, \partial_t \zeta}{r\zeta}, \tag{11.101}$$

or

$$D\left(\tilde{\boldsymbol{x}}_m, \tilde{\boldsymbol{x}}_h, t\right) = \frac{1}{8\pi\rho c} \int d^2x \frac{\dot{\mathcal{R}}\left(\tilde{\boldsymbol{x}}, z\right) \cos^2\gamma}{\zeta^2} \left(1 - \frac{16\left(\left(\tilde{\boldsymbol{x}} - \tilde{\boldsymbol{x}}_m\right) \cdot \tilde{\boldsymbol{x}}_h\right)^2}{c^4t^4}\right). \tag{11.102}$$

This modeling operator is seen to have the form (11.17), with

$$\begin{aligned} B\left(\tilde{\boldsymbol{x}}, \tilde{\boldsymbol{x}}_m, \tilde{\boldsymbol{x}}_h, t\right) &= \frac{1}{8\pi\rho c} \frac{\cos^2\gamma}{\zeta^2} \left(1 - \frac{16\left(\left(\tilde{\boldsymbol{x}} - \tilde{\boldsymbol{x}}_m\right) \cdot \tilde{\boldsymbol{x}}_h\right)^2}{c^4t^4}\right) \\ &= \frac{1}{8\pi\rho c} \frac{1}{\zeta^2} \left(1 - \frac{4\left|\tilde{\boldsymbol{x}}_h\right|^2}{c^2t^2}\right). \end{aligned} \tag{11.103}$$

That the data, depth, and amplitude functions depend on more variables than earlier in this chapter is not a concern – to define an inverse operation, one restricts the data to a subset of three independent coordinates. There is, of course, more than one definable inverse operation, depending on which subset of the data is chosen.

11.6.4 Constant-velocity constant-offset elastic prestack migration

Consider a constant-offset data set determined by the modeling operator (11.102). According to (11.20), the migration operation inverse to this equation is

$$\mathcal{R}(\tilde{\boldsymbol{x}}, z) = -\int d^2 x_m \dot{D}\left(\tilde{\boldsymbol{x}}_m, \tilde{\boldsymbol{x}}_h, \tau\left(\tilde{\boldsymbol{x}}_m, \tilde{\boldsymbol{x}}_h, \tilde{\boldsymbol{x}}, z\right)\right) A\left(\tilde{\boldsymbol{x}}_m, \tilde{\boldsymbol{x}}_h, \tilde{\boldsymbol{x}}, z\right), \quad (11.104)$$

where the traveltime function τ is given by

$$\begin{aligned}\tau\left(\tilde{\boldsymbol{x}}_m, \tilde{\boldsymbol{x}}, z\right) &= \frac{r}{c} \\ &= \frac{\sqrt{z^2 + |\tilde{\boldsymbol{x}} - \tilde{\boldsymbol{x}}_m - \tilde{\boldsymbol{x}}_h|^2} + \sqrt{z^2 + |\tilde{\boldsymbol{x}} - \tilde{\boldsymbol{x}}_m + \tilde{\boldsymbol{x}}_h|^2}}{c}, \quad (11.105)\end{aligned}$$

and the amplitude factor A by (11.34), or

$$A\left(\tilde{\boldsymbol{x}}_m, \tilde{\boldsymbol{x}}_h, \tilde{\boldsymbol{x}}, z\right) = \frac{(2\pi)^{-2}\left|J_\lambda\left(\tilde{\boldsymbol{x}}_m, \tilde{\boldsymbol{x}}_h, \tilde{\boldsymbol{x}}, z\right)\right|}{B\left(\tilde{\boldsymbol{x}}_m, \tilde{\boldsymbol{x}}, \tau\left(\tilde{\boldsymbol{x}}_m, \tilde{\boldsymbol{x}}_h, \tilde{\boldsymbol{x}}, z\right)\right)\dot{\zeta}\left(\tilde{\boldsymbol{x}}, \tilde{\boldsymbol{x}}_m, \tilde{\boldsymbol{x}}_h, \tau\left(\tilde{\boldsymbol{x}}_m, \tilde{\boldsymbol{x}}_h, \tilde{\boldsymbol{x}}, z\right)\right)}, \quad (11.106)$$

with the depth function ζ as given in equation (11.100).

$$\begin{aligned}J_\lambda\left(\tilde{\boldsymbol{x}}_m, \tilde{\boldsymbol{x}}, z\right) &= \left|\operatorname{Det}\begin{pmatrix}\frac{\partial^2\zeta}{\partial x \partial x_m} & \frac{\partial^2\zeta}{\partial x \partial y_m} \\ \frac{\partial^2\zeta}{\partial y \partial x_m} & \frac{\partial^2\zeta}{\partial y \partial y_m}\end{pmatrix}\right| \\ &= \frac{c^2 t^2}{4\zeta\left(\tilde{\boldsymbol{x}}, \tilde{\boldsymbol{x}}_m, t\right)^4}\left(1 - \frac{4\left|\tilde{\boldsymbol{x}}_h\right|^2}{c^2 t^2}\right)^2, \quad (11.107)\end{aligned}$$

and

$$\dot{\zeta}\left(\tilde{\boldsymbol{x}}, \tilde{\boldsymbol{x}}_m, t\right) = \frac{c}{2}\frac{ct}{2\zeta}\left(1 - \frac{16\left(\left(\tilde{\boldsymbol{x}} - \tilde{\boldsymbol{x}}_m\right)\cdot\tilde{\boldsymbol{x}}_h\right)^2}{c^4 t^4}\right). \quad (11.108)$$

After some algebra, A evaluates to

$$A\left(\tilde{\boldsymbol{x}}_m, \tilde{\boldsymbol{x}}, z\right) = \frac{2\rho}{\pi}\frac{c\tau}{z}\cos^2\gamma. \quad (11.109)$$

For constant-offset migration of constant-velocity data, ignoring the dependence of reflectivity upon angle, one uses the operator (11.104) with the amplitude factor (11.109).

11.6.5 Constant-velocity small-offset elastic migration and modeling

At small offsets, the expression (11.109) for A reduces to

$$A\left(\tilde{\boldsymbol{x}}_m, \tilde{\boldsymbol{x}}, z\right) = \frac{2\rho}{\pi}\frac{c\tau}{z}. \quad (11.110)$$

Similarly, equation (11.103) for B reduces to

$$B(\tilde{x}, \tilde{x}_m, t) = \frac{1}{8\pi\rho c}\frac{1}{\zeta^2}. \tag{11.111}$$

These formulas differ slightly from the exploding-reflector formulas (11.78) and (11.86), as of course they should, since they are based upon a different physical model.

Exercises

11.1 Using stationary phase, evaluate the k_x and k_y wavenumber integrals in equation (11.80) to obtain (11.81).

11.2 Confirm that the reciprocal relation (11.34) between the forward and inverse migration amplitudes hold for the zero-offset constant-velocity amplitude functions (11.77) and (11.85).

11.3 From the constant-velocity finite-offset two-way traveltime formula, derive the corresponding depth function (11.100).

11.4 Derive expressions for the two one-way traveltime components in terms of total traveltime τ, midpoint displacement $\tilde{\boldsymbol{x}} - \tilde{\boldsymbol{x}}_m$, and offset $\tilde{\boldsymbol{x}}_h$.

11.5 Derive expressions for the derivative of traveltime with respect to depth and for the derivative of the depth function with respect to traveltime and confirm equation (11.3).

11.6 From the prestack modeling formula (11.102) and the reciprocal relation (11.35), derive the constant-offset migration operator (11.104).

12
Residual asymptotic migration

12.1 Combining modeling and migration

The idea behind residual migration is to combine forward and inverse mappings into a single operation (Rothman *et al.*, 1985; Fomel, 2003; Sava, 2003; Schleicher *et al.*, 2007). First one undoes with a forward operation an initial migration performed with (presumably) the wrong velocity, then does a new migration with a (presumably) better velocity structure. For forward and inverse time migrations, combining the two operators is relatively simple. For constant velocity, the *f–k* formulas presented in Section 10.3 showed how simple it could be. If one or both is a depth migration, things are more complicated, though still doable, as seen below.

Suppose one has (for now, single-valued) migration and modeling algorithms for two distinct image functions:

$$\mathcal{R}_f\left(\tilde{\boldsymbol{x}}_f, z_f\right) = \int d\tilde{\boldsymbol{x}} A_f\left(\tilde{\boldsymbol{x}}, \tilde{\boldsymbol{x}}_f, z_f\right) \bar{D}\left(\tilde{\boldsymbol{x}}, \tau_f\left(\tilde{\boldsymbol{x}}, \tilde{\boldsymbol{x}}_f, z_f\right)\right), \tag{12.1}$$

and

$$D\left(\tilde{\boldsymbol{x}}, t\right) = \int d\tilde{\boldsymbol{x}}_i B_i\left(\tilde{\boldsymbol{x}}_i, \tilde{\boldsymbol{x}}, t\right) \bar{\mathcal{R}}_i\left(\tilde{\boldsymbol{x}}_i, \zeta_i\left(\tilde{\boldsymbol{x}}_i, \tilde{\boldsymbol{x}}, t\right)\right). \tag{12.2}$$

Here, ζ_i and B_i are the depth and migration-amplitude functions appropriate to the initial image function $\mathcal{R}_i$, while τ_f and A_f are the traveltime and modeling-amplitude functions appropriate to the final image function $\mathcal{R}_f$. The integrals in either equation are either one- or two-dimensional, depending on whether the underlying medium is two- or three-dimensional. Note that both operations (12.1) and (12.2) share the same data function D. The bars over D and $\mathcal{R}_i$ indicate a filter with respect to the last argument in their list of arguments. For the migration operation (12.1), the filter is most commonly one or two applications of the half-derivative adjoint, or $d_{h-}{}^n$. For the modeling operation (12.2), the most common filter is $d_{h+}{}^n$. For specificity, these filters will be assumed in the following discussion.

For the traveltime function τ_f, there is a corresponding depth function ζ_f, as defined in Section 11.2:

$$\zeta_f\left(\tilde{\boldsymbol{x}}_f, \tilde{\boldsymbol{x}}, \tau_f\left(\tilde{\boldsymbol{x}}, \tilde{\boldsymbol{x}}_f, z_f\right)\right) = z_f. \tag{12.3}$$

Likewise, for the depth function ζ_i there is a corresponding time function τ_i which obeys

$$\tau_i\left(\tilde{\boldsymbol{x}}, \tilde{\boldsymbol{x}}_i, \zeta_i\left(\tilde{\boldsymbol{x}}_i, \tilde{\boldsymbol{x}}, t\right)\right) = t. \tag{12.4}$$

We note, however, that $\tau_i \neq \tau_f$ and $\zeta_i \neq \zeta_f$.

Note that (12.1) and (12.2) have almost the same functional form, with the role of time and depth interchanged. We could, if desired, express the image functions in units of traveltime rather than depth, which would enhance the similarities of the two transformations.

To express the final migration $\mathcal{R}_f$ in terms of the initial migration $\mathcal{R}_i$, insert equation (12.2) into (12.1):

$$\begin{aligned}\mathcal{R}_f\left(\tilde{\boldsymbol{x}}_f, z_f\right) = \int d\tilde{\boldsymbol{x}} A_f\left(\tilde{\boldsymbol{x}}, \tilde{\boldsymbol{x}}_f, z_f\right) d_{h-}{}^n * \int d\tilde{\boldsymbol{x}}_i B_i\left(\tilde{\boldsymbol{x}}_i, \tilde{\boldsymbol{x}}, \tau_f\left(\tilde{\boldsymbol{x}}, \tilde{\boldsymbol{x}}_f, z_f\right)\right) \\ \times d_{h+}{}^n * \mathcal{R}_{\mathrm{i}}\left(\tilde{\boldsymbol{x}}_i, \zeta_i\left(\tilde{\boldsymbol{x}}_{\mathrm{i}}, \tilde{\boldsymbol{x}}, \tau_f\left(\tilde{\boldsymbol{x}}, \tilde{\boldsymbol{x}}_f, z_f\right)\right)\right).\end{aligned} \tag{12.5}$$

The filter $d_{h-}{}^n$ applies to the time coordinate τ_f. Using the formula (E.26), this filter can be converted to a depth-filter and brought inside the second integral, to produce (changing the order of integration)

$$\begin{aligned}\mathcal{R}_f\left(\tilde{\boldsymbol{x}}_f, z_f\right) = \int d\tilde{\boldsymbol{x}}_i \int d\tilde{\boldsymbol{x}} A_f\left(\tilde{\boldsymbol{x}}, \tilde{\boldsymbol{x}}_f, z_f\right) B_i\left(\tilde{\boldsymbol{x}}_i, \tilde{\boldsymbol{x}}, \tau_f\left(\tilde{\boldsymbol{x}}, \tilde{\boldsymbol{x}}_f, z_f\right)\right) \\ \times \dot{\zeta}_i\left(\tilde{\boldsymbol{x}}_i, \tilde{\boldsymbol{x}}, \tau_f\left(\tilde{\boldsymbol{x}}, \tilde{\boldsymbol{x}}_f, z_f\right)\right)^{n/2} d_{h-}{}^n * d_{h+}{}^n * \mathcal{R}_i\left(\tilde{\boldsymbol{x}}_i, \zeta_i\left(\tilde{\boldsymbol{x}}_i, \tilde{\boldsymbol{x}}, \tau_f\left(\tilde{\boldsymbol{x}}, \tilde{\boldsymbol{x}}_f, z_f\right)\right)\right),\end{aligned} \tag{12.6}$$

with $\dot{\zeta}_i$ the derivative of ζ_i with respect to its last argument τ_f. The integral over $\tilde{\boldsymbol{x}}$ is an expression of the diffraction integral (G.1), with the correspondence

$$\begin{aligned} \boldsymbol{s} &\leftrightarrow \tilde{\boldsymbol{x}}, \\ \rho &\leftrightarrow z_f, \\ f &\leftrightarrow d_{h-}{}^n, \\ I &\leftrightarrow \overline{\mathcal{R}}_i, \\ C &\leftrightarrow A_f B_i \dot{\zeta}_i{}^{n/2}, \\ \gamma(\rho, \boldsymbol{s}) &\leftrightarrow \zeta_i\left(\tilde{\boldsymbol{x}}_i, \tilde{\boldsymbol{x}}, \tau_f\left(\tilde{\boldsymbol{x}}_f, \tilde{\boldsymbol{x}}, z_f\right)\right). \end{aligned} \tag{12.7}$$

Two parameters $\tilde{\boldsymbol{x}}_f$ and $\tilde{\boldsymbol{x}}_i$ appear in (12.7) but not (G.1). These extra parameters can be considered constants during the integration over $\tilde{\boldsymbol{x}}$.

Invoking Appendix G, the $\tilde{\boldsymbol{x}}$-integral in (12.6) can be evaluated. The result is one or more events at locations $\tilde{\boldsymbol{x}} = \tilde{\boldsymbol{x}}_e$, depending upon the detailed behavior of the trajectory function $\zeta_i\left(\tilde{\boldsymbol{x}}, \tilde{\boldsymbol{x}}_i, \tau_f\left(\tilde{\boldsymbol{x}}_f, \tilde{\boldsymbol{x}}, z_f\right)\right)$. Minima, maxima, saddle points, and inflection points in ζ_i all build events through constructive interference. Thus, the general result is a summation over events

$$\mathcal{R}_f\left(\tilde{\boldsymbol{x}}_f, z_f\right) \simeq \sum_e \int d\tilde{\boldsymbol{x}}_i \frac{C_e\left(\tilde{\boldsymbol{x}}_i, \tilde{\boldsymbol{x}}_f, z_f\right)}{\mathcal{F}_e\left(\tilde{\boldsymbol{x}}_i, \tilde{\boldsymbol{x}}_f, z_f\right)} \times f_e * d_{h-}{}^n * d_{h+}{}^n * \mathcal{R}_i\left(\tilde{\boldsymbol{x}}_i, \zeta_i\left(\tilde{\boldsymbol{x}}_i, \tilde{\boldsymbol{x}}_e, \tau_f\left(\tilde{\boldsymbol{x}}_e, \tilde{\boldsymbol{x}}_f, z_f\right)\right)\right). \tag{12.8}$$

In this expression, $C_e = A_f B_i \dot{\zeta}_i{}^{n/2}$, evaluated at $\tilde{\boldsymbol{x}} = \tilde{\boldsymbol{x}}_e$, $\mathcal{F}_e$ is the amplitude characteristic of the event, and f_e is the filter characteristic of the event, as discussed in Appendix G.

If the average velocity in the velocity function used in constructing the initial image $\mathcal{R}_i$ is lower than the average velocity of the function used to construct the final image $\mathcal{R}_f$, then the raypaths associated with $\mathcal{R}_i$ will be shorter than the raypaths associated with $\mathcal{R}_f$. For a given z_f, it follows that, for integration variable $\tilde{\boldsymbol{x}}$ sufficiently far from $\tilde{\boldsymbol{x}}_i$ and $\tilde{\boldsymbol{x}}_f$, ζ_i must go to zero. Hence, within the domain of the $\tilde{\boldsymbol{x}}$-integral, ζ_i must undergo a maximum. Conversely, if the initial velocity is higher on average than the final velocity, then ζ_i must undergo a minimum.

There is nothing to prevent additional minima, maxima, saddles, and inflections from occurring, but in the simplest case, one can anticipate that where the initial velocity is too low, the trajectory function will generate a maximum-type event, with a migration-like residual operator. Where the initial velocity is too high, a minimum-type event will occur, with a modeling-like residual operator.

Assume that ζ_i goes through an extremal point at $\tilde{\boldsymbol{x}} = \tilde{\boldsymbol{x}}_e$, and define $\mathrm{Det}\left(\zeta_i''\right)$ to be the determinant of the matrix of second derivatives of ζ_i with respect to the components of $\tilde{\boldsymbol{x}}$ at the point $\tilde{\boldsymbol{x}} = \tilde{\boldsymbol{x}}_e$. (If $n = 1$, the determinant is just the second derivative of ζ_i with respect to the single component of $\tilde{\boldsymbol{x}}$.) Then, the contribution to (12.8) from the vicinity of $\tilde{\boldsymbol{x}}_e$ is

$$\mathcal{R}_f\left(\tilde{\boldsymbol{x}}_f, z_f\right) \simeq 2\pi S_e \int d\tilde{\boldsymbol{x}}_i \frac{C\left(\tilde{\boldsymbol{x}}_i, \tilde{\boldsymbol{x}}_f, z_f\right)}{\sqrt{\left|\mathrm{Det}\left(\zeta_i''\right)\right|}} d_{h+}{}^n * \mathcal{R}_i\left(\tilde{\boldsymbol{x}}_i, \gamma\left(\tilde{\boldsymbol{x}}_i, \tilde{\boldsymbol{x}}_f, z_f\right)\right), \tag{12.9}$$

where $S_e = -1$ for a maximum and $S_e = +1$ for a minimum. The amplitude function is

$$C\left(\tilde{\boldsymbol{x}}_i, \tilde{\boldsymbol{x}}_f, z_f\right) = A_f\left(\tilde{\boldsymbol{x}}_e, \tilde{\boldsymbol{x}}_f, z_f\right) B_i\left(\tilde{\boldsymbol{x}}_i, \tilde{\boldsymbol{x}}_e, \tau_f\left(\tilde{\boldsymbol{x}}_e, \tilde{\boldsymbol{x}}_f, z_f\right)\right) \times \left(\partial_t \zeta_i\left(\tilde{\boldsymbol{x}}_i, \tilde{\boldsymbol{x}}_e, \tau_f\left(\tilde{\boldsymbol{x}}_e, \tilde{\boldsymbol{x}}_f, z_f\right)\right)\right)^{n/2}, \tag{12.10}$$

and the trajectory function is

$$\gamma\left(\tilde{\boldsymbol{x}}_i, \tilde{\boldsymbol{x}}_f, z_f\right) = \zeta_i\left(\tilde{\boldsymbol{x}}_i, \tilde{\boldsymbol{x}}_e, \tau_f\left(\tilde{\boldsymbol{x}}_e, \tilde{\boldsymbol{x}}_f, z_f\right)\right). \tag{12.11}$$

Thus, in the case of a single extremum, the residual migration operator becomes (12.9), a simple modeling or migration operation. Unfortunately, there is no guarantee that things will remain that simple. Even if the original modeling and migration operations (12.1) and (12.2) are single-valued, the residual operation (12.5) may well be multivalued. In this case, Appendix G provides the tools to calculate the multiple events.

On the other hand, under some circumstances it may be possible to simplify things considerably, by performing two (or more) single-path migration operations in lieu of one multipath operation. A multipath Green's function by definition exhibits triplications and diffractions, which appear in the data as roughly hyperbolic events with effective velocities lower than the medium velocity. By performing an initial migration at or near this effective velocity, the triplications in the effective Green's function can be reduced or eliminated, permitting a single-path residual operation to complete the migration.

An example using residual migration to convert a time migration into a depth migration is illustrated in Figures 12.1 to 12.3. Reflecting points have been placed at 1000 ft depth intervals in a medium of linearly increasing velocity

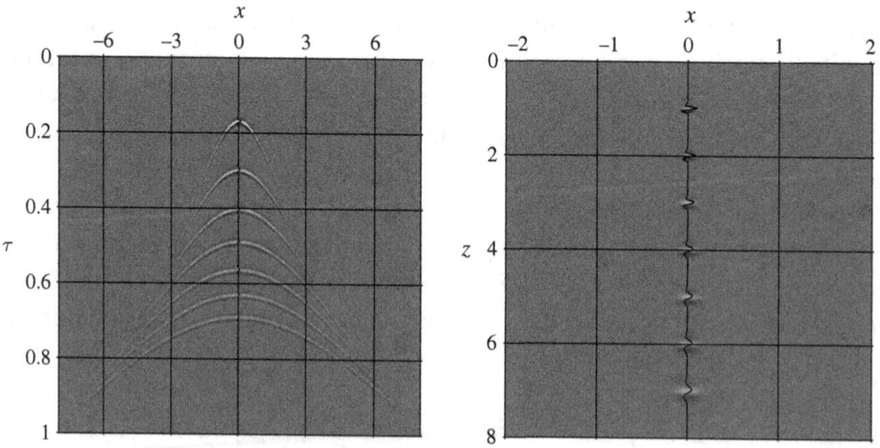

Figure 12.1 The left panel shows the reflection response to a sequence of point reflectors in a medium where velocity increases linearly with depth. The right panel shows a depth migration of the reflection response.

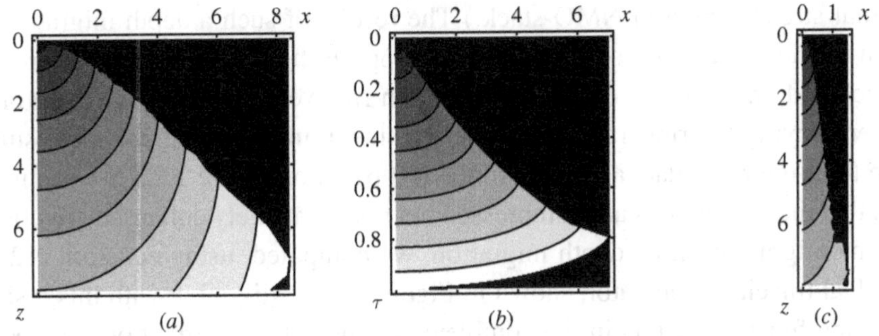

Figure 12.2 Traveltime functions for migration in a medium where $c(z) = 5 + 1.875z$. (a), The depth migration traveltime function needed to collapse the diffraction curves in Figure 12.1 to points. (b), An rms-velocity time migration traveltime function which partially collapses the curves as per Figure 12.3. (c), A residual traveltime function which, when applied to the time-migrated data in Figure 12.3, produces the depth-migration equivalent in the right panel of Figure 12.3.

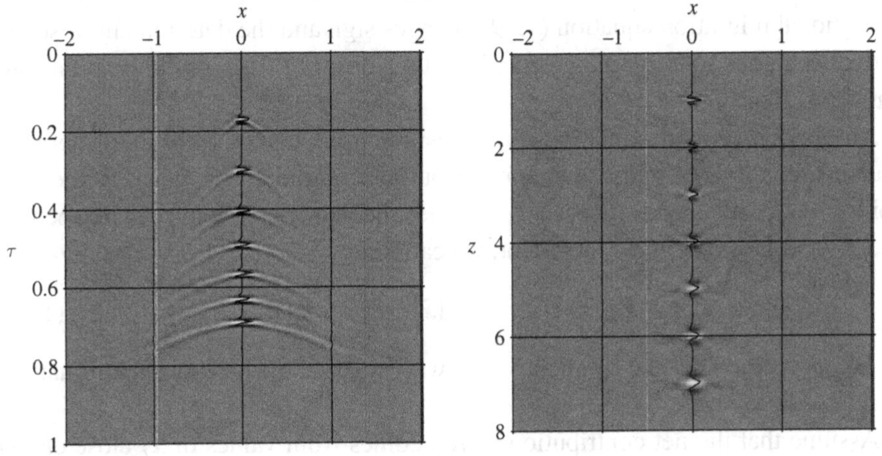

Figure 12.3 On the left is a time migration of the reflecting point data, resulting in the partial collapse of the diffraction curves seen in the left panel of Figure 12.1. The right panel is a residual migration of the left panel, which completes the collapse of the diffractions into (bandlimited) points.

($c(z) = 5 + 1.875z$), resulting in the sequence of diffractions shown in the left panel of Figure 12.1.

The traveltime function needed to perform a depth migration of this data is shown in Figure 12.2a. (Note that the function is truncated at a maximum offset which generally increases with depth. Past the truncation point, data stretch becomes too large for the algorithm to handle. This effect is related to the normal moveout stretch found in NMO-stack.) The results of such a depth migration are shown in the right panel of Figure 12.1. Suppose, however, that initial processing included, instead of a depth migration, an rms-velocity time migration of the data, employing the rms traveltime function shown in Figure 12.2b, and resulting in the time-migrated data as shown in the left panel of Figure 12.3. We see in this panel that the diffractions are compressed but not completely collapsed. To convert the time migration into a depth migration, we computed, using equation (12.11), a residual traveltime operator, shown in Figure 12.2c. Migration with this residual traveltime function results in the complete collapse of the diffractions, as seen in the right panel of Figure 12.3.

12.2 Transition zones

For complicated traveltime functions, one could expect regions where the residual operator is migration-like, and regions where it may be modeling-like. Managing the transition between these regions could be sticky, since at the transition point the residual migration equation (12.9) changes sign and the determinant of second derivatives goes to zero, suggesting that the form (12.9) is not appropriate in a transition zone.

An alternative approximation in a transition zone comes from the generalized Radon transform. Suppose that for the output coordinate $\left(\tilde{\boldsymbol{x}}_f, z_f\right)$, at the input point $\tilde{\boldsymbol{x}}_i = \tilde{\boldsymbol{x}}_e$, the depth function ζ_i assumes the value z_e, where z_e is independent of $\tilde{\boldsymbol{x}}$. For values of $\tilde{\boldsymbol{x}}_i$ near this point, we can write

$$\zeta_{\mathrm{i}}\left(\tilde{\boldsymbol{x}}_i, \tilde{\boldsymbol{x}}, \tau_f\left(\tilde{\boldsymbol{x}}, \tilde{\boldsymbol{x}}_f, z_f\right)\right) \simeq z_e + \left(\tilde{\boldsymbol{x}}_i - \tilde{\boldsymbol{x}}_e\right) \cdot \lambda\left(\tilde{\boldsymbol{x}}, \tilde{\boldsymbol{x}}_f, z_f\right). \qquad (12.12)$$

The n-vector $\boldsymbol{\lambda}$ is the gradient of ζ_i with respect to the input coordinate $\tilde{\boldsymbol{x}}_i$, evaluated at $\tilde{\boldsymbol{x}}_i = \tilde{\boldsymbol{x}}_e$.

Assume that the net contribution to $\mathcal{R}_f$ comes from values of $\tilde{\boldsymbol{x}}_i$ close enough to $\tilde{\boldsymbol{x}}_e$ that (12.12) holds. Then, equation (12.6) becomes

$$\begin{aligned}\mathcal{R}_f\left(\tilde{\boldsymbol{x}}_f, z_f\right) = &-\int d\tilde{\boldsymbol{x}}_i \int d\boldsymbol{x}\, A_f\left(\tilde{\boldsymbol{x}}, \tilde{\boldsymbol{x}}_f, z_f\right) \\ &\times B_{\mathrm{i}}\left(\tilde{\boldsymbol{x}}_i, \tilde{\boldsymbol{x}}, \tau_f\left(\tilde{\boldsymbol{x}}, \tilde{\boldsymbol{x}}_f, z_f\right)\right) \dot{\zeta}_{\mathrm{i}}\left(\tilde{\boldsymbol{x}}_i, \tilde{\boldsymbol{x}}, \tau_f\left(\tilde{\boldsymbol{x}}, \tilde{\boldsymbol{x}}_f, z_f\right)\right)^{n/2} \\ &\times \rho_0{}^n * \mathcal{R}_i\left(\tilde{\boldsymbol{x}}_i, z_e + \left(\tilde{\boldsymbol{x}}_i - \tilde{\boldsymbol{x}}_e\right) \cdot \lambda\left(\tilde{\boldsymbol{x}}, \tilde{\boldsymbol{x}}_f, z_f\right)\right). \end{aligned} \qquad (12.13)$$

Interchanging the order of integration, and replacing $\tilde{\boldsymbol{x}}_i$ with $\tilde{\boldsymbol{x}}_e$ in the slowly varying amplitude terms, the integral over $\tilde{\boldsymbol{x}}_i$ is recognizable as a double inverse Radon transform. Denoting

$$\mathcal{R}_i(\boldsymbol{p}, z) \equiv \frac{1}{(2\pi)^n} \int d\tilde{\boldsymbol{x}}_i \rho_0{}^n * \mathcal{R}_i\left(\tilde{\boldsymbol{x}}, z + \boldsymbol{p} \cdot \tilde{\boldsymbol{x}}\right), \qquad (12.14)$$

equation (12.13) can be written

$$\begin{aligned}\mathcal{R}_f\left(\tilde{\boldsymbol{x}}_f, z_f\right) &= (2\pi)^n \int\int d\tilde{\boldsymbol{x}} A_f\left(\tilde{\boldsymbol{x}}, \tilde{\boldsymbol{x}}_f, z_f\right) \\ &\times B_i\left(\tilde{\boldsymbol{x}}, \tilde{\boldsymbol{x}}_e, \tau_f\left(\tilde{\boldsymbol{x}}, \tilde{\boldsymbol{x}}_f, z_f\right)\right) \dot{\zeta}_i\left(\tilde{\boldsymbol{x}}, \tilde{\boldsymbol{x}}_e, \tau_f\left(\tilde{\boldsymbol{x}}, \tilde{\boldsymbol{x}}_f, z_f\right)\right)^{n/2} \\ &\times \mathcal{R}_i\left(\boldsymbol{\lambda}, z_e - \tilde{\boldsymbol{x}}_e \cdot \lambda\left(\tilde{\boldsymbol{x}}, \tilde{\boldsymbol{x}}_f, z_f\right)\right). \end{aligned} \tag{12.15}$$

The next step is to change integration variable from $\tilde{\boldsymbol{x}}$ to $\boldsymbol{\lambda}$. Provided that $\boldsymbol{\lambda}$ is monotonic in x, the integral remains single-valued, and

$$\begin{aligned}\mathcal{R}_f\left(\tilde{\boldsymbol{x}}_f, z_f\right) = &(2\pi)^n \int d\boldsymbol{\lambda} \boldsymbol{\mathcal{R}}_i\left(\boldsymbol{\lambda}, z_e - \boldsymbol{x}_e \cdot \lambda\right) A_f\left(\tilde{\boldsymbol{x}}, \tilde{\boldsymbol{x}}_f, z_f\right) \\ &\times B_i\left(\tilde{\boldsymbol{x}}, \tilde{\boldsymbol{x}}_e, \tau_f\left(\tilde{\boldsymbol{x}}, \tilde{\boldsymbol{x}}_f, z_f\right)\right) \frac{\dot{\zeta}_i\left(\tilde{\boldsymbol{x}}, \tilde{\boldsymbol{x}}_e, \tau_f\left(\tilde{\boldsymbol{x}}, \tilde{\boldsymbol{x}}_f, z_f\right)\right)^{n/2}}{\left|J_\lambda\left(\tilde{\boldsymbol{x}}, \tilde{\boldsymbol{x}}_f, z_f\right)\right|}, \end{aligned} \tag{12.16}$$

where, for $n = 2$,

$$J_\lambda\left(\tilde{\boldsymbol{x}}, \tilde{\boldsymbol{x}}_f, z_f\right) = \operatorname{Det}\begin{pmatrix} \frac{\partial\lambda_x}{\partial x} & \frac{\partial\lambda_x}{\partial y} \\ \frac{\partial\lambda_y}{\partial x} & \frac{\partial\lambda_y}{\partial y} \end{pmatrix}. \tag{12.17}$$

For $n = 1$, J_λ is just the derivative of λ with respect to x.

Exercises

12.1 Develop a 3-D residual exploding-reflector, constant-velocity Kirchhoff time-migration operator which carries an initial velocity c_i to a final velocity c_f.

13
Asymptotic data mapping and continuation

13.1 Combining forward and inverse migration

Data mapping is an operation complementary to residual migration. Restricting initially to single-valued modeling and migration, for data mapping one has an input and an output data set, related by a single migrated image (Stolt, 2002; Alkhalifah & Bagaini, 2006; Schleicher *et al.*, 2007; Ramirez & Weglein, 2009):

$$\mathcal{R}(\tilde{\boldsymbol{x}}, z) = \int d\tilde{\boldsymbol{x}}_i A_i(\tilde{\boldsymbol{x}}_i, \tilde{\boldsymbol{x}}, z)\, d_{h-}{}^n * D_i(\tilde{\boldsymbol{x}}_i, \tau_i(\tilde{\boldsymbol{x}}_i, \tilde{\boldsymbol{x}}, z)), \tag{13.1}$$

and

$$D_f(\tilde{\boldsymbol{x}}_f, t_f) = \int d\tilde{\boldsymbol{x}} B_f(\tilde{\boldsymbol{x}}, \tilde{\boldsymbol{x}}_f, t_f)\, d_{h+}{}^n * \mathcal{R}(\tilde{\boldsymbol{x}}, \zeta_f(\tilde{\boldsymbol{x}}, \tilde{\boldsymbol{x}}_f, t_f)). \tag{13.2}$$

In these formulas, n is one less than the dimensionality of the system. The coordinates $\tilde{\boldsymbol{x}}$, $\tilde{\boldsymbol{x}}_i$, and $\tilde{\boldsymbol{x}}_f$ are n-vectors, and $d_{h\pm}{}^n$ are the half-derivative operators (see Appendix E) applied n times to the last (time or depth) coordinate of the data or image function.

D_i and D_f represent data at different physical locations. They may represent data at different offsets, different source locations, different receiver locations, or different depths. For both data sets, the underlying velocity structure should be the same, though the traveltime functions τ_i and τ_f (and depth functions ζ_i and ζ_f) are different.

One develops a continuation operator analogously to residual migration. Substituting the migration equation (13.1) into the modeling equation (13.2),

$$\begin{aligned} D_f(\tilde{\boldsymbol{x}}_f, t_f) = \int d\tilde{\boldsymbol{x}} B_f(\tilde{\boldsymbol{x}}, \tilde{\boldsymbol{x}}_f, t_f)\, d_{h+}{}^n * \int d\tilde{\boldsymbol{x}}_i A_i(\tilde{\boldsymbol{x}}_i, \tilde{\boldsymbol{x}}, \zeta_f(\tilde{\boldsymbol{x}}, \tilde{\boldsymbol{x}}_f, t_f)) \\ \times d_{h-}{}^n * D_i(\tilde{\boldsymbol{x}}_i, \tau_i(\tilde{\boldsymbol{x}}_i, \tilde{\boldsymbol{x}}, \zeta_f(\tilde{\boldsymbol{x}}, \tilde{\boldsymbol{x}}_f, t_f))). \end{aligned} \tag{13.3}$$

The filter $d_{h+}{}^n$ in this formula operates on the depth coordinate ζ_f. Invoking its scaling property, this filter can be brought inside the $\tilde{\boldsymbol{x}}_i$ integral and converted to a filter operating on the time coordinate τ_i of D_i. Changing the order of integration, this results in

$$D_f\left(\tilde{\boldsymbol{x}}_f, t_f\right) = \int d\tilde{\boldsymbol{x}}_i \int d\tilde{\boldsymbol{x}} B_f\left(\tilde{\boldsymbol{x}}, \tilde{\boldsymbol{x}}_f, t_f\right) A_i\left(\tilde{\boldsymbol{x}}_i, \tilde{\boldsymbol{x}}, \zeta_f\left(\tilde{\boldsymbol{x}}, \tilde{\boldsymbol{x}}_f, t_f\right)\right) \tag{13.4}$$
$$\times \dot{\tau}_i\left(\tilde{\boldsymbol{x}}_i, \tilde{\boldsymbol{x}}, \zeta_f\left(\tilde{\boldsymbol{x}}, \tilde{\boldsymbol{x}}_f, t_f\right)\right)^{n/2} d_{h+}{}^n * d_{h-}{}^n * D_i\left(\tilde{\boldsymbol{x}}_i, \tau_i\left(\tilde{\boldsymbol{x}}_i, \tilde{\boldsymbol{x}}, \zeta_f\left(\tilde{\boldsymbol{x}}, \tilde{\boldsymbol{x}}_f, t_f\right)\right)\right).$$

The integral over $\tilde{\boldsymbol{x}}$ is an expression of the diffraction integral (G.1), with the correspondence

$$\begin{aligned}
s &\leftrightarrow \tilde{\boldsymbol{x}}, \\
\rho &\leftrightarrow t_f, \\
I &\leftrightarrow d_{h-}{}^n * D_i, \\
C &\leftrightarrow B_f A_i \dot{\tau}_i{}^{n/2}, \\
f &\leftrightarrow d_{h+}{}^n, \\
\gamma(\rho, \boldsymbol{x}) &\leftrightarrow \tau_i\left(\tilde{\boldsymbol{x}}_i, \tilde{\boldsymbol{x}}, \zeta_f\left(\tilde{\boldsymbol{x}}, \tilde{\boldsymbol{x}}_f, t_f\right)\right).
\end{aligned} \tag{13.5}$$

The $\tilde{\boldsymbol{x}}$-integral in (13.4) can be evaluated as per Appendix G. The result is one or more events, depending upon the detailed behavior of the trajectory function $\tau_i\left(\tilde{\boldsymbol{x}}_i, \tilde{\boldsymbol{x}}, \zeta_f\left(\tilde{\boldsymbol{x}}, \tilde{\boldsymbol{x}}_f, t_f\right)\right)$ with respect to $\tilde{\boldsymbol{x}}$. Minima, maxima, saddle points, and inflection points in τ_i all build events through constructive interference. The general result takes the form (G.4), or

$$D_f\left(\tilde{\boldsymbol{x}}_f, t_f\right) \simeq \sum_e \int d\tilde{\boldsymbol{x}}_i \frac{C_e\left(\tilde{\boldsymbol{x}}_i, \tilde{\boldsymbol{x}}_f, t_f\right)}{\mathcal{F}_e\left(\tilde{\boldsymbol{x}}_i, \tilde{\boldsymbol{x}}_f, t_f\right)} f_e * D_i\left(\tilde{\boldsymbol{x}}_i, \tau_e\left(\tilde{\boldsymbol{x}}_i, \tilde{\boldsymbol{x}}_f, t_f\right)\right), \tag{13.6}$$

with $C_e = B_f A_i \dot{\tau}_i{}^{n/2}$and $\tau_e = \tau_i$, all evaluated at the event location $\tilde{\boldsymbol{x}} = \tilde{\boldsymbol{x}}_e$. $\mathcal{F}_e$ is an amplitude and f_e a time filter, determined by the type and other features of the event as discussed in Appendix G.

In the simplest case, one would expect a single event prompted by an extremum in τ_i. The type of extremal point one encounters will depend on circumstances. For vertical continuation, where one extrapolates a wavefield from one depth to another, one would expect in the simplest case for τ_i to go through a minimum. For azimuthal continuation, which maps data of one offset azimuth onto data of another offset azimuth, one would expect a highly asymmetric traveltime function which could well result in a saddle point.

Suppose τ_i goes through an extremal point with respect to $\tilde{\boldsymbol{x}}$ at the point $\tilde{\boldsymbol{x}}_e = \tilde{\boldsymbol{x}}_e\left(\tilde{\boldsymbol{x}}_i, \tilde{\boldsymbol{x}}_f, t_f\right)$. If $\operatorname{Det}\left(\tau_i''\right)$ is the determinant of the matrix of second derivatives of τ_i with respect to x and y at $(x, y) = \tilde{\boldsymbol{x}}_e$, then according to Appendix G,

$$D_f\left(\tilde{\boldsymbol{x}}_f, t_f\right) = \int d\tilde{\boldsymbol{x}}_i \frac{C_e\left(\tilde{\boldsymbol{x}}_i, \tilde{\boldsymbol{x}}_f, t_f\right)}{\sqrt{\left|\operatorname{Det}\left(\tau_i''\right)\right|}} f_e * D_i\left(\tilde{\boldsymbol{x}}_i, \tau_e\right), \tag{13.7}$$

where, if

$$z_e = z_e\left(\tilde{\boldsymbol{x}}_i, \tilde{\boldsymbol{x}}_f, t_f\right) = \zeta_f\left(\tilde{\boldsymbol{x}}_e, \tilde{\boldsymbol{x}}_f, t_f\right), \tag{13.8}$$

then

$$\tau_e = \tau_e\left(\tilde{\boldsymbol{x}}_i, \tilde{\boldsymbol{x}}_f, t_f\right) = \tau_i\left(\tilde{\boldsymbol{x}}_i, \tilde{\boldsymbol{x}}_e, z_e\right), \tag{13.9}$$

$$C_e\left(\tilde{\boldsymbol{x}}_i, \tilde{\boldsymbol{x}}_f, t_f\right) = B_f\left(\tilde{\boldsymbol{x}}_e, \tilde{\boldsymbol{x}}_f, t_f\right) A_i\left(\tilde{\boldsymbol{x}}_i, \tilde{\boldsymbol{x}}_e, z_e\right) \dot{\tau}_i\left(\tilde{\boldsymbol{x}}_i, \tilde{\boldsymbol{x}}_e, z_e\right)^{n/2}, \tag{13.10}$$

and

$$\begin{aligned} f_e(t) &= d_{h+}{}^n(t), \text{ at a minimum in } \tau_i, \\ &= d_{h-}{}^n(t), \text{ at a maximum in } \tau_i, \\ &= \rho(t), \text{ at a 3-D saddle point in } \tau_i. \end{aligned} \tag{13.11}$$

In two dimensions, Det $\left(\tau_i''\right)$ reduces to a single second derivative, and the filter f_e is the half-derivative or its adjoint. In three dimensions, Det $\left(\tau_i''\right)$ is the determinant of the 2×2 matrix of second derivatives, and f_e is plus or minus a full derivative in the case of a full minimum or maximum, or the "rho" filter in the case of a saddle point.

13.2 Constant-velocity depth extrapolation in 2-D

Suppose that D_i and D_f represent 2-D constant-velocity wavefields, measured at depths z_i and z_f respectively. Equation (13.1) represents migration of the data D_i at z_i to an image at points (x,z), while equation (13.2) reformulates the data at the new depth z_f. The amplitude functions A_i and B_f depend on the propagation–reflection model, to be specified later. The traveltime function for the first operation (13.1) is in the constant-velocity case

$$\tau_i\left(x_i, z_i, x, z\right) = \frac{2}{c}\sqrt{\left(x - x_i\right)^2 + \left(z - z_i\right)^2}, \tag{13.12}$$

as illustrated in Figure 2.25. We have added an extra parameter z_i to the argument list of τ_i to remind ourselves that the input data surface is $z = z_i$. Since z_i remains constant within the mapping (13.1), it is not, strictly, a variable, but since the mapping formula depends on its value, it does no harm to treat it as such.

The depth function in (13.2) is

$$\zeta_f\left(x, x_f, z_f, t_f\right) = z_f + \sqrt{\frac{c^2 t_f{}^2}{4} - \left(x - x_f\right)^2}, \tag{13.13}$$

corresponding to the mapping operation also plotted in Figure 2.25. We have added z_f to the argument list of ζ_f because that is the depth of the output data surface.

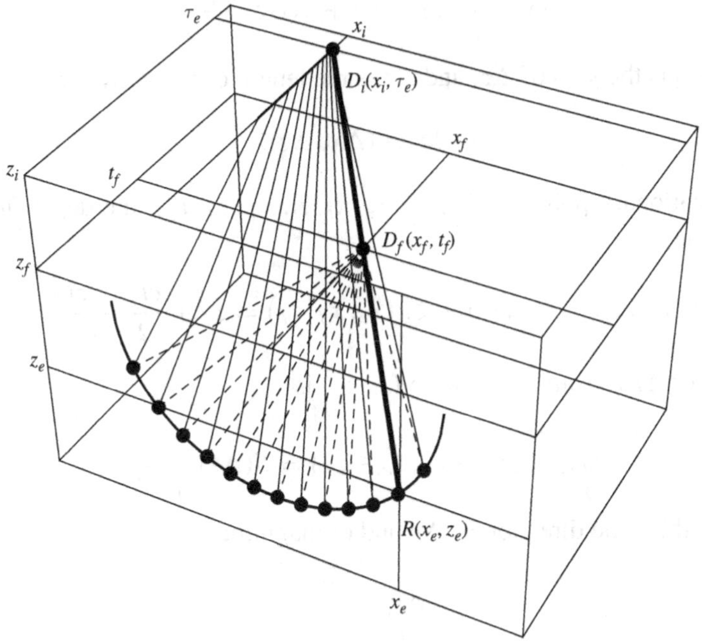

Figure 13.1 The composite mapping (13.4) from z_i to z_f in a constant-velocity 2-D medium. In this illustration, $z_f > z_i$. The input data plane $z = z_i$ is the upper horizontal surface. The output data plane $z = z_f$ is the horizontal surface interior to the volume, and the reflectivity image plane is the front vertical surface. For a fixed x_f, t_f, and x_i, there is on the input data plane a range of contributing τ_i, varying with the integration variable x in (13.4). For $z_f > z_i$, τ_i goes through a maximum τ_e at the image point $\left(x, \zeta_f, 0\right) = (x_e, z_e, 0)$, which lies on an extension of the line (shown here as a wide black line) from (x_i, z_i, τ_e) to $\left(x_f, z_f, t_f\right)$.

The composite mapping operation (13.4) is shown in Figure 13.1. The operation is a double integral representing two mappings, the first mapping from input data D_i to an image function R, and the second mapping from R to the output data D_f. The composite trajectory function appearing in (13.4) is

$$\tau_i\left(x_i, z_i, x, \zeta_f\left(x, x_f, z_f, t_f\right)\right)$$
$$= \frac{2}{c}\sqrt{\left(x - x_f + \Delta x\right)^2 + \left(\Delta z + \sqrt{c^2\, t_f{}^2/4 - \left(x - x_f\right)^2}\right)^2}, \quad (13.14)$$

with $\Delta x = x_f - x_i$ and $\Delta z = z_f - z_i$. As illustrated in Figure 13.1, for a given x_i, x_f, and t_f, as one integrates over x, the traveltime τ_i varies, assuming an extremal value τ_e at

$$x_e\left(x_i, x_f, t_f\right) = x_f + S(\Delta z)\frac{ct_f}{2}\frac{\Delta x}{\Delta r}, \tag{13.15}$$

where $S(\Delta z)$ is the sign of Δz, and Δr is the length of the vector

$$\Delta r \equiv (\Delta x, \Delta z). \tag{13.16}$$

At the stationary point $x = x_e = x_e\left(x_i, z_i, x_f, z_f, t_f\right)$, the depth function ζ_f has the value

$$z_e{=}z_e\left(x_i, z_i, x_f, z_f, t_f\right) \equiv \zeta_f\left(x_e, x_f, t_f\right) = z_f + \frac{ct_f}{2}\frac{|\Delta z|}{\Delta r}. \tag{13.17}$$

The vector $\Delta \boldsymbol{r}_{ef}$ from $\left(x_f, z_f\right)$ to (x_e, z_e) is

$$\Delta r_{ef} \equiv \left(x_e - x_f, z_e - z_f\right) = S(\Delta z)\frac{ct_f}{2}\frac{\Delta r}{\Delta r}, \tag{13.18}$$

which is in the same direction as $\Delta \boldsymbol{r}$ and of magnitude

$$\Delta r_{ef} = \frac{ct_f}{2}. \tag{13.19}$$

The traveltime function τ_i becomes at the extremal image point

$$\tau_e = \tau_e\left(x_i, z_i, x_f, z_f, t_f\right) = t_f + S(\Delta z)\frac{2}{c}\Delta r, \tag{13.20}$$

which is just the sum (or, if $z_f < z_i$, the difference) of the output traveltime t_f and the traveltime between the input and output data points. For $z_f > z_i$, τ_e is a maximum. For $z_f < z_i$, τ_e is a minimum.

The three points (x_i, z_i, τ_e), $\left(x_f, z_f, t_f\right)$, and $(x_e, z_e, 0)$ lie on the same line, as depicted in Figure 13.1. This means, among other things, that the distance ratios are equal:

$$\frac{x_e - x_i}{z_e - z_i} = \frac{x_e - x_f}{z_e - z_f} = \frac{x_f - x_i}{z_f - z_i} = \frac{\Delta x}{\Delta z} \equiv \tan\alpha. \tag{13.21}$$

Equation (13.7) gives the effective mapping operator from z_i to z_f in terms of the extremal image point. In this case, (13.7) has the 2-D form (see Figure 13.2)

$$D_f\left(x_f, t_f\right) = \int dx_i \sqrt{\frac{\dot{\tau}_i\left(x_i, x_e, z_e\right)}{\left|\tau_{ixx}\right|_{x\to x_e}}} B_f\left(x_e, x_f, z_f, t_f\right)\, A_i\left(x_i, z_i, x_e, z_e\right) \\ \times \left[d_{h\mp} * D_i\right]\left(x_i, \tau_e\right). \tag{13.22}$$

The input data in this formula are filtered either by the half-derivative operator d_{h+} or its adjoint d_{h-}, depending on whether z_f is less than or greater than z_i. To

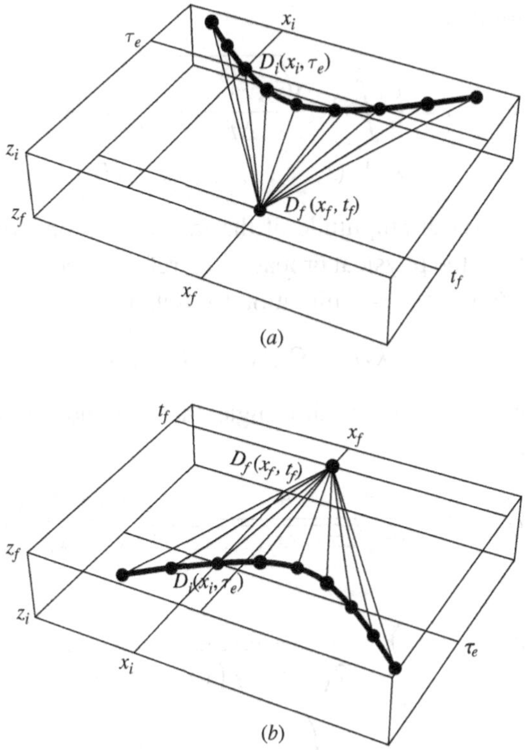

Figure 13.2 The direct mapping (13.22) of data from z_i to z_f in a constant-velocity 2-D medium. The input data plane $z = z_i$ is the upper (lower) horizontal surface in panel (a) (panel (b)). The output data plane $z = z_f$ is the lower (upper) horizontal surface in panel (a) (panel (b)). For the depicted output point $D_f\left(x_f, t_f\right)$, input data is summed along the heavy black trajectory. For $z_f > z_i$ (panel (a)), the transformation maps from later times to earlier times, and is accompanied by a half-derivative adjoint filter d_{h-}. For $z_f < z_i$ (panel (b)), the transformation maps earlier times onto later times, and includes a half-derivative filter d_{h+}.

calculate the amplitude term in (13.7) one needs the second x-derivative and the first depth derivative of τ_i at the extremal point. The second x-derivative is

$$\tau_{ixx}|_{x \to x_e} = -\frac{8\Delta r^3}{c^3 \Delta z^2 t_f \tau_e}. \tag{13.23}$$

The depth derivative of τ_i at the extremal image point is

$$\dot{\tau}_i\left(x_i, x_e, z_e\right) = \frac{2}{c}\frac{|\Delta z|}{\Delta r}. \tag{13.24}$$

Thus, from equation (13.22),

$$D_f\left(x_f, t_f\right) = \frac{c}{2}\int dx_i \frac{\sqrt{|\Delta z|^3 t_f \tau_e}}{\Delta r^2} B_f\left(x_e, x_f, z_f, t_f\right) \\ \times A_i\left(x_i, z_i, x_e, z_e\right)\left[d_{h\mp} * D_i\right]\left(x_i, \tau_e\right). \tag{13.25}$$

To determine the overall amplitude of the transformation, we must specify A_i and B_f, which is to say the physical propagation–reflection model. Borrowing from Section 2.6.5 and 2.6.6, if the *f–k* migration formula is

$$R\left(k_x, k_z\right) = \Omega\left(k_x / k_z\right) D\left(k_x, \omega\right), \tag{13.26}$$

with Ω a function of tangent of dip angle that depends on the propagation–reflection model, then

$$A_i\left(x_i, z_i, x, z\right) = \frac{4}{c^2\sqrt{2\pi \tau_i\left(x_i, z_i, x, z\right)}}\Omega\left(\frac{x - x_i}{z - z_i}\right), \tag{13.27}$$

and

$$B_f\left(x, x_f, z_f, t_f\right) = \frac{c}{2\sqrt{2\pi\left(\zeta_f\left(x, x_f, z_f, t_f\right) - z_f\right)}}\Omega^{-1} \\ \times\left(\frac{x - x_f}{\zeta_f\left(x, x_f, z_f, t_f\right) - z_f}\right). \tag{13.28}$$

Thus,

$$A_i\left(x_i, z_i, x_e, z_e\right) = \frac{4}{c^2\sqrt{2\pi \tau_e}}\Omega\left(\frac{x_e - x_i}{z_e - z_i}\right), \tag{13.29}$$

$$B_f\left(x_e, x_f, z_f, t_f\right) = \frac{c}{2\sqrt{2\pi\left(z_e - z_f\right)}}\Omega^{-1}\left(\frac{x_e - x_f}{z_e - z_f}\right) \\ = \frac{\sqrt{c\Delta r}}{2\sqrt{\pi t_f|\Delta z|}}\Omega^{-1}\left(\frac{x_e - x_f}{z_e - z_f}\right). \tag{13.30}$$

But we saw above that $\left(x_e - x_i\right) / \left(z_e - z_i\right) = \left(x_e - x_f\right) / \left(z_e - z_f\right)$, so, independently of our choice of Ω,

$$A_i\left(x_i, z_i, x_e, z_e\right) B_f\left(x_e, x_f, z_f, t_f\right) = \frac{1}{\pi}\sqrt{\frac{2\Delta r}{c^3 \tau_e t_f |\Delta z|}}, \tag{13.31}$$

and

$$D_f\left(x_f, t_f\right) = \frac{|\Delta z|}{2\pi}\sqrt{\frac{2}{c}}\int \frac{dx_i}{\Delta r^{3/2}}\left[d_{h\mp} * D_i\right]\left(x_i, t_f + 2S(\Delta z)\Delta r / c\right). \tag{13.32}$$

In three dimensions, data extrapolation from z_i to z_f becomes

$$D_f\left(\tilde{\boldsymbol{x}}_f, t_f\right) = \frac{|\Delta z|}{2\pi}\frac{2}{c}\int d\tilde{\boldsymbol{x}}_i \frac{\dot{D}_i\left(\tilde{\boldsymbol{x}}_i, t_f + 2S(\Delta z)\Delta r/c\right)}{\Delta r^2}, \tag{13.33}$$

where $\Delta r = \sqrt{\left(x_f - x_i\right)^2 + \left(y_f - y_i\right)^2 + \left(z_f - z_i\right)^2}$.

13.3 MZO and DMO

Migration to zero offset MZO can be considered a constant-offset migration followed by a zero-offset modeling. Working in two dimensions, the imaging and modeling operators are

$$\mathcal{R}(x, z) = -\int dx_i \dot{D}_i\left(x_i, \tau_i\left(x_i, x, z\right)\right) A_i\left(x_i, x, z\right), \tag{13.34}$$

and

$$D_0\left(x_0, t_0\right) = \int dx \dot{\mathcal{R}}\left(x, \zeta_0\left(\left(x, x_0, t_0\right)\right)\right) B_0\left(x, x_0, t_0\right), \tag{13.35}$$

with (if the initial half-offset is x_h) traveltime and depth functions

$$\tau_i\left(x_i, x, z\right) = \frac{r}{c} = \frac{r_g + r_s}{c}, \tag{13.36}$$

$$r_g = \sqrt{z^2 + \left(x - x_i + x_h\right)^2}, \qquad r_s = \sqrt{z^2 + \left(x - x_i - x_h\right)^2}, \tag{13.37}$$

$$\zeta_0\left(x, x_0, t_0\right) = \sqrt{\frac{c^2 t_0^2}{4} - \left(x - x_0\right)^2}. \tag{13.38}$$

The resulting composite operator (13.4) is

$$\begin{aligned} D_0\left(x_0, t_0\right) = & \int dx_i \int dx B_0\left(x, x_0, t_0\right) A_i\left(x_i, x, \zeta_0\left(x, x_0, t_0\right)\right) \\ & \times \tau_i\left(x_i, x, \zeta_0\left(x, x_0, t_0\right)\right)^{1/2}\left[d_{h+} * d_{h-} * D_i\right]\left(x_i, \tau_i\left(x_i, x, \zeta_0\left(x, x_0, t_0\right)\right)\right). \end{aligned} \tag{13.39}$$

The composite traveltime function is

$$\begin{aligned} \tau_i\left(x_i, x, \zeta_0\left(x, x_0, t_0\right)\right) = & \frac{1}{c}\sqrt{c^2 t_0^2/4 - \left(x - x_0\right)^2 + \left(x - x_i + x_h\right)^2} \\ & + \frac{1}{c}\sqrt{c^2 t_0^2/4 - \left(x - x_0\right)^2 + \left(x - x_i - x_h\right)^2}. \end{aligned} \tag{13.40}$$

We look for an extremal value of this function with respect to x. The x-derivative of this function is

$$\partial_x \tau_i \left(x_i, x, \zeta_0 \left(x, x_0, t_0\right)\right) = \frac{1}{c} \left(\frac{x_0 - x_i + x_h}{r_g} + \frac{x_0 - x_i - x_h}{r_s} \right). \tag{13.41}$$

This derivative can be zero only if $x_0 - x_i + x_h$ has opposite sign to $x_0 - x_i - x_h$. This requires that $|x_0 - x_i| < |x_h|$. With this constraint, the value of x at which the derivative is zero is

$$x \to x_e = x_0 + \frac{c^2 t^2}{4} \frac{x_0 - x_i}{{x_h}^2 - \left(x_0 - x_i\right)^2}. \tag{13.42}$$

At this point,

$$r_g = \sqrt{\left|\frac{x_h - x_i + x_0}{x_h + x_i - x_0}\right| \left(\frac{c^2 {t_0}^2}{4} + {x_h}^2 - \left(x_0 - x_i\right)^2\right)},$$

$$r_s = \sqrt{\left|\frac{x_h + x_i - x_0}{x_h - x_i + x_0}\right| \left(\frac{c^2 {t_0}^2}{4} + {x_h}^2 - \left(x_0 - x_i\right)^2\right)}, \tag{13.43}$$

the derivative of traveltime is zero:

$$\begin{aligned} &\partial_x \tau_i \left(x_i, x_e, \zeta_0 \left(x_e, x_0, t_0\right)\right) = \\ &\left(\frac{x_0 - x_i + x_h}{|x_h - x_i + x_0|} + \frac{x_0 - x_i - x_h}{|x_h + x_i - x_0|} \right) \times \frac{\sqrt{{x_h}^2 - \left(x_0 - x_i\right)^2}}{c\sqrt{c^2\, {t_0}^2 / 4 + {x_h}^2 - \left(x_0 - x_i\right)^2}} \\ &= 0, \end{aligned} \tag{13.44}$$

while traveltime assumes the form

$$\begin{aligned} \tau_e \left(x_i, x_0, t_0\right) &\equiv \tau_i \left(x_i, x_e, \zeta_0 \left(x_e, x_0, t_0\right)\right) \\ &= \frac{|x_h|}{\sqrt{{x_h}^2 - \left(x_0 - x_i\right)^2}} \sqrt{{t_0}^2 + \frac{4}{c^2} \left({x_h}^2 - \left(x_0 - x_i\right)^2\right)}. \end{aligned} \tag{13.45}$$

The second derivative of traveltime is

$$\partial^2{}_x \tau_i \left(x_i, x, \zeta_0 \left(x, x_0, t_0\right)\right) = -\frac{1}{c} \left(\frac{\left(x_0 - x_i + x_h\right)^2}{{r_g}^3} + \frac{\left(x_0 - x_i - x_h\right)^2}{{r_s}^3} \right), \tag{13.46}$$

becoming, at the extremal point,

$$\begin{aligned} \tau_e'' \left(x_i, x_0, t_0\right) &\equiv \partial^2{}_x \tau_i \left(x_i, x_e, \zeta_0 \left(x_e, x_0, t_0\right)\right) \\ &= -\frac{2\,|x_h| \sqrt{{x_h}^2 - \left(x_0 - x_i\right)^2}}{c\left(c^2\, {t_0}^2/4 + {x_h}^2 - \left(x_0 - x_i\right)^2\right)^{3/2}}. \end{aligned} \tag{13.47}$$

The second derivative is negative, indicating a maximum in traveltime at the extremal point. Borrowing formula (13.7),

$$D_0(x_0, t_0) = \int dx_i \frac{C_e(x_i, x_0, t_0)}{\sqrt{|\tau_e''|}} d_{h-} * D_i(x_i, \tau_e(x_i, x_0, t_0)), \tag{13.48}$$

with

$$C_e(x_i, x_0, t_0) = B_0(x_e, x_0, t_0)\, A_i\left(x_i, x_e, \zeta_f(x_e, x_0, t_0)\right) \times \dot{\tau}_i(x_i, x_e, \zeta_0(x_e, x_0, t_0))^{1/2}. \tag{13.49}$$

For DMO, the initial input is moveout-corrected data. That is, define

$$D_{\mathrm{NMO}}(x_i, t_{\mathrm{NMO}}) = D_i(x_i, t_i), \tag{13.50}$$

where

$$t_{\mathrm{NMO}} = \sqrt{t_i^2 - 4x_h^2/c^2}. \tag{13.51}$$

At the point $t_i \to \tau_e$, we have

$$t_{\mathrm{NMO}} \to \sqrt{\tau_e^2 - 4x_h^2/c^2} = t_0 \frac{|x_h|}{\sqrt{x_h^2 - (x_0 - x_i)^2}}. \tag{13.52}$$

This is the familiar velocity-independent DMO mapping trajectory. See, for example, equation 51 of Stolt (2002). In terms of moveout-corrected data, the transformation equation reads

$$D_0(x_0, t_0) = \int dx_i \frac{C_e(x_i, x_0, t_0)}{\sqrt{|\tau_e''|}} \int dt' d_{h-}(\tau_e - t')\, D_{\mathrm{NMO}}\left(x_i, \sqrt{t'^2 - 4x_h^2/c^2}\right). \tag{13.53}$$

In this formula, the filter integral has been written explicitly because the convolution is over moveout uncorrected time. Invoking the placement formula (E.26), and the 3/2 power law (E.45) for the half-derivative filter, the filter can be converted to an operation on moveout corrected time. The result is

$$D_0(x_0, t_0) = \int dx_i \frac{C_e(x_i, x_0, t_0)}{\sqrt{|\tau_e''|}} \sqrt{\frac{\tau_e}{t_{\mathrm{NMO}}}} d_{h-} * D_{\mathrm{NMO}}(x_i, t_{\mathrm{NMO}}). \tag{13.54}$$

Exercises

13.1 Starting from a generalized 2-D f–k migration formula $R(k_x, k_z) = \Omega(k_x, \omega) D(k_x, \omega)$, with $k_z = -\mathrm{Sign}(\omega)\sqrt{4\omega^2/c^2 - k_x^2}$, use stationary phase to find the amplitude coefficients (a) $A_i(x_i, x, z)$ and (b) $B_f(x, x_f, t_f)$ of the corresponding space-time imaging and modeling operations (13.1) and (13.2).

13.2 For a constant-velocity background, derive the extrapolation formula for moving 3-D zero-offset data from depth z_i to depth z_f, using (a) the exploding reflector model (as per equations (11.77) and (11.85)), and (b) a zero-offset model (equations (11.110) and (11.111)).

14

Least-squares asymptotic migration

14.1 Direct versus least-squares inversion

As discussed in Section 2.1, migration presupposes a linear model between data $\mathcal{D}$ and an image model $\mathcal{R}$:

$$\mathcal{D} = \mathfrak{D}\mathcal{R}. \tag{14.1}$$

For some restricted data sets, this equation can be directly inverted to produce a migration equation

$$\mathcal{R} = \mathfrak{D}^{-1}\mathcal{D}. \tag{14.2}$$

However, seismic data are noisy, suggesting that noise should be added to the modeling equation:

$$\mathcal{D} = \mathfrak{D}\mathcal{R} + \mathcal{N}. \tag{14.3}$$

This, plus the high potential redundancy in seismic data suggests a statistical estimation of $\mathcal{R}$. Minimizing a weighted least-squares estimate of noise leads to (Operto *et al.*, 2000; Ronen & Liner, 2000; Lambar *et al.*, 2003; Tang, 2009; Aveni & Biondi, 2010)

$$\mathcal{R} \simeq \left[\mathfrak{D}^{\dagger}\mathcal{W}\mathfrak{D}\right]^{-1}\mathfrak{D}^{\dagger}\mathcal{W}\mathcal{D}, \tag{14.4}$$

with $\mathcal{W}$ a suitable weighting operator. When a unique inverse $\mathfrak{D}^{-1}$ exists, the least-squares inverse (14.4) reduces to it. It is, however, more generally defined, and offers greater flexibility in practice.

To complete the discussion, one should mention a third option, which chooses as an imaging operator the transpose of the forward operator:

$$\mathcal{R} \sim \mathfrak{D}^{\dagger}\mathcal{D}. \tag{14.5}$$

While this operation will generally form an image, it is unfaithful to amplitudes. It is simplest to implement, but is inappropriate for amplitude-preserving processing.

14.2 Multipathing

Multipathing occurs when more than one possible raypath connects two points. For complicated velocity structures, it is easy to see how this could occur. Consider the ensemble of all conceivable paths connecting two points. Starting from a straight line between the two points, for any bounded velocity structure, one can find a path of arbitrarily large traveltime simply by bulging the path enough in any direction. It follows that one of the paths in the ensemble must have the minimum traveltime. In addition, there may be other raypaths where traveltime is a local minimum, maximum, or (in three dimensions) a saddle point. Such extremal paths are the raypaths predicted by Snell's law and computable from the Eikonal equation (6.26).

The situation for transmitted rays is analogous to that for reflected ones (see Section 9.5.1). A traveltime minimum does not add to the KMAH index of the ray. A traveltime saddle point corresponds to a ray which has gone through a focus point or caustic. This adds one to the KMAH index, adding 90° of phase, just like a reflection saddle point. A 2-D traveltime maximum adds two to the KMAH index, changing the sign of the corresponding Green's function, again just like the corresponding reflected wave.

Also analogous to reflected waves, the Snell raypaths do not tell the whole story. Moving from a region where a triplication (min–max–min) exists to a region with a single minimum, ray theory predicts an abrupt transition. In truth, the transition is gradual, with a diffractive event corresponding to a traveltime inflection point replacing the disappearing max–min. Additionally, isolated scattering points produce forward scattering diffractions not predictable from Snell's law and not computable from the Eikonal equation. Boundary terminations (faults, pinchouts, etc.) also produce forward scattering diffractions that appear to emanate from the termination point.

For reflections, migration algorithms need the diffractive events to form a complete image. For transmitted waves, the need is less clear. One can argue that diffracted energy in the transmitted wave contains little information on the location of reflectors, because as source and receiver move, the diffraction point tends to stay relatively fixed, making it difficult to form a reflector image through a diffraction. Regardless of the validity of this argument, purely ray-theoretical migration algorithms as a rule ignore diffracted energy in the transmitted wave while incorporating it in the reflected wave.

Multipathing can usually be sorted out by describing rays not in terms of starting and ending points, but in terms of either starting point and starting direction, or ending point and ending direction. There may be many rays with the same starting and ending point, but the Eikonal equation pretty well ensures that, given a starting point and initial direction, a unique ray can be computed. Thus, provided diffractive energy in the transmitted wave can be ignored, use of an angle-based coordinate system can eliminate many of the problems of multipathing.

14.3 The asymptotic modeling operator in three and two dimensions

14.3.1 Three dimensions

Allowing for multipathing, the general 3-D reflected impulse response (9.143) can be written as

$$D\left(\tilde{\boldsymbol{x}}_g|\tilde{\boldsymbol{x}}_s;\omega\right) \simeq -i\omega\int\!\!\int\!\!\int d^3x \sum_{l=1}^{L} \mathcal{R}(\boldsymbol{x},\sigma_l)\mathfrak{A}_l\left(\tilde{\boldsymbol{x}}_g,\tilde{\boldsymbol{x}}_s|\,\boldsymbol{x}\right) e^{i\omega\tau_l\left(\tilde{\boldsymbol{x}}_g,\tilde{\boldsymbol{x}}_s|\boldsymbol{x}\right)} f_l\left(\tilde{\boldsymbol{x}}_g,\tilde{\boldsymbol{x}}_s|\,\boldsymbol{x},\omega\right), \tag{14.6}$$

where $\mathfrak{A}_l$ is a combined amplitude factor for the two-way raypath from $\tilde{\boldsymbol{x}}_s$ to $\boldsymbol{x}$ to $\tilde{\boldsymbol{x}}_g$ denoted by the index l,

$$\mathfrak{A}_l\left(\tilde{\boldsymbol{x}}_g,\tilde{\boldsymbol{x}}_s|\,\boldsymbol{x}\right) = 2\frac{C_r{}^2\cos\gamma_{rl}}{c_{rl}}\left|g_{il}\left(\boldsymbol{x}\,|\tilde{\boldsymbol{x}}_s\right)g_{rl}\left(\tilde{\boldsymbol{x}}_g|\,\boldsymbol{x}\right)\right|, \tag{14.7}$$

and τ_l is the total traveltime on the same path:

$$\tau_l\left(\tilde{\boldsymbol{x}}_g,\tilde{\boldsymbol{x}}_s|\,\boldsymbol{x}\right) = \tau_{il}\left(\boldsymbol{x}\,|\tilde{\boldsymbol{x}}_s\right) + \tau_{rl}\left(\tilde{\boldsymbol{x}}_g|\,\boldsymbol{x}\right). \tag{14.8}$$

The quantity $f_l(\tilde{\boldsymbol{x}}_g,\tilde{\boldsymbol{x}}_s|\,\boldsymbol{x},\omega)$ is a filter characteristic of the raypath. If only extremal (that is, non-diffractive) paths are considered, then the filter is purely a phase:

$$f_l\left(\tilde{\boldsymbol{x}}_g,\tilde{\boldsymbol{x}}_s|\,\boldsymbol{x},\omega\right) \rightarrow e^{-i\frac{\pi}{2}a_l\left(\tilde{\boldsymbol{x}}_g,\tilde{\boldsymbol{x}}_s|\boldsymbol{x}\right)\mathrm{Sign}(\omega)}, \tag{14.9}$$

where $a_l\left(\tilde{\boldsymbol{x}}_g,\tilde{\boldsymbol{x}}_s|\,\boldsymbol{x}\right)$ is the KMAH index of the two-way raypath. For diffractive raypaths, the filter f_l has amplitude as well as phase.

$\mathcal{R}(\boldsymbol{x},\sigma_l)$ is the image function at point $\boldsymbol{x} = (x,y,z)$ and opening angle σ_l. Though not explicitly written, σ_l depends on coordinates $\tilde{\boldsymbol{x}}_s$, $\tilde{\boldsymbol{x}}_g$, and $\boldsymbol{x}$. $D\left(\tilde{\boldsymbol{x}}_g|\tilde{\boldsymbol{x}}_s;\omega\right)$ is the data function representing a full-bandwidth ($\mathfrak{w} = 1$) impulse for source coordinate $\tilde{\boldsymbol{x}}_s$, receiver coordinate $\tilde{\boldsymbol{x}}_g$, and frequency ω. The number L of paths may be spatially variable.

14.3.2 Two dimensions

Some formulas for migration of 2-D data sets use a 2-D model of wave propagation. Whether applied to a 2-D or a 3-D data set, an amplitude-preserving imaging operator needs to acknowledge that we live in three dimensions. When working with point sources and receivers, one should always use a 3-D Green's function, even if propagation is to be confined to two dimensions. A 2.5-D forward modeling operator differs both in phase and amplitude from 2-D modeling operators.

If $\mathcal{R}(\boldsymbol{x}, \sigma_l)$ depends on x and z only, and if $\tilde{\boldsymbol{x}}_g$ and $\tilde{\boldsymbol{x}}_s$ are confined to the x–z plane, then the third (y) dimension in (14.6) can be removed via stationary phase, leaving (ignoring factors of 2π)

$$\dot{D}\left(\tilde{\boldsymbol{x}}_g|\tilde{\boldsymbol{x}}_s;\omega\right) \simeq \sqrt{-i\omega}\int\int d^2x \sum_{l=1}^{L} \mathcal{R}(\tilde{\boldsymbol{x}}, \sigma_l) \times \left(\frac{\mathfrak{A}_l\left(\tilde{\boldsymbol{x}}_g, \tilde{\boldsymbol{x}}_s|\tilde{\boldsymbol{x}}\right)}{\sqrt{\partial_{yy}\tau_l\left(\tilde{\boldsymbol{x}}_g, \tilde{\boldsymbol{x}}_s|\tilde{\boldsymbol{x}}\right)}}\right) e^{i\omega\tau_l\left(\tilde{\boldsymbol{x}}_g,\tilde{\boldsymbol{x}}_s|\tilde{\boldsymbol{x}}\right)} f_l\left(\tilde{\boldsymbol{x}}_g, \tilde{\boldsymbol{x}}_s \tilde{\boldsymbol{x}}, \omega\right). \quad (14.10)$$

In (14.10) the quantity $\tilde{\boldsymbol{x}}$ is the 2-vector (x, z). Compared with a 2-D modeling equation, the 2.5-D expression differs by the factor $\sqrt{i\omega/\tau_{lyy}}$. Additionally, the Green's function amplitudes within the amplitude function $\mathfrak{A}_l$ in (14.10) satisfy the 3-D rather than 2-D transport equation.

The 2.5-D formula (14.10) requires knowledge of the curvature of the rays in the y-direction at the image point $\boldsymbol{x} = (x, 0, z)$, which is a measure of leakage of energy into the third dimension. Without the curvature information, one cannot preserve relative amplitudes during a migration. At $y = 0$, the second traveltime derivative can be expressed in terms of integrals along the path as

$$\begin{aligned} \partial_{yy}\tau_l\left(\tilde{\boldsymbol{x}}_g, \tilde{\boldsymbol{x}}_s|\boldsymbol{x}\right) &= \partial_{yy}\tau_{il}\left(\boldsymbol{x}\,|\tilde{\boldsymbol{x}}_s\right) + \partial_{yy}\tau_{rl}\left(\tilde{\boldsymbol{x}}_g|\,\boldsymbol{x}\right), \\ &= \frac{1}{\int_0^{\tau_{il}} c_{il}{}^2\, d\tau} + \frac{1}{\int_0^{\tau_{rl}} c_{rl}{}^2\, d\tau}. \end{aligned} \quad (14.11)$$

14.4 Least-squares migration in 2.5-D

To keep things as simple as possible, we will not consider converted waves in what follows. This means that at any point in the subsurface, there is a single associated velocity which we will call c. Traveltime functions will be governed by a single Eikonal equation, which we write as $|\nabla\tau|=1/c$. The extension to converted waves is straightforward.

14.4.1 2.5-D modeling operator with noise

Start with the 2.5-D modeling operator (14.10) and assume, for simplicity, that sources and receivers are at $z = 0$. With the addition of a noise term, we have

$$D\left(x_g|x_s;\omega\right) \simeq \sqrt{-i\omega}\int\int d^2x \sum_{l=1}^{L} \mathcal{R}(\tilde{\boldsymbol{x}},\sigma_l) \tag{14.12}$$
$$\times \left(\frac{\mathfrak{A}_l\left(x_g,x_s|\tilde{\boldsymbol{x}}\right)}{\sqrt{\tau_{lyy}\left(x_g,x_s|\tilde{\boldsymbol{x}}\right)}}\right) e^{i\omega\tau_l(x_g,x_s|\tilde{\boldsymbol{x}})} f_l\left(x_g,x_s|\tilde{\boldsymbol{x}},\omega\right) + \mathcal{N}\left(x_g|x_s,\omega\right).$$

14.4.2 Linearized inverse migration

To find a migration formula, we must invert equation (14.12) for reflectivity. Since reflectivity is a function of opening angle σ_l, and σ_l a function of x_g, x_s, and $\tilde{\boldsymbol{x}}$, equation (14.12) is not quite the simple linear operation it first appears. One way to achieve a linear form is to use a linearized model of reflectivity, i.e. (8.63), or

$$\mathcal{R}(\boldsymbol{x},\sigma) = \sum_i b_i(\boldsymbol{x})\mathbb{B}_i(\boldsymbol{x},\sigma), \tag{14.13}$$

This separates the angle-dependence of reflectivity into a small set $b_i\left(\tilde{\boldsymbol{x}}\right)$ of angle-independent kernels as the unknowns to solve for, each multiplied by a known, angle-dependent coefficient $\mathbb{B}_i(x,\sigma)$. This adds an additional discrete index to the forward and inverse problems. Though straightforward enough, this approach will not be developed here.

14.4.3 Constant-angle migration

Another way is to use only data corresponding to a small range $\mathfrak{R}_\sigma$ of opening angles, for which reflectivity is essentially constant. Following Operto *et al.* (2000), we do this by defining a restriction operator Λ_0, whose value is constant over the "pass" region of σ, and zero for all other σ. That is,

$$\begin{aligned}\Lambda_0(\sigma) &= \frac{1}{2\Delta\sigma}, \quad \sigma_0-\Delta\sigma<\sigma<\sigma_0+\Delta\sigma,\\ &= 0, \qquad \text{otherwise.}\end{aligned} \tag{14.14}$$

A collection of all the data under the above restriction could be referred to as a (nearly) constant-angle section. It is analogous to a constant-offset section, except that to find the data belonging to it one must build raypaths and compute opening angles.

Invoking the weighted least-squares inversion formula (14.4) with the weight factor $\mathcal{W} = \Lambda_0\mathcal{Q}$ ($\mathcal{Q}$ yet to be specified),

$$\mathfrak{O}^{\dagger}\Lambda_0\mathcal{Q}\mathfrak{O}\left(\tilde{\boldsymbol{x}}|\tilde{\boldsymbol{x}}'\right)$$
$$= \int\int\int dx_g dx_s d\omega \sum_{l,l'} \Lambda_0\mathcal{Q}\cdot\left(\frac{|\omega|\mathfrak{A}_l\left(x_g, x_s|\tilde{\boldsymbol{x}}\right)\mathfrak{A}_{l'}\left(x_g, x_s|\tilde{\boldsymbol{x}}'\right)}{\sqrt{\left|\tau_{l'yy}\left(x_g, x_s|\tilde{\boldsymbol{x}}'\right)\tau_{lyy}\left(x_g, x_s|\tilde{\boldsymbol{x}}\right)\right|}}\right)$$
$$\times f_{l'}\left(x_g, x_s|\tilde{\boldsymbol{x}}, \omega\right) f_l^*\left(x_g, x_s|\tilde{\boldsymbol{x}}, \omega\right) e^{i\omega\left(\tau_{l'}\left(x_g,x_s|\tilde{x}'\right)-\tau_l\left(x_g,x_s|\tilde{x}\right)\right)}, \tag{14.15}$$

and

$$\mathfrak{O}^{\dagger}\Lambda_0\mathcal{Q}\left(\tilde{\boldsymbol{x}}|x_g, x_s, \omega\right) = \sqrt{i\omega}\sum_l\left(\frac{\mathfrak{A}_l\left(x_g, x_s|\tilde{\boldsymbol{x}}\right) f_l^*\left(x_g, x_s|\tilde{\boldsymbol{x}}, \omega\right)}{\sqrt{\left|\tau_{lyy}\left(x_g, x_s|\tilde{\boldsymbol{x}}\right)\right|}}\right)$$
$$\times \Lambda_0\mathcal{Q}e^{-i\omega\tau_l\left(x_g,x_s|\tilde{x}\right)}. \tag{14.16}$$

In the 2-D case, the receiver and source positions x_g and x_s are assumed to be scalars. We now assume that the operation $\mathfrak{O}^{\dagger}\Lambda_0\mathcal{Q}\mathfrak{O}$ is nearly diagonal, so that attention may be confined to $l' = l,$ and $\tilde{\boldsymbol{x}}'$ near $\tilde{\boldsymbol{x}}$:

$$\mathfrak{O}^{\dagger}\Lambda_0\mathcal{Q}\mathfrak{O}\left(\tilde{\boldsymbol{x}}|\tilde{\boldsymbol{x}}'\right) = \int\int\int dx_g dx_s d\omega \sum_l \Lambda_0\mathcal{Q} \tag{14.17}$$
$$\times\left(\frac{|\omega|\,|f_l\left(x_g, x_s|\tilde{\boldsymbol{x}}, \omega\right)|^2\mathfrak{A}_l{}^2\left(x_g, x_s|\tilde{\boldsymbol{x}}\right)}{\left|\tau_{lyy}\left(x_g, x_s|\tilde{\boldsymbol{x}}\right)\right|}\right)e^{i\omega\nabla\tau_l\left(x_g,\, x_s|\tilde{x}\right)\cdot\left(\tilde{x}'-\tilde{x}\right)}.$$

The multipath sum can be eliminated by a change of integration variable to horizontal slowness of the source and receiver rays at $\tilde{\boldsymbol{x}}$. If β_i and β_r are the angles made with respect to the vertical at the reflector by the incident and reflected waves, define

$$\tilde{\boldsymbol{p}}_r = \left(p_{rx}, p_{rz}\right) = \frac{1}{c\left(\tilde{\boldsymbol{x}}\right)}\left(\sin\beta_r, -\cos\beta_r\right),$$
$$\tilde{\boldsymbol{p}}_i = \left(p_{ix}, p_{iz}\right) = \frac{1}{c\left(\tilde{\boldsymbol{x}}\right)}\left(\sin\beta_i, \cos\beta_i\right). \tag{14.18}$$

Choosing p_{ix} and p_{rx} as integration variables in place of source and receiver position,

$$\mathfrak{O}^{\dagger}\Lambda_0\mathcal{Q}\mathfrak{O}\left(\tilde{\boldsymbol{x}}|\tilde{\boldsymbol{x}}'\right) = \int\int\int dp_{rx}dp_{ix}d\omega\left|\omega\frac{\partial x_g}{\partial p_{rx}}\frac{\partial x_s}{\partial p_{ix}}\right|\Lambda_0\mathcal{Q}\cdot$$
$$\frac{\mathfrak{A}_l{}^2\left(x_g, x_s|\tilde{\boldsymbol{x}}\right)|f_l\left(x_g, x_s|\tilde{\boldsymbol{x}}, \omega\right)|^2}{\left|\tau_{lyy}\left(x_g, x_s|\tilde{\boldsymbol{x}}\right)\right|}e^{i\omega\nabla\tau_l\left(x_g,\, x_s|\tilde{x}\right)\cdot\left(\tilde{x}'-\tilde{x}\right)}. \tag{14.19}$$

To simplify the exponential, define a new integration variable $\tilde{\boldsymbol{k}}$ as

$$\begin{aligned}\tilde{\boldsymbol{k}} &= -\omega \nabla_{\tilde{\boldsymbol{x}}} \tau_l \left(x_g, x_s | \tilde{\boldsymbol{x}}\right) \\ &= \omega \left(\tilde{\boldsymbol{p}}_r - \tilde{\boldsymbol{p}}_i\right) \\ &= -\frac{\omega}{c\left(\tilde{\boldsymbol{x}}\right)} \left(\sin\beta_i - \sin\beta_r, \cos\beta_i + \cos\beta_r\right). \end{aligned} \tag{14.20}$$

The variable $\tilde{\boldsymbol{k}}$ can be considered the local wavenumber of the reflectivity at $\tilde{\boldsymbol{x}}$. Its direction bisects the source and receiver rays at $\tilde{\boldsymbol{x}}$, hence is normal to a reflecting surface observing Snell's law at that point. Equation (14.20) can be written

$$\begin{aligned}\tilde{\boldsymbol{k}} &= -\frac{2\omega}{c\left(\tilde{\boldsymbol{x}}\right)} \cos\left(\frac{\beta_i - \beta_r}{2}\right)\left(\sin\left(\frac{\beta_i + \beta_r}{2}\right), \cos\left(\frac{\beta_i + \beta_r}{2}\right)\right) \\ &= -\frac{2\omega}{c\left(\tilde{\boldsymbol{x}}\right)} \cos\left(\frac{\sigma}{2}\right)(\sin\alpha, \cos\alpha), \end{aligned} \tag{14.21}$$

with $\alpha = (\beta_i + \beta_r)/2$.

We can also specify the opening angle σ in terms of the dot and cross product of $\tilde{\boldsymbol{p}}_i$ and $\tilde{\boldsymbol{p}}_r$:

$$\cos\sigma = -\tilde{\boldsymbol{p}}_r \cdot \tilde{\boldsymbol{p}}_i c^2, \tag{14.22}$$

$$\sin\sigma = \left(\tilde{\boldsymbol{p}}_r \times \tilde{\boldsymbol{p}}_i\right) \cdot \hat{\boldsymbol{y}} c^2. \tag{14.23}$$

With $\tilde{\boldsymbol{k}}$ and σ as the integration variables, the Jacobian of the transformation is (see exercise 14.1)

$$\left|\frac{\partial\left(p_{ix}, p_{rx}, \omega\right)}{\partial\left(\tilde{\boldsymbol{k}}, \sigma\right)}\right| = \left|\frac{\cos\beta_r \cos\beta_i}{4\omega\, \cos^2(\sigma/2)}\right|. \tag{14.24}$$

In terms of $\tilde{\boldsymbol{k}}$ and σ, equation (14.19) becomes

$$\begin{aligned}\mathfrak{D}^{\dagger}\Lambda_0 \mathcal{Q}\mathfrak{D}\left(\tilde{\boldsymbol{x}}|\tilde{\boldsymbol{x}}'\right) = \frac{1}{2\Delta\sigma} \int\int d\tilde{\boldsymbol{k}} \int_{\sigma_0-\Delta\sigma}^{\sigma_0+\Delta\sigma} d\sigma \left|\frac{\partial x_g}{\partial p_{rx}} \frac{\partial x_s}{\partial p_{ix}} \left(\frac{\cos\beta_r \cos\beta_i}{4\cos^2(\sigma/2)}\right)\right| \\ \mathcal{Q}\left(\frac{\mathfrak{A}_l{}^2\left(x_g, x_s|\tilde{\boldsymbol{x}}\right) | f_l\left(x_g, x_s|\tilde{\boldsymbol{x}}, \omega\right)|^2}{\left|\tau_{lyy}\left(x_g, x_s|\tilde{\boldsymbol{x}}\right)\right|}\right) e^{-i\tilde{\boldsymbol{k}}\cdot(\tilde{\boldsymbol{x}}'-\tilde{\boldsymbol{x}})}. \end{aligned} \tag{14.25}$$

Equation (14.25) becomes a delta function in $\tilde{\boldsymbol{x}}'-\tilde{\boldsymbol{x}}$, provided $\mathcal{Q}$ takes the form

$$\mathcal{Q} = \left(\frac{\left|\tau_{lyy}\left(x_g, x_s|\tilde{\boldsymbol{x}}\right)\right|}{\mathfrak{A}_l{}^2\left(x_g, x_s|\tilde{\boldsymbol{x}}\right) | f_l\left(x_g, x_s|\tilde{\boldsymbol{x}}, \omega\right)|^2}\right) \left|\frac{\partial p_{rx}}{\partial x_g} \frac{\partial p_{ix}}{\partial x_s} \frac{4\cos^2(\sigma/2)}{\cos\beta_r \cos\beta_i}\right|. \tag{14.26}$$

With choice (14.26) for $\mathcal{Q}$, the formula (14.4) for $\mathcal{R}$ becomes

$$\mathcal{R}(\tilde{\boldsymbol{x}}, \sigma_0) = \int \frac{d\omega}{\sqrt{-i\omega}} \int\int dx_g dx_s \Lambda_0(\sigma) \left|\frac{\partial p_{rx}}{\partial x_g}\frac{\partial p_{ix}}{\partial x_s}\right| \left|\frac{\partial\left(\tilde{\boldsymbol{k}}, \sigma\right)}{\partial\left(p_{ix}, p_{rx}, \omega\right)}\right| \times \left(\frac{\sqrt{\left|\tau_{lyy}\left(x_g, x_s|\tilde{\boldsymbol{x}}\right)\right|}}{\mathfrak{A}_l\left(x_g, x_s|\tilde{\boldsymbol{x}}\right)}\right) e^{-i\omega\tau_l\left(x_g, x_s|\tilde{\boldsymbol{x}}\right)} \frac{D\left(x_g|x_s; \omega\right)}{f_l\left(x_g, x_s|\tilde{\boldsymbol{x}}, \omega\right)}. \tag{14.27}$$

There is no sum over multiple paths because there can be only one path with the correct σ_0. The path number l appearing in this formula is thus determined by σ_0.

The partial derivatives in the above formula effect a change of integration variable to $\tilde{\boldsymbol{k}}$ and σ:

$$\mathcal{R}(\tilde{\boldsymbol{x}}, \sigma_0) = \frac{1}{2\Delta\sigma} \int\int d\tilde{\boldsymbol{k}} \int_{\sigma_0-\Delta\sigma}^{\sigma_0+\Delta\sigma} d\sigma \left(\frac{\sqrt{\left|\tau_{lyy}\left(x_g, x_s|\tilde{\boldsymbol{x}}\right)\right|}}{\sqrt{-i\omega}\mathfrak{A}_l\left(x_g, x_s|\tilde{\boldsymbol{x}}\right)}\right) \times e^{-i\omega\tau_l\left(x_g, x_s|\tilde{\boldsymbol{x}}\right)} f_l^*\left(x_g, x_s|\tilde{\boldsymbol{x}}, \omega\right) D\left(x_g|x_s; \omega\right). \tag{14.28}$$

It is convenient to retain ω as an integration variable. Recalling (14.21), we can substitute for $\tilde{\boldsymbol{k}}$ the integration variables ω and α:

$$\mathcal{R}(\tilde{\boldsymbol{x}}, \sigma_0) = \frac{2}{\Delta\sigma c\left(\tilde{\boldsymbol{x}}\right)^2} \int d\alpha \int_{\sigma_0-\Delta\sigma}^{\sigma_0+\Delta\sigma} d\sigma \cos^2\frac{\sigma}{2} \left(\frac{\sqrt{\left|\tau_{lyy}\left(x_g, x_s|\tilde{\boldsymbol{x}}\right)\right|}}{\mathfrak{A}_l\left(x_g, x_s|\tilde{\boldsymbol{x}}\right)}\right) \times \int d\omega\sqrt{i\omega} e^{-i\omega\tau_l\left(x_g, x_s|\tilde{\boldsymbol{x}}\right)} f_l^*\left(x_g, x_s|\tilde{\boldsymbol{x}}, \omega\right) D\left(x_g|x_s; \omega\right). \tag{14.29}$$

The frequency integral can be performed, leaving a convolutional filter in the time domain of the data D at time $t = \tau_l\left(x_g, x_s|\tilde{\boldsymbol{x}}\right)$:

$$\mathcal{R}(\tilde{\boldsymbol{x}}, \sigma_0) = \frac{2}{\Delta\sigma c\left(\tilde{\boldsymbol{x}}\right)^2} \int d\alpha \int_{\sigma_0-\Delta\sigma}^{\sigma_0+\Delta\sigma} d\sigma \cos^2\frac{\sigma}{2} \frac{\sqrt{\left|\tau_{lyy}\left(x_g, x_s|\tilde{\boldsymbol{x}}\right)\right|}}{\left(\mathfrak{A}_l\left(x_g, x_s|\tilde{\boldsymbol{x}}\right)\right)} \times w_l * D\left(x_g|x_s; \tau_l\left(x_g, x_s|\tilde{\boldsymbol{x}}\right)\right), \tag{14.30}$$

with w_l the 2.5-D filter expressed in the frequency domain

$$w_l(\omega) = \sqrt{i\omega} f_l^*\left(x_g, x_s|\tilde{\boldsymbol{x}}, \omega\right). \tag{14.31}$$

In the absence of caustics or diffractions, w_l is a half-derivative filter.

Alternatively, we could leave the migration equation (14.27) in terms of the integration variables x_g and x_s. Evaluating the frequency integral,

$$\mathcal{R}(\tilde{\boldsymbol{x}},\sigma_0)=\int\int dx_g dx_s \Lambda_0(\sigma)\left|\frac{\partial p_{rx}}{\partial x_g}\frac{\partial p_{ix}}{\partial x_s}\right|\left|\left(\frac{4\cos^2(\sigma/2)}{\cos\beta_r\cos\beta_i}\right)\right|\left(\frac{\sqrt{\left|\tau_{lyy}\left(x_g,x_s|\tilde{\boldsymbol{x}}\right)\right|}}{\mathfrak{A}_l\left(x_g,x_s|\tilde{\boldsymbol{x}}\right)}\right)$$
$$\times w_l * D\left(x_g|x_s;\tau_l\left(x_g,x_s|\tilde{\boldsymbol{x}}\right)\right), \tag{14.32}$$

The two migration formulas (14.30) and (14.32) are formally equivalent, though they would likely be implemented differently. Both can be described as common-opening-angle or common-incident-angle migration. In (14.30), one starts at output point $\tilde{\boldsymbol{x}}$ and output opening angle σ_0. For a small range $\pm\Delta\sigma$ around σ_0, and for the range of allowable dip angles θ_k, calculate the corresponding incident and reflected angles β_i and β_r, then trace rays to the surface, locating x_s and x_g, determining $\tau_l\left(x_g,x_s|\tilde{\boldsymbol{x}}\right)$, and computing the amplitude factors $\mathfrak{A}_l$ and τ_{lyy}. For each of the θ_k and σ, the corresponding filtered data point is selected, multiplied by the amplitude factor and added to the sum to form $\mathcal{R}$. Since x_g and x_s are unlikely to lie exactly on realized data points, some interpolation is necessarily involved.

For formula (14.32), one starts with the available data at points x_g and x_s. Rays are computed to subsurface points $\tilde{\boldsymbol{x}}$, looking for points where the opening angle between the two rays falls in the desired range $\sigma_0\pm\Delta\sigma$. For such points, calculate the amplitudes and traveltime and add the weighted filtered data at that traveltime to $\mathcal{R}(\tilde{\boldsymbol{x}},\sigma_0)$. Alternatively, given $\tilde{\boldsymbol{x}}$, find the values of x_g and x_s with σ in the desired range. One does not know a priori which values of $\left(x_g,x_s,\tilde{\boldsymbol{x}}\right)$ have the right σ. To avoid wasted computation, one would likely want to calculate $\mathcal{R}$ for all ranges of σ simultaneously.

14.5 Three-dimensional common-angle least-squares migration

As in 2.5 dimensions, we will not attempt a multiparameter linearized inverse, but rather solve for reflectivity at a fixed opening angle.

Start with a possibly multivalued modeling operation (14.3) in three dimensions. Expanded, it is equation (14.6) with the addition of a noise term:

$$D\left(\tilde{\boldsymbol{x}}_g|\tilde{\boldsymbol{x}}_s;\omega\right)\simeq -i\omega\int\int\int d^3x\sum_{l=1}^{L}\mathcal{R}(\boldsymbol{x},\sigma_l)\mathfrak{A}_l\left(\tilde{\boldsymbol{x}}_g,\tilde{\boldsymbol{x}}_s|\,\boldsymbol{x}\right)$$
$$\times e^{i\omega\tau_l\left(\tilde{\boldsymbol{x}}_g,\tilde{\boldsymbol{x}}_s|\boldsymbol{x}\right)}e^{-i\frac{\pi}{2}a_l\left(\tilde{\boldsymbol{x}}_g,\tilde{\boldsymbol{x}}_s|\boldsymbol{x}\right)}+\mathcal{N}\left(\tilde{\boldsymbol{x}}_g|\tilde{\boldsymbol{x}}_s;\omega\right). \tag{14.33}$$

To find a migration formula, we must invert the above equation. Similarly to the 2-D case, we do this by defining a restriction operator $\boldsymbol{\Lambda}_0\left(\tilde{\boldsymbol{\lambda}}\right)$ that projects onto a small region $\mathfrak{R}_\lambda$ of a two-variable space $\tilde{\boldsymbol{\lambda}} = (\lambda_1, \lambda_2)$. This may define a subset of the available data, and also reflect the fact that the data set as a whole may be limited. $\boldsymbol{\Lambda}_0$ assumes a constant value over the region $\mathfrak{R}_\lambda$ of $\tilde{\boldsymbol{\lambda}}$, and is zero outside the region. We have, if

$$\Gamma = \int_{\mathfrak{R}_\lambda} d\tilde{\boldsymbol{\lambda}}, \tag{14.34}$$

then

$$\begin{aligned} \Lambda_0\left(\tilde{\boldsymbol{\lambda}}\right) &= \frac{1}{\Gamma}, \qquad \tilde{\boldsymbol{\lambda}} \in \mathfrak{R}_\lambda, \\ &= 0, \qquad \text{otherwise.} \end{aligned} \tag{14.35}$$

14.5.1 Restriction to common opening angle and azimuth

For specificity and simplicity, we choose as our restriction variables $\tilde{\lambda} = (\sigma, \chi_{ir})$, where σ is opening angle between incident and reflected rays at the reflection or scattering point, and χ_{ir} is reflection point azimuth. Further, we will assume that the region $\mathfrak{R}_\lambda$ covers the σ-interval $\sigma_0 \pm \Delta\sigma$, and the azimuth point $\chi_{ir} = \chi_{ir0} \pm \Delta\chi_{ir}$. Thus,

$$\begin{aligned} \Lambda_0\left(\sigma, \chi_{ir}\right) &= \frac{1}{4\Delta\sigma\Delta\chi_{ir}}, \qquad \sigma \in \sigma_0 \pm \Delta\sigma, \qquad \chi_{ir} \in \chi_{ir0} \pm \Delta\chi_{ir} \\ &= 0, \qquad \text{otherwise.} \end{aligned} \tag{14.36}$$

If there is anything weird about our choice of restriction operator, it is that azimuthal angle is constrained at the reflector rather than at the source–receiver surface, where we can directly measure it. That, however, can be overcome. If we allow for the possibility that χ_{ir0} and $\Delta\chi_{ir}$ vary with $\boldsymbol{k}_m$ and σ, then a transformation from χ_{ir} to χ_{gs} can be made.

14.5.2 The least-squares inverse

Incorporating the restriction operator, we minimize the weighted least-squares noise, or

$$\mathcal{E} = \int\int\int d\tilde{\boldsymbol{x}}_g d\tilde{\boldsymbol{x}}_s d\omega |\mathcal{N}\left(\tilde{\boldsymbol{x}}_g|\tilde{\boldsymbol{x}}_s;\omega\right)|^2 \Lambda_0\left(\tilde{\lambda}\right) \mathcal{Q}(\cdot). \tag{14.37}$$

In this expression $\mathcal{Q}$ is a weighting function, the exact functional dependence of which is left unspecified for now.

Setting partial derivatives of $\mathcal{E}$ with respect to $\boldsymbol{\mathcal{R}}$ to zero, one obtains the relation

$$\boldsymbol{\mathfrak{D}}^\dagger \boldsymbol{\Lambda}_0 \mathcal{Q} \boldsymbol{\mathfrak{D}} \boldsymbol{\mathcal{R}} = \boldsymbol{\mathfrak{D}} \boldsymbol{\Lambda}_0 \mathcal{Q} \boldsymbol{\mathcal{D}}, \tag{14.38}$$

with

$$\mathfrak{O}^{\dagger}\Lambda_0\mathcal{Q}\mathfrak{O}(\boldsymbol{x}|\boldsymbol{x}') = -\sum_{l,l'}\int\int\int d\tilde{x}_g d\tilde{\boldsymbol{x}}_s d\omega\omega^2 \mathfrak{A}_l\left(\tilde{\boldsymbol{x}}_g, \tilde{\boldsymbol{x}}_s|\boldsymbol{x}\right)\mathfrak{A}_{l'}\left(\tilde{\boldsymbol{x}}_g, \tilde{\boldsymbol{x}}_s|\boldsymbol{x}'\right)$$
$$\times\Lambda_0\mathcal{Q}e^{i\omega(\tau_{l'}(\tilde{x}_g,\tilde{x}_s|x')-\tau_l(\tilde{x}_g,\tilde{x}_s|x))-\frac{i\pi}{2}(\alpha_{l'}(\tilde{x}_g,\tilde{x}_s|x')-\alpha_l(\tilde{x}_g,\tilde{x}_s|x))}, \tag{14.39}$$

and

$$\mathfrak{O}^{\dagger}\Lambda_0\mathcal{Q}\left(\boldsymbol{x}\left|\{\tilde{\boldsymbol{x}}\}_g, \{\tilde{\boldsymbol{x}}\}_s; \omega\right.\right) = i\omega\sum_l \mathfrak{A}_l\left(\{\tilde{\boldsymbol{x}}\}_g, \{\tilde{\boldsymbol{x}}\}_s|\boldsymbol{x}\right)\Lambda_0$$
$$\times\mathcal{Q}e^{-i\omega\tau_l(\{\tilde{x}\}_g,\{\tilde{x}\}_s|x)+\frac{i\pi}{2}\alpha_l(\{\tilde{x}\}_g,\{\tilde{x}\}_s|x)}. \tag{14.40}$$

The indices l and l' each run over all possible combinations of incident and reflected raypaths linking $\tilde{\boldsymbol{x}}_g$, $\tilde{\boldsymbol{x}}_s$, and $\tilde{\boldsymbol{x}}$.

The traveltime τ_l and the KMAH index α_l are the sums of the traveltimes and KMAH indexes for the incoming and reflected waves; i.e.

$$\tau_l\left(\{\tilde{\boldsymbol{x}}\}_g, \{\tilde{\boldsymbol{x}}\}_s|\boldsymbol{x}\right) = \tau_{il}\left(\boldsymbol{x}\left|\{\tilde{\boldsymbol{x}}\}_s\right.\right) + \tau_{rl}\left(\{\tilde{\boldsymbol{x}}\}_g\middle|\boldsymbol{x}\right), \alpha_l\left(\{\tilde{\boldsymbol{x}}\}_g, \{\tilde{\boldsymbol{x}}\}_s|\boldsymbol{x}\right)$$
$$= \alpha_{il}\left(\boldsymbol{x}\left|\{\tilde{\boldsymbol{x}}\}_s\right.\right) + \alpha_{rl}\left(\{\tilde{\boldsymbol{x}}\}_g\middle|\boldsymbol{x}\right). \tag{14.41}$$

We assume that the operation $\mathfrak{O}^{\dagger}\Lambda_0\mathcal{Q}\mathfrak{O}$ is nearly diagonal, so that

$$\mathfrak{O}^{\dagger}\Lambda_0\mathcal{Q}\mathfrak{O}(\boldsymbol{x}|\boldsymbol{x}') \simeq -\sum_l\int d\{\tilde{x}_g\}\int d\{\tilde{\boldsymbol{x}}_s\}\int d\omega\,\omega^2$$
$$\times\mathfrak{A}_l^{\,2}\left(\{\tilde{\boldsymbol{x}}\}_g, \{\tilde{\boldsymbol{x}}\}_s|\boldsymbol{x}\right)\Lambda_0\mathcal{Q}e^{i\omega\nabla\tau_l(\{\tilde{x}\}_g,\{\tilde{x}\}_s|x)\cdot(x'-x)}. \tag{14.42}$$

This assumption is not likely to be violated for a complete data set, where $\Lambda_0 = 1$. For a restricted data set, violation is possible, in which case some "cross-talk" between the various paths may show up in the inversion. We expect this to be less of a problem for the restriction we have chosen (i.e. $\tilde{\boldsymbol{\lambda}} = (\sigma, \chi_{ir})$), than for, say, a constant-offset data set.

The sum over multiple paths is a bother. To get rid of it, we can change integration variables from source and receiver location to parameters that define a unique ray. In three dimensions, there are a number of ways to do that. Here, as in Section 11.5.1, we choose to describe the rays in terms of slowness vectors at the scattering or reflection point. Define slowness 3-vectors $\boldsymbol{p}_i$ and $\boldsymbol{p}_r$ for the incident and reflected rays at $\boldsymbol{x}$, as

$$\boldsymbol{p}_i = \nabla_{\boldsymbol{x}}\tau_{il}\left(\boldsymbol{x}\left|\tilde{\boldsymbol{x}}_s\right.\right), \tag{14.43}$$

and

$$\boldsymbol{p}_r = -\nabla_{\boldsymbol{x}}\tau_{rl}\left(\tilde{\boldsymbol{x}}_g|\boldsymbol{x}\right), \tag{14.44}$$

with the gradient in both cases with respect to the scattering point $\boldsymbol{x}$. As defined, both of these slowness vectors point from $\boldsymbol{x}$ back up the ray towards the surface. We expect only one ray leaving $\boldsymbol{x}$ in any given direction, so have not tagged the slowness vectors with the multipathing parameter l.

Thanks to the Eikonal equation, the slowness vectors are uniquely determined by their first two components. Define 2-vectors $\tilde{\boldsymbol{p}}_i$ and $\tilde{\boldsymbol{p}}_r$, equal to the horizontal components of the slowness 3-vectors $\boldsymbol{p}_i$ and $\boldsymbol{p}_r$ of the incident and reflected rays at $\boldsymbol{x}$. We wish to make a change of integration variable in (14.42) from $\tilde{\boldsymbol{x}}_g$ and $\tilde{\boldsymbol{x}}_s$ to $\tilde{\boldsymbol{p}}_i$ and $\tilde{\boldsymbol{p}}_r$. Even though there may be more than one ray from $\boldsymbol{x}$ to $\tilde{\boldsymbol{x}}_g$ or $\tilde{\boldsymbol{x}}_s$, there should be only one ray leaving $\boldsymbol{x}$ with a given slowness vector. Every ray between $\tilde{\boldsymbol{x}}_s$ and $\boldsymbol{x}$ must correspond to some $\tilde{\boldsymbol{p}}_i$, and every ray between $\tilde{\boldsymbol{x}}_g$ and $\boldsymbol{x}$ must correspond to some $\tilde{\boldsymbol{p}}_r$. Each component l of the sum in (14.42) must correspond to a distinct range of slowness directions. Consequently, with the slowness vectors as integration variables, the sum over multiple paths disappears:

$$\mathfrak{D}^{\dagger}\Lambda\mathcal{Q}\mathfrak{D}(\boldsymbol{x}|\boldsymbol{x}') \simeq -\int d\{\tilde{\boldsymbol{p}}\}_r \int d\{\tilde{\boldsymbol{p}}\}_s \left|\frac{\partial\{\tilde{x}_g\}}{\partial\{\tilde{\boldsymbol{p}}\}_r}\right| \left|\frac{\partial\{\tilde{x}_s\}}{\partial\{\tilde{\boldsymbol{p}}\}_i}\right| \int d\omega \Lambda_0\left(\{\tilde{\boldsymbol{\lambda}}\}\right)$$
$$\times\, \mathcal{Q}\omega^2\mathfrak{A}_l{}^2\left(\{\tilde{\boldsymbol{x}}\}_g, \{\tilde{\boldsymbol{x}}\}_s|\,\boldsymbol{x}\right) e^{i\omega\nabla\tau_l\left(\{\tilde{\boldsymbol{x}}\}_g, \{\tilde{\boldsymbol{x}}\}_s|\boldsymbol{x}\right)\cdot(\boldsymbol{x}'-\boldsymbol{x})}. \quad (14.45)$$

The two Jacobians in this expression that transform coordinates from source and receiver positions to incident and reflected wavenumber depend, like the rays themselves, upon the velocity structure of the medium. They can be computed from ray information in any of a number of ways.

Because each path corresponds to a different range of slowness vectors, the subscript l has been removed from them. However, it remains on some of the parameters in this equation because they are expressed as functions of $\tilde{\boldsymbol{x}}_g$ and $\tilde{\boldsymbol{x}}_s$. The parameters $\tilde{\boldsymbol{x}}_g$, $\tilde{\boldsymbol{x}}_s$, and l are dependent variables in this equation, functions of the slowness vectors.

To simplify the exponential term define the 3-vector variable (see Section 6.5)

$$\boldsymbol{k}_m = -\omega\,\nabla\tau_l\left(\tilde{\boldsymbol{x}}_g, \tilde{\boldsymbol{x}}_s|\,\boldsymbol{x}\right), \quad (14.46)$$

related to the slowness vectors as (see equation H.36)

$$\boldsymbol{k}_m = \omega\left(\boldsymbol{p}_r - \boldsymbol{p}_i\right). \quad (14.47)$$

It is convenient to define another vector $\boldsymbol{q}_\perp$ normal to $\boldsymbol{k}_m$ as

$$\boldsymbol{q}_\perp = \boldsymbol{p}_r \times \boldsymbol{p}_i. \quad (14.48)$$

and its 2-vector counterpart

$$\tilde{\boldsymbol{q}}_\perp = \left(q_x, q_y\right). \quad (14.49)$$

Between ω, $\boldsymbol{p}_i$, and $\boldsymbol{p}_r$, there are five independent quantities. If one considers $\boldsymbol{k}_m$ and the first two components of $\boldsymbol{q}_\perp$ as independent variables, then the third component of $\boldsymbol{q}_\perp$ must be determinable from the other variables. Indeed, since $\boldsymbol{q}_\perp$ is normal to $\boldsymbol{k}_m$:

$$q_{\perp z} = -\frac{q_x k_{mx} + q_y k_{my}}{k_{mz}}. \tag{14.50}$$

From Appendix H, $\boldsymbol{q}_\perp$ is related to opening angle σ and azimuthal angle χ_{ir} at the reflection point (see Figure H.1):

$$\tilde{\boldsymbol{q}}_\perp \equiv \frac{k_{mz}\sin\sigma}{c_i c_r \sqrt{k_{mz}{}^2 + (k_{my}\cos\chi_{ir} - k_{mx}\sin\chi_{ir})^2}} \begin{pmatrix} +\sin\chi_{ir} \\ -\cos\chi_{ir} \end{pmatrix}. \tag{14.51}$$

The reflection-point azimuth can be expressed in terms of $\tilde{\boldsymbol{q}}_\perp as$

$$\tan\chi_{ir} = -\frac{q_{\perp x}}{q_{\perp y}}. \tag{14.52}$$

Source–receiver azimuthal angle χ_{gs} does not relate so easily to $\boldsymbol{k}_m$ and $\boldsymbol{q}_\perp$, since it is measured on the other end of the raypaths. In a constant-velocity medium, χ_{ir} and χ_{gs} would be the same angle, though in a general medium they are different, and ray-tracing information is required to determine one from the other.

The variables $\boldsymbol{k}_m$ and $\tilde{\boldsymbol{q}}_\perp$ can replace the integration variables ω, $\tilde{\boldsymbol{p}}_r$, and $\tilde{\boldsymbol{p}}_s$ in (14.45). That is,

$$\begin{aligned}\mathfrak{D}^\dagger \boldsymbol{\Lambda} \mathfrak{Q}\mathfrak{D}(\boldsymbol{x}|\boldsymbol{x}') \simeq -\int d\tilde{\boldsymbol{q}}_\perp \int d\boldsymbol{k}_m \left|\frac{\partial \tilde{\boldsymbol{x}}_g}{\partial \tilde{\boldsymbol{p}}_r}\right| \left|\frac{\partial \tilde{\boldsymbol{x}}_s}{\partial \tilde{\boldsymbol{p}}_s}\right| \left|\frac{\partial\left(\omega, \tilde{\boldsymbol{p}}_r, \tilde{\boldsymbol{p}}_s\right)}{\partial\left(\boldsymbol{k}_m, \tilde{\boldsymbol{q}}_\perp\right)}\right| \\ \times \Lambda_0\left(\tilde{\boldsymbol{\lambda}}\right) \mathfrak{Q}\omega^2 \mathfrak{A}_l{}^2\left(\tilde{\boldsymbol{x}}_g, \tilde{\boldsymbol{x}}_s | \boldsymbol{x}\right) e^{-i\boldsymbol{k}_m\cdot(\boldsymbol{x}'-\boldsymbol{x})}.\end{aligned} \tag{14.53}$$

The Jacobian of this change of variables is

$$\left|\frac{\partial\left(\omega, \tilde{\boldsymbol{p}}_r, \tilde{\boldsymbol{p}}_s\right)}{\partial\left(\boldsymbol{k}_m, \tilde{\boldsymbol{q}}_\perp\right)}\right| = \left|\frac{c_i{}^3 c_r{}^3 p_{rz} p_{iz}}{4\omega^2 c_e(\sigma)^2 (p_{rz} - p_{iz})\cos\sigma}\right|. \tag{14.54}$$

Now make another change of integration variable from $\tilde{\boldsymbol{q}}$ to the restriction 2-vector $\tilde{\boldsymbol{\lambda}} = (\sigma, \chi_{ir})$.

$$\begin{aligned}\mathfrak{D}^\dagger \boldsymbol{\Lambda} \mathfrak{Q}\mathfrak{D}(\boldsymbol{x}|\boldsymbol{x}') \simeq -\frac{1}{\Gamma}\int d\boldsymbol{k}_m \int_{\mathcal{R}_\lambda} d\tilde{\boldsymbol{\lambda}}\omega^2 \left|\frac{\partial \tilde{\boldsymbol{x}}_g}{\partial \tilde{\boldsymbol{p}}_r}\right| \left|\frac{\partial \tilde{\boldsymbol{x}}_s}{\partial \tilde{\boldsymbol{p}}_s}\right| \left|\frac{\partial\left(\omega, \tilde{\boldsymbol{p}}_r, \tilde{\boldsymbol{p}}_s\right)}{\partial\left(\boldsymbol{k}_m, \tilde{\boldsymbol{q}}_\perp\right)}\right| \\ \times \left(\left|\frac{\partial \tilde{\boldsymbol{q}}_\perp}{\partial(\sigma, \chi_{ir})}\right|\right)_{\boldsymbol{k}_m} \mathfrak{A}_l{}^2\left(\tilde{\boldsymbol{x}}_g, \tilde{\boldsymbol{x}}_s | \boldsymbol{x}\right) \mathfrak{Q} e^{-i\boldsymbol{k}_m\cdot(\boldsymbol{x}'-\boldsymbol{x})}.\end{aligned} \tag{14.55}$$

The Jacobian between $\tilde{\boldsymbol{q}}_\perp$ and $\tilde{\boldsymbol{\lambda}} = (\sigma, \chi_{ir})$ is easily derived:

$$\left(\left|\frac{\partial \tilde{\boldsymbol{q}}_\perp}{\partial (\sigma, \chi_{ir})}\right|\right)_{\boldsymbol{k}_m} = \frac{{k_z}^2 |\sin\sigma \cos\sigma|}{{c_i}^2 {c_r}^2 \left({k_{mz}}^2 + (k_{my}\cos\chi_{ir} - k_{mx}\sin\chi_{ir})^2\right)}. \tag{14.56}$$

Now we are positioned to pick the weight factor $\mathcal{Q}$. If we take

$$\mathcal{Q} = \frac{-1}{\omega^2 \mathfrak{A}_l{}^2 \left(\tilde{\boldsymbol{x}}_g, \tilde{\boldsymbol{x}}_s | \boldsymbol{x}\right)} \left|\frac{\partial \tilde{\boldsymbol{p}}_r}{\partial \tilde{\boldsymbol{x}}_g} \frac{\partial \tilde{\boldsymbol{p}}_s}{\partial \tilde{\boldsymbol{x}}_s}\right| \cdot \left|\frac{\partial \left(\boldsymbol{k}_m, \tilde{\boldsymbol{q}}_\perp\right)}{\partial \left(\omega, \tilde{\boldsymbol{p}}_r, \tilde{\boldsymbol{p}}_s\right)}\right| \left(\left|\frac{\partial (\sigma, \chi_{ir})}{\partial \tilde{\boldsymbol{q}}_\perp}\right|\right)_{\boldsymbol{k}_m}, \tag{14.57}$$

then

$$\mathfrak{D}^\dagger \boldsymbol{\Lambda} \mathcal{Q} \mathfrak{D}(\boldsymbol{x}|\boldsymbol{x}') \simeq \delta(\boldsymbol{x} - \boldsymbol{x}'). \tag{14.58}$$

With this choice for $\mathcal{Q}$, the formula (14.4) for $\boldsymbol{\mathcal{R}}$ becomes

$$\mathcal{R}(\boldsymbol{x}, \sigma_0, \chi_{ir0}) = \sum_l \int\int\int d\tilde{\boldsymbol{x}}_g d\tilde{\boldsymbol{x}}_s d\omega \Lambda_0\left(\tilde{\boldsymbol{\lambda}}\right) \left|\frac{\partial \tilde{\boldsymbol{p}}_r}{\partial \tilde{\boldsymbol{x}}_g}\right| \cdot \left|\frac{\partial \tilde{\boldsymbol{p}}_s}{\partial \tilde{\boldsymbol{x}}_s}\right| \left(\left|\frac{\partial (\sigma, \chi_{ir})}{\partial \tilde{\boldsymbol{q}}_\perp}\right|\right)_{\boldsymbol{k}_m}$$
$$\times \left|\frac{\partial \left(\boldsymbol{k}_m, \tilde{\boldsymbol{q}}_\perp\right)}{\partial \left(\omega, \tilde{\boldsymbol{p}}_r, \tilde{\boldsymbol{p}}_s\right)}\right| \frac{e^{-i\omega\tau_l(\tilde{\boldsymbol{x}}_g, \tilde{\boldsymbol{x}}_s|\boldsymbol{x}) + \frac{i\pi}{2}\alpha_l(\tilde{\boldsymbol{x}}_g, \tilde{\boldsymbol{x}}_s|\boldsymbol{x})}}{-i\omega\mathfrak{A}_l\left(\tilde{\boldsymbol{x}}_g, \tilde{\boldsymbol{x}}_s | \boldsymbol{x}\right)} D\left(\tilde{\boldsymbol{x}}_g, \tilde{\boldsymbol{x}}_s, \omega\right). \tag{14.59}$$

14.5.3 *The elegant formula*

The determinants in formula (14.59) are in effect a change of integration variables from ω, $\tilde{\boldsymbol{x}}_g$ and $\tilde{\boldsymbol{x}}_s$ to $\boldsymbol{k}_m$ and $\tilde{\boldsymbol{\lambda}}$:

$$\mathcal{R}(\boldsymbol{x}, \sigma_0, \chi_{ir0}) = \int d^3\boldsymbol{k}_m \int_{\mathcal{R}_\lambda} \frac{d\tilde{\boldsymbol{\lambda}}}{\Gamma} \frac{e^{-i\omega\tau_l(\tilde{\boldsymbol{x}}_g, \tilde{\boldsymbol{x}}_s|\boldsymbol{x}) + \frac{i\pi}{2}\alpha_l(\tilde{\boldsymbol{x}}_g, \tilde{\boldsymbol{x}}_s|\boldsymbol{x})}}{-i\omega\mathfrak{A}_l\left(\tilde{\boldsymbol{x}}_g, \tilde{\boldsymbol{x}}_s | \boldsymbol{x}\right)} D\left(\tilde{\boldsymbol{x}}_g, \tilde{\boldsymbol{x}}_s, \omega\right). \tag{14.60}$$

In this expression the multipath sum disappears because $\boldsymbol{k}_m$ and $\tilde{\boldsymbol{\lambda}}$ together specify a unique path. If we compare this least-squares multipath formula with the deterministic formula (11.62), we find they closely correspond, the differences being that the present formula (a) explicitly includes possible KMAH factors, (b) forms an average over a range of opening and azimuthal angles, and (c) ignores amplitude factors of 2π. This close correspondence is to be anticipated, since we have engineered the least-squares weight factors to produce, in the absence of redundancy or averaging, the deterministic result.

It would be helpful to retain ω as an integration variable, in the hope of reducing the ω integration to a simple filter in the time domain. This can be done by writing $\boldsymbol{k}_m$ in spherical coordinates, i.e.

$$
\begin{aligned}
k_{mx} &= k_m \sin\alpha \cos\phi_k, \\
k_{my} &= k_m \sin\alpha \sin\phi_k, \\
k_{mz} &= k_m \cos\alpha,
\end{aligned}
\tag{14.61}
$$

and noting, from its definition (14.47),

$$
\begin{aligned}
k_m &= \frac{\omega}{c_i(\boldsymbol{x})c_r(\boldsymbol{x})}\sqrt{c_r(\boldsymbol{x})^2 + c_i(\boldsymbol{x})^2 + 2c_i(\boldsymbol{x})c_r(\boldsymbol{x})\cos\sigma} \\
&= \frac{2\omega c_e(\sigma, \boldsymbol{x})}{c_i(\boldsymbol{x})c_r(\boldsymbol{x})}.
\end{aligned}
\tag{14.62}
$$

Thus, given our choice of restriction operator,

$$
\begin{aligned}
\mathcal{R}(\boldsymbol{x}, \sigma_0, \chi_{ir0}) &= \int d\phi_k \int d\alpha \sin\alpha \int dk_m {k_m}^2 \\
&\times \int_{\mathcal{R}_\lambda} \frac{d\tilde{\boldsymbol{\lambda}}}{\Gamma} \frac{e^{-i\omega\tau_l(\tilde{\boldsymbol{x}}_g, \tilde{\boldsymbol{x}}_s|\boldsymbol{x}) + \frac{i\pi}{2}\alpha_l(\tilde{\boldsymbol{x}}_g, \tilde{\boldsymbol{x}}_s|\boldsymbol{x})}}{i\omega\mathfrak{A}_l\left(\tilde{\boldsymbol{x}}_g, \tilde{\boldsymbol{x}}_s|\boldsymbol{x}\right)} D\left(\tilde{\boldsymbol{x}}_g, \tilde{\boldsymbol{x}}_s, \omega\right),
\end{aligned}
\tag{14.63}
$$

or, with a change of variable from k_m to ω,

$$
\begin{aligned}
\mathcal{R}(\boldsymbol{x}, \sigma_0, \chi_{ir0}) &= \frac{8{c_e}^3(\sigma_0, \boldsymbol{x})}{{c_r}^3(\boldsymbol{x}){c_i}^3(\boldsymbol{x})} \int d\phi_k \int d\alpha \sin\alpha \int_{\mathcal{R}_\lambda} \frac{d\tilde{\boldsymbol{\lambda}}}{\Gamma} \\
&\times \int d\omega i\omega \frac{e^{-i\omega\tau_l(\tilde{\boldsymbol{x}}_g, \tilde{\boldsymbol{x}}_s|\boldsymbol{x}) + \frac{i\pi}{2}\alpha_l(\tilde{\boldsymbol{x}}_g, \tilde{\boldsymbol{x}}_s|\boldsymbol{x})}}{\mathfrak{A}_l\left(\tilde{\boldsymbol{x}}_g, \tilde{\boldsymbol{x}}_s|\boldsymbol{x}\right)} D\left(\tilde{\boldsymbol{x}}_g, \tilde{\boldsymbol{x}}_s, \omega\right).
\end{aligned}
\tag{14.64}
$$

In the time domain, this becomes

$$
\begin{aligned}
\mathcal{R}(\boldsymbol{x}, \sigma_0, \chi_{ir0}) &= \frac{-8{c_e}^3(\sigma_0, \boldsymbol{x})}{{c_r}^3(\boldsymbol{x}){c_i}^3(\boldsymbol{x})} \int d\phi_k \int d\alpha \sin\alpha \int_{\mathcal{R}_\lambda} \frac{d\tilde{\boldsymbol{\lambda}}}{\Gamma} \\
&\times \left(\frac{w_l * D\left(\tilde{\boldsymbol{x}}_g, \tilde{\boldsymbol{x}}_s, \tau_l\left(\tilde{\boldsymbol{x}}_g, \tilde{\boldsymbol{x}}_s|\boldsymbol{x}\right)\right)}{\mathfrak{A}_l\left(\tilde{\boldsymbol{x}}_g, \tilde{\boldsymbol{x}}_s|\boldsymbol{x}\right)} \right),
\end{aligned}
\tag{14.65}
$$

with w_l a filter on the time component of D, with form in the frequency domain

$$
w_l(\omega) = -i\omega e^{i\frac{\pi}{2}\alpha_l(\tilde{\boldsymbol{x}}_g, \tilde{\boldsymbol{x}}_s|\boldsymbol{x})}. \tag{14.66}
$$

In the absence of caustics, the migration filter w_l is a simple time derivative.

To rephrase (14.65) in terms of source–receiver azimuth, note that the restriction integral is invariant under a change of variable from χ_{ir} to χ_{gs}:

$$
\begin{aligned}
\int_{\mathcal{R}_\lambda} \frac{d\tilde{\boldsymbol{\lambda}}}{\Gamma} &= \frac{1}{4\Delta\sigma\Delta\chi_{ir}} \int_{\sigma_0-\Delta\sigma}^{\sigma_0+\Delta\sigma} d\sigma \int_{\chi_{ir0}-\Delta\chi_{ir}}^{\chi_{ir0}+\Delta\chi_{ir}} d\chi_{ir} \\
&= \frac{1}{4\Delta\sigma\Delta\chi_{gs}} \int_{\sigma_0-\Delta\sigma}^{\sigma_0+\Delta\sigma} d\sigma \int_{\chi_{gs0}-\Delta\chi_{gs}}^{\chi_{gs0}+\Delta\chi_{gs}} d\chi_{gs} \\
&= 1.
\end{aligned}
\tag{14.67}
$$

This leaves us free to rewrite (14.65) as

$$\mathcal{R}\left(\boldsymbol{x}, \sigma_0, \chi_{gs0}\right) = \frac{-8c_e{}^3\left(\sigma_0, \boldsymbol{x}\right)}{c_r{}^3(\boldsymbol{x})c_i{}^3(\boldsymbol{x})} \int d\phi_k \int d\alpha \sin\alpha \times \int_{\mathcal{R}_\lambda} \frac{d\sigma d\chi_{gs}}{\Gamma} \left(\frac{w_l * D\left(\tilde{\boldsymbol{x}}_g, \tilde{\boldsymbol{x}}_s, \tau_l\left(\tilde{\boldsymbol{x}}_g, \tilde{\boldsymbol{x}}_s | \boldsymbol{x}\right)\right)}{\mathfrak{A}_l\left(\tilde{\boldsymbol{x}}_g, \tilde{\boldsymbol{x}}_s | \boldsymbol{x}\right)} \right), \tag{14.68}$$

The formula (14.68) looks very simple, but is not necessarily easy to implement. For every possible subsurface scattering or reflection point $\boldsymbol{x}$, one puts together ray pairs corresponding to the ranges of σ, α, and ϕ_k. The source–receiver azimuthal integration variable χ_{gs} can be mapped into the reflection-point azimuth χ_{ir} by computing the partial derivative of one with respect to the other with σ, α, and ϕ_k fixed. However, that does not alleviate the responsibility of figuring out which subsurface azimuth χ_{ir0} corresponds to the realized surface azimuth $\chi_{gs} = \chi_{gs0}$.

14.5.4 The messy formula

The less elegant formula is to leave surface parameters as the integration variables. The advantage of this approach is that one only needs to deal with rays that are realized in the data.

With expressions (14.54) and (14.56) for the determinants, one can write (14.59) as

$$\mathcal{R}\left(\boldsymbol{x}, \sigma_0, \chi_{gs0}\right) = \sum_l \int\int\int d\tilde{\boldsymbol{x}}_g d\tilde{\boldsymbol{x}}_s d\omega \Lambda_0\left(\tilde{\boldsymbol{\lambda}}\right) \left|\frac{\partial \tilde{\boldsymbol{p}}_r}{\partial \tilde{\boldsymbol{x}}_g}\right| \left|\frac{\partial \tilde{\boldsymbol{p}}_s}{\partial \tilde{\boldsymbol{x}}_s}\right| \left(\left|\frac{\partial \chi_{gs}}{\partial \chi_{ir}}\right|\right)_{\boldsymbol{k}_m, \sigma} \times \left(\left|\frac{\partial\left(\sigma, \chi_{ir}\right)}{\partial \tilde{\boldsymbol{q}}_\perp}\right|\right)_{\boldsymbol{k}} \left|\frac{\partial\left(\boldsymbol{k}_m, \tilde{\boldsymbol{q}}_\perp\right)}{\partial\left(\omega, \tilde{\boldsymbol{p}}r, \tilde{\boldsymbol{p}}_s\right)}\right| \frac{e^{-i\omega\tau_l\left(\tilde{\boldsymbol{x}}_g, \tilde{\boldsymbol{x}}_s | \boldsymbol{x}\right) + \frac{i\pi}{2}\alpha_l\left(\tilde{\boldsymbol{x}}_g, \tilde{\boldsymbol{x}}_s | \boldsymbol{x}\right)}}{-i\omega\mathfrak{A}_l\left(\tilde{\boldsymbol{x}}_g, \tilde{\boldsymbol{x}}_s | \boldsymbol{x}\right)} \times D\left(\tilde{\boldsymbol{x}}_g, \tilde{\boldsymbol{x}}_s, \omega\right), \tag{14.69}$$

or

$$\mathcal{R}\left(\boldsymbol{x}, \sigma_0, \chi_{gs0}\right) = \sum_l \int d\tilde{\boldsymbol{x}}_g \int d\tilde{\boldsymbol{x}}_s \int d\omega i\omega \Lambda_0\left(\sigma, \chi_{gs}\right) \left|\frac{\partial \chi_{gs}}{\partial \chi_{ir}}\right| \left|\frac{\partial \tilde{\boldsymbol{p}}_r}{\partial \tilde{\boldsymbol{x}}_g}\right| \left|\frac{\partial \tilde{\boldsymbol{p}}_s}{\partial \tilde{\boldsymbol{x}}_s}\right| \times \frac{k_{mz}{}^2 + \left(k_{my}\cos\chi_{ir} - k_{mx}\sin\chi_{ir}\right)^2}{k_{mz}{}^2} \left|\frac{4c_e(\sigma)^2\left(p_{rz} - p_{iz}\right)}{c_i c_r p_{rz} p_{iz} \sin\sigma}\right| \times \frac{e^{-i\omega\tau_l\left(\tilde{\boldsymbol{x}}_g, \tilde{\boldsymbol{x}}_s | \boldsymbol{x}\right) + \frac{i\pi}{2}\alpha_l\left(\tilde{\boldsymbol{x}}_g, \tilde{\boldsymbol{x}}_s | \boldsymbol{x}\right)}}{\mathfrak{A}_l\left(\tilde{\boldsymbol{x}}_g, \tilde{\boldsymbol{x}}_s | \boldsymbol{x}\right)} D\left(\tilde{\boldsymbol{x}}_g, \tilde{\boldsymbol{x}}_s, \omega\right). \tag{14.70}$$

The time-domain version of this formula is

$$\mathcal{R}\left(\boldsymbol{x}, \sigma_0, \chi_{gs0}\right) = -\sum_l \int\int d\tilde{\boldsymbol{x}}_g d\tilde{\boldsymbol{x}}_s \Lambda\left(\sigma, \chi_{gs}\right) \left|\frac{\partial \chi_{gs}}{\partial \chi_{ir}}\right| \left|\frac{\partial \tilde{\boldsymbol{p}}_r}{\partial \tilde{\boldsymbol{x}}_g}\right| \left|\frac{\partial \tilde{\boldsymbol{p}}_s}{\partial \tilde{\boldsymbol{x}}_s}\right|$$
$$\times \left|\frac{4c_e(\sigma)^2 \left(p_{rz} - p_{iz}\right)}{c_i c_r p_{rz} p_{iz} \sin\sigma}\right| \frac{{k_{mz}}^2 + \left(k_{my}\cos\chi_{ir} - k_{mx}\sin\chi_{ir}\right)^2}{{k_{mz}}^2}$$
$$\times \frac{w_l * D\left(\tilde{\boldsymbol{x}}_g, \tilde{\boldsymbol{x}}_s, \tau_l\left(\tilde{\boldsymbol{x}}_g, \tilde{\boldsymbol{x}}_s \middle| \boldsymbol{x}\right)\right)}{\mathfrak{A}_l\left(\tilde{\boldsymbol{x}}_g, \tilde{\boldsymbol{x}}_s \middle| \boldsymbol{x}\right)}. \tag{14.71}$$

Zero azimuth. For zero azimuth, the restriction parameter Λ essentially involves a delta function in source–receiver azimuth. With a change of integration variables from source and receiver coordinates $\tilde{\boldsymbol{x}}_g$ and $\tilde{\boldsymbol{x}}_s$ to midpoint and offset coordinates $\tilde{\boldsymbol{x}}_m = \left(\tilde{\boldsymbol{x}}_g + \tilde{\boldsymbol{x}}_s\right)/2$, $\tilde{\boldsymbol{x}}_h = \left(\tilde{\boldsymbol{x}}_g - \tilde{\boldsymbol{x}}_s\right)/2$, the delta function can be used to eliminate the y-component of the offset integral:

$$\mathcal{R}(\boldsymbol{x}, \sigma) = -\sum_l \int d\tilde{\boldsymbol{x}}_m \int d\boldsymbol{x}_h \left|x_h\right| \Lambda_0(\sigma) \left|\frac{\partial \chi_{gs}}{\partial \chi_{ir}}\right| \left|\frac{\partial \tilde{\boldsymbol{p}}_r}{\partial \tilde{\boldsymbol{x}}_g}\right| \left|\frac{\partial \tilde{\boldsymbol{p}}_s}{\partial \tilde{\boldsymbol{x}}_s}\right| \left|\frac{4c_e(\sigma)^2 \left(p_{rz} - p_{iz}\right)}{c_i c_r p_{rz} p_{iz} \sin\sigma}\right|$$
$$\times \frac{{k_{mz}}^2 + \left(k_{my}\cos\chi_{ir} - k_{mx}\sin\chi_{ir}\right)^2}{{k_{mz}}^2} \left(\frac{w_l * D\left(\tilde{\boldsymbol{x}}_g, \tilde{\boldsymbol{x}}_s, \tau_l\left(\tilde{\boldsymbol{x}}_g, \tilde{\boldsymbol{x}}_s \middle| \boldsymbol{x}\right)\right)}{\mathfrak{A}_l\left(\tilde{\boldsymbol{x}}_g, \tilde{\boldsymbol{x}}_s \middle| \boldsymbol{x}\right)}\right). \tag{14.72}$$

The reduced restriction parameter $\Lambda_0(\sigma)$ has the same definition as the 2-D parameter (14.14). The formula (14.72) isn't as pretty as (14.68), but is relatively straightforward. One starts with the full, 3-D prestack data set. An input trace is determined by its coordinates $(\tilde{\boldsymbol{x}}_g, \tilde{\boldsymbol{x}}_s)$. For a given output point $\boldsymbol{x}$, one produces the incident and reflected rays. With $\boldsymbol{p}_r$ and $\boldsymbol{p}_s$ known, one can pick out that portion of the data that falls within the support $\sigma_0 \pm \Delta\sigma$ of the restriction operator Λ_0. (Alternatively, one can partition out the entire data set to all relevant ranges of σ.)

Exercises

14.1 Derive the 2-D Jacobian (14.24).

14.2 Derive the Jacobian $\left|\mathrm{Det}\left(\frac{\partial(\boldsymbol{k}, \tilde{\boldsymbol{q}}_\perp)}{\partial(\omega, \tilde{\boldsymbol{p}}_r, \tilde{\boldsymbol{p}}_i)}\right)\right|$.

14.3 Evaluate the Jacobian $\left[\left|\mathrm{Det}\left(\frac{\partial(\tilde{\boldsymbol{q}}_\perp)}{\partial(\sigma,\chi)}\right)\right|\right]_k$.

14.4 Express the Jacobian $\left|\mathrm{Det}\left(\frac{\partial(\boldsymbol{k}, \sigma, \chi)}{\partial(\omega, \tilde{\boldsymbol{p}}_r, \tilde{\boldsymbol{p}}_i)}\right)\right|$ in terms of the magnitude of $\tilde{q}_\perp$.

Appendix A
Conventions and glossary of terms

A.1 Conventions

Vector conventions

Vector quantities appear in bold-face type (e.g. $\boldsymbol{x}$). Two-dimensional projections of three-dimensional vectors are given a $\sim$ overhead, as in $\tilde{\boldsymbol{x}} = (x, y)$, as a projection of $\boldsymbol{x} = (x, y, z)$. Matrices are represented as bold-face capitals.

Operator argument conventions

Some quantities can be viewed as operations which map a set of *input coordinates* onto a set of *output coordinates*. We represent such quantities with the form $D\left(\boldsymbol{x}_g|\boldsymbol{x}_s; t\right)$. The $\boldsymbol{x}_s$ is placed right of the vertical bar to indicate that it is an input coordinate, while the $\boldsymbol{x}_g$ is placed left of the vertical bar to indicate that it is an output coordinate. The t is placed on the far right to separate it from the input and output coordinates. We could define a source time as an input coordinate and a measurement time as an output coordinate but, since we consider only the difference between these two times to be important, have chosen not to. Our notation is analogous to a generalized matrix notation in which the receiver locations $\boldsymbol{x}_g$ indicate rows and the source locations $\boldsymbol{x}_s$ indicate columns. Were we to employ Dirac notation (commonly used in physics), $D\left(\boldsymbol{x}_g|\boldsymbol{x}_s; t\right)$ would become $\langle\boldsymbol{x}_g|\mathbf{D}(t)|\boldsymbol{x}_s\rangle$. Though better in many respects, Dirac notation is not in common use within the geophysics community, so we refrain from using it in this volume.

Derivative convention

Functions of multiple variables will typically list the last variable as t or z $\left(\text{e.g. } D\left(\boldsymbol{x}_g|\boldsymbol{x}_s; t\right)\right)$. Derivatives of such a function with respect to its last variable may be indicated with a dot above the function $\left(\text{e.g. } \dot{D}\left(\boldsymbol{x}_g|\boldsymbol{x}_s; t\right)\right)$.

Transform nomenclature

Generally, we represent both a space or time function (e.g. $f(t)$) and its Fourier transform (e.g. $f(\omega)$) by the same symbol, letting the argument signify the domain. Philosophically, we consider the function (e.g. f) to be the same abstract entity in either case, expressible in e.g. time as $f(t)$ and in frequency as $f(\omega)$. Were we to succumb to Dirac notation, f

would be the "ket" $|f\rangle$, with $f(t) = \langle t|f\rangle$, and $f(\omega) = \langle\omega|f\rangle$. Use of the same symbol for both a function and its transform will not please everyone. Our feeling is that when dealing with multidimensional functions and many possible combinations of domain, this is the most practical alternative.

Stationary points

In this volume, extensive use is made of the stationary phase approximation, in which the stationary point of a function $f(x)$ is the point at which the derivative of f with respect to x is zero. Where practical, we denote such a stationary point as $\underset{\wedge}{x}$. When dealing with multidimensional functions, we reserve the right to adjust notation as necessary to minimize ambiguities.

A.2 Glossary

$\boldsymbol{A}$	An amplitude function of one or more variables
$\mathbf{A}$	The shear-wave component of an elastic wavefunction
$\mathcal{A}$	An amplitude-phase coefficient
a_c	Velocity perturbation
a_α	P-velocity perturbation
a_β	S-velocity perturbation
a_ρ	Density perturbation
a_λ	Bulk modulus perturbation
a_μ	Shear modulus perturbation
B	Modeling amplitude function
$b(\boldsymbol{x})$	A function that defines a surface at $b(\boldsymbol{x}) = 0$
b_c	Normal gradient of a velocity perturbation
b_α	Normal gradient of a P-velocity perturbation
b_β	Normal gradient of an S-velocity perturbation
b_ρ	Normal gradient of a density perturbation
C_a	An acoustic scale factor
C_P	A P-wave scale factor
C_S	An S-wave scale factor
c	Propagation velocity for a wave of unspecified type
c_0	A background or reference velocity
c_i, c_r	Velocities of incident and reflected wave of unspecified type
$c_e(\sigma)$	An angle-dependent average velocity
$\underline{c}$	Summed slowness
$\boldsymbol{D}(\tilde{\boldsymbol{x}}_g\|\tilde{\boldsymbol{x}}_s;t)$	A seismic data set, expressed as a function of receiver location, source location, and time
$\boldsymbol{\mathcal{D}}$	A seismic reflection data set, represented abstractly
$\mathcal{D}$	An integration domain
$d_{h\pm}(t)$	Causal and anticausal half-derivative operator
$\boldsymbol{E}$	Rotation matrix into P, SV, and SH components
$\hat{\mathbf{e}}_{\mathbf{SH}}$	A unit vector in the SH-direction
$\hat{\mathbf{e}}_{\mathbf{SV}i}$	A unit vector in the incident-wave SV-direction
$\hat{\mathbf{e}}_{\mathbf{SV}r}$	A unit vector in the reflected-wave SV-direction

e	The exponential constant
$\boldsymbol{f}$–$\boldsymbol{k}$	Shorthand for frequency–wavenumber domain
$\boldsymbol{G}$	A Green's function, expressed abstractly
$\boldsymbol{G}_{\pm}$	Causal and anticausal Green's functions
$G\left(\boldsymbol{x}_g \vert \boldsymbol{x}_s; t\right)$	A Green's function, expressed as a function of receiver position, source position, and time
$\boldsymbol{G}_0$	An approximate or background Green's function, expressed abstractly
$G_0\left(\boldsymbol{x}_g \vert \boldsymbol{x}_s; t\right)$	A background Green's function, expressed as a function of receiver position, source position, and time
G_a	An acoustic Green's function
G_{scalar}	A scalar Green's function
G_P	A P-wave Green's function
G_S	An S-wave Green's function
g_a	Acoustic Green's function amplitude
g_{scalar}	Scalar Green's function amplitude
g_P	P-wave Green's function amplitude
g_S	S-wave Green's function amplitude
i	The unit imaginary number
$\boldsymbol{I}$	The unit operator, in one or more dimensions
$\boldsymbol{J}$	The Jacobian of a coordinate transformation
J_0	A Bessel function of order zero
J_1	A Bessel function of order one
j_1	A spherical Bessel function of order one
$\boldsymbol{K}$	Matrix of second derivatives of $\boldsymbol{k}$
$\boldsymbol{k} = \left(k_x, k_y, k_z\right)$	The Fourier wavenumber vector conjugate to cartesian spatial coordinates $\boldsymbol{x} = (x,y,z)$. Also, local asymptotic wavenumbers
$\hat{\boldsymbol{k}}$	A unit vector in the direction of $\boldsymbol{k}$
$\boldsymbol{k}$	Wavenumber magnitude
$\tilde{\boldsymbol{k}}_g$	Receiver wavenumber in one or two dimensions
$\tilde{\boldsymbol{k}}_h$	Offset wavenumber in one or two dimensions
$\tilde{\boldsymbol{k}}_m$	Horizontal midpoint wavenumber in one or two dimensions
$\boldsymbol{k}_m$	Midpoint wavenumber
$\tilde{\boldsymbol{k}}_s$	Source wavenumber in one or two dimensions
$\mathcal{L}$	A wave operator
$\boldsymbol{\mathcal{L}}_D$	A diagonalized elastic wave operator
$\boldsymbol{\mathcal{L}}_E$	An elastic wave operator
$\boldsymbol{\mathcal{L}}_{E0}$	An approximate or background elastic wave operator
$\mathcal{L}_P$	The P-component of a background elastic wave operator
$\mathcal{L}_S$	The S-component of a background elastic wave operator
$\boldsymbol{\mathcal{M}}$	Abstract representation of a migration or imaging operator
$\boldsymbol{n}$	A coordinate in the direction normal to a surface
n_t	Number of time samples in a data set
$\hat{\boldsymbol{n}}_g$	Receiver-ray unit vector
$\tilde{\boldsymbol{n}}_g$	Horizontal components of the receiver-ray unit vector
$\hat{\boldsymbol{n}}_s$	Source-ray unit vector

$\tilde{\boldsymbol{n}}_s$	Horizontal components of the source-ray unit vector
$\mathcal{O}$	Abstract representation of a modeling operator
$\boldsymbol{P}(\boldsymbol{x}, t)$	A wave function in terms of location and time
$\boldsymbol{p} = (p_x, p_y, p_z)$	Slowness vector
$\boldsymbol{p}_i = (p_{ix}, p_{iy}, p_{iz})$	Incident ray slowness vector
$\boldsymbol{p}_r = (p_{rx}, p_{ry}, p_{rz})$	Reflected ray slowness vector
$\tilde{\boldsymbol{p}} = (p_x, p_y)$	Components of the slowness vector p. Also, Radon or slant-stack parameters conjugate to (x, y)
$\tilde{\boldsymbol{p}}_g = (p_{gx}, p_{gy})$	x–y slowness vector of geophone ray
$\tilde{\boldsymbol{p}}_s = (p_{sx}, p_{sy})$	x–y slowness vector of source ray
$\tilde{\boldsymbol{p}}_h = (p_{hx}, p_{hy})$	Offset slant-stack parameter
$\tilde{\boldsymbol{p}}_m = (p_{mx}, p_{my})$	Midpoint slant-stack parameter
$\mathcal{Q}$	A weighting function
q	Tangent of dip angle
$\boldsymbol{q} = (q_x, q_y)$	Two-dimensional dip angle tangent
q_x, q_h	Slant stack parameters in midpoint-offset-depth
$\boldsymbol{r}$	A distance in three dimensions
r_s	Length of source ray
r_g	Length of receiver ray
R	A reflection coefficient
$\boldsymbol{\mathcal{R}}$	An image or reflectivity function, represented abstractly
$\mathcal{R}(\boldsymbol{x})$	An image or reflectivity function, expressed as a function of subsurface location $\boldsymbol{x}$
$\mathcal{R}_p(\boldsymbol{x}; \boldsymbol{x}_0)$	A point reflector centered at $\boldsymbol{x}_0$
$\mathcal{R}_{\mathcal{S}}(\boldsymbol{x}; \boldsymbol{x}_r; \tilde{\boldsymbol{p}})$	A line reflector passing through point $\boldsymbol{x}_r$ with dip tangents $\tilde{\boldsymbol{p}} = (p_x, p_y)$
S	The sign function
$\mathcal{S}$	A surface
$\boldsymbol{s}$	A surface coordinate
$\mathrm{sinc}(\boldsymbol{x})$	The sinc function sin $(x)/x$
t	Time
T	A transmission coefficient
T_f	Maximum time in a time window
$\boldsymbol{u}$	Displacement vector for elastic waves
$U(z)$	Divergence factor in a depth-variable medium
$\mathcal{V}$	A three- or two-dimensional region
$\mathcal{V}$	The scattering potential
$\boldsymbol{\mathcal{V}}_E$	The elastic scattering potential
$\boldsymbol{\mathcal{V}}_W$	The elastic scattering potential rotated into P, SV, and SH components
$\boldsymbol{\mathcal{V}}_{\mathrm{PP}}$	The P to P wave scattering potential
$\boldsymbol{\mathcal{V}}_{\mathrm{SVSV}}$	The SV to SV wave scattering potential
$\boldsymbol{\mathcal{V}}_{\mathrm{SHSH}}$	The SH to SH wave scattering potential
$\boldsymbol{\mathcal{V}}_{\mathrm{SVP}}$	The P to SV wave scattering potential
$\boldsymbol{\mathcal{V}}_{\mathrm{PSV}}$	The SV to P wave scattering potential
$\boldsymbol{\mathcal{V}}_{\mathrm{PSH}}$	The SH to P wave scattering potential (equals zero)

$\mathcal{V}_{\mathrm{SHP}}$	The P to SH wave scattering potential (equals zero)
$\mathcal{V}_{\mathrm{SVSH}}$	The SH to SV wave scattering potential (equals zero)
$\mathcal{V}_{\mathrm{SHSV}}$	The SV to SH wave scattering potential (equals zero)
$\mathbb{V}$	The frequency-reduced scattering potential
$\mathcal{W}$	A weighting operator
$\mathfrak{w}$	A source wavelet
$\boldsymbol{x} = (\boldsymbol{x}, \boldsymbol{y}, \boldsymbol{z})$	Cartesian spatial coordinates
$\tilde{\boldsymbol{x}}_g = (x_g, y_g)$	Receiver x–y-location on the receiving surface
$\tilde{\boldsymbol{x}}_h$	Source–receiver half-offset in one or two dimensions
$\tilde{\boldsymbol{x}}_m$	Source–receiver midpoint in one or two dimensions
$\boldsymbol{x}_r$	Location of a reflecting point
$\tilde{\boldsymbol{x}}_s = (x_s, y_s)$	Source x–y-location on the source surface
Z_f	Maximum depth in a depth window
α	Reflector dip angle
α	P-wave velocity
α_0	Approximate or background P-wave velocity
β	S-wave velocity
β_0	Approximate or background S-wave velocity
γ	P-modulus
γ, γ_i	Angle of incidence
γ_r	Angle of reflection
Δx	Midpoint displacement $x_m - x$
$\delta()$	A Dirac delta function
$\Theta(t)$	The unit step function or integration operator
$\Theta_{h\pm}(t)$	Causal/anticausal half-integral operator
θ	An angular coordinate
$\boldsymbol{\Lambda}$	A restriction operator
λ, μ	Lamé parameters
$\boldsymbol{\Pi}$	Diagonalization operator for elastic waves
ρ	Density
ρ_0	Approximate or background density
ρ	A distance in two dimensions
$\rho^{(n)}(x)$	An n-dimensional "rho" filter in the coordinate x
σ	Opening angle between an incident and reflected wave or ray
τ	Traveltime
ϕ_i, ϕ_r	Propagation angles, in 2-D, of incident and reflected waves relative to the vertical
χ_{gs}	Source–receiver azimuthal angle
χ_{ir}	Incident-reflected ray azimuthal angle
ψ	A wavefunction
ψ_a	An acoustic wavefunction
ψ_{scalar}	A scalar wavefunction
$\boldsymbol{\Psi}$	An elastic wavefunction
ω	Angular frequency in radians per second
$\varsigma(\boldsymbol{x}, \boldsymbol{y})$	Boundary depth function
ζ	Traveltime depth function

Appendix B

Coordinates, vectors, and identities

B.1 Coordinates

We would like to maintain a right-handed coordinate system, in which a right-hand screw rotated from the x-axis towards the y-axis advances in the positive z-direction. In geophysics, we prefer to define the z-axis as positive downward, which encourages defining the three orthogonal axes as shown in Figure B.1. Cartesian coordinates of a point are given as (x, y, z). Alternatively, we can describe the point in cylindrical coordinates as (ρ, ϕ, z), with

$$\begin{aligned} x &= \rho \cos\phi, \\ y &= \rho \sin\phi, \end{aligned} \tag{B.1}$$

or in spherical coordinates (r, ϕ, θ), with

$$\begin{aligned} x &= r \cos\phi \sin\theta, \\ y &= r \sin\phi \sin\theta, \\ z &= r \cos\theta. \end{aligned} \tag{B.2}$$

Unit vectors

We can define unit vectors for our coordinate system as follows: $\hat{\boldsymbol{x}}$, $\hat{\boldsymbol{y}}$, and $\hat{\boldsymbol{z}}$ are vectors of unit length pointing in the positive x, y, and z directions respectively. In cylindrical coordinates, we take the cartesian unit vector $\hat{\boldsymbol{z}}$ and add $\hat{\boldsymbol{\rho}}$ and $\hat{\boldsymbol{\phi}}$ as the unit vectors pointing in the direction of increasing ρ and ϕ. For spherical coordinates, we retain $\hat{\boldsymbol{\phi}}$ and add $\hat{\boldsymbol{r}}$ and $\hat{\boldsymbol{\theta}}$ as unit vectors pointing in the direction of increasing r and θ. We could, if desired, define arbitrary orthogonal coordinates as (u_1, u_2, u_3), with corresponding unit vectors $\hat{\boldsymbol{u}}_1$, $\hat{\boldsymbol{u}}_2$, and $\hat{\boldsymbol{u}}_3$.

Dot and cross products

A dot product of two vectors $\boldsymbol{p}$ and $\boldsymbol{q}$ is defined as

$$\boldsymbol{p} \cdot \boldsymbol{q} = pq \cos\sigma, \tag{B.3}$$

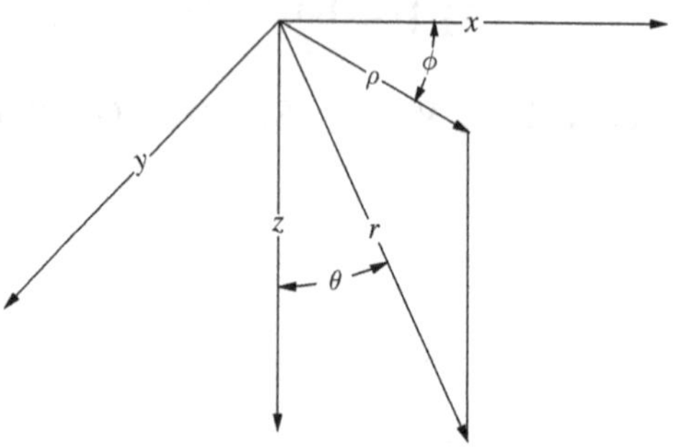

Figure B.1 A right-handed coordinate system with the z-axis pointing down. Cartesian coordinates of a point are denoted by (x, y, z), cylindrical coordinates by (ρ, ϕ, z), and spherical coordinates by (r, ϕ, θ), all as depicted in the figure.

where p and q are the magnitudes of $\boldsymbol{p}$ and $\boldsymbol{q}$, and σ is the angle between them. For a set of orthogonal unit vectors $\hat{\boldsymbol{u}}_i$,

$$\hat{\boldsymbol{u}}_i \cdot \hat{\boldsymbol{u}}_j = \delta_{ij}. \tag{B.4}$$

The cross product of $\boldsymbol{p}$ and $\boldsymbol{q}$ is a vector $\boldsymbol{p} \times \boldsymbol{q}$. Its magnitude is

$$|\boldsymbol{p} \times \boldsymbol{q}| = pq|\sin\sigma|. \tag{B.5}$$

Its direction is orthogonal to $\boldsymbol{p}$ and $\boldsymbol{q}$. Following the right-hand rule, it is the direction of advance for a right-hand screw rotated from $\boldsymbol{p}$ towards $\boldsymbol{q}$. In a right-handed coordinate system, we have

$$\begin{aligned} \hat{\boldsymbol{x}} \times \hat{\boldsymbol{y}} &= \hat{\boldsymbol{z}}, \\ \hat{\boldsymbol{y}} \times \hat{\boldsymbol{z}} &= \hat{\boldsymbol{x}}, \\ \hat{\boldsymbol{z}} \times \hat{\boldsymbol{x}} &= \hat{\boldsymbol{y}}. \end{aligned} \tag{B.6}$$

The cross product can be expressed as a matrix operation. In a right-handed cartesian coordinate system,

$$\boldsymbol{a} \times \boldsymbol{b} = \begin{pmatrix} a_y b_z - a_z b_y \\ a_z b_x - a_x b_z \\ a_x b_y - a_y b_x \end{pmatrix} = \begin{pmatrix} 0 & -a_z & a_y \\ a_z & 0 & -a_x \\ -a_y & a_x & 0 \end{pmatrix} \begin{pmatrix} b_x \\ b_y \\ b_z \end{pmatrix}. \tag{B.7}$$

Vector identities

For any vector $\boldsymbol{a}$,

$$\begin{aligned} \boldsymbol{a} \cdot \boldsymbol{a} &= a^2, \\ \boldsymbol{a} \times \boldsymbol{a} &= \boldsymbol{0}. \end{aligned} \tag{B.8}$$

These simple relations hold for any two vectors $\boldsymbol{a}$ and $\boldsymbol{b}$:

$$\begin{aligned} \boldsymbol{a} \cdot \boldsymbol{b} &= \boldsymbol{b} \cdot \boldsymbol{a}, \\ \boldsymbol{a} \times \boldsymbol{b} &= -\boldsymbol{b} \times \boldsymbol{a}. \end{aligned} \tag{B.9}$$

The following relations hold for any three vectors $\boldsymbol{a}, \boldsymbol{b}$, and $\boldsymbol{c}$:

$$\begin{aligned}
&\boldsymbol{a} \times (\boldsymbol{b} \times \boldsymbol{c}) = \boldsymbol{b}(\boldsymbol{a} \cdot \boldsymbol{c}) - \boldsymbol{c}(\boldsymbol{a} \cdot \boldsymbol{b}),\\
&(\boldsymbol{a} \times \boldsymbol{b}) \times \boldsymbol{c} = \boldsymbol{b}(\boldsymbol{a} \cdot \boldsymbol{c}) - \boldsymbol{a}(\boldsymbol{b} \cdot \boldsymbol{c}),\\
&\boldsymbol{a} \times \boldsymbol{b} \cdot \boldsymbol{c} = \boldsymbol{a} \cdot \boldsymbol{b} \times \boldsymbol{c} = \mathrm{Det}\begin{pmatrix} a_x & a_y & a_z \\ b_x & b_y & b_z \\ c_x & c_y & c_z \end{pmatrix} = \mathrm{Det}\begin{pmatrix} \boldsymbol{a}^T \\ \boldsymbol{b}^T \\ \boldsymbol{c}^T \end{pmatrix}.
\end{aligned} \tag{B.10}$$

This last expression is the volume of the parallelepiped defined by vectors $\boldsymbol{a}, \boldsymbol{b}$, and $\boldsymbol{c}$.

B.2 Differential operations

Cartesian coordinates

If the location of a point changes infinitesimally, the change can be expressed in cartesian coordinates as

$$d\boldsymbol{r} = \hat{\boldsymbol{x}}dx + \hat{\boldsymbol{y}}dy + \hat{\boldsymbol{z}}dz. \tag{B.11}$$

The gradient of a scalar function $\Psi(x, y, z)$ is

$$\nabla\Psi = \hat{\boldsymbol{x}}\frac{\partial\Psi}{\partial x} + \hat{\boldsymbol{y}}\frac{\partial\Psi}{\partial y} + \hat{\boldsymbol{z}}\frac{\partial\Psi}{\partial z}. \tag{B.12}$$

The divergence of a vector function $\boldsymbol{A}(x, y, z)$ is

$$\nabla \cdot \boldsymbol{A} = \frac{\partial A_x}{\partial x} + \frac{\partial A_y}{\partial y} + \frac{\partial A_z}{\partial z}. \tag{B.13}$$

The curl of $\boldsymbol{A}$ is

$$\begin{aligned}
\nabla \times \boldsymbol{A} &= \hat{\boldsymbol{x}}\left(\frac{\partial A_z}{\partial y} - \frac{\partial A_y}{\partial z}\right) + \hat{\boldsymbol{y}}\left(\frac{\partial A_x}{\partial z} - \frac{\partial A_z}{\partial x}\right) + \hat{\boldsymbol{z}}\left(\frac{\partial A_y}{\partial x} - \frac{\partial A_x}{\partial y}\right)\\
&= \begin{pmatrix} 0 & -\partial_z & \partial_y \\ \partial_z & 0 & -\partial_x \\ -\partial_y & \partial_x & 0 \end{pmatrix}\begin{pmatrix} A_x \\ A_y \\ A_z \end{pmatrix}.
\end{aligned} \tag{B.14}$$

The Laplacian of Ψ is

$$\nabla^2\Psi = \frac{\partial^2\Psi}{\partial x^2} + \frac{\partial^2\Psi}{\partial y^2} + \frac{\partial^2\Psi}{\partial z^2}. \tag{B.15}$$

Cylindrical coordinates

For cylindrical coordinates, the corresponding quantities are

$$d\boldsymbol{r} = \hat{\boldsymbol{\rho}}d\rho + \hat{\boldsymbol{\phi}}\rho d\phi + \hat{\boldsymbol{z}}dz, \tag{B.16}$$

$$\nabla\Psi = \hat{\boldsymbol{\rho}}\frac{\partial\Psi}{\partial\rho} + \hat{\boldsymbol{\phi}}\frac{1}{\rho}\frac{\partial\Psi}{\partial\phi} + \hat{\boldsymbol{z}}\frac{\partial\Psi}{\partial z}, \tag{B.17}$$

$$\nabla \cdot \boldsymbol{A} = \frac{1}{\rho}\frac{\partial\left(\rho A_\rho\right)}{\partial\rho} + \frac{1}{\rho}\frac{\partial A_\phi}{\partial\phi} + \frac{\partial A_z}{\partial z}, \tag{B.18}$$

$$\nabla \times \boldsymbol{A} = \hat{\boldsymbol{\rho}}\left(\frac{1}{\rho}\frac{\partial A_z}{\partial \phi} - \frac{\partial A_\phi}{\partial z}\right) + \hat{\boldsymbol{\phi}}\left(\frac{\partial A_\rho}{\partial z} - \frac{\partial A_z}{\partial \rho}\right) + \hat{\boldsymbol{z}}\left(\frac{1}{\rho}\frac{\partial\left(\rho A_\phi\right)}{\partial \rho} - \frac{1}{\rho}\frac{\partial A_\rho}{\partial \phi}\right), \tag{B.19}$$

$$\nabla^2 \Psi = \frac{1}{\rho}\frac{\partial}{\partial \rho}\left(\rho\frac{\partial \Psi}{\partial \rho}\right) + \frac{1}{\rho^2}\frac{\partial^2 \Psi}{\partial \phi^2} + \frac{\partial^2 \Psi}{\partial z^2}. \tag{B.20}$$

Spherical coordinates

In spherical coordinates, we have

$$d\boldsymbol{r} = \hat{\boldsymbol{r}}dr + \hat{\boldsymbol{\phi}}r\sin\theta d\phi + \hat{\boldsymbol{\theta}}rd\theta, \tag{B.21}$$

$$\nabla \Psi = \hat{\boldsymbol{r}}\frac{\partial \Psi}{\partial r} + \hat{\boldsymbol{\theta}}\frac{1}{\rho}\frac{\partial \Psi}{\partial \theta} + \hat{\boldsymbol{\phi}}\frac{1}{r\sin\theta}\frac{\partial \Psi}{\partial \phi}, \tag{B.22}$$

$$\nabla \cdot \boldsymbol{A} = \frac{1}{r^2}\frac{\partial\left(r^2 A_r\right)}{\partial r} + \frac{1}{r\sin\theta}\frac{\partial\left(A_\theta \sin\theta\right)}{\partial \theta} + \frac{1}{r\sin\theta}\frac{\partial A_\phi}{\partial \phi}. \tag{B.23}$$

$$\begin{aligned}\nabla \times \boldsymbol{A} = \hat{\boldsymbol{r}}&\frac{1}{r\sin\theta}\left(\frac{\partial\left(A_\phi \sin\theta\right)}{\partial \theta} - \frac{\partial A_\theta}{\partial \phi}\right)\\ &+ \hat{\boldsymbol{\theta}}\frac{1}{r}\left(\frac{1}{\sin\theta}\frac{\partial A_r}{\partial \phi} - \frac{\partial\left(rA_\phi\right)}{\partial r}\right)\\ &+ \hat{\boldsymbol{\phi}}\frac{1}{r}\left(\frac{\partial\left(rA_\theta\right)}{\partial r} - \frac{\partial A_r}{\partial \theta}\right),\end{aligned} \tag{B.24}$$

$$\nabla^2 \Psi = \frac{1}{r^2}\frac{\partial}{\partial r}\left(r^2\frac{\partial \Psi}{\partial r}\right) + \frac{1}{r^2\sin\theta}\frac{\partial}{\partial \theta}\left(\sin\theta\frac{\partial \Psi}{\partial \theta}\right) + \frac{1}{r^2\sin^2\theta}\frac{\partial^2 \Psi}{\partial \phi^2}. \tag{B.25}$$

General orthogonal coordinates

For general orthogonal coordinates,

$$d\boldsymbol{r} = \hat{\boldsymbol{u}}_1 h_1 du_1 + \hat{\boldsymbol{u}}_2 h_2 du_2 + \hat{\boldsymbol{u}}_3 h_3 du_3. \tag{B.26}$$

$$\nabla \Psi = \hat{\boldsymbol{u}}_1\frac{1}{h_1}\frac{\partial \Psi}{\partial u_1} + \hat{\boldsymbol{u}}_2\frac{1}{h_2}\frac{\partial \Psi}{\partial u_2} + \hat{\boldsymbol{u}}_3\frac{1}{h_3}\frac{\partial \Psi}{\partial u_3}. \tag{B.27}$$

The expression for the gradient follows from the requirement that $\nabla \Phi \cdot d\boldsymbol{r} = d\Phi$.

$$\nabla \cdot \boldsymbol{A} = \frac{1}{h_1 h_2 h_3}\left[\frac{\partial\left(h_2 h_3 A_1\right)}{\partial u_1} + \frac{\partial\left(h_1 h_3 A_2\right)}{\partial u_2} + \frac{\partial\left(h_1 h_2 A_3\right)}{\partial u_3}\right]. \tag{B.28}$$

The expression for the divergence follows from the relation $\nabla \cdot \boldsymbol{A} dV = \boldsymbol{A} \cdot d\boldsymbol{S}$, with dV an elemental volume bounded by the elemental surface $d\boldsymbol{S}$. The volume element

$dV = h_1 h_2 h_3 du_1 du_2 du_3$, and $d\boldsymbol{S}$ has six components of the form $\hat{\boldsymbol{u}}_i h_j h_k du_j du_k$, with i, j, and k a cyclic permutation of 1, 2, and 3.

$$\nabla \times \boldsymbol{A} = \hat{\boldsymbol{u}}_1 \frac{1}{h_2 h_3}\left(\frac{\partial (h_3 A_3)}{\partial u_2} - \frac{\partial (h_2 A_2)}{\partial u_3}\right) + \hat{\boldsymbol{u}}_2 \frac{1}{h_1 h_3}\left(\frac{\partial (h_1 A_1)}{\partial u_3} - \frac{\partial (h_3 A_3)}{\partial u_1}\right) + \hat{\boldsymbol{u}}_3 \frac{1}{h_1 h_2}\left(\frac{\partial (h_2 A_2)}{\partial u_1} - \frac{\partial (h_1 A_1)}{\partial u_2}\right), \tag{B.29}$$

The expression for the curl derives from the relation $\nabla \times \boldsymbol{A} \cdot d\boldsymbol{S} = \boldsymbol{A} \cdot d\boldsymbol{l}$, with $d\boldsymbol{S}$ an elemental surface and $\boldsymbol{A} \cdot d\boldsymbol{l}$ an elemental line integral around the perimeter of the surface.

$$\nabla^2 \Psi = \frac{1}{h_1 h_2 h_3}\left[\frac{\partial}{\partial u_1}\left(\frac{h_2 h_3}{h_1}\frac{\partial \Psi}{\partial u_1}\right) + \frac{\partial}{\partial u_2}\left(\frac{h_1 h_3}{h_2}\frac{\partial \Psi}{\partial u_2}\right) + \frac{\partial}{\partial u_3}\left(\frac{h_1 h_2}{h_3}\frac{\partial \Psi}{\partial u_3}\right)\right]. \tag{B.30}$$

The Laplacian can be obtained by combining the expressions for the gradient and the divergence.

Differential vector identities

$$\nabla \cdot (\Psi \boldsymbol{A}) = \Psi \nabla \cdot \boldsymbol{A} + \boldsymbol{A} \cdot \nabla \Psi, \tag{B.31}$$

$$\nabla \times (\Psi \boldsymbol{A}) = \Psi \nabla \times \boldsymbol{A} - \boldsymbol{A} \times \nabla \Psi, \tag{B.32}$$

$$\nabla \cdot (\boldsymbol{A} \times \boldsymbol{B}) = \boldsymbol{B} \cdot \nabla \times \boldsymbol{A} + \boldsymbol{A} \cdot \nabla \times \boldsymbol{B}, \tag{B.33}$$

$$\nabla \cdot (\nabla \times \boldsymbol{A}) = 0, \tag{B.34}$$

$$\nabla \times (\nabla \times \boldsymbol{A}) = \nabla(\nabla \cdot \boldsymbol{A}) - \nabla^2 \boldsymbol{A}, \tag{B.35}$$

$$\nabla \cdot (\Psi \nabla \Psi) = \Psi \nabla^2 \Psi + (\nabla \Psi)^2, \tag{B.36}$$

$$\nabla \times (\nabla \Psi) = 0. \tag{B.37}$$

Appendix C
Fourier and Radon transforms

Transformation of seismic data into domains other than space–time is often useful both conceptually and computationally. Of particular note are the Fourier transform and its close relative the Radon transform.

C.1 The Fourier transform

C.1.1 The forward and inverse temporal Fourier transform

Given a temporal function $a(t)$, we define its Fourier transform to be

$$a(\omega) = \int dt\, a(t)e^{i\omega t}, \tag{C.1}$$

and its inverse Fourier transform to be

$$a(t) = \frac{1}{2\pi}\int d\omega\, a(\omega)e^{-i\omega t}. \tag{C.2}$$

If t is time, then its Fourier conjugate ω is the corresponding *angular frequency*. (Generally, we will use the same symbol (a, in this case) to denote a function and its Fourier transform, allowing the argument (t or ω) of the function to tell us which domain the function is in. The justification for this is that both the function and its Fourier transform contain the same information, just represented in different domains.)

That the forward and inverse Fourier transform are indeed each other's inverses is easily confirmed using the orthogonality identity

$$\frac{1}{2\pi}\int dte^{i(\omega-\omega')t} = \delta(\omega-\omega'), \tag{C.3}$$

or the completeness identity

$$\frac{1}{2\pi}\int d\omega e^{i\omega(t-t')} = \delta(t-t'). \tag{C.4}$$

We have chosen to normalize the transform in such a way that a delta function in the t-domain transforms into the unit function in the ω-domain. That is, if $a(t) = \delta(t)$, then $a(\omega) = 1$. The reciprocal relation is the same except for scale: if $a(\omega) = \delta(\omega)$, then $a(t) = 1/(2\pi)$. If perfect symmetry between the forward and inverse transforms is desired, it can be achieved by rescaling the conjugate variable to cyclical frequency $\nu = \omega/(2\pi)$. Then the forward and inverse transforms would be

$$a(\nu) = \int dt a(t) e^{i2\pi\nu t}, \tag{C.5}$$

and

$$a(t) = \int d\nu a(\nu) e^{-i2\pi\nu t}. \tag{C.6}$$

In these expressions, the factor of 2π has been transferred to the exponent. No matter what one does, a factor of 2π is going to pop up somewhere, and we are content to leave it in the denominator of the inverse transform as per equation (C.2).

C.1.2 The Fourier transform of convolutions and linear operators

With our chosen normalization, a convolution in the t-domain becomes a product in the ω-domain. Suppose $f(t)$ is a convolutional filter applied to the function $a(t)$, producing $b(t)$:

$$b(t) = \int dt' f(t - t') a(t'). \tag{C.7}$$

Then

$$\begin{aligned} b(\omega) &= \int dt b(t) e^{i\omega t} \\ &= \int dt \int dt' f(t - t') a(t') e^{i\omega t} \\ &= \int dt \int dt' f(t - t') e^{i\omega(t - t')} a(t') e^{i\omega t'}. \end{aligned} \tag{C.8}$$

Changing integration variable from t to $t'' = t - t'$, and rearranging,

$$\begin{aligned} b(\omega) &= \int dt' a(t') e^{i\omega t'} \int dt'' f(t'') e^{i\omega t''} \\ &= a(\omega) f(\omega). \end{aligned} \tag{C.9}$$

For a general linear operator, the situation is a little more complicated. Suppose

$$b(t) = \int dt' \mathcal{O}(t|t') a(t'). \tag{C.10}$$

For the operator $\mathcal{O}$, arguments to the right of the vertical bar (in this case, t') are input variables, while arguments to the left (here, t) are the output variables. A Fourier transform of an output variable is straightforward:

$$\begin{aligned} b(\omega) &= \int dt b(t) e^{i\omega t} \\ &= \int dt' \int dt e^{i\omega t} \mathcal{O}(t|t') a(t') \\ &= \int dt' \mathcal{O}(\omega|t') a(t'), \end{aligned} \tag{C.11}$$

provided we define

$$\mathcal{O}(\omega|t') \equiv \int dt \mathcal{O}(t|t') e^{i\omega t}. \tag{C.12}$$

For the input variable, we have (expanding $a(t')$ into its Fourier components)

$$\begin{aligned} b(\omega) &= \int dt' \mathcal{O}(\omega|t') \frac{1}{2\pi} \int d\omega' a(\omega') e^{-i\omega' t'} \\ &= \frac{1}{2\pi} \int d\omega' a(\omega') \int dt' \mathcal{O}(\omega|t') e^{-i\omega' t'} \\ &= \frac{1}{2\pi} \int d\omega' \mathcal{O}(\omega|\omega') a(\omega'), \end{aligned} \tag{C.13}$$

provided we choose

$$\mathcal{O}(\omega|\omega') \equiv \int dt' \mathcal{O}(\omega|t') e^{-i\omega t'}. \tag{C.14}$$

That is, the phase of the Fourier transform of an input variable of a linear operator is chosen opposite in sign to the phase of the Fourier transform of an input variable. With this convention, the integral form of the linear operation is the same in either domain, except for the scale factor of $1/(2\pi)$ in the frequency-domain form (C.13). Because of the normalization chosen for the Fourier transform, we can expect this scale factor to appear whenever there is an integral over ω to perform. Had we chosen to parameterize the transform in terms of frequency ν rather than angular frequency ω, the factor of $1/(2\pi)$ would be absent.

C.1.3 The forward and inverse spatial Fourier transform

For Fourier transforms over spatial coordinates, we give the phase term the opposite sign. The spatial Fourier transform from cartesian coordinate x to *wavenumber* or *spatial frequency* k is

$$a(k) = \int dx a(x) e^{-ikx}. \tag{C.15}$$

The sign of the exponent for the forward transform of the spatial variables has been chosen to be negative. The Fourier transform effects a decomposition into components of all possible spatial frequencies.

The inverse spatial Fourier transform is

$$a(x) = \frac{1}{(2\pi)} \int dk a(k) e^{+ikx}. \tag{C.16}$$

As for the temporal transform, one could, if desired, eliminate the asymmetry in scale between the forward and inverse transforms by parameterizing the transform in terms of $q = k/(2\pi)$.

The reason for choosing opposite signs for the phase terms in the spatial and temporal transforms is apparent if we look at a plane wave in space and time. Suppose $a(\boldsymbol{x}, t)$ corresponds to a wave of a particular (positive) frequency ω_0 and wavenumber $\boldsymbol{k}_0$, so that its Fourier transform is proportional to

$$a(\boldsymbol{k}, \omega) \propto \delta(\omega - \omega_0) \delta(\boldsymbol{k} - \boldsymbol{k}_0). \tag{C.17}$$

Then, in the space–time domain,

$$a(\boldsymbol{x}, t) \propto e^{+i(\boldsymbol{k}_0 \cdot \boldsymbol{x} - \omega_0 t)}. \tag{C.18}$$

This is the expression for a plane wave of frequency ω_0 and wavenumber vector $\boldsymbol{k}_0$, traveling in the direction of $\boldsymbol{k}_0$. Had we chosen the same sign convention for the spatial and temporal transform, the travel would have been in the direction opposite to the direction of $\boldsymbol{k}_0$.

Linear operations work in space like they do in time. If $\mathcal{O}(\boldsymbol{x}|\boldsymbol{x}')$ represents a linear operation in the space domain, then

$$\mathcal{O}(\boldsymbol{k}|\boldsymbol{x}') \equiv \int d\boldsymbol{x} \mathcal{O}(\boldsymbol{x}|\boldsymbol{x}') e^{-i\boldsymbol{k}\cdot\boldsymbol{x}}, \tag{C.19}$$

but

$$\mathcal{O}(\boldsymbol{x}|\boldsymbol{k}') \equiv \int d\boldsymbol{x}' \mathcal{O}(\boldsymbol{x}|\boldsymbol{x}') e^{+i\boldsymbol{k}'\cdot\boldsymbol{x}'}. \tag{C.20}$$

C.1.4 Discrete Fourier transforms

Under certain assumptions, the Fourier integral can be reduced to a discrete sum. Consider a function of one variable $a(x)$ with Fourier transform $a(k)$. Suppose, for instance, that $a(k)$ is non-zero only in the region $(-k_f, k_f)$, so that

$$a(x) = \frac{1}{2\pi} \int_{-k_f}^{k_f} dk a(k) e^{+ikx}. \tag{C.21}$$

In this case, $a(k)$ can be built from a discrete set $a(x_i)$, with $x_i = i\Delta x = i\pi / k_f$. That is, define a discrete transform $a_{\text{dis}}(k)$ to be

$$a_{\text{dis}}(k) \equiv \Delta x \sum_{i=-\infty}^{\infty} a(x_i) e^{-ikx_i}, \quad -k_f < k < k_f. \tag{C.22}$$

Writing $a(x_i)$ in terms of $a(k)$, we can write $a_{\text{dis}}(k)$ in terms of $a(k)$:

$$a_{\text{dis}}(k) = \Delta x \sum_{i=-\infty}^{\infty} \frac{1}{2\pi} \int_{-k_f}^{k_f} dk' a(k') e^{+i(k'-k)x_i}. \tag{C.23}$$

Rearranging the integrals,

$$a_{\text{dis}}(k) = \frac{\Delta x}{2\pi} \int_{-k_f}^{k_f} dk' a(k') \sum_{i=-\infty}^{\infty} e^{+i(k'-k)\Delta x i}. \tag{C.24}$$

Invoke the delta-function identity

$$\frac{1}{2\pi} \sum_{i=-\infty}^{\infty} e^{-iui} = \sum_{m=-\infty}^{\infty} \delta(u - 2\pi m). \tag{C.25}$$

Then

$$a_{\text{dis}}(k) = \Delta x \int_{-k_f}^{k_f} dk' a(k') \sum_{m=-\infty}^{\infty} \delta((k'-k)\Delta x - 2\pi m). \tag{C.26}$$

Both k and k' have magnitudes less than k_f, so $|k' - k|\Delta x < 2\pi$, and the delta function can be satisfied only for $m = 0$. Integrating over k', we get

$$a_{\text{dis}}(k) = a(k); \tag{C.27}$$

i.e.,

$$a(k) = \Delta x \sum_{i=-\infty}^{\infty} a(x_i) e^{-ikx_i}. \tag{C.28}$$

For values of x other than x_i, we can write the inverse transform

$$a(x) = \frac{1}{2\pi} \int_{-k_f}^{k_f} dk a(k) e^{+ikx}, \tag{C.29}$$

or

$$a(x) = \frac{\Delta x}{2\pi} \int_{-k_f}^{k_f} dk \sum_{i=-\infty}^{\infty} a(x_i)\, e^{+ik\,(x-x_i)}. \tag{C.30}$$

Performing the integral over k,

$$\begin{aligned} a(x) &= \frac{\Delta x}{\pi} \sum_{i=-\infty}^{\infty} a(x_i) \frac{\sin\left(k_f\,(x - x_i)\right)}{(x - x_i)} \\ &= \sum_{i=-\infty}^{\infty} a(x_i)\, \frac{\sin\left(k_f\,(x - x_i)\right)}{k_f\,(x - x_i)}. \end{aligned} \tag{C.31}$$

That is, $a(x)$ for all values of x and $a(k)$ for all values of k both can be generated from the discrete set of values $a\,(x_i)$.

The discrete Fourier transform and its inverse can be implemented very efficiently using algorithms such as the fast Fourier transform, or FFT.

C.1.5 Doubly discrete Fourier transforms

What happens if $a\,(x_i)$ is also non-zero only over a finite range of x_i? Well, strictly speaking, that can't happen. An annoying little fact prevents both a function and its Fourier transform from having compact support. However, we can do something almost as good. Suppose that $a(x)$ repeats itself over intervals $2X = M\Delta x$, so that $a(x + 2X) = a(x)$. Then the discrete Fourier transform (C.28) becomes

$$a(k) = \Delta x \sum_{i=0}^{M-1} a(x_i) e^{-ikx_i} \sum_{r=-\infty}^{\infty} e^{-ik\Delta x Mr}. \tag{C.32}$$

Invoking the identity (C.25),

$$a(k) = 2\Delta x \sum_{i=0}^{M-1} a(x_i) e^{-ikx_i} \sum_{m=-\infty}^{\infty} \delta\left(Mk/k_f - 2m\right). \tag{C.33}$$

Thus, $a(k)$ is zero except at discrete values $k_m = 2m\, k_f / M$. Since the magnitude of k_m is bounded by k_f, the magnitude of m is bounded by $M/2$. Because of the delta functions, $a\,(k_m)$ is strictly infinite. However, if we define a core quantity

$$A_m = \frac{1}{M} \sum_{i=0}^{M-1} a(x_i) e^{-i2\pi mi/M}, \quad m = -M/2 + 1, M/2, \tag{C.34}$$

then

$$a(k) = 2\pi \sum_m A_m \delta\left(k - 2m\, k_f / M\right). \tag{C.35}$$

The discrete set $a\,(x_i)$ is obtainable from the set A_m from the inverse relation

$$a\,(x_i) = \sum_{m=-M/2+1}^{M/2} A_m e^{i2\pi mi/M}. \tag{C.36}$$

The function appears finite and discrete in both the x and k domains. However, the interpretations are not the same. In the x domain, we assumed that the function repeats itself at regular intervals, leading to a fundamentally discrete set in k-space. Other than at the discrete values k_i, $a(k) = 0$. In the k-domain, we assumed that the function is non-zero only over a finite interval, which implies that the x-space function exists for all x values, and is obtainable from the discrete set via a sinc-function interpolation.

If one does interpolate between the discrete k-space values, one is implicitly altering the assumptions from which the discrete expressions derive. Specifically, one is replacing the assumption of a repeating x-space function with something else, that something else depending on the method of interpolation. The upshot is that interpolation in k-space is less straightforward than interpolation in x-space.

C.2 The Radon transform

C.2.1 The forward Radon transform

A close relative of the Fourier transform is the Radon transform or slant stack. The Radon transform is commonly performed on functions of one or two space coordinates and one time coordinate. Given a function $f\,(\tilde{x}, t)$ of n (= 1 or 2) spatial variables and a time variable, we define the n-dimensional Radon transform as

$$f\,(\tilde{p}, \tau) = \int d\tilde{x}\, f\,(\tilde{x}, t = \tau + \tilde{p} \cdot \tilde{x}). \tag{C.37}$$

The Radon transform (C.37) leaves the time argument τ in its original domain, though shifted by the terms $\tilde{p} \cdot \tilde{x}$; the remaining arguments are transformed from the $\tilde{x}$ domain to the $\tilde{p}$ domain. The Radon conjugate variables $\tilde{p}$ denote slopes in particular directions. The Radon transform is sometimes referred to as a slant stack, or, bowing to nomenclature, a p–τ transform.

C.2.2 Relating the Radon and Fourier transforms

For the relation between the Radon and Fourier transforms, take a Fourier transform over τ:

$$f(\tilde{\boldsymbol{p}}, \omega) = \int d\tau f(\tilde{\boldsymbol{p}}, \tau) e^{i\omega\tau}. \tag{C.38}$$

Substitute the definition (C.37):

$$f(\tilde{\boldsymbol{p}}, \omega) = \int d\tau e^{i\omega\tau} \int d\tilde{\boldsymbol{x}} f(\tilde{\boldsymbol{x}}, t = \tau + \tilde{\boldsymbol{p}} \cdot \tilde{\boldsymbol{x}}). \tag{C.39}$$

Add and subtract the terms $\tilde{\boldsymbol{p}} \cdot \tilde{\boldsymbol{x}}$ to and from the exponent:

$$f(\tilde{\boldsymbol{p}}, \omega) = \int d\tilde{\boldsymbol{x}} e^{-i\omega\tilde{\boldsymbol{p}}\cdot\tilde{\boldsymbol{x}}} \int d\tau e^{i\omega} (\tau + \tilde{\boldsymbol{p}} \cdot \tilde{\boldsymbol{x}}) f(\tilde{\boldsymbol{x}}, \tau + \tilde{\boldsymbol{p}} \cdot \tilde{\boldsymbol{x}}). \tag{C.40}$$

Change the last integration variable from τ to $t = \tau + \tilde{\boldsymbol{p}} \cdot \tilde{\boldsymbol{x}}$:

$$f(\tilde{\boldsymbol{p}}, \omega) = \int d\tilde{\boldsymbol{x}} e^{-i\omega\tilde{p}\cdot\tilde{x}} \int dt e^{i\omega t} f(\tilde{\boldsymbol{x}}, t). \tag{C.41}$$

All the integrals are now recognizable as Fourier transforms. If $f\left(\tilde{\boldsymbol{k}}, \omega\right)$ is the $n + 1$-fold Fourier transform of $f(\tilde{\boldsymbol{x}}, t)$, then the Radon–Fourier transform is just the full Fourier transform at $\tilde{\boldsymbol{k}} = \omega\tilde{\boldsymbol{p}}$:

$$f(\tilde{\boldsymbol{p}}, \omega) = f\left(\tilde{\boldsymbol{k}} = \omega\tilde{\boldsymbol{p}}, \omega\right). \tag{C.42}$$

C.2.3 The inverse Radon transform

Since the Fourier and Radon transforms are related, the inverse Radon transform is readily calculated from the inverse Fourier transform. Starting from the definition (C.2) of the inverse Fourier transform, and changing integration variables from $\tilde{\boldsymbol{k}}$ to $\tilde{\boldsymbol{p}} = \tilde{\boldsymbol{k}}/\omega$,

$$\begin{aligned} f(\tilde{\boldsymbol{x}}, t) &= \frac{1}{(2\pi)^{n+1}} \int d\tilde{\boldsymbol{k}} \int d\omega f\left(\tilde{\boldsymbol{k}}, \omega\right) e^{+i\left(\tilde{k}\cdot\tilde{x} - \omega t\right)} \\ &= \frac{1}{(2\pi)^{n+1}} \int d\tilde{\boldsymbol{p}} \int d\omega |\omega|^n f\left(\tilde{\boldsymbol{k}} = \omega\tilde{\boldsymbol{p}}, \omega\right) e^{+i\omega(\tilde{p}\cdot\tilde{x} - t)}. \end{aligned} \tag{C.43}$$

Invoking the relation (C.42) between the radon and Fourier transforms,

$$f(\tilde{\boldsymbol{x}}, t) = \frac{1}{(2\pi)^{n+1}} \int d\tilde{\boldsymbol{p}} \int d\omega |\omega|^n f(\tilde{\boldsymbol{p}}, \omega) e^{+i\omega(\tilde{p}\cdot\tilde{x} - t)}. \tag{C.44}$$

Define the filtered function

$$\tilde{f}^{(n\rho)}(\tilde{\boldsymbol{p}}, \omega) \equiv |\omega|^n f(\tilde{\boldsymbol{p}}, \omega). \tag{C.45}$$

The filter $|\omega|^n$ is recognizable as a multidimensional "rho" filter $\rho(\omega)^n$ (see Appendix E).

With this definition, (C.44) can be written

$$f(\tilde{\boldsymbol{x}}, t) = \frac{1}{(2\pi)^{n+1}} \int d\tilde{\boldsymbol{p}} \int d\omega \tilde{f}^{(n\rho)}(\tilde{\boldsymbol{p}}, \omega) e^{-i\omega(t - \tilde{p}\cdot\tilde{x})}. \tag{C.46}$$

The integral over ω is an inverse Fourier transform, returning f from p–ω space to p–τ space:

$$f(\tilde{\boldsymbol{x}}, t) = \frac{1}{(2\pi)^n} \int d\tilde{\boldsymbol{p}}\, \tilde{f}^{(n\rho)}(\tilde{\boldsymbol{p}}, \tau = t - \tilde{\boldsymbol{p}} \cdot \tilde{\boldsymbol{x}}) \,. \tag{C.47}$$

The inverse Radon transform (C.46), as defined and normalized, has almost the same form as the forward transform (C.37). Besides the switch in integration variable, it differs in the signs of the delay terms $\tilde{\boldsymbol{p}} \cdot \tilde{\boldsymbol{x}}$, the normalization factor $(2\pi)^{-n}$, and the rho-filter $|\omega|^n$, where n, the number of spatial coordinates in the transform, may be one or two.

Appendix D

Surface and pointwise reflectivity

Which is the more fundamental concept, specular surface reflectivity or pointwise reflectivity? Rather than resolve this issue, we demonstrate here that either quantity can be decomposed in terms of the other.

D.1 Specular reflecting surfaces and point reflectors

Consider a specular reflecting surface passing through the point $\boldsymbol{x}_r = (x_r, y_r, z_r)$ with slope tangents $\boldsymbol{p} = \left(p_x, p_y\right)$ in the x and y directions. It may be expressed as

$$\mathcal{R}_{\mathcal{S}}\left(\boldsymbol{x}; \boldsymbol{x}_r; \boldsymbol{p}\right) = R\sqrt{1 + {p_x}^2 + {p_y}^2}\delta\left(z - z_r - \left(x - x_r\right)p_x - \left(y - y_r\right)p_y\right). \quad \text{(D.1)}$$

This expression is normalized to a delta function in the direction normal to the surface $z = z_r + (x - x_r)\,p_x + (y - y_r)\,p_y$. The set of all possible reflecting surfaces can be specified from a single value each for x_r and y_r; multiple values are included in (D.1) for convenience.

In contrast, a point reflector at $\boldsymbol{x}_0 = (x_0, y_0, z_0)$ has the form

$$\mathcal{R}_{\mathcal{P}}\left(\boldsymbol{x}; \boldsymbol{x}_0\right) = R\delta\left(z - z_0\right)\delta\left(y - y_0\right)\delta\left(x - x_0\right). \quad \text{(D.2)}$$

D.2 Decomposition of a specular reflector into point reflectors

It should be clear that the specular reflection (D.1) can be formed from a superposition of point reflections (D.2). Starting with (D.1) with $x_r = y_r = 0$, add integrals over x_0 and y_0:

$$\begin{aligned}\mathcal{R}_{\mathcal{S}}\left(\boldsymbol{x};\, z_r;\, \boldsymbol{p}\right) = {} & R\sqrt{1 + {p_x}^2 + {p_y}^2} \\ & \times \int\!\!\int dx_0 dy_0 \delta\left(z - z_r - x_0 p_x - y_0 p_y\right)\delta\left(y - y_0\right)\delta\left(x - x_0\right). \quad \text{(D.3)}\end{aligned}$$

The integrand is proportional to the point reflector (D.2):

$$\begin{aligned}\mathcal{R}_{\mathcal{S}}\left(\boldsymbol{x}; z_r; \boldsymbol{p}\right) = {} & \sqrt{1 + {p_x}^2 + {p_y}^2}\int\!\!\int dx_0\, dy_0 \\ & \times \mathcal{R}_{\mathcal{P}}\left(\boldsymbol{x}; \left(x_0, y_0, z_r + x_0 p_x + y_0 p_y\right)\right). \quad \text{(D.4)}\end{aligned}$$

That is, the line reflector is just a collection of point reflectors along the reflecting surface.

D.3 Decomposition in k-space of specular and point reflectors

D.3.1 Line reflector

Invoking (C.1), the Fourier transform of a line reflector becomes

$$\mathcal{R}_{\mathcal{S}}(\boldsymbol{k}, \boldsymbol{x}_r; \boldsymbol{p}) = R\sqrt{1 + {p_x}^2 + {p_y}^2} \int dx \int dy \int dz \times e^{-i(k_x x + k_y y + k_z z)} \delta\left(z - z_r - (x - x_r)\, p_x - (y - y_r)\, p_y\right). \quad \text{(D.5)}$$

Evaluating the z-integral,

$$\mathcal{R}_{\mathcal{S}}(\boldsymbol{k}, \boldsymbol{x}_r; \boldsymbol{p}) = R\sqrt{1 + {p_x}^2 + {p_y}^2} \int dx \times \int dy e^{-i(k_x x + k_y y + k_z (z_r + (x - x_r) p_x + (y - y_r) p_y))}. \quad \text{(D.6)}$$

Collecting terms within the exponential,

$$\mathcal{R}_{\mathcal{S}}(\boldsymbol{k}, \boldsymbol{x}_r; \boldsymbol{p}) = R\sqrt{1 + {p_x}^2 + {p_y}^2} e^{-ik_z(z_r - x_r p_x - y_r p_y)} \times \int dx e^{-i(k_x + k_z p_x)x} \int dy\, e^{-i(k_y + k_z p_y)y}. \quad \text{(D.7)}$$

The x- and y-integrals evaluate to delta functions:

$$\mathcal{R}_{\mathcal{S}}(\boldsymbol{k}, \boldsymbol{x}_r; \boldsymbol{p}) = (2\pi)^2 R\sqrt{1 + {p_x}^2 + {p_y}^2} e^{-ik_z(z_r - x_r p_x - y_r p_y)} \times \delta\left(k_x + k_z p_x\right) \delta\left(k_y + k_z p_y\right). \quad \text{(D.8)}$$

Substituting k_x for $k_z p_x$, and likewise for the y-term, we are left with

$$\mathcal{R}_{\mathcal{S}}(\boldsymbol{k}, \boldsymbol{x}_r; \boldsymbol{p}) = (2\pi)^2 R\sqrt{1 + {p_x}^2 + {p_y}^2} e^{-i(k_z z_r + k_x x_r + k_y y_r)} \times \delta\left(k_x + k_z p_x\right) \delta\left(k_y + k_z p_y\right). \quad \text{(D.9)}$$

D.3.2 Point reflector

The k-space expression for the point reflector is somewhat simpler:

$$\mathcal{R}_{\mathcal{P}}(\boldsymbol{k}; \boldsymbol{x}_0) = Re^{-i(k_x x_0 + k_y y_0 + k_z z_0)}. \quad \text{(D.10)}$$

D.4 Decomposition of the point reflector into surface reflections

Exploiting the properties of the delta function, the expression (D.10) for a point reflector in k-space can be written as integrals over p_x and p_y:

$$\mathcal{R}_{\mathcal{P}}(\boldsymbol{k}; \boldsymbol{x}_0) = R|k_z|^2 \int dp_x \int dp_y \times e^{-i(k_x x_0 + k_y y_0 + k_z z_0)} \delta\left(k_x + k_z p_x\right) \delta\left(k_y + k_z p_y\right). \quad \text{(D.11)}$$

Invoking the expression (D.9) for the line reflector, (D.11) becomes

$$\mathcal{R}_{\mathcal{P}}(\boldsymbol{k};\,\boldsymbol{x}_0) = \frac{|k_z|^2}{(2\pi)^2}\int dp_x \int dp_y \frac{\mathcal{R}_{\mathcal{S}}(\boldsymbol{k},\boldsymbol{x}_0;\,\boldsymbol{p})}{\sqrt{1+{p_x}^2+{p_y}^2}}. \tag{D.12}$$

The factor of $|k_z|^2$ appearing in this formula can be considered a filter acting on the reflector $\mathcal{R}_{\mathcal{S}}$. Define

$$\tilde{\mathcal{R}}_{\mathcal{S}}^{(2)}\left(\boldsymbol{k},\boldsymbol{x}_0;\,p_x,p_y\right) \equiv |k_z|^2\mathcal{R}_{\mathcal{S}}(\boldsymbol{k},\boldsymbol{x}_0;\,\boldsymbol{p}). \tag{D.13}$$

The final step is to perform the inverse transform to spatial coordinates:

$$\mathcal{R}_{\mathcal{P}}(\boldsymbol{x};\,\boldsymbol{x}_0) = \frac{1}{(2\pi)^2}\int dp_x \int dp_y \frac{\tilde{\mathcal{R}}_{\mathcal{S}}^{(2)}(\boldsymbol{x},\boldsymbol{x}_0;\,\boldsymbol{p})}{\sqrt{1+{p_x}^2+{p_y}^2}}. \tag{D.14}$$

The point reflector at $\boldsymbol{x}_0$ is here expressed as a superposition of specular surface reflectors passing through $\boldsymbol{x}_0$ at all possible angles, each convolved with a 2-D "rho" filter. The factor $\sqrt{1+{p_x}^2+{p_y}^2}$ appearing in the denominator of this expression is a consequence of the normalization chosen for the surface reflections.

The conclusion to be drawn is that the point reflector is not a more fundamental entity than a line or surface reflector, rather just an alternate basis for describing all possible reflectors.

Appendix E
Useful filters

This appendix contains a brief discussion of some of the filters and functions often encountered in seismic imaging. Not all the filters are completely domesticated, particularly at very small and very large times or frequencies, and occasionally at $t = 0$. Where infinities are encountered, the poor behavior is more a theoretical than a practical problem, in that the functions on which they will be applied are in practice bandlimited, with no very low- or very high-frequency components. Readers with a disposition towards rigor ought to avoid this section, and may instead refer to Appendix A of Bleistein *et al.* (2001). Alternatively, Claerbout (1976) offers a more intuitive, if limited, discussion.

E.1 Time and frequency expressions

We define the forward Fourier transform $f(\omega)$ of a temporal filter $f(t)$ as per Appendix C, equation (C.1), and the inverse transform as per (C.2). The phase term $e^{i\omega t}$ in the forward transform is chosen positive for time functions.

A temporal filter by definition is multiplicative in the frequency domain:

$$\overline{D}(\omega) \equiv f(\omega)D(\omega), \tag{E.1}$$

for any frequency function $D(\omega)$. Then, in the time domain, the filter is a convolution:

$$\begin{aligned}\overline{D}(t) &\equiv \frac{1}{2\pi}\int d\omega e^{-i\omega t} f(\omega)D(\omega)\\ &= \frac{1}{2\pi}\int dt' D(t')\int d\omega e^{-i\omega(t-t')} f(\omega),\end{aligned} \tag{E.2}$$

or

$$\begin{aligned}\overline{D}(t) &= [f * D](t)\\ &= \int dt' D(t') f(t-t').\end{aligned} \tag{E.3}$$

The phase term in Fourier transforms over spatial coordinates are given the opposite sign as for temporal coordinates; i.e. the forward transform from $f(z)$ to $f\,(k_z)$ has phase $e^{-ik_z z}$.

With this convention, the wavenumber-domain expression of a filter applied to a spatial coordinate will be the same as the frequency-domain expression of the same filter applied to the time coordinate, provided frequency is replaced by the negative of the wavenumber.

Positive and negative frequencies

If a filter is real in the time domain, then its values at negative frequencies are complex conjugates of its values at corresponding positive frequencies. That is,

$$f(-\omega) = f(\omega)^*. \tag{E.4}$$

Exploiting this fact, for filters real in the time domain, we will often write frequency-domain expressions only for positive frequencies, with the understanding that expressions at negative frequencies can be found by taking the complex conjugate.

Adjoints and inverses

Most filters will have an adjoint and an inverse. For a real convolutional filter in the time domain, the adjoint is formed by reversing the sign of the argument; i.e.

$$f^{\dagger}(t) = f(-t). \tag{E.5}$$

For a filter multiplicative in the frequency domain, the adjoint is the complex conjugate. The inverse filter undoes the forward filter; i.e.

$$\begin{aligned} \left[f^{-1} * f\right](t) &= \left[f * f^{-1}\right](t) \\ &= \int dt' f^{-1}(t') f(t - t') = \delta(t). \end{aligned} \tag{E.6}$$

In the frequency domain, the inverse filter is the simple arithmetic inverse of the forward filter.

Causal and anticausal filters

A filter that vanishes at all negative times is said to be one-sided or causal. If a filter is causal, its adjoint is necessarily anticausal. A numerical approximation to a causal filter may or may not be causal.

Commutivity

Convolutional filters are indifferent to the order in which they are applied. That is, if f and g are two filters, then

$$[f * g](t) = [g * f](t).$$

In the frequency domain, this property is obvious: Multiplications commute, therefore, multiplicative filters commute. Since the filters commute in the frequency domain, they must commute in the time domain also. Convolutional filters are special in this regard. Linear operators in general do not have this convenient commutivity property.

Scalability

Many of the filters discussed below are scalable in the sense that $f(at) = a^{-\nu} f(t)$. This is a useful property in imaging algorithms, which tend to stretch or squeeze the time and depth coordinates during the operation.

Scalable filters preserve this property when convolved with other scalable filters. If $f(at) = a^{-\nu} f(t)$, and $g(at) = a^{-\mu} g(t)$, and if $h(t) = [f * g](t)$, then

$$\begin{aligned} h(at) &= [f * g](at) \\ &= \int dt' g(t') * f(at - t') \\ &= a \int du g(au) * f(a(t-u)) \\ &= a^{-(\mu+\nu-1)} h(t). \end{aligned} \tag{E.7}$$

The scaling power of the composite filter is the sum of the scaling powers of the two constituent filters, less one.

Non-scalable filters

Though scalability is a useful property, it is not possessed by all filters. Examples of non-scalable filters include sinc functions, Gaussian filters, and Ricker wavelets, as well as the family of inflection filters discussed below. Stretching the argument of a non-scalable filter makes it fatter or leaner, altering its basic shape.

To retain some of the convenience of scalable filters, we adopt the following convention. If $f_c(u)$ is a non-scalable filter, we introduce an additional variable t_d, which we call a dilation factor, and define a class of non-scalable filters as

$$f(t_d, t) \equiv t_d^{-1} f_c\left(t / t_d\right). \tag{E.8}$$

The time dilation factor t_d is a measure of the width of the filter in the time domain. Alternatively, we can define a characteristic frequency $\omega_d = \pi / t_d$ which is a measure of the width of the filter in the frequency domain. Note that f_c is the class member with unit dilation; i.e. $f(1, t) = f_c(t)$.

The factor of t_d^{-1} has been placed in the definition (E.8) for convenience in normalization. The maximum value of $f(t_d, t)$ varies with t_d, while integrals of $f(t_d, t)$ over intervals of t proportional to t_d are invariant.

We define the Fourier transform of a non-scalable filter as we would any other filter. The Fourier transform of the core filter $f_c(u)$ is

$$f_c(q) \equiv \int du e^{iqu} f_c(u). \tag{E.9}$$

In the frequency domain, a dilated filter becomes

$$
\begin{aligned}
f(t_d, \omega) &= \int dt e^{i\omega t} f(t_d, t) \\
&= \frac{1}{t_d} \int dt e^{-i\omega t} f_c(t/t_d) \\
&= \int du e^{i\omega t_d u} f_c(u) \\
&= f_c(\omega t_d) = f_c(\pi\omega/\omega_d)
\end{aligned}
\tag{E.10}
$$

The same dilation factor t_d (or characteristic frequency ω_d) that stretches or compresses the time-domain wavelet also compresses or stretches the frequency-domain wavelet. With the chosen normalization, peak amplitude in the frequency domain is independent of t_d or ω_d.

With the definition (E.8), we can retain our concept of convolution in the presence of a changing dilation factor:

$$
\begin{aligned}
[f(t_d) * D](t) &\equiv \int dt' D(t-t') f(t_d, t') \\
&\equiv t_d^{-1} \int dt' D(t-t') f_c(t'/t_d).
\end{aligned}
\tag{E.11}
$$

The convolution of a scalable filter $g(t)$ with a non-scalable filter $f(t_d, t)$ is non-scalable:

$$
\begin{aligned}
[f(t_d) * g](t) &= t_d^{-1} \int dt' g(t-t') f_c(t'/t_d) \\
&= \int du g(t_d(t/t_d - u)) f_c(u).
\end{aligned}
\tag{E.12}
$$

If $g(at) = a^{-\nu} g(t)$, then

$$
\begin{aligned}
[f(t_d) * g](t) &= t_d^{-\nu} \int du g(t/t_d - u) f_c(u) \\
&= t_d^{-\nu} [f_c * g](t/t_d).
\end{aligned}
\tag{E.13}
$$

We can define a new class of non-scalable filters with the normalization of (E.8) based on the convolution of f_c with g:

$$
h_c(u) \equiv [f_c * g](u). \tag{E.14}
$$

$$
h(t_d, t) \equiv t_d^{-1} h_c(t/t_d). \tag{E.15}
$$

In terms of the new class of filters, equation (E.13) can be written

$$
[f(t_d) * g](t) = t_d^{-\nu} h_c(t/t_d) = t_d^{1-\nu} h(t_d, t). \tag{E.16}
$$

Equation (E.13) or (E.16) provides a general formula for convolving a scalable filter with a class of non-scalable filters.

E.2 Filters combined with linear operators

Although convolutional filters commute with each other, they do not commute in general with other linear operators. Abstractly, if $\boldsymbol{f}$ is a filter and $\boldsymbol{L}$ a linear operator, then in general $\boldsymbol{fL} \neq \boldsymbol{Lf}$.

In seismic imaging, a very common type of linear operation on a data set $D(x,t)$ can be expressed as the integral form

$$LD\,(x,z) = \int dx' L(x,x',z) D(x', \tau(x,x',z)), \tag{E.17}$$

with the quantities $\tau(x, x', z)$ and $L(x, x', z)$ well-behaved functions of z, slowly varying over distances comparable to the support of a typical filter f. The input function $D(x, t)$ on the other hand is assumed to oscillate rapidly with respect to its variable t, and the output function $LD(x, z)$ likewise varies rapidly with respect to its variable z. The integral over x' may be one- or two-dimensional, depending on whether the mapping described by (E.17) is two- or three-dimensional.

Suppose f is a scalable filter obeying the power law

$$f(az) = a^{-\nu} f(z). \tag{E.18}$$

If we apply the filter to the output function $\boldsymbol{LD}$, we obtain

$$I(x,z) = \int dz' f(z{-}z') \int dx' L(x,x',z') D(x', \tau(x,x',z')). \tag{E.19}$$

Abstractly, we have the situation

$$I = fLD. \tag{E.20}$$

Suppose we wish to find an equivalent operator $\boldsymbol{L}'$ such that

$$I = L' f D. \tag{E.21}$$

We first change the order of integration, and approximate $L(x, x', z'$ with its value $L(x, x', z)$ at z, so that it can be brought outside the z' integral:

$$I(x,z) \simeq \int dx' L(x,x',z) \int dz' f(z{-}z') D(x', \tau(x,x',z')). \tag{E.22}$$

Because of the rapid variation of D with respect to τ, we cannot approximate τ within D by its value at $z' = z'$. However, because the support of the wavelet f is assumed small, one can make a linear approximation to τ, valid within the support of f:

$$\tau(x,x',z') \simeq \tau(x,x',z) + (z' - z)\partial_z \tau(x,x',z). \tag{E.23}$$

Make a change of integration variable from z' to $u = (z - z')\partial_z \tau$. Then

$$I(x,z) \simeq \int dx' L(x,x',z) \int \frac{du}{|\partial_z \tau|} f\left(\frac{u}{\partial_z \tau}\right) D(x', \tau(x,x',z) - u). \tag{E.24}$$

Invoking the power law (E.18), I becomes

$$I(x,z) \simeq \int dx' \,(\partial_z \tau)^{\nu - 1}\, L(x,x',z) \int du\, f(\tau(x,x',z) - u) D(x',u). \tag{E.25}$$

Thus, for filters that obey the power law (E.18), one can apply a filter either before or after the integration operator. If before, the amplitude of the integration operator is multiplied by a power of the derivative of the old vertical coordinate with respect to the new. That is,

$$\int dz' f(z-z') \int dx' L(x,x',z') D(x', \tau(x,x',z')) \simeq$$
$$\int dx' \, (\partial_z \tau)^{\nu-1} \, L(x,x',z) \int du \, f(\tau(x,x',z) - u) D(x',u). \qquad \text{(E.26)}$$

On the left-hand side of this expression, the filter f is applied to the coordinate z. On the right-hand side, f is applied to the coordinate τ. The scaling factor $(\partial_z \tau)^{\nu-1}$ renders the two sides of the equation dimensionally consistent.

E.3 Bandpass filters

Bandpass filters are by nature non-scalable. We present a few simple filters here, namely sinc, Gaussian, and Ricker wavelets.

E.3.1 The sinc function

We define the core non-scalable sinc function to be

$$\text{sinc}_c(u) \equiv \frac{\sin(\pi u)}{\pi u}, \qquad \text{(E.27)}$$

in terms of which the general sinc function is

$$\text{sinc}\,(t_d, t) = \frac{\text{sinc}_c\left(t \,/ t_d\right)}{t_d} = \frac{\sin\left(\pi t \,/ t_d\right)}{\pi t}. \qquad \text{(E.28)}$$

The maximum amplitude of this function is $1/t_d$. It is symmetric about $t = 0$, and oscillatory, the amplitude of its envelope falling as $1/t$. The first zero occurs at time $\pm t_d$, with succeeding zeros at intervals of t_d. The Fourier transform of the core sinc function is

$$\text{sinc}_c(q) = 1, -\pi < q < \pi, \; = 0, |q| > \pi. \qquad \text{(E.29)}$$

In the frequency domain, the general sinc function is a block of unit amplitude for frequencies within $\pm\omega_d = \pm\pi / t_d$, and zero otherwise. For a discretized function sampled at intervals Δt, the maximum represented frequency is the Nyquist frequency $\omega_N = \pi/\Delta t$, and $\omega_d/\omega_N = \Delta t / t_d$. The left-hand panel of Figure E.1 shows a 64-point frequency-domain representation of a sinc function, with $\omega_d = \omega_N/2$ or $t_d = 2\Delta t$. The middle panel is a 64-point approximation to the time-domain representation of the same function, formed as a discrete inverse FFT of the frequency-domain representation.

This time-domain approximation clearly dies out more quickly than the exact sinc function, whose envelope fades as $1/t$ at large times. This is convenient if the filter is to be implemented in the time domain, since there are practical limits as to the length of a convolution operator. It is reasonable to ask, however, how much distortion is being introduced by the approximation. If we extend the length of the approximate time-domain operator by padding zeros and Fourier transform, we obtain the frequency-domain operator shown in the right panel of Figure E.1. This function, more finely sampled in frequency than

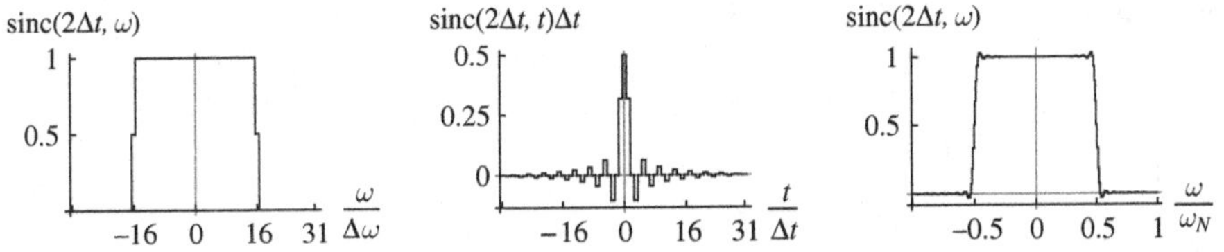

Figure E.1 A numerical (64 point) implementation of the sinc function in frequency and time. In frequency, shown on the left, the function is unity for frequencies of magnitude less than π divided by the dilation time t_d. The time function, shown in the center panel, was formed as an inverse discrete Fourier transform of the frequency-domain function. It is seen to be an oscillating function of peak magnitude $1/t_d$. The exact sinc function does not die out within the time window, but continues forever, dying as $1/t$. The effect of truncating the filter in time can be gauged in frequency, by padding the time function with zeros and performing a longer Fourier transform. This produces a more finely sampled frequency-domain function, as shown in the panel on the right. This function is near to the exact sinc function, with the addition of some low-amplitude oscillatory behavior near the cutoff frequency $\omega_d = \pi/t_d$.

the original on the left, shows some small Gibbs-effect oscillations near the truncation frequency ω_d, other than which the spectrum of the sinc function remains faithful.

Where the sinc filter is to be applied once, the Gibbs-effect oscillations are of little consequence. However, there may be occasions when caution need be exercised. If a filter is applied many times (as is possible within a finite-difference algorithm), even tiny inaccuracies can become significant, and might destabilize an operation.

E.3.2 The Gaussian

We define the non-scalable class of Gaussian filters $\text{gau}(t_d, t)$ to be

$$\text{gau}(t_d, t) \equiv \frac{\sqrt{\pi}}{t_d} e^{-(\pi t/t_d)^2}. \tag{E.30}$$

These filters are symmetric about the point $t = 0$, tapering smoothly and monotonically towards zero for positive and negative times.

Figure E.2 shows a 32-point numerical approximation to $\text{gau}(10\Delta t, t)\Delta t$, where Δt is the time increment. The filter has been multiplied by the time increment Δt. This is convenient for discrete approximations, in that when integral operations such as convolutions and transforms are replaced by sums, the discrete increments Δt are automatically included.

The Fourier transform of a Gaussian is also a Gaussian:

$$\text{gau}(t_d, \omega) = e^{-\left(\frac{\omega t_d}{2\pi}\right)^2}. \tag{E.31}$$

The filter has been defined so that its peak amplitude in the frequency domain is unity. In the time domain, peak amplitude decreases as t_d increases, but area under the curve is the same for all t_d. Had we defined the Gaussian to have unit peak amplitude in the time domain, its frequency-domain peak amplitude would be proportional to t_d.

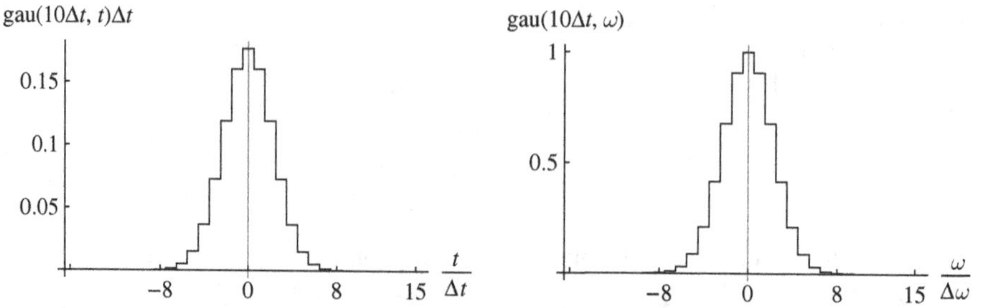

Figure E.2 A Gaussian in time and frequency. On the left is the discretized Gaussian filter $\frac{\sqrt{\pi}}{10}\, e^{-(\pi i t/10)^2}$, and on the right its discrete Fourier transform. For this example, the frequency increment is $\Delta\omega = \pi\ /\ (16\ \Delta t)$.

Technically, the Gaussian wavelet goes on forever, though as a practical matter it dies out rapidly. At time $t = t_d$, the time-domain wavelet is only $e^{-\pi^2} \simeq 5 \times 10^{-5}$ times its peak value. For the numerical example in Figure E.2, the maximum time in the window of the filter is 1.6 t_d, for which the wavelet amplitude is 10^{-11} times the peak amplitude, close enough to zero for most geophysical applications.

A numerical Fourier transform of the filter on the left in Figure E.2 produces the frequency-domain result shown on the right. The numerical result is a close approximation to the theoretical result in equation (E.31). For the example in E.2, theoretical filter amplitude at the maximum, or Nyquist, frequency is $e^{-25} \simeq 1.4 \times 10^{-11}$, also effectively zero for most applications.

E.3.3 The Ricker wavelet

We define the non-scalable class of Ricker wavelets rck (t_d, t) to be proportional to the second derivative of the Gaussian. Specifically,

$$\text{rck}\,(t_d, t) \equiv \frac{\pi^{1/2}}{2t_d}\left(1 - 2\left(\pi t\ /t_d\right)^2\right) e^{1-(\pi t/t_d)^2}. \tag{E.32}$$

The Ricker wavelet is also symmetric about $t = 0$, starting positive, becoming negative, then asymptotically approaching zero at large times.

A numerical approximation to this function is shown in the left-hand side of Figure E.3. Like the Gaussian, the filter technically goes on forever, but as a practical matter it quickly becomes negligibly small. The Ricker wavelet has zero crossings at $t = \pm t_d/\sqrt{2}\pi$.

In the frequency domain, this filter becomes

$$\text{rck}\,(t_d, \omega) = \left(\frac{\omega t_d}{2\pi}\right)^2 e^{1-\left(\frac{\omega t_d}{2\pi}\right)^2}, \tag{E.33}$$

as illustrated in the right-hand side of Figure E.3. We defined the Ricker wavelet to have unit peak amplitude in the frequency domain. Peak amplitude occurs at $\omega = \pm\frac{2\pi}{t_d} = \pm 2\omega_d$.

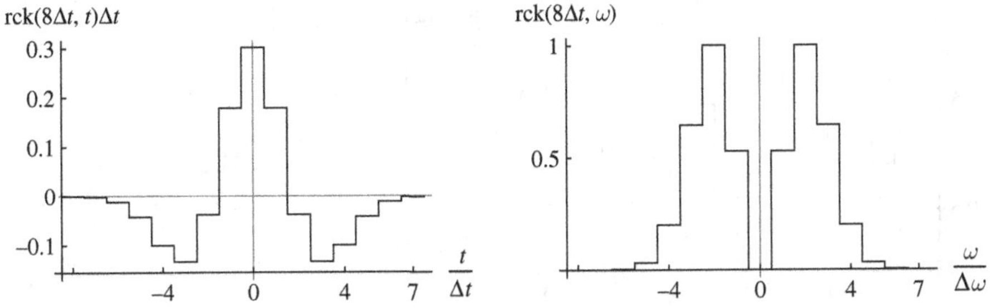

Figure E.3 A Ricker wavelet in time and frequency. The Ricker wavelet (E.32)–(E.33) with $t_d = 8\Delta t$.The amplitude of the wavelet (E.32) has been multiplied by Δt for convenience in performing integral operations. On the right is the discrete Fourier transform of the Ricker wavelet on the left, which closely matches the exact expression (E.33). As defined, the wavelet has unit peak amplitude in the frequency domain. In the time domain, peak amplitude is $\pi^{1/2} e \,/(2t_d)$.

E.4 The delta function and relatives

Perhaps the most useful filter of all is the delta function $\delta(t)$. As a filter, the delta function is the identity filter, in that its output is the same as its input:

$$D(t) = \int dt' D(t')\delta(t-t'). \tag{E.34}$$

In the frequency domain, the delta function becomes unity:

$$1 = \int dt e^{i\omega t}\delta(t). \tag{E.35}$$

The delta function is scalable with scaling power one:

$$\delta(at) = \frac{1}{a}\delta(t). \tag{E.36}$$

The delta function is its own inverse and its own adjoint.

The obvious numerical approximation to $\delta(t)$ is a spike $\delta_i/\Delta t$ centered on zero time.

Derivatives of the delta function are also useful. The first derivative $\delta'(t)$ as a filter is the differentiation operator

$$D'(t) = \int dt' D(t')\delta'(t-t'), \tag{E.37}$$

as can be confirmed by integration by parts. The frequency-domain expression of the delta-derivative filter is $-i\omega$. It has the scaling power two:

$$\delta'(a\,t) = \frac{1}{a^2}\delta'(t). \tag{E.38}$$

The adjoint of δ' is its negative.

The most common implementation of the differentiation operator is the difference operator $\Delta_i = (\delta_{i0} - \delta_{i1})/\,\Delta t$. This filter is centered on $t = \Delta t/2$ rather than zero.

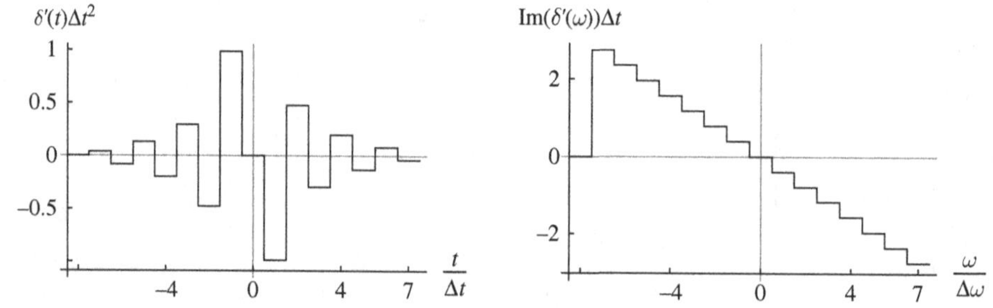

Figure E.4 A zero-centered 16-point approximation to the derivative operator in time and frequency. In the frequency domain, the filter is purely imaginary and linear in frequency except at Nyquist frequency. The filter has zero amplitude at Nyquist, which is necessary to keep the time-domain filter real.

Its frequency-domain expression is $-2ie^{i\omega\Delta t/2}\sin(\omega\Delta t/2)/\Delta t$. At low frequencies, this reduces to the ideal $-i\omega$, but at high frequencies accuracy is lost. One can also construct numerical approximations to this filter that are centered on zero time. Figure E.4 shows on the left a 16-point zero-centered approximation to the differentiation operator, constructed as an inverse discrete transform of the function on the right.

This filter, as one might expect, is a little "ringy". It is reasonable to infer that the oscillations are due in part to the discontinuity in the frequency domain at Nyquist frequency $\omega_N = \pi/\Delta t$. If instead we center the filter at $\Delta t/2$, so that its frequency-domain expression is $-i\omega\, e^{i\omega\Delta t/2}$, then the discontinuity at $\pm\omega_N$ disappears, and a more compact time-domain filter results. A 16-point implementation of this filter is illustrated in Figure E.5. Though very similar in appearance to a simple difference operator, this filter retains its accuracy over the full bandwidth of the data. The outlying side lobes, already small, could be reduced further by very slightly attenuating the amplitude of the filter at Nyquist frequency.

Integrals of the delta function are also often encountered as filters. The first integral of the delta function is a step function

$$\begin{aligned}\Theta(t) &= \int_{-\infty}^{t} dt'\delta(t') - \frac{1}{2},\\ &= +\frac{1}{2}, \qquad t > 0,\\ &= -\frac{1}{2}, \qquad t < 0.\end{aligned} \tag{E.39}$$

The additive factor of $-1/2$ might be considered arbitrary, in that the derivative of Θ will be the delta function regardless. With the additive factor of $-1/2$, the step function in the frequency domain has the form $-1/(i\omega)$. This form is antisymmetric in frequency, hence antisymmetric in time, as observed in (E.39). When acting upon functions or filters with no zero-frequency content, the presence or absence of the additive factor is immaterial. Step functions appear throughout this document, for the most part not symmetrized as is (E.39). Hopefully, this inconsistency will not add to the confusion.

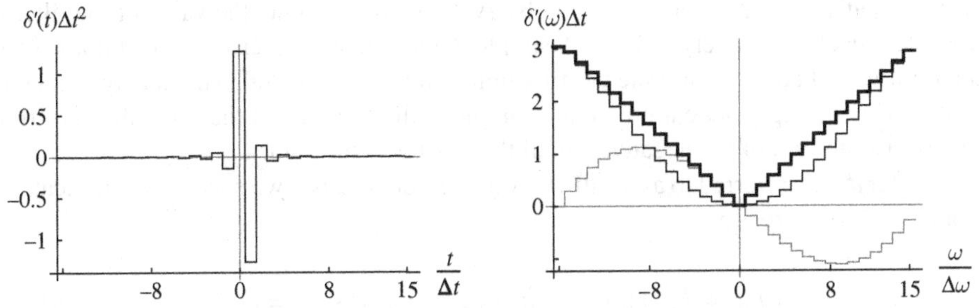

Figure E.5 A 16-point approximation to the derivative operator centered on $\Delta t/2$. The frequency domain filter is shown on the right, its real component in single-thickness black, its imaginary component in gray, and its absolute value in double-thickness black. It differs from the filter in Figure E.4 by the factor $e^{i\omega\Delta t/2}$. With this phase factor, at Nyquist frequency the imaginary component of the filter becomes zero and the real component finite but continuous. The time-domain filter, shown on the left, is antisymmetric about the point $\Delta t/2$, and dies quickly with distance from the center.

The step function has scaling power zero:

$$\Theta(at) = \Theta(t). \tag{E.40}$$

The adjoint of the step function is a downward step; that is,

$$\Theta^\dagger(t) = -\Theta(t). \tag{E.41}$$

The inverse of the step function is the δ' function.

A time-domain implementation of the step function filter is awkward in that it goes on forever.

E.5 Half-derivatives

Imaging algorithms often add or subtract 45° of phase in the form of one of the following filters.

The causal half-derivative filter (see Figure E.6)

$$d_{h+}(\omega) \equiv \sqrt{-i\omega}, \tag{E.42}$$

has the causal time-domain expression

$$\begin{aligned} d_{h+}(t) &= \frac{-1}{2\sqrt{\pi}} t^{-3/2}, \quad t > 0; \\ &= 0, \qquad\qquad\quad t < 0; \\ &= +\infty, \qquad\qquad t = 0. \end{aligned} \tag{E.43}$$

There is a reason why this filter must be positive infinity at t=0. The frequency-domain expression (E.42) for d_{h+} is clearly zero at zero frequency, which implies that the integral of the time-domain expression must be zero. However, the integral of (E.43) over positive times is negative infinity. For these two observations to reconcile, the value of the filter at zero time must be positively infinite. A simple delta function at zero time will not suffice, because it would add only a finite contribution to the filter integral. This theoretical complication is not a big practical problem: to numerically implement the filter, the zero-time value is just the sum of the negatives of all the positive-time values.

The filter d_{h+} is referred to as a half-derivative filter because two successive applications amount to a first derivative:

$$\left[d_{h+} * d_{h+}\right](t) = \int dt' d_{h+}(t') d_{h+}(t-t') = \delta'(t). \tag{E.44}$$

While it would be more correct to call d_{h+} a square-root derivative filter, it is hard to fight common usage.

There are any number of ways to implement the filter. We might, for instance start with the frequency-domain formula (E.42) and do a discrete inverse transform to the time domain. As illustrated in Figure E.6, the result is of the expected size and shape, with the addition of a high-frequency oscillation.

The oscillation is related to a discontinuity in the filter at the Nyquist frequency. When acting on bandlimited wavefunctions with no near-Nyquist frequency components, the oscillations have no effect. Even so, one might prefer an implementation of the filter which does not exhibit oscillatory behavior.

One possibility is just to move the center of the filter from $t = 0$ to $t = \Delta t/4$. At $\pm$ the Nyquist frequency $\omega_N = \pi/\Delta t$, $d_h(\pm\omega_N) = \sqrt{\mp i\pi/\Delta t}$. Shifting the filter center to $\Delta t/4$ multiplies the frequency-domain filter by $e^{i\omega\Delta t/4}$, so that $d_h(\pm\omega_N) \to \sqrt{\mp i\pi/\Delta t}e^{\pm i\pi/4}$

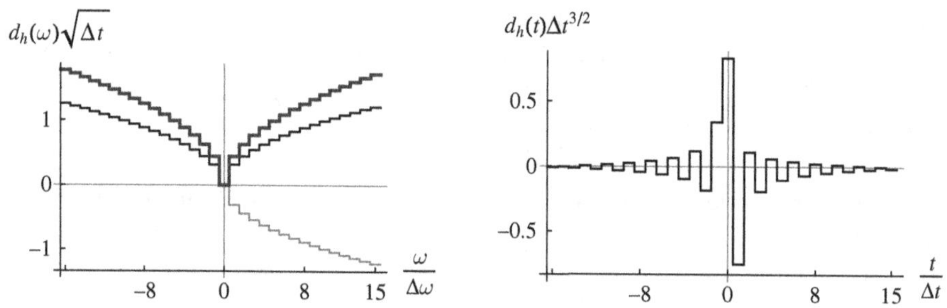

Figure E.6 Half-derivative in time and frequency. A numerical (32 point) implementation of the half-derivative filter in the frequency and time domains. On the left are the real (in black) and imaginary (in gray) components of the filter in the frequency domain, and also (in thick black) its absolute value. For negative frequencies, the real and imaginary parts of the filter overlay. For positive frequencies, the imaginary component is the negative of the real component. The time-domain filter, shown on the right, is formed by an inverse FFT of the frequency-domain form on the left. The time-domain filter has a high-frequency oscillatory component. In the frequency domain, as the filter wraps from positive to negative Nyquist frequency, the imaginary part is discontinuous, imposing a Nyquist-frequency oscillation in the time domain.

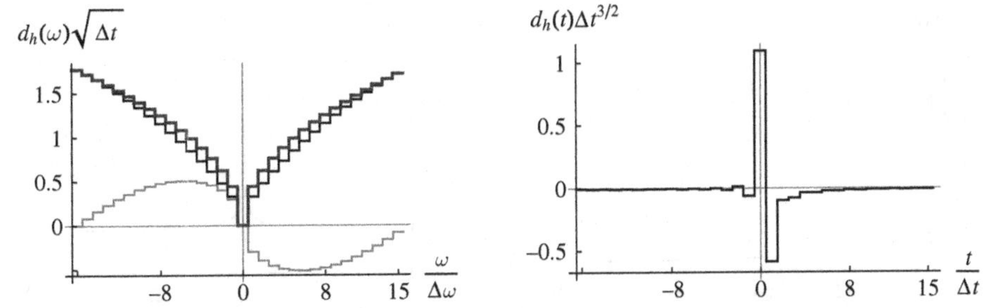

Figure E.7 Half-derivative filter centered on $\Delta t/4$. A numerical (32 point) implementation of the half-derivative filter in the frequency and time domains. On the left are the real (thin black) and imaginary (gray) components of the filter in the frequency domain, and also (thick black) its absolute value. This filter was formed by multiplying the filter in Figure E.6 by the phase factor $e^{i\omega\Delta t/4}$, corresponding to a time shift of $\Delta t/4$. This produces a figure that is continuous at Nyquist frequency. On the right is the time-domain filter, which is seen to be much less oscillatory than the filter in Figure E.6.

$= \sqrt{+\pi/\Delta t}$. That is, the time-shifted filter at the Nyquist frequency becomes continuous and real, as illustrated in Figure E.7.

The time-shifted filter in the time domain is certainly less oscillatory, and is nearly (but not perfectly) one-sided. True one-sided implementations of the filter can be constructed, generally with some loss of fidelity at high frequencies.

The half-derivative filter has the scaling property

$$d_{h+}(at) = a^{-3/2} d_{h+}(t). \tag{E.45}$$

The adjoint of d_{h+} is

$$d_{h-}(\omega) \equiv \sqrt{+i\omega}, \tag{E.46}$$

which has the anticausal time-domain expression

$$\begin{aligned} d_{h-}(t) &= -\frac{1}{2\sqrt{\pi}}(-t)^{-3/2}, \quad t < 0; \\ &= 0, \qquad\qquad\qquad\quad t > 0. \end{aligned} \tag{E.47}$$

The filter d_{h-} has the same scaling properties as its adjoint. Two successive applications of d_{h-} yield the adjoint of the derivative operator, or $-\delta'$.

E.6 Half-integral filters

The inverse of d_{h+} is

$$\Theta_{h+}(\omega) \equiv \frac{1}{\sqrt{-i\omega}}. \tag{E.48}$$

This filter enjoys the causal time-domain expression

$$\Theta_{h+}(t) = +\frac{1}{\sqrt{\pi t}}, t > 0; \ = 0, t < 0. \tag{E.49}$$

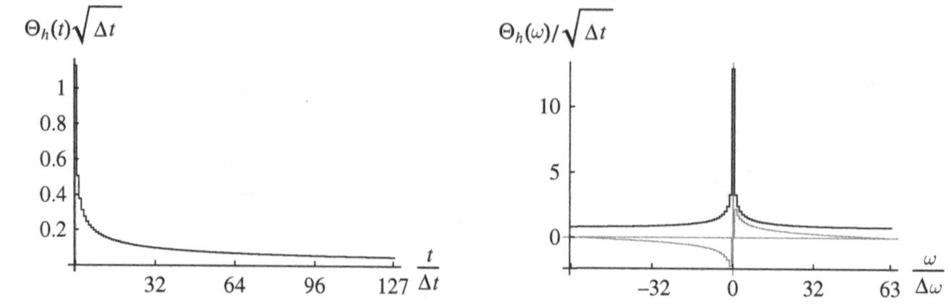

Figure E.8 Half-integral filter in frequency and time. A 128-point numerical expression of the filter Θ_{h+}. The filter is plotted in the time domain on the left, and in the frequency domain (real = black, imaginary = gray) on the right. The filter was created in the time domain from equation (E.49), evaluated at points $t_i = (i + 0.25)\Delta t$, consistent with placing the zero time of the filter at $-\Delta t/4$ rather than at time zero. The frequency-domain filter was obtained via a discrete Fourier transform of the time-domain filter. This filter differs from the theoretical frequency-domain expression (E.48) in that the imaginary component drops to zero at the Nyquist frequency. This can be explained as a phase shift of $e^{-i\omega\Delta t/4}$, corresponding to a time shift of $-\Delta t/4$ in the time-domain expression.

A straightforward implementation of equation (E.49) is shown in Figure E.8 on the left, with the corresponding frequency-domain filter on the right. To keep the first filter value finite, the zero point has been shifted to the left by $\Delta t/4$. In the frequency domain, the resulting filter is continuous at $\pm$ Nyquist frequency. With this implementation, convolution of this filter with an implementation of the half-derivative filter centered at $+\Delta t/4$ produces an approximation to the unit operator centered on $t = 0$.

The half-integral filter is seen to decay very slowly in both time and frequency, which poses a problem in direct implementation. However, if we discretize the filter in the frequency domain at intervals $\Delta\omega$, its length in the time domain is automatically limited to $\pm\pi/\Delta\omega$. Figure E.9 shows a 64-point numerical implementation of Θ_{h+}. To keep the frequency-domain filter continuous at the Nyquist frequency, an $e^{-i\omega\Delta t/4}$phase shift has been incorporated, centering the time-domain filter on the point $-\Delta t/4$. Zero frequency has also been adjusted so that the time-domain filter is zero at the edges of the time window.

The resulting filter is not causal, but it is reasonably compact. Provided $\Delta\omega$ is lower than the lowest frequency in the data, application of this filter should have the same result as the full-bandwidth filter.

Note that in the time domain, the derivative of Θ_{h+} is d_{h+}:

$$\frac{d\Theta_{h+}(t)}{dt} = d_{h+}(t). \tag{E.50}$$

In the frequency domain, the time derivative becomes a multiplication by $-i\,\omega$:

$$-i\omega\Theta_{h+}(\omega) = d_{h+}(\omega). \tag{E.51}$$

In the same sense that d_{h+} can be thought of as a half-derivative filter, Θ_{h+} can be thought of as a half-integral filter. Applying Θ_{h+} to itself,

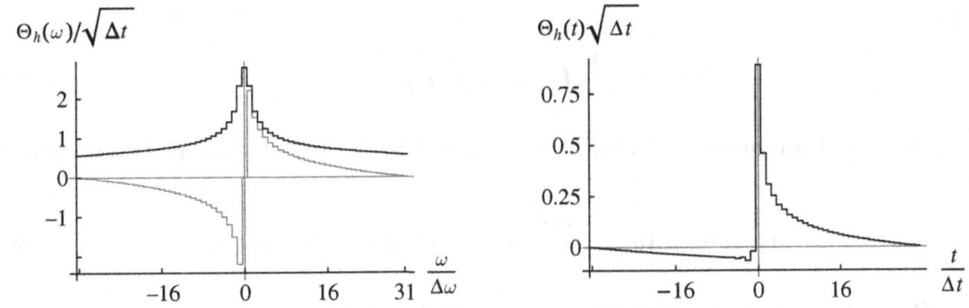

Figure E.9 Half integral filter centered at $-\Delta t/4$. A 64-point numerical implementation of $\Theta_{h+}(\omega)$ and $\Theta_{h+}(t)$, formed in the frequency domain and phase-shifted by $e^{-i\omega\Delta t/4}$, centering the time-domain implementation at $t = \Delta t/4$, and adjusting the zero-frequency value so that the time-domain filter is zero at the edge of its time window. This operator is more compact than the pure half-integral filter, which dies off rather slowly as $1/\sqrt{t}$.

$$\left[\Theta_{h+} * \Theta_{h+}\right](t) = \int dt' \Theta_{h+}(t')\Theta_{h+}(t-t') = \Theta(t). \tag{E.52}$$

The adjoint of Θ_{h+}, and the inverse of d_{h-} is

$$\Theta_{h-}(\omega) \equiv \frac{1}{\sqrt{i\omega}}. \tag{E.53}$$

The time-derivative of Θ_{h-} is -d_{h-}:

$$-i\omega\Theta_{h-}(\omega) = -\sqrt{i\omega} = -d_{h-}(\omega), \tag{E.54}$$

or

$$\frac{d\Theta_{h-}(t)}{dt} = -d_{h-}(t). \tag{E.55}$$

Θ_{h-} has the anti-causal time-domain expression

$$\Theta_{h-}(t) = \frac{1}{\sqrt{\pi}}(-t)^{-1/2}, t < 0; \;\; = 0, \; t > 0. \tag{E.56}$$

Θ_{h-} and Θ_{h+} share the scaling property

$$\Theta_{h\pm}(at) = \frac{\Theta_{h\pm}(t)}{\sqrt{a}}. \tag{E.57}$$

E.7 The Hilbert transform

The combination of two filters Θ_{h-} and d_{h+} results in the Hilbert transform filter. In frequency,

$$\text{hlb}(\omega) = \Theta_{h-}(\omega)d_{h+}(\omega) = \frac{\sqrt{-i\omega}}{\sqrt{+i\omega}} = -i\text{Sign}(\omega), \tag{E.58}$$

and in time,

$$\mathrm{hlb}(t) = \frac{-1}{2\pi}\int d\omega i \ \mathrm{Sign}(\omega)e^{-i\omega t} = \frac{-1}{\pi t}. \tag{E.59}$$

Likewise, the combination of Θ_{h+} and d_{h-} results in the adjoint of the Hilbert transform filter:

$$\Theta_{h+}(\omega)d_{h-}(\omega) = \frac{\sqrt{+i\omega}}{\sqrt{-i\omega}} = +i \ \mathrm{Sign}(\omega) = \mathrm{hlb}^{\dagger}(\omega). \tag{E.60}$$

hlb(t) has the same scaling property as the delta function:

$$\mathrm{hlb}(at) = \frac{\mathrm{hlb}(t)}{a}. \tag{E.61}$$

The effect of applying hlb(t) is to alter phase by 90° while leaving the frequency amplitude spectrum intact. As an antisymmetric function, the adjoint of hlb(t) is its negative.

Implementation

We can use equation (E.58) to define the Hilbert transform filter in the frequency domain, and perform the inverse transform digitally to produce a time-domain filter. For a 64-point filter, the result is shown in Figure E.10. The frequency-domain filter is shown on the left, and the time-domain filter on the right. The envelope of the time-domain filter has the expected $1/t$ behavior, with some high-frequency oscillatory behavior imposed. The even-numbered $(0, \pm 2, \ldots)$ values are in fact all zero. When convolved with a low-frequency function, the oscillations are immaterial. For high-frequency fidelity, they would appear to be necessary.

One could argue that the oscillations arise because we have placed the center of the filter at time zero, rather than at the "natural" center $\Delta t/2$ of this filter. Indeed, if we multiply the frequency-domain filter by $e^{i\omega\Delta t/2}$, then at the Nyquist frequency $\pm\omega_N = \pm\pi/\Delta t$,

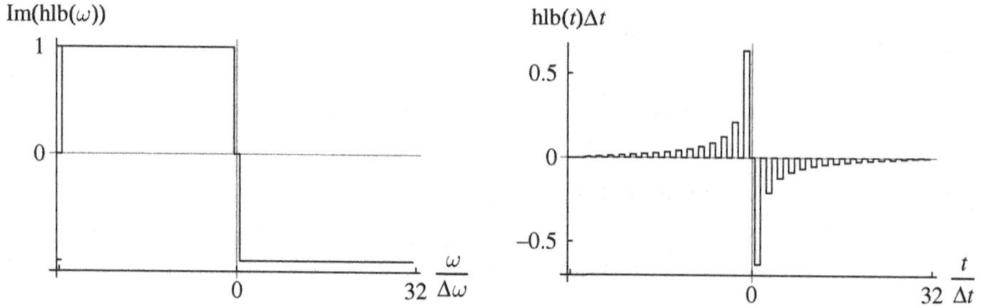

Figure E.10 A 64-point approximation to the Hilbert transform filter hlb(t), formed by evaluating (E.58) in the frequency domain, and performing an inverse FFT. The function at even-numbered times (counting from time zero) is zero, the oscillations arising because of the abrupt change in sign of the frequency-domain function at $\omega/\Delta\omega = \pm 32$. Since half of the filter values are zero, we could argue that the filter is more compact than it appears.

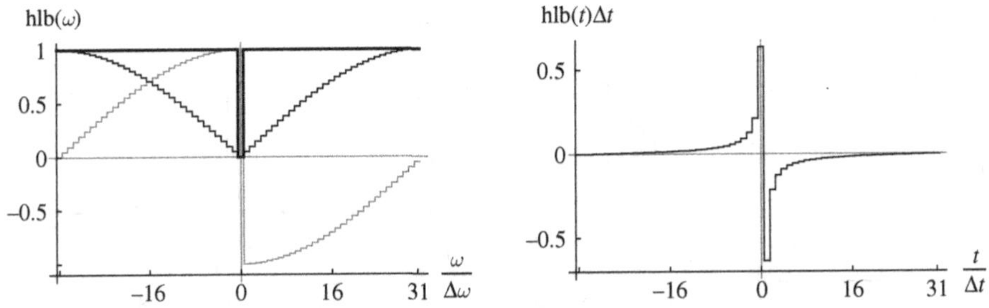

Figure E.11 Hilbert transform filter centered at $\Delta t/2$. A 64-point approximation to the Hilbert transform filter hlb(t), formed by evaluating (E.58) in the frequency domain and adding a phase shift of $e^{i\omega\Delta t/2}$. The real part of the function is shown in black, the imaginary in gray, and the absolute value in double-thickness. The time-domain function, found by performing an inverse FFT, is seen to be an anti-symmetric filter centered about $\Delta t/2$. The frequency-domain filter is continuous at Nyquist frequency, leading to a time-domain filter without the oscillations seen in the zero-centered filter.

the filter becomes real and continuous, as illustrated in Figure E.11 on the left. The corresponding time-domain filter, shown on the right, is a non-oscillatory function centered on the point $t = \Delta t/2$.

E.8 The rho filter and its inverse

Combining the two 45° filters $d_{h\pm}$ yields the zero-phase "rho" filter ρ:

$$\rho(\omega) \equiv d_{h-}(\omega) d_{h+}(\omega) = |\omega|. \tag{E.62}$$

The "rho" filter has the spectrum of a derivative filter, without the phase. The time-domain expression of this filter is

$$\rho(t) = -\frac{1}{\pi t^2}, \qquad t \neq 0. \tag{E.63}$$

Since the filter has no zero-frequency component, its integral must be zero, which means that at $t = 0$, the filter must be positive infinity. Figure E.12 shows a 32-point approximation to the rho filter, formed in the frequency domain from its defining equation, and transformed to the time domain with an FFT.

The rho filter gets its name from its occurrence in the inverse Radon transform.

The rho filter is self-adjoint. It has the same scaling property as δ':

$$\rho(at) = \frac{\rho(t)}{a^2}. \tag{E.64}$$

The inverse rho filter

The inverse of ρ is

$$P(\omega) \equiv \Theta_{h-}(\omega)\Theta_{h+}(\omega) = \frac{1}{|\omega|}. \tag{E.65}$$

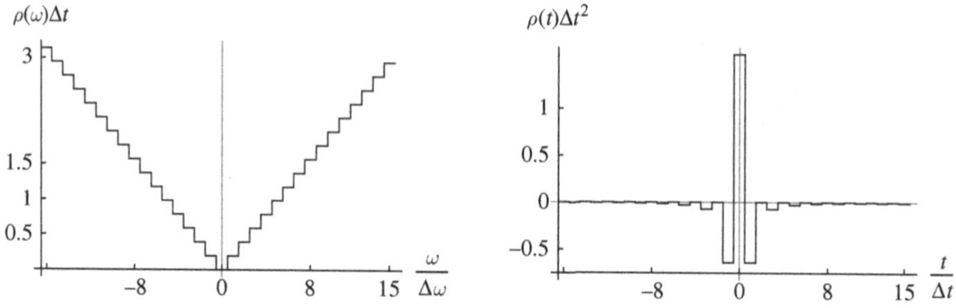

Figure E.12 The rho filter. A 32-point discretization of the rho filter ρ, formed in the frequency domain from equation (E.62), and transformed to the time domain.

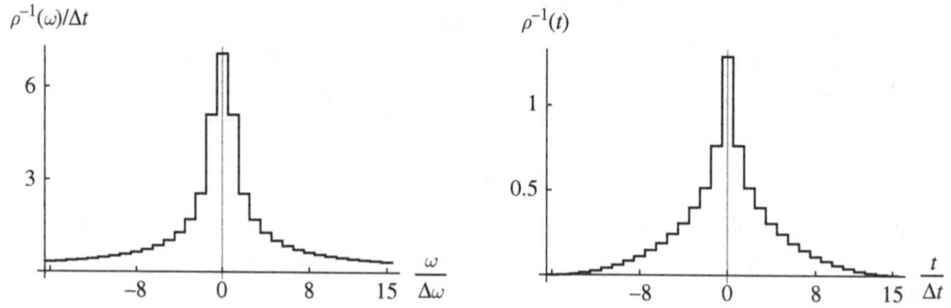

Figure E.13 The inverse rho filter. A 32-point approximation to the inverse rho filter $P(t)$, formed in the frequency domain from (E.65) and transformed to time. The zero-frequency component of the filter was chosen so that the time-domain filter approaches zero at the edges of the time window.

Its time-domain expression is (within an additive constant)

$$P(t) = -\frac{\ln(|t|)}{\pi}. \tag{E.66}$$

When applied to bandlimited functions, the constant is immaterial. Figure E.13 shows a numerical approximation to the inverse rho filter in frequency and time.

The derivative of the inverse rho filter is the Hilbert transform filter. This is easily verified in the frequency domain, where

$$-\, i\omega P(\omega) = -i = \mathrm{hlb}(\omega). \tag{E.67}$$

E.9 The inflection filter and its relatives

This next group of filters lacks some of the simple analytic and scaling properties of the filters presented above.

The inflection filter W_0 (so called because of its association with inflection points in the trajectory function), we define as

$$W_0(t_d, t) = t_d^{-1} W_{0c}\,(t/t_d)\,, \tag{E.68}$$

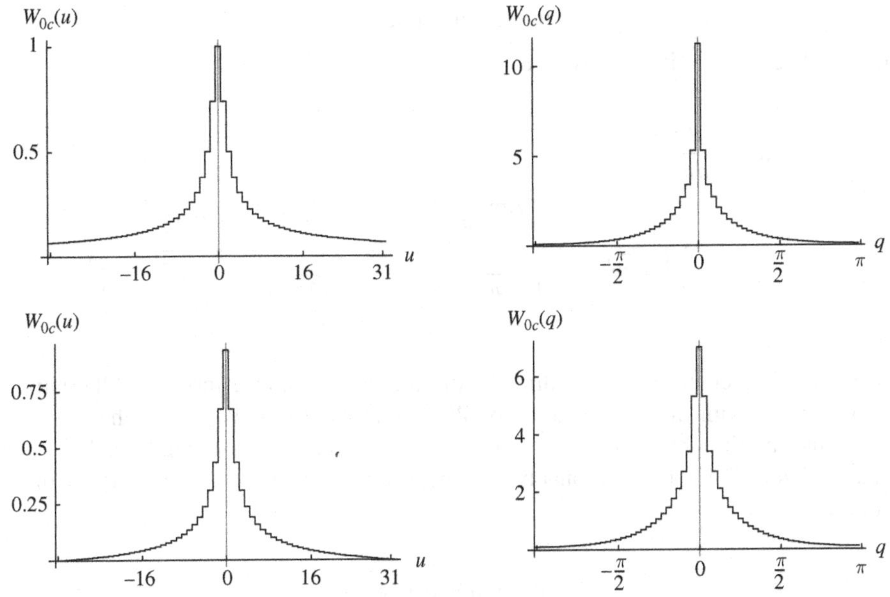

Figure E.14 The inflection filter. A 64-point approximation to the inflection filter $W_{0c}(u)$ (top left) and its Fourier transform $W_{0c}(q)$(top right). The discretization interval for u is $\Delta u = 1$, corresponding to a discretization interval in q of $\pi/32$. The exact function is symmetric and long-lived in time, whereas this approximation is truncated outside its time window. Because of the abrupt cutoff at the edges of the time window, convolution with the time-domain filter would produce artifacts. This problem is dealt with by subtracting from the filter its value at the edge of the time window, as depicted in the lower-left panel. In the frequency domain, shown in the lower-right panel, this reduces the zero-frequency component, while leaving all other values of the filter unchanged.

where

$$W_{0c}(u) = \frac{1}{\left(\sqrt{1+u^2}+u\right)^{2/3} + \left(\sqrt{1+u^2}-u\right)^{2/3} - 1}. \tag{E.69}$$

The filter W_0 does not retain its shape as t_d changes; rather, the wavelet expands or contracts. The width of W_0 is proportional to the dilation factor t_d. W_0 as defined is of peak amplitude $1/t_d$, declining to half-amplitude at $t = 2t_d$. At large times, the filter dies quite slowly as $t^{-2/3}$, making a digital implementation of the filter awkward. An implementation of the filter W_0 is depicted in Figure E.14. In Figure E.14, the time-domain filter (upper left) was truncated outside a 32-point time window, then the value at the window's edge was subtracted from all filter values to produce the panel on the lower left. The filter in the frequency domain is shown in the right panels before (top) and after (bottom) the amplitude adjustment. Only zero frequency is affected by the amplitude adjustment.

Inflection derivative

Closely related to W_{0c} is its derivative

$$W_{dc}(u) \equiv \frac{d}{du} W_{0c}(u)$$
$$= \frac{2}{3\sqrt{1+u^2}} \frac{\left(\sqrt{1+u^2}-u\right)^{2/3} - \left(\sqrt{1+u^2}+u\right)^{2/3}}{\left(\left(\sqrt{1+u^2}-u\right)^{2/3} + \left(\sqrt{1+u^2}+u\right)^{2/3} - 1\right)^2}. \tag{E.70}$$

We could implement this filter directly, discretizing the above formula. Alternatively, we could begin with the discretization of $W_{0c}(u)$, transform to $W_{0c}(q)$ in the frequency domain and multiply by $-iqe^{iq\Delta u/2}$ to effect a derivative. The resulting filter W_{dc}, illustrated in Figure E.15, is a reasonably well-behaved 90° phase wavelet centered on the point $\Delta u/2$.

Inflection half-integral

Combining W_0 with the half-integral filter Θ_{h+} leads to a filter of the form

$$W_{h+}(t_d, t) = t_d^{-1/2}\left[\Theta_{h+} * W_0(t_d)\right](t) = \frac{1}{\sqrt{\pi t_d}} \int_{-\infty}^{t} dv \frac{W_0(t_d, v)}{\sqrt{t-v}}. \tag{E.71}$$

A closed-form solution to this equation can be had in terms of the elliptic integral of the first kind. Alternatively, we can just perform the convolution, obtaining the result shown in Figure E.16.

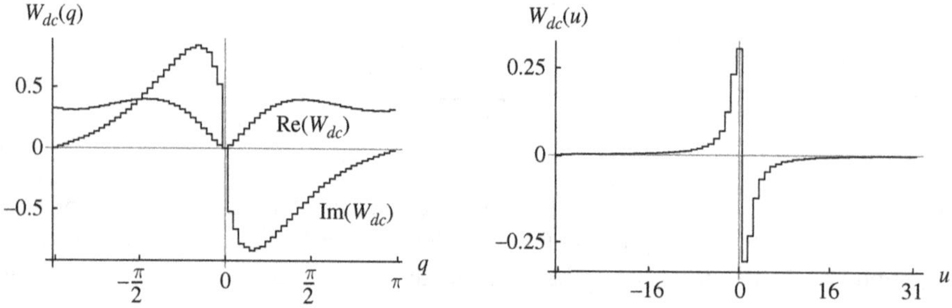

Figure E.15 The differentiated inflection filter in frequency and time. The differentiated inflection filter $W_{dc}(q)$ and $W_{dc}(u)$, discretized in units $\Delta q = \pi/32$ and $\Delta u = 1$. These filters were formed starting with the discretization of W_{0c} depicted in Figure E.14, and applying the discretized differentiation filter $-iqe^{iq\Delta u/2}$ as illustrated in Figure E.5. The resulting time-filter is antisymmetric about the point $u = \Delta u/2$.

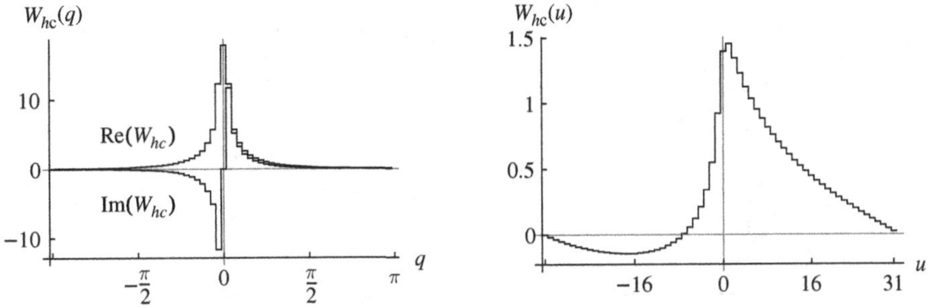

Figure E.16 The half-integrated inflection filter. The function W_{hc} is the inflection filter W_{0c} combined with the half-integral filter Θ_{h+}. This 64-point discretization was formed in the frequency domain and transformed to the time domain. The value at zero frequency was adjusted so that the time-domain filter is zero at the edge of the filter window.

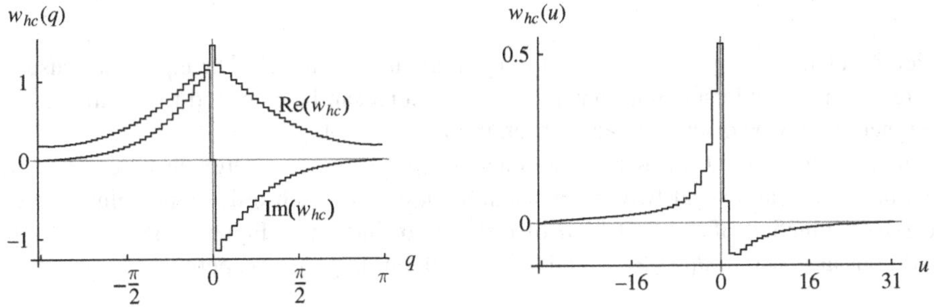

Figure E.17 The half-differentiated inflection filter. The inflection filter w_{h+}, or the convolution of W_0 with the half-derivative filter.

Inflection half-derivative

Another filter in the inflection family is the derivative of W_{h+}, or

$$\begin{aligned} w_{h+}(t_d, t) &\equiv t_d \frac{d}{dt} W_{h+}(t_d, t) \\ &= {t_d}^{-1/2} \left[\Theta_{h+} * W_d(t_d) \right](t) \\ &= {t_d}^{1/2} \left[d_{h+} * W_0(t_d) \right](t). \end{aligned} \tag{E.72}$$

This filter may be considered to be the convolution of W_d with a half-integral filter, or of W_0 with a half-derivative. It is shown in Figure E.17.

E.10 Bessel functions

In seismic processing, one encounters from time to time Bessel functions. A Bessel function $B_m(\rho)$ of order m is a solution to the equation

$$\left(\frac{d^2}{d\gamma^2} + \frac{1}{\gamma} \frac{d}{d\gamma} + 1 - \frac{m^2}{\gamma^2} \right) B_m(\gamma) = 0, \tag{E.73}$$

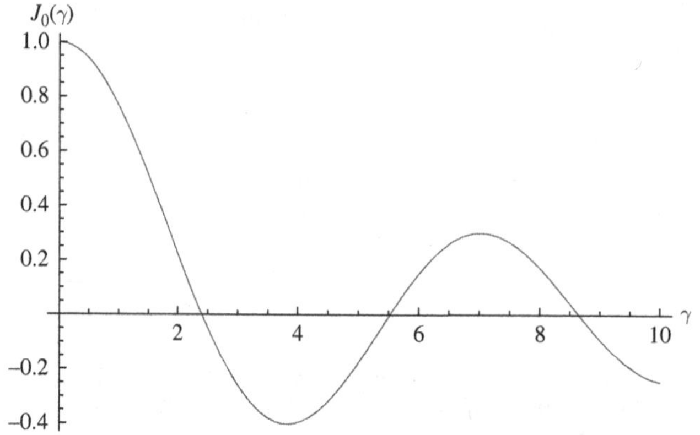

Figure E.18 Bessel function of the first kind of order zero.

which is related to Laplace's equation in cylindrical coordinates. The equation is usually written with the understanding that $\gamma > 0$. Solutions can be extended to negative γ, but the function may be discontinuous or even infinite at $\gamma = 0$.

There are many kinds of Bessel functions. Since the Bessel equation has second derivatives in γ, one can expect two independent solutions for each order. One solution is J_0, the Bessel function of the first kind of order zero. Shown in Figure E.18, it has initial conditions at $\gamma = 0$ of $J_0(0) = 1$, and $J_0{}'(0) = 0$. For large γ, J_0 approaches

$$J_0(\gamma) \to \sqrt{\frac{2}{\pi\gamma}}\cos(\gamma - \pi/4). \tag{E.74}$$

Another Bessel function of order zero is Y_0, depicted in Figure E.19. As $\gamma \downarrow 0$, Y_0 approaches $-\infty$ and Y_0' approaches $+\infty$; for $\gamma \gg 0$, it approaches

$$Y_0(\gamma) \to \sqrt{\frac{2}{\pi\gamma}}\sin(\gamma - \pi/4). \tag{E.75}$$

The Bessel functions of order zero can be related to solutions of the wave equation as follows. In cylindrical coordinates, the 2-D frequency-domain wave equation is

$$\left(\frac{\partial^2}{\partial\rho^2} + \frac{1}{\rho}\frac{\partial}{\partial\rho} + \frac{1}{\rho^2}\frac{\partial^2}{\partial\theta^2} + \frac{\omega^2}{c^2}\right)\psi(\rho, \theta, \omega) = 0. \tag{E.76}$$

Seeking a solution that depends on ρ and ω only,

$$\left(\frac{\partial^2}{\partial\rho^2} + \frac{1}{\rho}\frac{\partial}{\partial\rho} + \frac{\omega^2}{c^2}\right)\psi(\rho, \omega) = 0, \tag{E.77}$$

or, setting $\gamma = \rho c/\omega$,

$$\left(\frac{\partial^2}{\partial\gamma^2} + \frac{1}{\gamma}\frac{\partial}{\partial\gamma} + 1\right)\psi(\gamma) = 0, \tag{E.78}$$

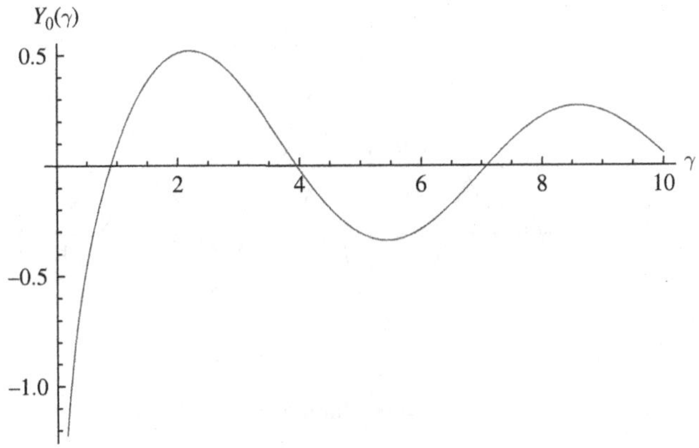

Figure E.19 Bessel function of the second kind of order zero.

which is the Bessel equation of order zero. Hence the Bessel functions of order zero are solutions to the 2-D wave equation with radial symmetry.

The Bessel function J_0 can be related to plane-wave solutions of the wave equation as follows. Look at the function

$$f(\omega\rho/c) = \int_{-\pi}^{\pi} d\theta e^{i\tilde{\mathbf{k}}\cdot\tilde{\mathbf{x}}} = \int_{-\pi}^{\pi} d\theta e^{i\frac{\omega\rho}{c}\cos\theta}. \tag{E.79}$$

As a superposition of plane waves, f obeys the wave equation (E.76). Since f does not depend on θ, it also obeys (E.77) or equivalently, (E.78), which means that f is a Bessel function of order zero. At $\omega\rho/c = 0$,

$$f(0) = \int_{-\pi}^{\pi} d\theta = 2\pi, \tag{E.80}$$

and

$$f'(0) = i\int_{-\pi}^{\pi} d\theta \cos\theta = 0. \tag{E.81}$$

These are the value and derivative at zero argument of $2\pi\ J_0$. We conclude that $f = 2\pi J_0$, or

$$J_0(\omega\rho/c) = \frac{1}{2\pi}\int_{-\pi}^{\pi} d\theta e^{i\tilde{\mathbf{k}}\cdot\tilde{\mathbf{x}}}. \tag{E.82}$$

Another function worthy of mention is H_0^1, the Hankel function of the first kind (or Bessel function of the third kind) of order zero. We single this function out from its Bessel relatives because it is proportional to the 2-D constant-background space–frequency domain Green's function. It is shown in Figure E.20. When its argument is positive, $H_0^1 = J_0 + iY_0$. The real part of H_0^1 is antisymmetric in its argument, and the imaginary part is symmetric, which implies that its inverse Fourier transform is imaginary.

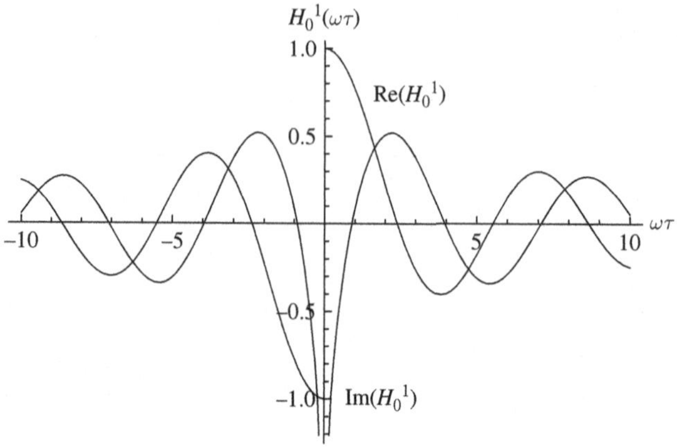

Figure E.20 Hankel function of the first kind of order zero. The real and imaginary parts of the Hankel function $H_0^1(\omega\tau)$. The real part of H_0^1 is antisymmetric in $\omega\tau$, while the imaginary part is symmetric.

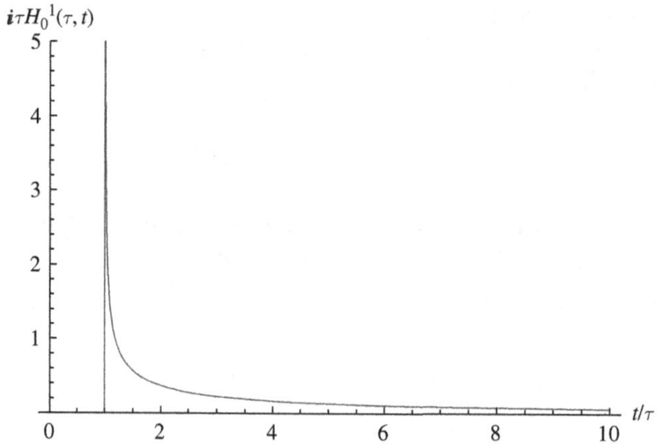

Figure E.21 The Hankel function $H_0^1(\tau, t)$ in the time domain.

In fact, the inverse Fourier transform of H_0^1 is

$$H_0^1(\tau, t) = -\frac{2i}{\pi} \frac{\Theta(t-\tau)}{\sqrt{t^2-\tau^2}}. \tag{E.83}$$

The time-domain function $i\, H_0^1(\tau, t)$ is shown in Figure E.21.
This function looks similar to the half-integral filter Θ_{h+}. In fact, for $t - \tau \ll t + \tau$,

$$H_0^1(\tau, t) \simeq -i\sqrt{\frac{2}{\pi\tau}} \frac{\Theta(t-\tau)}{\sqrt{\pi(t-\tau)}} = -i\sqrt{\frac{2}{\pi\tau}}\Theta_{h+}(t-\tau). \tag{E.84}$$

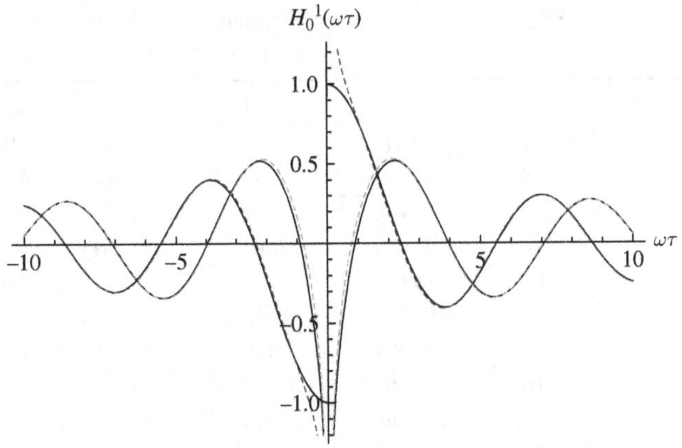

Figure E.22 A comparison between the Hankel function $H_0^1(\omega\tau)$(the solid lines) and its high-frequency approximation (dashed).

In the high-frequency approximation, the small values of $t - \tau$ are the important ones, and the half-integral approximation will be valid. The corresponding frequency-domain approximation is

$$H_0^1(\omega\tau) \simeq -i\sqrt{\frac{2}{\pi\tau}}e^{i\omega\tau}\Theta_{h+}(\omega) = \sqrt{\frac{2i}{\pi\omega\tau}}e^{i(\omega\tau-\pi/2)}. \tag{E.85}$$

Figure E.22 overlays the Hankel function with its high-frequency approximation (E.84). For $\omega\tau \gtrsim 2$, the match is close.

Higher-order Bessel functions can be constructed from lower orders using recursion relations. A general formula for either J_{m+1} or Y_{m+1} is

$$B_{m+1}(\gamma) = \frac{m}{\gamma}B_m(\gamma) - B_m{}'(\gamma). \tag{E.86}$$

For J_1, the recursion simplifies to $J_1 = -J_0{}'$. For J_1, we consequently have the integral relation

$$J_1(\gamma) = \frac{1}{\gamma}\int_0^\gamma d\lambda\lambda J_0(\lambda). \tag{E.87}$$

Shown in Figure E.23, J_1 has the asymptotic properties

$$\begin{aligned} J_1(\gamma) &\underset{\gamma\downarrow 0}{\to} \frac{\gamma}{2}, \\ &\underset{\gamma\to\infty}{\to} \sqrt{\frac{2}{\pi\gamma}}\cos(\gamma - 3\pi/4). \end{aligned} \tag{E.88}$$

Filter name	Symbol	Time formula	Frequency Formula	Adjoint filter	Inverse filter	Scaling power
Delta function	δ	$\delta(t)$	1	$\delta(t)$	$\delta(t)$	1
Derivative function	δ'	$\delta'(t)$	$i\omega$	$-\delta'(t)$	$\Theta(t)$	2
Step function	Θ	Sign(t)/2	$(i\omega)^{-1}$	$-\Theta(t)$	$\delta'(t)$	0
Half-derivative	d_{h+}	$-1/\sqrt{4\pi t^3}$	$\sqrt{i\omega}$	$d_{h-}(t)$	$\Theta_{h+}(t)$	3/2
Half-integral	Θ_{h+}	$1/\sqrt{\pi t}$	$(i\omega)^{-1/2}$	$\Theta_{h-}(t)$	$d_{h+}(t)$	1/2
Hilbert transform	hlb	$(\pi t)^{-1}$	$-i\text{Sign}(\omega)$	$-\text{hlb}(t)$	$-\text{hlb}(t)$	1
rho	ρ	$-(\pi t^2) - 1$	$\|\omega\|$	$\rho(t)$	$P(t)$	2
Inverse rho	P	$-\pi^{-1}\ln(t)$	$\|\omega\|^{-1}$	$P(t)$	$\rho(t)$	0
Inflection function	W_0	see text	Fig. E.14	$W_0(t)$	nd	na
Inflection derivative	W_d	see text	Fig. E.15	$-W_d(t)$	nd	na
Inflection half-integral	W_{h+}	see text	Fig. E.16	$W_{h-}(t)$	nd	na
Inflection half-derivative	w_{h+}	see text	Fig. E.17	$w_{h-}(t)$	nd	na

na, not applicable.

nd, not discussed.

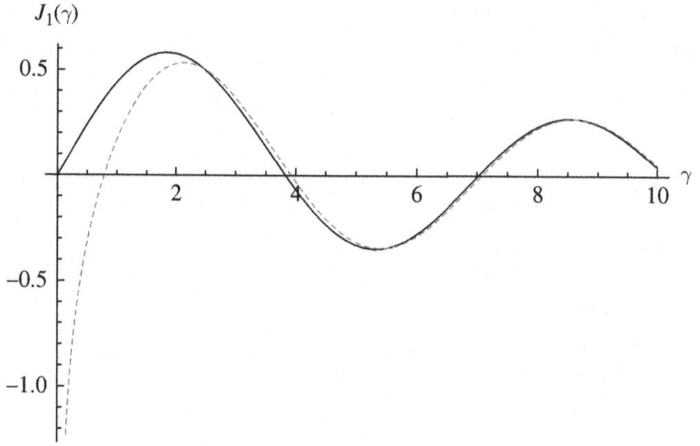

Figure E.23 The Bessel function J_1 (the solid curve) and its large-argument asymptote (the dashed curve).

E.11 Summary

The above table summarizes the properties of the filters discussed in the text. Though included for completeness, the family of inflection filters have properties that are too complex to be neatly tabulated.

Appendix F
The phase integral and the stationary phase approximation

The stationary phase approximation is a powerful tool for the evaluation of asymptotic integrals. Without digging deeply into the subject, we here present the basic idea. For more information, see Newton (1966), Bleistein and Handelsman (1986), and Bleistein *et al.* (2001).

F.1 Stationary phase in one dimension

F.1.1 The phase integral without end points

Suppose one has the integral

$$I = \int dx e^{if(x)} A(x), \tag{F.1}$$

in which the exponential varies "rapidly" while the function A varies "slowly". The oscillations of the exponential will cause the integrand to average to zero, except in regions where phase changes slowly. This will happen close to a stationary point; i.e. where

$$f'(x_0) \equiv \frac{df}{dx}(x_0) \equiv f_0' = 0. \tag{F.2}$$

Near such a point,

$$f(x) \simeq f_0 + f_0''(x - x_0)^2 \big/ 2. \tag{F.3}$$

Assuming that only points near x_0 contribute to the integral (see Figure F.1), we replace $f(x)$ within the integral with its quadratic approximation (F.3), and replace $A(x)$ with its value at the stationary point:

$$I \simeq e^{if_0} A(x_0) \int dx e^{if_0''(x-x_0)^2/2}. \tag{F.4}$$

To evaluate the phase integral, make a change of variable to

$$u = \sqrt{\frac{|f_0''|}{\pi}}(x - x_0), \tag{F.5}$$

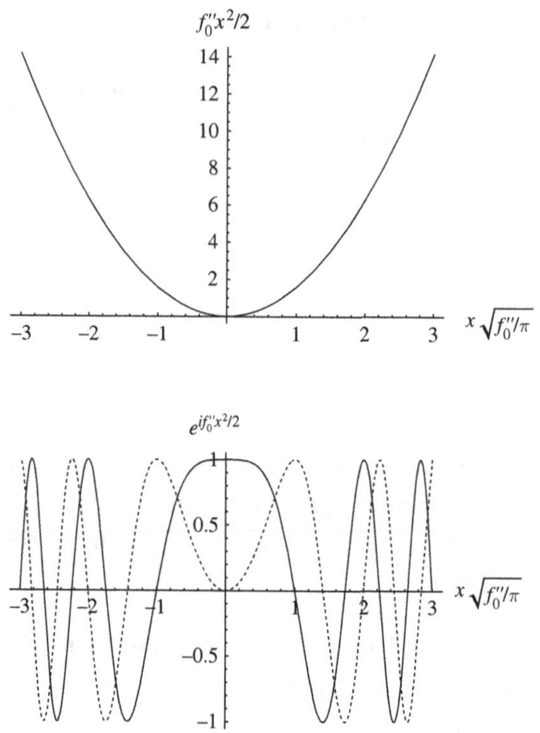

Figure F.1 A graphic depiction of the phase function. Above is the function $f(x) = f_0'' x^2/2$, plotted as a function of $x\sqrt{f_0''/\pi}$, for the case where the second derivative function f_0'' is positive. Below is the phase $e^{if_0'' x^2/2}$, also plotted as a function of $x\sqrt{f_0''/\pi}$, with the real part plotted as a solid line and the imaginary part a dotted line. Enough of the phase function has been plotted to reveal the oscillatory behavior at large x. The oscillations will damp out contributions to the phase integral $\int dx e^{if_0'' x^2/2}$ except from near the point $x = 0$, where both the real and imaginary parts of the phase have large positive lobes not balanced by adjacent negative lobes.

so that

$$I \simeq e^{if_0} A(x_0) \sqrt{\frac{\pi}{|f_0''|}} \int_{-\infty}^{\infty} du e^{i\mathrm{Sign}(f_0'')\pi u^2/2}. \tag{F.6}$$

In this form, the integral is related to the Fresnel integrals (Figure F.2)

$$C(x) \equiv \int_0^x \cos\left(\frac{\pi}{2}t^2\right) dt, \tag{F.7}$$

and

$$S(x) \equiv \int_0^x \sin\left(\frac{\pi}{2}t^2\right) dt, \tag{F.8}$$

evaluated at $\pm\infty$. These values are

$$C(\pm\infty) = S(\pm\infty) = \pm\frac{1}{2}. \tag{F.9}$$

Consequently, the integral I becomes

$$I \simeq e^{i f_0} A\left(x_0\right) \sqrt{\frac{\pi}{|f_0"|}} \left(1 + i \text{ Sign}\left(f_0''\right)\right), \tag{F.10}$$

or

$$I \simeq e^{i f_0} A\left(x_0\right) \sqrt{\frac{2\pi i}{f_0''}}. \tag{F.11}$$

F.1.2 Stationary phase with boundaries – a Fresnel function solution

Suppose we have a phase integral with end points, i.e.

$$I(a,b) = \int_a^b e^{i f(x)} dx. \tag{F.12}$$

Replacing $f(x)$ with its quadratic approximation, we have

$$\begin{aligned} I(a,b) &\simeq e^{i f_0} \int_a^b e^{i(x-x_0)^2 f_0''/2} dx \\ &= \sqrt{\frac{\pi}{|f_0''|}} e^{i f_0} \int_{u_a}^{u_b} e^{i \text{Sign}(f_0'') \frac{\pi}{2} u^2} du, \end{aligned} \tag{F.13}$$

where

$$u = \sqrt{\frac{|f_0''|}{\pi}} \left(x - x_0\right), \tag{F.14}$$

$$u_a = \sqrt{\frac{|f_0''|}{\pi}} \left(a - x_0\right), \tag{F.15}$$

and

$$u_b = \sqrt{\frac{|f_0''|}{\pi}} \left(b - x_0\right). \tag{F.16}$$

The integral is separable into four pieces:

$$\begin{aligned} I(a,b) \simeq \sqrt{\frac{\pi}{|f_0''|}} e^{i f_0} &\left\{ \int_0^{u_b} \cos\left(\frac{\pi}{2}u^2\right) du + \int_0^{-u_a} \cos\left(\frac{\pi}{2}u^2\right) du \right. \\ &\left. + i \text{Sign}\left(f_0''\right) \left(\int_0^{u_b} \sin\left(\frac{\pi}{2}u^2\right) du + \int_0^{-u_a} \sin\left(\frac{\pi}{2}u^2\right) du \right) \right\}. \end{aligned} \tag{F.17}$$

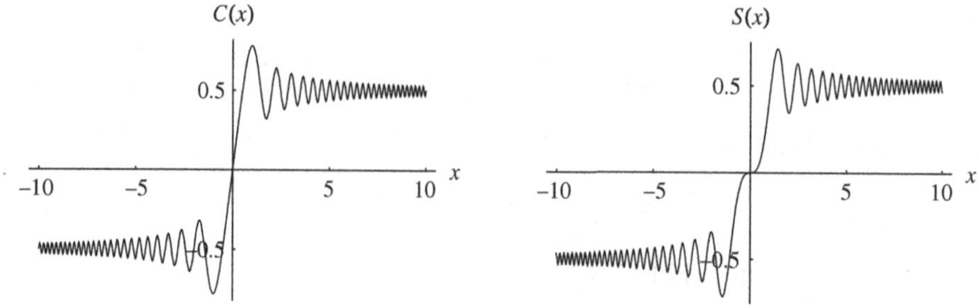

Figure F.2 The Fresnel integral functions *C(x)* and *S(x)*. Both functions are zero at the origin and asymptotically approach $\pm 1/2$ as $x \to \pm\infty$.

Each of these pieces is recognizable as a Fresnel function, as defined in equations (F.7) and (F.8). We write

$$I(a,b) \simeq \sqrt{\frac{\pi}{|f_0''|}} e^{i f_0} \left(E(u_b) + E(-u_a)\right), \tag{F.18}$$

where

$$E(x) \equiv C(x) + i \,\mathrm{Sign}\left(f_0''\right) S(x). \tag{F.19}$$

Since the Fresnel functions are zero at the origin and approach $\pm\frac{1}{2}$ at $\pm\infty$, the E-function has the properties

$$\begin{aligned} E(0) &= 0, \\ E(\pm\infty) &= \pm\sqrt{\frac{i \,\mathrm{Sign}\left(f_0''\right)}{2}}. \end{aligned} \tag{F.20}$$

If the boundaries are at $a = -\infty$ and $b = +\infty$, then the phase integral (F.18) becomes the stationary phase result

$$I(-\infty,\ \infty) \simeq \sqrt{\frac{2\pi i}{f_0''}} e^{i f_0}. \tag{F.21}$$

If one of the boundaries is at $\pm\infty$, and the other at the stationary points x_0, then the result is 1/2 the stationary phase value:

$$I(-\infty,\ x_0) \simeq I(x_0, \infty) \simeq \sqrt{\frac{\pi i}{2 f_0''}} e^{i f_0}. \tag{F.22}$$

More generally, the Fresnel functions have far- and near-field asymptotic expansions. For large x,

$$\begin{aligned} C(x) \underset{x\to\pm\infty}{\to} & \pm\frac{1}{2} + \sin\left(\frac{\pi}{2}x^2\right)\left(\frac{1}{\pi x} - \frac{3}{\pi^3 x^5} + \frac{105}{\pi^5 x^9} - \cdots\right) \\ & - \cos\left(\frac{\pi}{2}x^2\right)\left(\frac{1}{\pi^2 x^3} - \frac{15}{\pi^4 x^7} + \frac{945}{\pi^6 x^{11}} - \cdots\right) \end{aligned} \tag{F.23}$$

and

$$S(x) \underset{x\to\pm\infty}{\to} \pm\frac{1}{2} - \cos\left(\frac{\pi}{2}x^2\right)\left(\frac{1}{\pi x} - \frac{3}{\pi^3 x^5} + \frac{105}{\pi^5 x^9} - \cdots\right)$$
$$- \sin\left(\frac{\pi}{2}x^2\right)\left(\frac{1}{\pi^2 x^3} - \frac{15}{\pi^4 x^7} + \frac{945}{\pi^6 x^{11}} - \cdots\right). \quad \text{(F.24)}$$

In the limit of small x, the Fresnel functions have the asymptotic forms

$$C(x) \underset{x\to 0}{\to} x - \frac{\pi^2 x^5}{40} + \frac{\pi^4 x^9}{3456} - \cdots, \quad \text{(F.25)}$$

$$S(x) \underset{x\to 0}{\to} \frac{\pi x^3}{6} - \frac{\pi^3 x^7}{336} + \frac{\pi^5 x^{11}}{42240} - \cdots. \quad \text{(F.26)}$$

If the absolute values of u_a and u_b in (F.18) are large enough, we can approximate the Fresnel integrals with the first-order terms in the asymptotic expansions (F.23) and (F.24):

$$C(x) \sim \frac{\text{Sign}(x)}{2} + \frac{\sin\left(\pi x^2/2\right)}{\pi x}, \quad \text{(F.27)}$$

$$S(x) \sim \frac{\text{Sign}(x)}{2} - \frac{\cos\left(\pi x^2/2\right)}{\pi x}. \quad \text{(F.28)}$$

The phase integral (F.18) then becomes

$$I(a,b) \simeq \sqrt{\frac{\pi}{|f_0''|}} e^{i f_0} \left\{ \frac{\text{Sign}(u_b) - \text{Sign}(u_a)}{2} \left(1 + i\text{Sign}\left(f_0''\right)\right) \right.$$
$$+ \frac{\sin\left(\pi\ u_b{}^2/2\right) - i\ \text{Sign}\left(f_0''\right)\cos\left(\pi\ u_b{}^2/2\right)}{\pi u_b}$$
$$\left. - \frac{\sin\left(\pi\ u_a{}^2/2\right) - i\ \text{Sign}\left(f_0''\right)\cos\left(\pi\ u_a{}^2/2\right)}{\pi u_a} \right\}$$
$$= \sqrt{\frac{2\pi i}{f_0''}} e^{i f_0} \frac{\text{Sign}(b - x_0) - \text{Sign}(a - x_0)}{2}$$
$$- \frac{i e^{i\left(f_0 + f_0''(b-x_0)^2/2\right)}}{f_0''(b - x_0)} + \frac{i e^{i\left(f_0 + f_0''(a-x_0)^2/2\right)}}{f_0''(a - x_0)}. \quad \text{(F.29)}$$

Where $a < x_e$ and $b > x_e$, the first term is just the stationary phase approximation. If both a and b lie on the same side of x_0, the first term becomes zero. In either case, the remaining two terms describe the first-order perturbations in the phase integral due to the boundaries at a and b.

Stationary point near a boundary

If either of the boundary points a or b is at the stationary point x_0, the approximation (F.29) becomes infinite. The actual value of both C and S at the stationary point, however, is zero. We consequently conclude that for a boundary at or near a stationary point, the far-field extension (F.29) of the stationary phase approximation is inadequate.

Where one of the boundary points (suppose a, for specificity) is near the stationary point x_0, so that $|u_a| = \sqrt{|f_0"|/\pi}\,|a - x_0| \ll 1$, the small-argument expansions (F.25) and (F.26) of the Fresnel functions are more appropriate. A simple approximation to the Fresnel functions is to take the first term in the small-argument expansion for $|u_a|$ smaller than some breakover point, and the first term in the large-argument expansion otherwise:

$$\begin{aligned} \hat{C}(x) &= x, & |x| < x_{\text{tr}}, \\ &= \frac{\text{Sign}(x)}{2} + \frac{\sin\left(\pi x^2/2\right)}{\pi x}, & |x| > x_{\text{tr}}, \end{aligned} \tag{F.30}$$

$$\begin{aligned} \hat{S}(x) &= \frac{\pi x^3}{6}, & |x| < x_{\text{tr}}, \\ &= \frac{\text{Sign}(x)}{2} - \frac{\cos\left(\pi x^2/2\right)}{\pi x}, & |x| > x_{\text{tr}}, \end{aligned} \tag{F.31}$$

and, from (F.19),

$$\begin{aligned} \hat{E}(x) &= x\left(1 + i\,\text{Sign}\left(f_0''\right)\frac{\pi x^2}{6}\right), & |x| < x_{\text{tr}}, \\ &= \text{Sign}(x)\sqrt{\frac{i\,\text{Sign}\left(f_0''\right)}{2}} - \frac{i\,\text{Sign}\left(f_0''\right)}{\pi x}e^{i\,\text{Sign}(f_0'')\pi x^2/2}, & |x| > x_{\text{tr}}. \end{aligned} \tag{F.32}$$

The near- and far-field approximations to $C(x)$ coincide near the value 0.8391. Choosing $x_{\text{tr}} = 0.8391$ as the breakover point for both C and S results in Figure F.3. The resulting approximation, though not perfect, is not too bad. We could achieve a closer fit near the breakover point by fitting a higher-order polynomial approximation to the near-field function, but question whether the improvement would be worth the effort.

With the near/far approximations to C and S, (F.18) becomes

$$I(a,b) \simeq \sqrt{\frac{\pi}{|f_0''|}}\,e^{i f_0}\left(\hat{E}\left(u_b\right) + \hat{E}\left(-u_a\right)\right), \tag{F.33}$$

with the complex function $\hat{E}$ as given in equation (F.32).

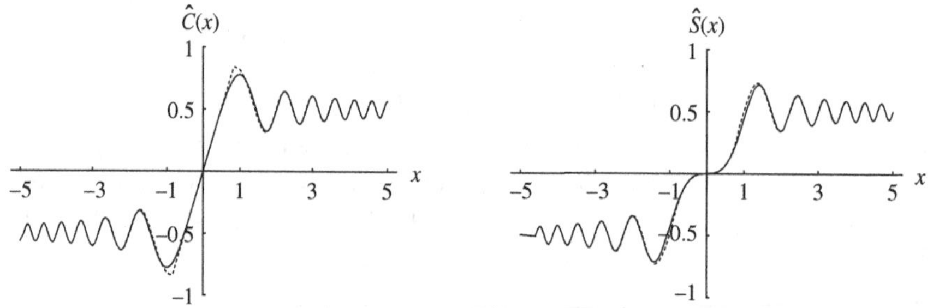

Figure F.3 The Fresnel integral functions *C(x)* and *S(x)* are shown as solid lines, and the first-order near/far-field approximations $\hat{C}(x)$ and $\hat{S}(x)$ shown as dashed lines. For $|x| < 0.8391$, *C(x)* is approximated by x and *S(x)* by $\pi x^3/6$. For $|x| > 0.8391$, *C(x)* is approximated by (F.27) and *S(x)* by (F.28).

Linearized approximation

If a and b are far enough from the stationary point that $f(x)$ can be considered linear in x near a and b, then

$$\begin{aligned} I(a,b) &= \int_a^b e^{if(x)}\,dx \\ &= \int_{-\infty}^{\infty} e^{if(x)}\,dx - \int_b^{\infty} e^{if(x)}\,dx - \int_{-\infty}^{a} e^{if(x)}\,dx \\ &\simeq \sqrt{\frac{2\pi i}{f_0''}}e^{if_0} - \frac{e^{if_a}}{if_a'} + \frac{e^{if_b}}{if_b'}. \end{aligned} \tag{F.34}$$

If x_0 does not lie between a and b, the first term in this expression does not contribute. Note that

$$f_a \simeq f_0 + f_0''\frac{(a-x_0)^2}{2}, \tag{F.35}$$

$$f_a' \simeq f_0''(a-x_0), \tag{F.36}$$

$$f_b \simeq f_0 + f_0''\frac{(b-x_0)^2}{2}, \tag{F.37}$$

$$f_b' \simeq f_0''(b-x_0). \tag{F.38}$$

If the quadratic approximation to f is exact within the interval (a, b), the linearly modified stationary phase integral (F.34) is identical to the first-order far-field approximation (F.29) to the Fresnel representation of the phase integral. In general, one would expect (F.34) to be slightly more accurate. Similarly, for a phase integral with a slowly varying component $A(x)$ (see equation (F.1)), it would be most accurate to use the local value of A in each term:

$$\begin{aligned} I(a,b) &= \int_a^b A(x)e^{if(x)}dx \\ &\simeq A(x_0)\sqrt{\frac{2\pi i}{f_0''}}e^{if_0} - A(a)\frac{e^{if_a}}{if_a'} + A(b)\frac{e^{if_b}}{if_b'}. \end{aligned} \tag{F.39}$$

F.2 Two-dimensional stationary phase

F.2.1 The 2-D phase integral

Suppose I is the 2-D integral

$$I = \int dx \int dy e^{if(x,y)}A(x,y). \tag{F.40}$$

In this case, if (x_0, y_0) is a stationary point where $\partial_x f$ and $\partial_y f$ are zero,

$$I \simeq e^{if(x_0,y_0)}A(x_0,y_0)\int dx \int dy e^{i\left[f_{xx}\frac{(x-x_0)^2}{2} + f_{yy}\frac{(y-y_0)^2}{2} + f_{xy}(x-x_0)(y-y_0)\right]}. \tag{F.41}$$

Shifting the origin to (x_0, y_0), this becomes

$$I \simeq e^{if(x_0,y_0)} A\left(x_0, y_0\right) \int dx \int dy e^{i\left[f_{xx}\frac{x^2}{2}+f_{yy}\frac{y^2}{2}+f_{xy}xy\right]}. \tag{F.42}$$

One can decouple the two integrals by rotating the x–y coordinate system so that the cross-term in the exponential disappears. That is, define rotated coordinates (r_1, r_2) such that

$$f_{xx}\frac{x^2}{2} + f_{yy}\frac{y^2}{2} + f_{xy}xy = \left(f_{11}{r_1}^2 + f_{22}{r_2}^2\right)/2\,. \tag{F.43}$$

Then,

$$I \simeq e^{if(x_0,y_0)} A\left(x_0, y_0\right) \int dr_1 e^{if_{11}{r_1}^2/2} \int dr_2 e^{if_{22}{r_2}^2/2}\,. \tag{F.44}$$

Each of the integrals can now be evaluated independently, yielding

$$I \simeq e^{if(x_0,y_0)} A\left(x_0, y_0\right) \sqrt{\frac{2\pi i}{f_{11}}}\sqrt{\frac{2\pi i}{f_{22}}}\,. \tag{F.45}$$

The overall phase of this expression depends upon the signs of f_{11} and f_{22}. If S_{11} is the sign of f_{11} and S_{22} the sign of f_{22},we can write

$$I \simeq 2\pi e^{i[f(x_0,y_0)+(S_{11}+S_{22})\pi/4]} \frac{A\left(x_0, y_0\right)}{\sqrt{|f_{11}f_{22}|}}\,. \tag{F.46}$$

If f goes through a double minimum at (x_0, y_0), the net additional phase amounts to $e^{i\pi/2}$ or i. If f goes through a double maximum, the additional phase is $e^{-i\pi/2}$ or $-$i. If f goes through a saddle point, the additional phase is zero.

The product of the two second derivatives is the determinant of the second-derivative matrix, which is invariant under rotations:

$$f_{11}f_{22} = \mathrm{Det}\begin{pmatrix} f_{xx} & f_{xy} \\ f_{yx} & f_{yy} \end{pmatrix} = f_{xx}f_{yy} - {f_{xy}}^2. \tag{F.47}$$

Hence, one need not actually perform the rotation to (r_1, r_2) to calculate the magnitude of $f_{11}f_{22}$. The additional phase term is $\pi/4$ times the sum of the signs of the eigenvalues of the second-derivative matrix, which usually is easy to determine also.

F.3 Useful examples

F.3.1 A one-dimensional phase integral

Suppose that the integral to evaluate is

$$I = \int dk \mathcal{A}(k) e^{if(k)}, \tag{F.48}$$

with $\mathcal{A}$ a smooth and slowly varying function of k, and

$$f(k) = kx + S_a z\sqrt{\lambda^2 + S_b k^2}, \tag{F.49}$$

where λ and z are assumed positive and S_a and S_b are signs ± 1. The stationary point of f is at $\underset{\wedge}{k}$, where

$$f'\left(\underset{\wedge}{k}\right) = 0 = x + \frac{S_a S_b z \underset{\wedge}{k}}{\sqrt{\lambda^2 + S_b \underset{\wedge}{k}^2}}, \tag{F.50}$$

which implies that

$$\underset{\wedge}{k} = -S_a S_b \lambda \frac{x}{\rho}, \tag{F.51}$$

and

$$\sqrt{\lambda^2 + S_b \underset{\wedge}{k}^2} = \lambda \frac{z}{\rho}, \tag{F.52}$$

with

$$\rho = \sqrt{z^2 - S_b x^2}. \tag{F.53}$$

The function f at the stationary point is

$$f\left(\underset{\wedge}{k}\right) = S_a \lambda \rho, \tag{F.54}$$

while the second derivative of f is

$$f''\left(\underset{\wedge}{k}\right) = S_a S_b \frac{z\lambda^2}{(\lambda^2 + S_b \underset{\wedge}{k}^2)^{3/2}} = S_a S_b \frac{\rho^3}{\lambda z^2}. \tag{F.55}$$

With these values, the stationary phase expression (F.11) for I is

$$\begin{aligned} \int dk \mathcal{A}(k) e^{i\left(kx + S_a z\sqrt{\lambda^2 + S_b k^2}\right)} &\simeq z\mathcal{A}\left(\underset{\wedge}{k}\right) e^{i S_a \lambda \rho} \sqrt{\frac{2\pi i \lambda S_a S_b}{\rho^3}} \\ &= \sqrt{\frac{2\pi \lambda}{\rho^3}} z\mathcal{A}\left(-\frac{S_a S_b \lambda x}{\rho}\right) e^{i S_a (\lambda\rho + S_b \pi/4)}. \end{aligned} \tag{F.56}$$

The overall phase of the result depends on the signs S_a and S_b but not on the sign of x.

F.3.2 A two-dimensional example

Suppose that the integral to evaluate is

$$I = \int dk_x \int dk_y \mathcal{A}\left(k_x, k_y\right) e^{i f\left(k_x, k_y\right)}, \tag{F.57}$$

with

$$f\left(k_x, k_y\right) = k_x x + k_y y + S_a z \sqrt{\lambda^2 + S_b\left(k_x{}^2 + k_y{}^2\right)}, \tag{F.58}$$

where $\mathcal{A}$ is smoothly and slowly varying, and z and λ are positive. The stationary point of f is at $\left(\underset{\wedge}{k}_x, \underset{\wedge}{k}_y\right)$ where

$$f_{k_x}\left(\underset{\wedge}{k}_x, \underset{\wedge}{k}_y\right) = 0 = x + \frac{S_a S_b z \underset{\wedge}{k}_x}{\sqrt{\lambda^2 + S_b\left(\underset{\wedge}{k}_x{}^2 + \underset{\wedge}{k}_y{}^2\right)}}, \tag{F.59}$$

and

$$f_{k_y}\left(\underset{\wedge}{k}_x, \underset{\wedge}{k}_y\right) = 0 = y + \frac{S_a S_b z \underset{\wedge}{k}_y}{\sqrt{\lambda^2 + S_b\left(\underset{\wedge}{k}_x{}^2 + \underset{\wedge}{k}_y{}^2\right)}}, \tag{F.60}$$

from which it follows,

$$x^2 + y^2 = \frac{z^2\left(\underset{\wedge}{k}_x{}^2 + \underset{\wedge}{k}_y{}^2\right)}{\lambda^2 + S_b\left(\underset{\wedge}{k}_x{}^2 + \underset{\wedge}{k}_y{}^2\right)}, \tag{F.61}$$

or

$$\underset{\wedge}{k}_x{}^2 + \underset{\wedge}{k}_y{}^2 = \frac{\left(x^2 + y^2\right)\lambda^2}{\rho^2}, \tag{F.62}$$

and

$$\lambda^2 + S_b\left(\underset{\wedge}{k}_x{}^2 + \underset{\wedge}{k}_y{}^2\right) = \frac{\lambda^2 z^2}{\rho^2}, \tag{F.63}$$

where

$$\rho \equiv \sqrt{z^2 - S_b(x^2 + y^2)}. \tag{F.64}$$

Substituting this result into (F.59) and (F.60),

$$\begin{pmatrix} \underset{\wedge}{k}_x \\ \underset{\wedge}{k}_y \end{pmatrix} = \begin{pmatrix} x \\ y \end{pmatrix} \frac{-S_a S_b \lambda}{\rho}. \tag{F.65}$$

The function f at the stationary point is

$$f\left(\underset{\wedge}{k}_x, \underset{\wedge}{k}_y\right) = S_a \lambda \rho, \tag{F.66}$$

while the second derivatives of f are

$$f_{k_x k_x}\left(\underset{\wedge}{k_x}, \underset{\wedge}{k_y}\right) = \frac{z S_a S_b \left(\lambda^2 + S_b \underset{\wedge}{k_y}^2\right)}{\left(\lambda^2 + S_b\left(\underset{\wedge}{k_x}^2 + \underset{\wedge}{k_y}^2\right)\right)^{3/2}},$$

$$f_{k_y k_y}\left(\underset{\wedge}{k_x}, \underset{\wedge}{k_y}\right) = \frac{z S_a S_b \left(\lambda^2 + S_b \underset{\wedge}{k_x}^2\right)}{\left(\lambda^2 + S_b\left(\underset{\wedge}{k_x}^2 + \underset{\wedge}{k_y}^2\right)\right)^{3/2}},$$

$$f_{k_x k_y}\left(\underset{\wedge}{k_x}, \underset{\wedge}{k_y}\right) = -\frac{z S_a S_b \underset{\wedge}{k_x} \underset{\wedge}{k_y}}{\left(\lambda^2 + S_b\left(\underset{\wedge}{k_x}^2 + \underset{\wedge}{k_y}^2\right)\right)^{3/2}}, \tag{F.67}$$

and their determinant

$$f_{k_x k_x}\left(\underset{\wedge}{k_x}, \underset{\wedge}{k_y}\right) f_{k_y k_y}\left(\underset{\wedge}{k_x}, \underset{\wedge}{k_y}\right) - f_{k_x k_y}\left(\underset{\wedge}{k_x}, \underset{\wedge}{k_y}\right)^2 = \frac{z^2 \lambda^2}{\left(\lambda^2 + S_b\left(\underset{\wedge}{k_x}^2 + \underset{\wedge}{k_y}^2\right)\right)^2} = \frac{\rho^4}{z^2 \lambda^2}. \tag{F.68}$$

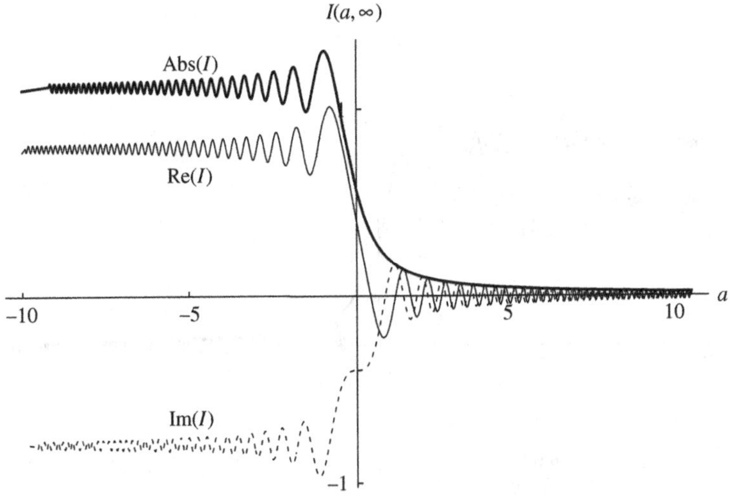

Figure F.4 Above is the Fresnel-function solution (F.72) for the phase integral $I(a, b)$, with the phase function (F.71), $b = +\infty$ and the lower boundary a ranging from -9 to $+9$. The upper curve is the absolute value, the middle curve the real part, and the bottom (dashed) curve the imaginary part of $I(a, \infty)$. For a much less than the stationary point $x_0 = 0$, the curves asymptotically approach constant values as a decreases. As a approaches the stationary point, the curves begin to drop towards zero. As a passes the stationary point, the curves asymptotically approach zero.

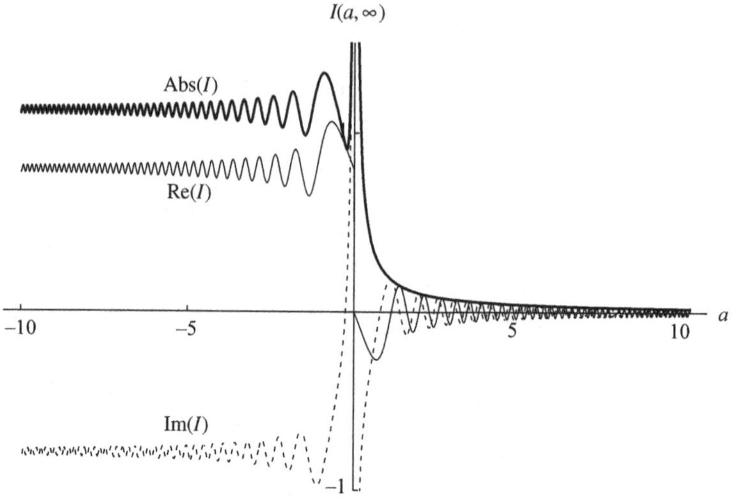

Figure F.5 Above is the stationary phase solution with linearized boundaries (F.34) for the phase integral $I(a, b)$, with the phase function (F.71), $b = +\infty$ and a ranging from -9 to $+9$. The upper curve is the absolute value, the middle curve the real part, and the bottom curve the imaginary part of $I(a, b)$. The linear approximation breaks down for a near the stationary point $x_0 = 0$; otherwise, we expect this to be a good approximation.

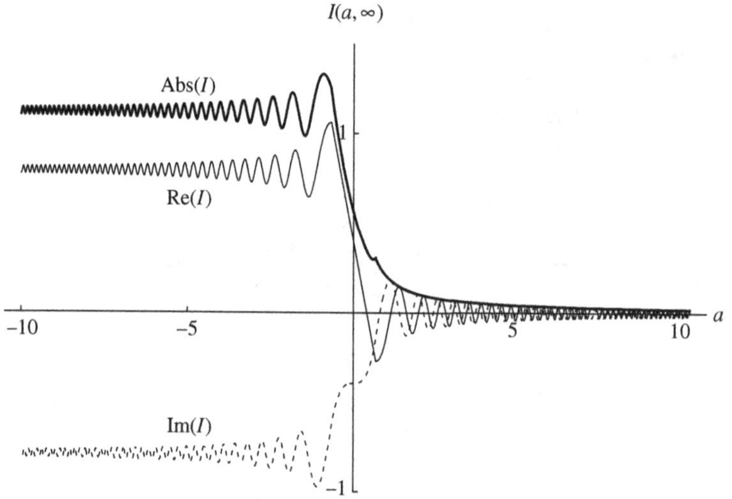

Figure F.6 Above is the stationary phase solution with the near/far approximation for the phase integral $I(a, b)$, with the phase function (F.71), $b = +\infty$ and a ranging from -10 to $+10$. The upper (bold) curve is the absolute value, the middle curve the real part, and the bottom (dashed) curve the imaginary part of $I(a, b)$. The curves closely match those in Figure F.4 using the exact C and S functions.

For $S_a S_b$ negative, f is clearly a double maximum at the stationary point, contributing a net phase of $-i$. For positive $S_a S_b$, f is a double minimum contributing a net phase of $+i$. In either case, the phase contribution is $i S_a S_b$.

With these values, the stationary phase expression (F.46) for the integral (F.57) is

$$\int dk_x \int dk_y A\left(k_x, k_y\right) e^{i\left(k_x x + k_y y + S_a z\sqrt{\lambda^2 + S_b\left(k_x{}^2 + k_y{}^2\right)}\right)} \simeq \frac{2\pi\lambda z A\left(-S_a S_b \lambda x/\rho, -S_a S_b \lambda y/\rho\right)}{\rho^2} e^{i S_a\left(\lambda\rho + S_b \frac{\pi}{2}\right)}. \tag{F.69}$$

We see that the result is not affected by the sign of x and y, but is affected by the sign of S_a and S_b.

F.3.3 A phase integral with a lower end point

We look at the following phase integral with a lower end point:

$$I(a, \infty) = \int_a^\infty e^{i f(x)}\, dx, \tag{F.70}$$

where

$$f(x) = f_0 + f_0'' \frac{(x - x_0)^2}{2}. \tag{F.71}$$

For specificity, choose $f_0 = 0$, $f_0'' = -5$, and $x_0 = 0$.

The exact Fresnel-function solution (F.18) in this case is

$$I(a, \infty) \simeq \sqrt{\frac{\pi}{5}}\left(\frac{1}{2} + C\left(-\sqrt{\frac{5}{\pi}}a\right) - i\left(\frac{1}{2} + S\left(-\sqrt{\frac{5}{\pi}}a\right)\right)\right), \tag{F.72}$$

We plot the Fresnel-function expression for $I(a, \infty)$ as a function of a in Figure F.4.

Where the lower integral boundary a is left of the stationary point ($x_0 = 0$), real and imaginary parts of the integral $I(a, \infty)$ approach constant values. For a right of the stationary point, both real and imaginary parts approach zero. At the stationary point, both real and imaginary parts drop to half their values at $a = -\infty$.

With this example, the far-field expression (F.34) becomes

$$I(a, \infty) \simeq \sqrt{\frac{-2\pi i}{5}}\Theta(-a) - \frac{i e^{-i2.5a^2}}{5a}. \tag{F.73}$$

We plot this expression in Figure F.5. Except near the stationary point, where the linear approximation is violated, the stationary-phase expression looks very much like the Fresnel-function expression plot in Figure F.4.

Finally, in Figure F.6 we plot the phase integral using the near/far approximations (F.30) and (F.31) to C and S.

Appendix G
The diffraction integral

For further discussion of diffractions and asymptotics, see Keller (1953) and Keller (1978).

G.1 The form of the diffraction integral

Integrals of the form (see e.g. equation (9.148))

$$\mathcal{O}(\rho) = \int_S d^n s C(\rho, s)\bar{\mathcal{I}}(\gamma(\rho, s)), \tag{G.1}$$

are found throughout this text. Integrals of this form are often encountered in high-frequency asymptotic expressions. We refer to them here as *diffraction integrals*. Such an integral may be one-dimensional or two, depending on whether one is dealing with a two- or three-dimensional medium. The output function $\mathcal{O}$ is a weighted integration over a line or surface $\mathcal{S}$ of the input function $\bar{\mathcal{I}}$. Position on this line or surface is indicated by the coordinate s, which is either a scalar or a two-vector.

The 'bar' over the $\mathcal{I}$ is meant to indicate that $\bar{\mathcal{I}}$ is a filtered version of another function $\mathcal{I}$:

$$\bar{\mathcal{I}}(\gamma) = [f * \mathcal{I}](\gamma) = \int d\gamma' f(\gamma - \gamma')\mathcal{I}(\gamma'). \tag{G.2}$$

For 3-D media, the filter is most commonly the derivative operator $\dot{\delta}(\gamma - \gamma')$, though others are also encountered. In two dimensions, the filter is typically one of the half-derivative filters $d_{h\pm}$ as described in Appendix E. The output coordinate ρ might be a depth or time coordinate, and is functionally related to the input coordinate γ, which may also be a depth or a time coordinate. If γ is a time coordinate, then the functional relation between ρ and γ can be considered to be a traveltime function, though it may or may not be associated with physical raypaths. If γ is a depth coordinate, then the function $\gamma(\rho,\ s)$ can be considered to be an inverse traveltime or depth function, which again may or may not be associated with physical raypaths. In either case, whether γ is a time or a depth coordinate, it may be generally referred to as a *trajectory function*.

In the high-frequency approximation, $\mathcal{I}$ varies rapidly. In contrast, the weighting coefficient C is very slowly varying. The function γ as a function of ρ is also slowly varying. These properties make it possible to approximately evaluate (G.1).

One particular instance of the form (G.1) is the Kirchhoff migration integral. That (G.1) is Kirchhoff-like is no accident, in that integral forms of imaging processes all have their roots in the divergence theorem, which relates fields within a volume to its properties on an enclosing surface.

The behavior of the diffraction integral (G.1) is determined by the behavior of its trajectory function γ as a function of position on the surface (or line) $\mathcal{S}$. Thinking three-dimensionally for a moment, suppose γ changes smoothly and monotonically across $\mathcal{S}$. Then one can contour the surface $\mathcal{S}$ into lines of constant γ. Along each line of constant γ, define a coordinate σ that measures path length along the contour from some reference point. Normal to the contour, let γ be the second coordinate. One can then make a change of integration variable in (G.1) from ds to $(d\gamma, d\sigma)$:

$$\mathcal{O}(\rho) = \int_{\mathcal{S}} d\gamma d\sigma \frac{C(\rho, s(\gamma, \sigma))}{|\nabla_s \gamma|} \bar{\mathcal{I}}(\gamma). \tag{G.3}$$

Now, $\overline{\mathcal{I}}$, as a high-frequency function, oscillates rapidly with γ. When integrated over a few of its oscillations, the result is essentially zero. The coefficient C varies little over an oscillation of $\overline{\mathcal{I}}$, hence does not affect the outcome of the integration. Unless the gradient of γ, which appears in the denominator of (G.3), varies significantly over an oscillation of $\overline{\mathcal{I}}$, then the integral over γ in (G.3) yields zero. This leads to the following observations:

The integral form (G.1) has negligible output, unless one or more of the following occurs:

1. The function $\gamma(\rho, s)$, considered as a function of s, undergoes a maximum, minimum, or saddle point at some point on the surface $\mathcal{S}$. If γ is not monotonic in s, then such points will exist. At such points, the gradient of γ goes to zero, and the form (G.3) becomes indeterminate. Near such points, if the coordinate transformation to (G.3) is applied, then the gradient in the denominator does vary significantly over an oscillation of the input function, and the result is non-zero.
2. The gradient of γ becomes small enough that changes in the gradient inverse become significant over a characteristic wavelength of $\overline{\mathcal{I}}$. The function $\gamma(\rho, s)$, considered as a function of s, may go through an inflection point at which the gradient of γ is a minimum. If the minimum is sufficiently small, then the denominator in (G.3) can change significantly over an oscillation of the input function, and the integral can yield a non-zero result. Events produced by other than a simple inflection point are possible also, though perhaps not quite in harmony with the assumed low-frequency character of γ.
3. The surface $\mathcal{S}$ terminates or suffers a discontinuity in slope. Oscillations in the input function near a termination may have nothing to cancel with, and (7.3) may yield a finite result.

All of this suggests that the surface integral can be replaced by effective contributions from a few discrete points. In the neighborhood of these points, constructive interference occurs, building an "event". Away from these points, the integral interferes destructively, and the net contribution is nearly zero.

In two dimensions, the situation is slightly simpler, in that the integral involves only one coordinate s. Analogously to (G.3), this coordinate may be changed to γ, and it is seen that the same observations apply.

In either two or three dimensions, one can anticipate that the diffraction integral will evaluate to an expression of the form

$$\begin{aligned} \mathcal{O}(\rho) &\simeq \sum_e \mathcal{O}_e(\rho) \\ &= \sum_e \frac{C(\rho, s_e)}{\mathcal{F}_e(\rho, s_e)} f_e * \bar{\mathcal{I}}(\gamma(\rho, s)), \end{aligned} \tag{G.4}$$

where $\mathcal{F}_e$ and f_e are, respectively, an amplitude function and a filter characteristic of the event at $s = s_e$.

Expressions are developed below for the events caused by extrema, inflection points, and surface terminations in γ. This is not meant to exhaust the list of possibilities, but does aspire to cover major, common ones.

G.2 Two dimensions

In two dimensions, the coordinate s becomes a scalar coordinate ℓ, and the diffraction integral takes the form

$$O(\rho) = \int d\ell C(\rho, \ell)\bar{\mathcal{I}}(\gamma(\rho, \ell)). \tag{G.5}$$

In this general form, the parameters of this equation may represent any of many physical quantities, depending upon the process being described. As an example, in a modeling process, ρ may represent time, and γ a traveltime function $t - \tau(\ell)$. In this case, $C(\rho, \ell)$ would represent a reflectivity amplitude combined with geometrical factors. Typically, a traveltime function would approach infinity as ℓ approaches plus or minus infinity, and go through at least one minimum somewhere on the line, possibly exhibiting other features as well. Figure G.1 illustrates a possible traveltime function over the interval $0 < \ell < 5000$. This particular function happens to be the polynomial $\tau(\ell) = 1.101\,03 - 0.000\,410\,981\ell$

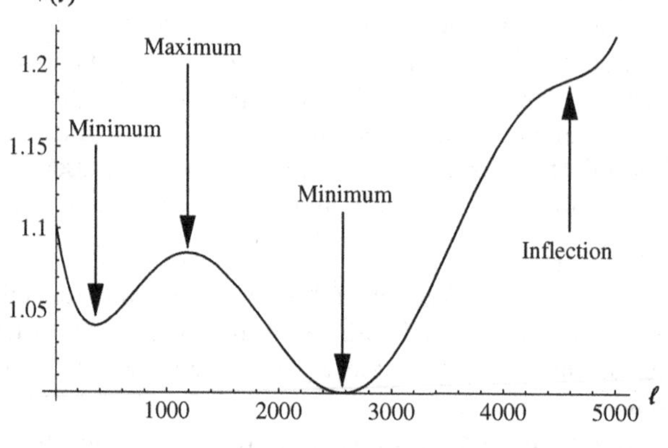

Figure G.1 Illustration of a traveltime function. Within the depicted range of ℓ, the function goes through two minima, one maximum, and one inflection point.

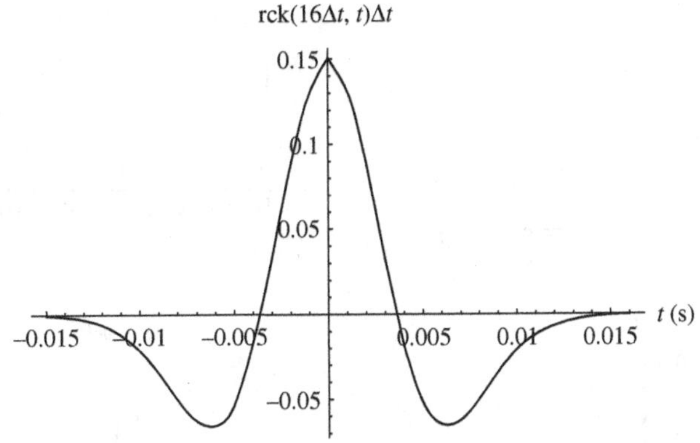

Figure G.2 The Ricker wavelet used in the diffraction integral (G.6). Sample rate is $\Delta t = .001$ sec. The peak frequency of this wavelet is 62.5 hz.

$+ 9.140\,76 \times 10^{-7}\ell^2 - 7.589\,03\times 10^{-10}\ell^3 + 2.800\,13\times 10^{-13}\ell^4 - 4.691\,48\times 10^{-17}\ell^5 + 2.930\,15\times 10^{-21}\ell^6$, and exhibits two minima, one maximum, and an inflection point.

Taking this sample traveltime function, we construct an example diffraction integral. We choose a time increment of $\Delta t = 0.001$ s. For simplicity, we set $C(\rho, \ell) = 1$. We take the wavelet $\mathcal{I}$ to be a Ricker wavelet rck$(16\Delta t, t)$, as defined in equation (E.32) with width $t_d = 16\Delta t$, corresponding to a peak frequency of 62.5 hz. This wavelet is depicted in Figure G.2. The filter f is taken to be the half-derivative function d_{h+} defined in Appendix E, equation (E.43). Implementations of this and other relevant filters are drawn in Figure G.3.

The diffraction integral in this example is thus

$$O(t) = \int_0^{5000} d\ell \left[d_{h+} * \mathrm{rck}(16\Delta t)\right](t - \tau(\ell)). \tag{G.6}$$

Numerical evaluation of equation (G.6) leads to the result shown in Figure G.4. The integral is seen to produce events at the traveltimes corresponding to the minima, maximum, inflection point, and end points.

G.2.1 Extrema

Suppose that $\gamma(\rho, \ell)$ undergoes an extremum at $\ell = \ell_e$, so that the derivative of γ is zero at ℓ_e. The contribution to the output function $\mathcal{O}$ from the neighborhood N_e of ℓ_e is

$$O_{e\pm}(\rho) \simeq C\left(\rho, \ell_e\right) \int_{N_e} d\ell \bar{\mathcal{I}}(\gamma(\rho, \ell)). \tag{G.7}$$

The coefficient C in this expression has been approximated by its value at the extremal point. In the neighborhood of the extremal point, one can make a Taylor series expansion of γ:

$$\gamma(\rho, \ell) \simeq \gamma_e(\rho) + \frac{(\ell - \ell_e)^2}{2}\gamma_e''(\rho), \tag{G.8}$$

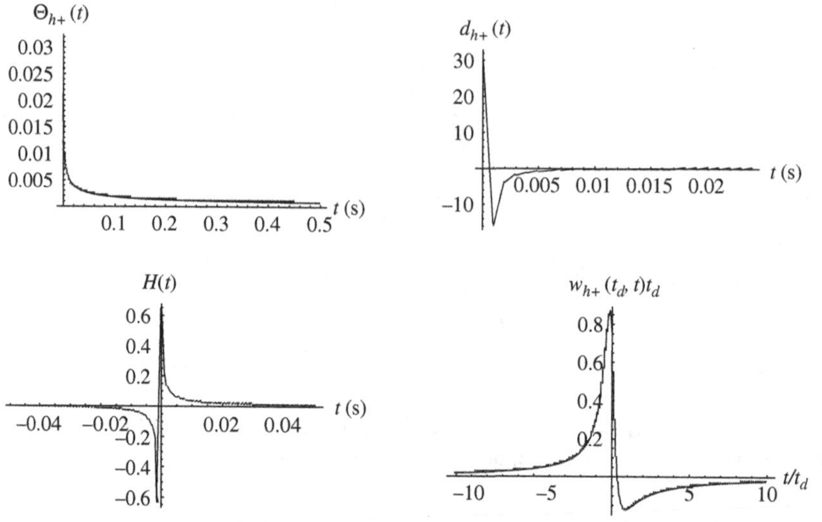

Figure G.3 Filters commonly used in constructing a diffraction integral. In the upper left is the half-integral filter, calculated by solving for the causal square root of the step function. In the upper right is a half-derivative filter, constructed by solving for the causal inverse of the half-integral filter. The filter on the lower left is the Hilbert transform, constructed as the convolution of the half-integral filter and the adjoint of the half-derivative filter. In the lower right is the half-derivative of the inflection filter W_0, as defined in Appendix E.

where

$$\gamma_e(\rho) \equiv \gamma(\rho, \ell_e) \tag{G.9}$$

and

$$\gamma_e''(\rho) \equiv \frac{\partial^2 \gamma}{\partial \ell^2}(\rho, \ell_e). \tag{G.10}$$

Now make a change of integration variable to

$$u = \frac{(\ell - \ell_e)^2}{2}\left|\gamma_e''(\rho)\right|. \tag{G.11}$$

Then,

$$\mathcal{O}_e(\rho) = \frac{\sqrt{2\pi}}{2}\frac{C(\rho, \ell_e)}{\sqrt{|\gamma_e''|}}\int_{N_e} du \sqrt{\frac{1}{\pi u}}\bar{\mathcal{I}}(\gamma_e + uS). \tag{G.12}$$

In this expression, S is the sign of the second derivative γ_e''.

The integral in (G.12) can now be extended to infinity in all directions. Doing so does not affect the result, because the rapid oscillations of $\mathcal{I}$ must average to zero away from the extremal point. Thus,

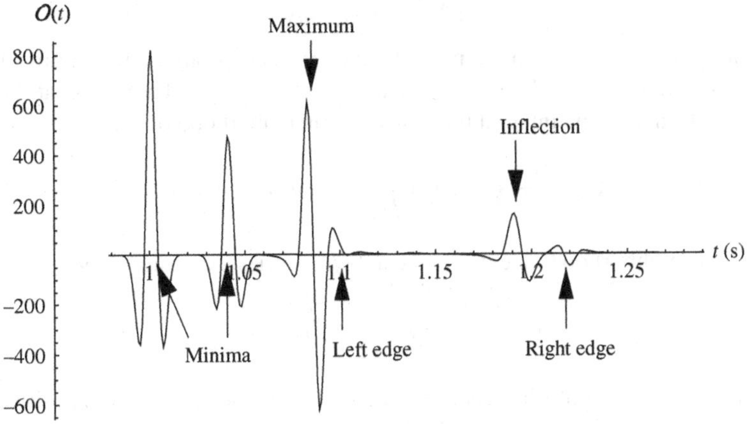

Figure G.4 Numerical evaluation of the integral (G.6) for the traveltime function in Figure G.1. Sample rate was chosen to be Δt = 0.001 s.

$$\mathcal{O}_e(\rho) = \sqrt{\frac{2\pi}{|\gamma_e''|}} C(\rho, \ell_e) \int_0^\infty \frac{du}{\sqrt{\pi u}} \bar{\mathcal{I}}(\gamma_e + uS) . \tag{G.13}$$

The integral over u is recognizable as the half-integral filter $\Theta_{h\mp}$ (see Appendix E), the sign $\mp$ depending on whether the sign S is positive or negative. If S is negative, corresponding to γ undergoing a maximum at ℓ_e, then the filter is Θ_{h+}. Likewise, a minimum at ℓ_e produces the filter Θ_{h-}. That is,

$$\begin{aligned} \mathcal{O}_e(\rho) &= \sqrt{\frac{2\pi}{|\gamma_e''|}} C(\rho, \ell_e) \left[\Theta_{h-} * f * \mathcal{I}\right](\gamma_e), \gamma \text{ a minimum,} \\ &= \sqrt{\frac{2\pi}{|\gamma_e''|}} C(\rho, \ell_e) \left[\Theta_{h+} * f * \mathcal{I}\right](\gamma_e), \gamma \text{ a maximum.} \end{aligned} \tag{G.14}$$

If f is a half-derivative filter, then the net filter is either the unit filter, or $\pm$ the Hilbert transform filter, as indicated by equation (E.60) or its adjoint.

For the traveltime example above, a minimum in γ corresponds to a maximum in τ, and vise versa. The traveltime has minima at $\tau = 1.0406$ and $\tau = 1.0000$ seconds, and a maximum at $\tau = 1.0854$ seconds. The second derivatives of τ at these points are 5.8429×10^{-7}, 2.2687×10^{-7}, and -2.3711×10^{-7}, producing events of amplitude 3279, 5263, and 5148. That is,

$$\mathcal{O}_1(t) = 3279\left[\Theta_{h+} * d_{h+}\right](t - 1.0406) = 3279\delta(t - 1.0406). \tag{G.15}$$

$$\mathcal{O}_2(t) = 5263\left[\Theta_{h+} * d_{h+}\right](t - 1) = 5263\delta(t - 1). \tag{G.16}$$

$$\mathcal{O}_3(t) = 5148\left[\Theta_{h-} * d_{h+}\right](t - 1) = -5148\text{hlb}(t - 1.0854). \tag{G.17}$$

G.2.2 Inflections

Suppose, at point ℓ_e, the derivative magnitude of γ goes through a minimum. If S is the sign of this derivative at γ_e, define coordinates $r = (\ell - \ell_e)\, S$. Then, look at the "event" $\mathcal{O}_e$ generated from the line integral (G.1) within a neighborhood of ℓ_e:

$$\mathcal{O}_e(\rho) = C\,(\rho, \ell_e) \int_{N_e} dr \bar{\mathcal{I}}\,(\gamma\,(\rho, r + \ell_e))\,. \tag{G.18}$$

Within this neighborhood, one can make a Taylor-series expansion of γ:

$$\gamma = \gamma_e + \gamma_e' r + \frac{1}{6}\gamma_e''' r^3 + \cdots\,. \tag{G.19}$$

Note that the direction of r has been defined so that γ_e' and γ_e''' are both positive. Within the neighborhood N_e of ℓ_e, the integral (G.18) becomes

$$\mathcal{O}_e(\rho) \simeq C\,(\rho, \ell_e) \int_{N_e} dr \bar{\mathcal{I}} \left(\gamma_e + \gamma_e' r + \frac{1}{6}\gamma_e''' r^3\right). \tag{G.20}$$

The argument of $\bar{\mathcal{I}}$ is monotonic in r, allowing a change of variable to

$$\nu = \sqrt{\frac{27}{4}} \left(\frac{r}{\delta r} + \frac{r^3}{\delta r^3}\right), \tag{G.21}$$

with δr a characteristic distance

$$\delta r = \sqrt{\frac{6\gamma_e'}{\gamma_e'''}}. \tag{G.22}$$

The event integral $\mathcal{O}_e$, in terms of ν, is

$$\mathcal{O}_e(\rho) \simeq C\,(\rho, \ell_e) \frac{2}{3}\sqrt{\frac{2\gamma_e'}{\gamma_e'''}} \int_{N_e} \frac{d\nu}{1 + 3r^2/\delta r^2} \bar{\mathcal{I}} \left(\gamma_e + \frac{2}{3}\sqrt{\frac{2\gamma_e'^3}{\gamma_e'''}}\,\nu\right). \tag{G.23}$$

Solving for r as a function of ν, we obtain

$$r = \frac{\delta r}{\sqrt{3}} \left(\left(\sqrt{1+\nu^2} + \nu\right)^{1/3} - \left(\sqrt{1+\nu^2} + \nu\right)^{-1/3}\right), \tag{G.24}$$

from which we see

$$\mathcal{O}_e(\rho) \simeq C\,(\rho, \ell_e) \frac{2}{3}\sqrt{\frac{2\gamma_e'}{\gamma_e'''}} \int_{N_e} d\nu W_{0c}(\nu) \bar{\mathcal{I}} \left(\gamma_e + \frac{2}{3}\sqrt{\frac{2\gamma_e'^3}{\gamma_e'''}}\,\nu\right), \tag{G.25}$$

with W_{0c} the inflection filter (E.69),

$$W_{0c}(\nu) = \frac{1}{\left(\sqrt{1+\nu^2} + \nu\right)^{2/3} + \left(\sqrt{1+\nu^2} + \nu\right)^{-2/3} - 1}, \tag{G.26}$$

or equivalently,

$$W_{0c}(\nu) = \frac{\left(\sqrt{1+\nu^2}+\nu\right)^{1/3} + \left(\sqrt{1+\nu^2}+\nu\right)^{-1/3}}{2\sqrt{1+\nu^2}}. \tag{G.27}$$

Extending the integral $\mathcal{O}_e$ to infinity, it can be written as the convolution

$$\mathcal{O}_e(\rho) \simeq \frac{C(\rho, \ell_e)\,\gamma_d}{\gamma_e'} \int_{-\infty}^{\infty} d\nu \bar{\mathcal{I}}(\gamma_e - \gamma_d \nu)\, W_{0c}(\nu), \tag{G.28}$$

with γ_d the characteristic dilation factor

$$\gamma_d = \frac{2}{3}\sqrt{\frac{2\gamma_e'^3}{\gamma_e'''}}. \tag{G.29}$$

Alternatively, the dilation factor can be placed in W_{0c}:

$$\begin{aligned}\mathcal{O}_e(\rho) &\simeq \frac{C(\rho, \ell_e)}{\gamma_e'} \int_{-\infty}^{\infty} du W_{0c}\left(\frac{u}{\gamma_d}\right) \bar{\mathcal{I}}(\gamma_e - u) \\ &= \frac{C(\rho, \ell_e)\,\gamma_d}{\gamma_e'} \int_{-\infty}^{\infty} du W_0(\gamma_d, u) \int_{-\infty}^{\infty} du' f(u') I\left(\gamma_e - u' - u\right) \\ &= \frac{C(\rho, \ell_e)\,\gamma_d}{\gamma_e'} \left[W_0(\gamma_d) * f * I\right](\gamma_e).\end{aligned} \tag{G.30}$$

$W_0(\gamma_d, u)$ is the symmetric inflection filter, as defined in Appendix E, equations (E.68) and (E.69), shown in Figure E.14. $W_0(\gamma_d, u)$ lacks the scaling invariance that would allow the factor γ_d to be removed from the filter. Instead, the factor expands or contracts the wavelet W_0.

If the filter $f(u)$ is scalable, so that $f(u\gamma_d) = \gamma_d^{-\eta_f} f(u)$, then the two filters f and W_0 can be combined. Defining

$$h(\gamma_d, \gamma) \equiv \frac{1}{\gamma_d}[W_{0c} * f]\left(\frac{\gamma}{\gamma_d}\right), \tag{G.31}$$

then, invoking equation (E.16),

$$[W_0(\gamma_d) * f](\gamma) = \gamma_d^{1-\eta_f} h(\gamma_d, \gamma), \tag{G.32}$$

and

$$\mathcal{O}_e(\rho) \simeq \frac{C(\rho, \ell_e)\,\gamma_d^{2-\eta_f}}{\gamma_e'} \left[h(\gamma_d) * \mathcal{I}\right](\gamma_e). \tag{G.33}$$

If f is a half-derivative filter $d_{h\pm}$, then $h(\gamma_d, \gamma)$ is the half-differentiated inflection filter $w_{h\pm}$, as shown in Figure E.17. In this case, $\eta_f = 3/2$, and

$$\mathcal{O}_e(\rho) \simeq \frac{C(\rho, \ell_e)\,\gamma_d^{1/2}}{\gamma_e'} \left[w_{h\pm}(\gamma_d) * \mathcal{I}\right](\gamma_e). \tag{G.34}$$

For the above traveltime example, an inflection point exists at $\tau_e = 1.191\,95$ s. At this point, the first derivative of traveltime is $\tau_e' = 0.000\,025\,252$, its second derivative is zero, and its third derivative is $\tau_e''' = 9.8425 \times 10^{-10}$. The dilation factor is $\gamma_d = \tau_d = \sqrt{8\,\tau_e'^3 / 9\tau_e'''} = 0.025\,739$ sec. The inflection produces the event

$$\mathcal{O}_4(t) \simeq 6353 w_{h+}(0.025\,739, t-1.19\,195). \tag{G.35}$$

G.2.3 Terminations

Suppose the line integral (G.5) terminates at a and b:

$$\mathcal{O}(\rho) = \int_a^b d\ell C(\rho,\ell)(\gamma(\rho,\ell)). \tag{G.36}$$

The total contribution to $\mathcal{O}$ can be divided into contributions from the interior of the interval plus contributions $\mathcal{O}_a$ and $\mathcal{O}_b$ from the end points. Assume that a and b are far enough from any extremal points that γ can be considered linear in ℓ near a and b. Then the two end-point contributions turn out to be step-function or integral filters:

$$\begin{aligned}\mathcal{O}_a(\rho) &\simeq C(\rho,a)\int_a^{\infty} d\ell\,(\gamma_a + \gamma_a'(\ell-a))\\ &= -\frac{C(\rho,a)}{\gamma_a'}\int_{-\infty}^{\gamma_a} du(u)\\ &= -\frac{C(\rho,a)}{\gamma_a'}\,[\Theta *]\,(\gamma_a)\,.\end{aligned} \tag{G.37}$$

$$\begin{aligned}\mathcal{O}_b(\rho) &\simeq C(\rho,b)\int_{-\infty}^{b} d\ell\,(\gamma_b + \gamma_b'(\ell-b))\\ &= \frac{C(\rho,b)}{\gamma_b'}\int_{-\infty}^{\gamma_b} d\ell(u)\\ &= \frac{C(\rho,b)}{\gamma_b'}\,[\Theta *]\,(\gamma_b)\,.\end{aligned} \tag{G.38}$$

This description of a termination is usually adequate, unless the termination point lies very close to an extremal point. If, for example, the termination point lies exactly on an extremal point, then the response $\mathcal{O}$ should be exactly one-half the extremal-point response. However, the response predicted by (G.37) or (G.38) is infinite.

Suppose the line integral (G.5) terminates at a point ℓ_0 which may be near an extremal point ℓ_e. Define a coordinate r to be the absolute value of the distance from the termination point ℓ_0:

$$r = |\ell - \ell_0|\,, \tag{G.39}$$

and γ_0' and γ_0'' to be the first and second derivatives of γ with respect to r, evaluated at the termination point. Then, near the termination point,

$$\begin{aligned}\gamma(\rho,\ell) \to \gamma(\rho,r) &\simeq \gamma_0 + \gamma_0' r + \gamma_0''\, r^2 / 2\\ &= \gamma_e + \gamma_0''\,(r - r_e)^2 / 2.\end{aligned} \tag{G.40}$$

We have parameterized the function γ two ways, with γ_0, γ_0', and γ_0'' the value of γ and its derivatives at the boundary point $r = 0$, and γ_e the value of γ at the extremal point r_e. The second derivative γ_0'' is, for a quadratic, the same at either point. The two sets of parameters have relations

$$r_e = -\gamma_0'/\gamma_0'', \tag{G.41}$$

and

$$\gamma_e = \gamma_0 + \Delta\gamma, \tag{G.42}$$

where

$$\Delta\gamma = -\frac{\gamma_0'' r_e^2}{2} = -\frac{\gamma_0'^2}{2\gamma_0''} \tag{G.43}$$

is the displacement in γ between the termination point $r = 0$ and the extremal point r_e. The sign of $\Delta\gamma$ is positive when the extremal point is a maximum, and negative for a minimum.

The line integral (G.5) is thus

$$\mathcal{O}_e(\rho) \simeq C(\rho, \ell_0) \int_0^\infty dr \bar{\mathcal{I}} \left(\gamma_0 + \gamma_0' r + \gamma_0'' r^2 \big/ 2\right). \tag{G.44}$$

The support of the function $\bar{\mathcal{I}}$ is concentrated near

$$\gamma_0 + \gamma_0' r + \gamma_0'' r^2 \big/ 2 = 0. \tag{G.45}$$

This may happen at most two points,

$$\begin{aligned} r_\pm &= r_e \pm |r_e| \sqrt{\gamma_e / \Delta\gamma} \\ &= r_e \pm \sqrt{\frac{-2\gamma_e}{\gamma_0''}}. \end{aligned} \tag{G.46}$$

For either of the roots $r_\pm$ to be real, γ_e and $\Delta\gamma$ must have the same sign, which means that γ_e and γ_0'' must have opposite signs. That is, γ_e must be positive when the neighboring event is a maximum, and negative for a minimum.

$r_e < 0$ Extremal point off boundary

If the point r_e is negative, then the extremal point lies outside the boundary (see Figure G.5a). In this case, only the root r_+ can lie within the boundaries of the curve, and then only provided γ_e is such that $\gamma_e / \Delta\gamma > 1$. For a maximum in γ, γ_0 must be positive, whereas for a minimum, γ_0 must be negative. In either case, the solution to the diffraction integral (G.44) is

$$\mathcal{O}_e(\rho) \simeq C(\rho, \ell_0) \frac{1}{|\gamma_0'|} \int d\mu \bar{\mathcal{I}}(\mu) \frac{\Theta\left((\gamma_0 - \mu)/\Delta\gamma\right)}{\sqrt{\left(1 + (\gamma_0 - \mu)/\Delta\gamma\right)}}. \tag{G.47}$$

This can be written as

$$\mathcal{O}_e(\rho) \simeq C(\rho, \ell_0) \frac{\Delta\gamma}{|\gamma_0'|} \int d\mu \bar{\mathcal{I}}(\mu) f_T(\Delta\gamma, \gamma_0 - \mu). \tag{G.48}$$

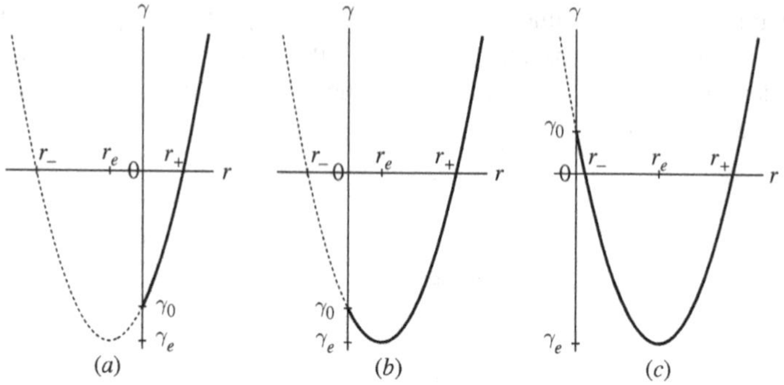

Figure G.5 Quadratic function left-terminated at zero. Example (a) shows the extremal point outside the end point ($r = 0$) of the curve. The curve has a single zero crossing inside the end point at $r = r_-$. In example (b), the extremal point is inside the end point, and there is a single zero crossing at $r = r_+$. In example (c), the extremal point has moved further to the right, and there are two zero crossings, at r_- and at r_+.

or

$$\mathcal{O}_e(\rho) \simeq \frac{\Delta\gamma}{|\gamma_0'|} C(\rho, \ell_0) \left[f_T(\Delta\gamma) * \bar{\mathcal{I}}\right](\gamma_0), \tag{G.49}$$

where f_T is the *termination filter*

$$f_T(\Delta\gamma, \gamma) = \frac{\Theta(\gamma/\Delta\gamma)}{\Delta\gamma\sqrt{(1+\gamma/\Delta\gamma)}}. \tag{G.50}$$

Where the termination point is near a maximum in γ, $\Delta\gamma$ is positive, and the filter is non-zero for positive γ. Near a minimum, $\Delta\gamma$ is negative and the filter is non-zero for negative γ.

The nature of the termination filter depends on the magnitude of the displacement $\Delta\gamma$. For $\gamma/\Delta\gamma << 1$ (i.e. the extremal point far from the end point), the termination filter is proportional to a step function:

$$f_T(\Delta\gamma, \gamma) \underset{\gamma/\Delta\gamma\to 0}{\to} \frac{\Theta(\gamma/\Delta\gamma)}{\Delta\gamma}. \tag{G.51}$$

In this case, $\mathcal{O}_e$ reduces to one of the linearized solutions (G.37) and (G.38).

For $\gamma/\Delta\gamma \gg 1$ (i.e. the extremal point very close to the end point), the termination filter is proportional to a half-integral operator:

$$f_T(\Delta\gamma, \gamma) \underset{\gamma/\Delta\gamma\to\infty}{\to} \frac{\Theta(\gamma/\Delta\gamma)}{\Delta\gamma\sqrt{\gamma/\Delta\gamma}} = \frac{\sqrt{\pi}\,\Theta_{h+}(\gamma/\Delta\gamma)}{\Delta\gamma}. \tag{G.52}$$

For data sampled at intervals $d\gamma$, the termination filter behaves like a step function when $|\Delta\gamma| \gg d\gamma$, and like a half-integral when $|\Delta\gamma| \ll d\gamma$.

Near the point $\gamma = 0$, the termination filter assumes its maximum value

$$f_T(\Delta\gamma, 0) = \frac{1}{\Delta\gamma}. \tag{G.53}$$

Combined with the factor $\Delta\gamma / |\gamma_0'| = -|\gamma_0'| / 2\gamma_0''$ appearing in (G.49), the magnitude of a termination event is seen to be proportional to $1/|\gamma_0'|$. The magnitude of the event depends inversely on the magnitude of the derivative γ_0' of γ at the termination point γ_0. If the surface terminates far from an extremum, γ_e' is likely to be large, and the corresponding event negligible.

$r_e > 0$ Extremal point inside boundary

If the point r_e is positive, then the extremal point lies inside the boundary. The root r_+ lies inside the boundary, and r_- will too, provided $\gamma_e / \Delta\gamma < 1$, or equivalently, $\gamma_0 / \Delta\gamma < 0$ (see Figure G.5b and c) . Writing the diffraction integral as

$$\begin{aligned} \mathcal{O}_e(\rho) \simeq \mathcal{O}_{e+}(\rho) + \mathcal{O}_{e-}(\rho) \simeq C(\rho, \ell_e) \int_{-\infty}^{\infty} dr \bar{\mathcal{I}} \left(\gamma_0 + \gamma_0' r + \gamma_0'' \, r^2 / 2\right) \\ - C(\rho, \ell_0) \int_{-\infty}^{0} dr \bar{\mathcal{I}} \left(\gamma_0 + \gamma_0' r + \gamma_0'' \, r^2 / 2\right), \end{aligned} \tag{G.54}$$

the first term is the expression for the extremum at γ_e. The second term can be written as an integral from 0 to ∞ by changing the sign of the integration variable:

$$\mathcal{O}_{e-}(\rho) \simeq -C(\rho, \ell_0) \int_0^{\infty} dr \bar{\mathcal{I}} \left(\gamma_0 - \gamma_0' r + \gamma_0'' \, r^2 / 2\right). \tag{G.55}$$

The argument of $\bar{\mathcal{I}}$ in this integral has a single positive root at $(\gamma_0'/\gamma_0'') \left(\sqrt{\gamma_e/\Delta\gamma} - 1\right)$, so that

$$\mathcal{O}_{e-}(\rho) \simeq -\frac{\Delta\gamma}{|\gamma_0'|} C(\rho, \ell_0) \left[f_T(\Delta\gamma) * \bar{\mathcal{I}}\right](\gamma_0), \tag{G.56}$$

with f_T given by (G.50). The difference between (G.56) for positive r_e and (G.49) for negative r_e is the overall sign. Thus, we can write an expression valid in either case as

$$\mathcal{O}_{e-}(\rho) \simeq -\text{Sign}(r_e) \frac{\Delta\gamma}{|\gamma_0'|} C(\rho, \ell_0) \left[f_T(\Delta\gamma) * \bar{\mathcal{I}}\right](\gamma_0). \tag{G.57}$$

G.2.4 Traveltime example

For our traveltime example, the line was allowed to terminate at points $\ell = 0$ and $\ell = 5000$, with corresponding traveltimes $\tau_e = 1.101\,03$ and $1.217\,98$. At these points, the derivatives of traveltime are $\tau_e' = -0.000\,410\,98$ and $-0.000\,150\,02$, respectively. The traveltime displacements are $\Delta\tau = 0.046\,195\,7$ s and $0.015\,66\,73$ s. Both displacements are much larger than the sample rate of 0.001 s, allowing the step-function approximation (G.51) to the termination filter. Thus, the terminations give rise to two events

$$\mathcal{O}_5(t) \simeq -2433\Theta_{h+}(t - 1.101\,03), \tag{G.58}$$

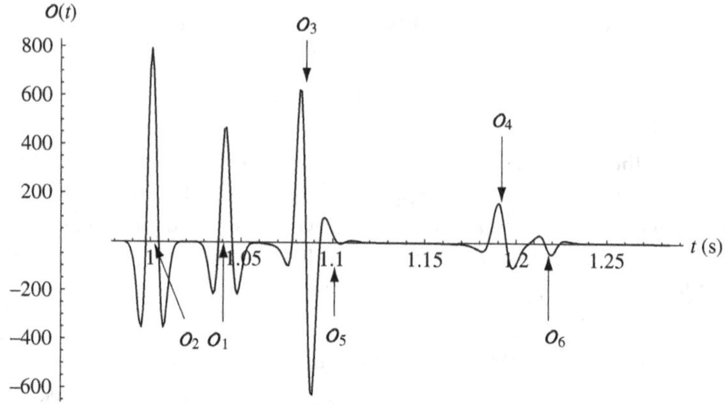

Figure G.6 The integral (G.6) constructed by calculating the individual events $\mathcal{O}_1 - \mathcal{O}_6$. Compare with the numerical integration of (G.6) in Figure G.4.

$$\mathcal{O}_6(t) \simeq -6666\Theta_{h+}(t - 1.217\,98). \tag{G.59}$$

For the example traveltime above, we plot the six events (G.15), (G.16), (G.17), (G.35), (G.58), and (G.59) in Figure G.6. The result closely matches (as it should) the numerical integration in Figure G.4.

G.2.5 Relation to the stationary phase approximation

The approximations made in constructing events from the diffraction integral are closely related to the stationary phase approximation. The diffraction integral (G.5)

$$\mathcal{O} = \int d\ell C(\ell)\bar{\mathcal{I}}(\gamma(\ell)) \tag{G.60}$$

can be decomposed into phase-integral components:

$$\mathcal{O} = \frac{1}{2\pi}\int dk\bar{\mathcal{I}}(k)\int d\ell e^{ik\gamma(\ell)}C(\ell), \tag{G.61}$$

each of which can be treated with the methods of Appendix F.

Diffraction integral without end points

Suppose that the variable γ goes through a minimum or maximum γ_e at point ℓ_e, so that

$$\gamma \simeq \gamma_e + (\ell - \ell_e)^2\frac{\gamma_e''}{2}. \tag{G.62}$$

For the moment, we have ignored the output variable ρ in (G.5) and assumed that the boundaries of the integral are at $\pm\infty$.

Addressing the phase-integral decomposition (G.61), the ℓ-integral is evaluable by the method of stationary phase; whence

$$\mathcal{O} \simeq \frac{C(\ell_e)}{\sqrt{2\pi}} \int dk \bar{\mathcal{I}}(k) \sqrt{\frac{i}{k\gamma_e''}} e^{ik\gamma_e}. \tag{G.63}$$

The k-integral is the inverse Fourier transform of a half-integral filter times $\overline{\mathcal{I}}(k)$. The phase of the filter depends upon the sign of γ_0''.That is,

$$\mathcal{O} \simeq C(\ell_e) \sqrt{\frac{2\pi}{|\gamma_e''|}} \left[\Theta_{h\mp} * \bar{\mathcal{I}}\right](\gamma_e), \tag{G.64}$$

where Θ_{h-} obtains if γ_e'' is positive (a minimum), and Θ_{h+} if γ_e'' is negative (a maximum).

Either way, $\mathcal{O}$ is proportional to a 45° phase-shifted "half integral" of the rapidly varying function $\overline{\mathcal{I}}$, evaluated at the stationary point γ_e. This is identical to the result (G.14) for extrema in the diffraction integral. Thus our construction of an extremal event in the diffraction integral is equivalent to a stationary phase approximation of the corresponding phase integral.

G.2.6 Diffraction integral with end points

Suppose now that the diffraction integral has an end point, so that

$$\mathcal{O} = \int_a^\infty d\ell C(\ell) \bar{\mathcal{I}}(\gamma(\ell)). \tag{G.65}$$

Suppose further that γ is nearly linear in ℓ near a. Then, performing a Fourier decomposition of $\overline{\mathcal{I}}$,

$$\begin{aligned}
\mathcal{O}_a &= \frac{1}{2\pi} \int dk \bar{\mathcal{I}}(k) \int_a^\infty d\ell e^{ik\gamma(\ell)} C(\ell) \\
&\simeq \frac{C(a)}{2\pi} \int dk \bar{\mathcal{I}}(k) e^{ik\gamma_a} \int_a^\infty d\ell e^{ik\gamma_a'(\ell-a)} \\
&= -\frac{C(a)}{2\pi\gamma_a'} \int dk \bar{\mathcal{I}}(k) \frac{e^{ik\gamma_a}}{ik} \\
&= -\frac{C(a)}{\gamma_a'} \left[\Theta * \bar{\mathcal{I}}\right](\gamma_a).
\end{aligned} \tag{G.66}$$

This is identical to the result (G.37) and consistent with the linearized extension of the stationary phase approximation in the presence of end points.

G.2.7 Diffraction integral with inflection points

Of the event types we have detected in the diffraction integral, the inflection point has no stationary-phase analog. This is because the phase at an inflection point is not, well, stationary. This is not to say that contributions from inflection points are not evident in the phase integral. Rather, it tells us that the stationary phase approximation to the phase integral may not be complete, even where the asymptotic assumptions behind stationary phase are otherwise well satisfied.

G.3 Three dimensions

The principal difference in three dimensions is that one now has a 2-D surface integral to evaluate. An event is characterized by something significant happening in both dimensions. More types of event are possible, since a minimum in one direction may be accompanied by a maximum in another, and so on.

G.3.1 extrema

Suppose that $\gamma(\rho, \boldsymbol{s})$ undergoes an extremum at $\boldsymbol{s} = \boldsymbol{s}_e$, so that the gradient of γ is zero at $\boldsymbol{s}_e$. The contribution to the output function $\mathcal{O}$ from the neighborhood N_e of $\boldsymbol{s}_e$ is

$$\mathcal{O}_e(\rho) \simeq C(\rho, \boldsymbol{s}_e) \int\int_{N_e} d\boldsymbol{s}\bar{\mathcal{I}}(\gamma(\rho, \boldsymbol{s})). \tag{G.67}$$

One can make a 2-D Taylor series expansion of γ about the extremal point γ_e:

$$\gamma(\rho, \boldsymbol{s}) \simeq \gamma_e(\rho) + (\boldsymbol{s} - \boldsymbol{s}_e) \cdot \boldsymbol{H} \cdot (\boldsymbol{s} - \boldsymbol{s}_e), \tag{G.68}$$

where $\gamma_e(\rho) \equiv \gamma(\rho, \boldsymbol{s}_e)$, and $\boldsymbol{H}$ is the matrix of second-derivatives of γ with respect to $\boldsymbol{s}$ at $\boldsymbol{s}_e$:

$$H_{xx} = \frac{1}{2}\frac{\partial^2 \gamma}{\partial s_x{}^2},$$

$$H_{xy} = H_{yx} = \frac{1}{2}\frac{\partial^2 \gamma}{\partial s_x \partial s_y},$$

$$H_{yy} = \frac{1}{2}\frac{\partial^2 \gamma}{\partial s_y{}^2}. \tag{G.69}$$

In general, the cross-terms H_{xy} and H_{yx} are non-zero. However, one can always rotate the coordinates until the cross-terms vanish. Define the vector $\boldsymbol{r} = \boldsymbol{s} - \boldsymbol{s}_e$, with elements r_1 and r_2 in the rotated coordinate system. In this system, $\boldsymbol{H}$ is a diagonal matrix with its eigenvalues on the diagonal:

$$\boldsymbol{H} = \begin{pmatrix} \gamma_{11}/2 & 0 \\ 0 & \gamma_{22}/2 \end{pmatrix}. \tag{G.70}$$

The diagonal elements are 1/2 the second derivatives of γ with respect to r_1 and r_2. The Taylor series expansion of γ assumes a simpler form in the rotated coordinates:

$$\gamma(\rho, \boldsymbol{r}) = \gamma_e + r_1{}^2 \frac{\gamma_{11}}{2} + r_2{}^2 \frac{\gamma_{22}}{2} + \cdots. \tag{G.71}$$

Now make another change of coordinates

$$u_1 = r_1{}^2 \left|\frac{\gamma_{11}}{2}\right|, \quad u_2 = r_2{}^2 \left|\frac{\gamma_{22}}{2}\right|, \tag{G.72}$$

and use them as the variables of integration. Then,

$$\mathcal{O}_e(\rho) = \frac{1}{2}\frac{C(\rho, \boldsymbol{s}_e)}{\sqrt{|\gamma_{11}\gamma_{22}|}} \int\int_{N_e} du_1 du_2 \sqrt{\frac{1}{u_1 u_2}} \bar{\mathcal{I}}(\gamma_e + u_1 S_1 + u_2 S_2). \tag{G.73}$$

In this expression, $S_1$is the sign of the second derivative γ_{11}, and S_2 is the sign of the eigenvalue γ_{22}.

The product of the two eigenvalues is the determinant of the matrix of second derivatives of γ. This determinant is invariant under rotations, so if desired, it may be expressed in the original (s_x, s_y) coordinate system. In general, this determinant will be referred to as $\det(\gamma'')$.

The integral in (G.73) can now be extended to infinity in all directions. Thus,

$$\mathcal{O}_e(\rho) = \frac{2\pi C(\rho, s_e)}{\sqrt{|\det(\gamma'')|}} \int_0^\infty \int_0^\infty \frac{du_1}{\sqrt{\pi u_1}} \frac{du_2}{\sqrt{\pi u_2}} \bar{\mathcal{I}}(\gamma_e + u_1 S_1 + u_2 S_2). \tag{G.74}$$

The integrals over u_1 and u_2 are recognizable as the half-integral filters $\Theta_{h\mp}$ (see Appendix E), the sign $\mp$ depending on whether the signs S_1 and S_2 are positive or negative. If both S_1 and S_2 are negative, corresponding to γ undergoing a two-dimensional maximum at s_e, then the net filter is $\Theta_{h+} * \Theta_{h+}$. According to equation (E.51), Two applications of the half-integral filter Θ_{h+} yield the integral or normalized step function Θ, defined in equation (E.38). Thus,

$$\mathcal{O}_e(\rho) = \frac{2\pi C(\rho, s_e)}{\sqrt{|\det(\gamma'')|}} [\Theta * f * \mathcal{I}](\gamma_e), \; \gamma_e \text{ a double maximum.} \tag{G.75}$$

If f is the differentiation filter, the net effect of the two filters in this case is the unit operator.

If both S_1 and S_2 are positive, then the net filter is the adjoint of Θ, which, since the normalized step function is antisymmetric, is $-\Theta$. Thus,

$$\mathcal{O}_e(\rho) = -\frac{2\pi C(\rho, s_e)}{\sqrt{|\det(\gamma'')|}} [\Theta * f * \mathcal{I}](\gamma_e), \; \gamma_e \text{ a double minimum.} \tag{G.76}$$

If one of S_1 and S_2 is positive and the other negative, corresponding to a saddle point in γ, then the net filter is $\Theta_{h-} * \Theta_{h+}$, or the inverse "rho" filter P, defined in equation (E.65).

$$\mathcal{O}_e(\rho) = \frac{2\pi C(\rho, s_e)}{\sqrt{|\det(\gamma'')|}} \left[P * f * \mathcal{I}\right](\gamma_e), \; \gamma_e \text{ a saddle point.} \tag{G.77}$$

If f is the derivative operator, the net filter is the Hilbert transform.

Whether a minimum, maximum, or saddle point is encountered, the result is a filtered version of the input function, evaluated at the extremal point γ_e. The amplitude in each case depends inversely on the square root of the determinant of the matrix of second derivatives of γ. If this determinant should approach zero at the extremal point, this formula will break down. In such a case, the second-order Taylor series approximation (G.68) is inadequate, and a higher-order approximation must be used.

The events produced by the 2- and 3-D operations both turn out to be proportional to the value of the input function or its Hilbert transform at the extremal point γ_e, provided that the diffraction integral filter f is a half-derivative in two dimensions, or a derivative operator in three dimensions.

G.3.2 Inflections

Suppose, at point $\boldsymbol{s}_e$, the gradient of γ goes through a minimum. Define coordinates $\boldsymbol{r} = \boldsymbol{s} - \boldsymbol{s}_e$, with r_1 aligned with the gradient of γ at $\boldsymbol{s}_e$, and r_2 normal to it. Then, look at the "event" $\mathcal{O}_e$ generated from the surface integral (G.1) within a neighborhood of $\boldsymbol{s}_e$:

$$\mathcal{O}_e(\rho) = C(\rho, \boldsymbol{s}_e) \int\int_{N_e} dr_1 dr_2 \bar{\mathcal{I}}(\gamma(\rho, \boldsymbol{r} + \boldsymbol{s}_e)). \tag{G.78}$$

Within this neighborhood, one can make a Taylor-series expansion of γ:

$$\gamma = \gamma_e + \sum_{i=1}^{2} \gamma_i r_i + \frac{1}{2} \sum_{i,j=1}^{2} \gamma_{ij} r_i r_j + \frac{1}{6} \sum_{i,j,k=1}^{2} \gamma_{ijk} r_i r_j r_k + \cdots \tag{G.79}$$

In this expression, γ_e is understood to be γ at $\boldsymbol{s}_e$, and the quantities $\gamma_i = \partial\gamma_e/\partial r_i$ and so on the derivatives of γ, evaluated at $\boldsymbol{s}_e$. As the r_1 and r_2 directions are defined, γ_1 is necessarily positive and γ_2 is zero.

For the gradient magnitude of γ to be a minimum at $\boldsymbol{s}_e$, one must have

$$\frac{\partial |\nabla\gamma|^2}{\partial r_i} = \frac{\partial}{\partial r_i}\left(\gamma_1{}^2 + \gamma_2{}^2\right) = 0, i = 1, 2. \tag{G.80}$$

This implies that

$$\begin{pmatrix} \gamma_{11} & \gamma_{12} \\ \gamma_{12} & \gamma_{22} \end{pmatrix} \begin{pmatrix} \gamma_1 \\ \gamma_2 \end{pmatrix} = \begin{pmatrix} 0 \\ 0 \end{pmatrix}. \tag{G.81}$$

That is to say, the gradient of γ is an eigenvector of the matrix of second derivatives of γ, with corresponding eigenvalue 0. Since γ_2 is zero, in order for (G.81) to be satisfied, γ_{11} and γ_{12} must also be zero.

The Taylor series expansion (G.79) reduces to

$$\gamma = \gamma_e + \gamma_1 r_1 + \gamma_{22} \frac{r_2{}^2}{2} + \frac{1}{6} \gamma_{111} r_1{}^3 + \cdots. \tag{G.82}$$

Terms in the expansion up to second order have been retained in the r_2 direction. In the r_1 direction, since the second-order term is zero, one may expect the range of r_1 values that contribute to the integral (G.78) to be somewhat larger than the range of r_2 values. Consequently, the term γ_{111} third-order in r_1 has been retained. Note that for the gradient of γ to be a minimum at γ_e, the third derivative γ_{111} must have the same sign as γ_1; i.e. $\gamma_{111} > 0$.

Within the neighborhood N_e of $\boldsymbol{s}_e$, the integral (G.78) becomes

$$\mathcal{O}_e(\rho) \simeq C(\rho, \boldsymbol{s}_e) \int\int_{N_e} dr_1 dr_2 \bar{\mathcal{I}}\left(\gamma_e + \gamma_1 r_1 + \gamma_{111} \frac{r_1{}^3}{6} + \gamma_{22} \frac{r_2{}^2}{2}\right). \tag{G.83}$$

To evaluate this integral, change integration variables to

$$\mu = \frac{r_2{}^2}{2} |\gamma_{22}|, \tag{G.84}$$

$$\nu = \sqrt{\frac{27}{4}}\left(\frac{r_1}{\delta r} + \frac{r_1{}^3}{\delta r^3}\right), \tag{G.85}$$

with δr a characteristic distance

$$\delta r = \sqrt{\frac{6\gamma_1}{\gamma_{111}}}. \tag{G.86}$$

The event integral $\mathcal{O}_e$, in terms of μ and ν, is

$$\mathcal{O}_e(\rho) \simeq \frac{2\sqrt{\pi}}{3}\sqrt{\frac{\gamma_1}{\gamma_{111}\,|\gamma_{22}|}}C(\rho, s_e)\int_0^\infty \frac{d\mu}{\sqrt{\pi\mu}}\int_{-\infty}^{\infty} d\nu W_{0c}(\nu)\bar{\mathcal{I}}(\gamma_e + \mu S_2 + \gamma_d\,\nu), \tag{G.87}$$

with S_2 the sign of γ_{22}, W_{0c} the inflection filter (G.26), and γ_d the dilation factor

$$\gamma_d = \frac{2}{3}\sqrt{\frac{2\gamma_1{}^3}{\gamma_{111}}}. \tag{G.88}$$

In this expression, the ranges of the integrals have been extended to infinity, in the expectation that only small values will contribute to the result.

From (E.68), this can be written in convolutional form as

$$\mathcal{O}_e(\rho) \simeq \frac{2\sqrt{\pi}}{3}\sqrt{\frac{\gamma_1}{\gamma_{111}\,|\gamma_{22}|}}C(\rho, s_e)\left[W_0(\gamma_d) * \Theta_{h(-S_2)} * f * \mathcal{I}\right](\gamma_e). \tag{G.89}$$

If f is the derivative operator, then $\Theta_{h(-S_2)} * f = -S_2 d_{h(-S_2)}$, a half-derivative operator. The convolution of this filter with W_0 then results in the half-differentiated inflection filter w_h, as defined by equation (E.72). This is the same filter observed in two dimensions, where the 2-D diffraction integral filter is a half-derivative.

Equation (G.89) states that the output event is proportional to a filtered version of the input function $\mathcal{I}$. The filter is a rescaled version of a convolution of the inflection filter W_0 with the two filters $\Theta_{h(-S_2)}$ and f. Since W_0 does not have the simple power-factor scaling properties of many of the useful filters in Appendix E, the factor γ_d stretches or compresses the wavelet $W_0 * \Theta_{h(-S_2)} * f$ accordingly. As the characteristic distance γ_d grows smaller, the wavelet grows sharper.

The amplitude of the event (G.89) is seen to depend on the derivatives of γ at the inflection point. For small γ_1, the amplitude is large and the filter is sharply peaked. As γ_1 increases, the event amplitude decreases and the filter becomes lower frequency, eventually smoothing over oscillations of $\mathcal{I}$ and further reducing the event amplitude.

G.3.3 Terminations

Suppose the integration surface ends at a line, which may be straight or, if curved, curving slowly relative to a wavelength of I. Looking at the trajectory function γ as a function of position on the line, suppose that at point $\boldsymbol{s}_e$ on the line, γ goes through a maximum, minimum, or inflection. Define local coordinates $\boldsymbol{r} = \boldsymbol{s} - \boldsymbol{s}_e$, with r_2 parallel to the line at $\boldsymbol{s}_e$, and r_1 pointing into the integration surface. To keep things simple, we assume that near $\boldsymbol{s}_e$, γ is linear in r_1. In the neighborhood of $\boldsymbol{s}_e$, the diffraction integral takes the form

$$\mathcal{O}_e(\rho) = C(\rho, \boldsymbol{s}_e)\int\int_{N_e} dr_1 dr_2 \bar{\mathcal{I}}(\gamma(\rho, \boldsymbol{r} + \boldsymbol{s}_e)). \tag{G.90}$$

An end-extremum

On the integration surface, in the neighborhood of s_e,

$$\gamma \simeq \gamma_0 + \gamma_1 r_1 + \gamma_{22}\frac{r_2^2}{2}, \tag{G.91}$$

and

$$\mathcal{O}_e(\rho) \simeq C(\rho, s_e) \int\int_{N_e} dr_1 dr_2 \bar{\mathcal{I}}\left(\gamma_0 + \gamma_1 r_1 + \gamma_{22}\frac{r_2^2}{2}\right). \tag{G.92}$$

With a change of integration variables to $\mu_1 = |\gamma_1| r_1$ and $\mu_2 = |\gamma_{22}| r_2^2/2$, this integral can be written

$$\mathcal{O}_e(\rho) \simeq \sqrt{\frac{2\pi}{|\gamma_{22}|}}\frac{C(\rho, s_e)}{|\gamma_1|} \int d\mu_1 \Theta(\mu_1) \int_0^\infty \frac{d\mu_2}{\sqrt{\pi \mu_2}} \bar{\mathcal{I}}(\gamma_0 + S_1\mu_1 + S_2\mu_2), \tag{G.93}$$

or

$$\mathcal{O}_e(\rho) \simeq -\sqrt{\frac{2\pi}{|\gamma_{22}|}}\frac{C(\rho, s_e)}{\gamma_1}\left[\Theta * \Theta_{h(-S_2)} * \bar{\mathcal{I}}\right](\gamma_0). \tag{G.94}$$

If $\bar{I} = dI/d\gamma$, then the integral reduces to

$$\mathcal{O}_e(\rho) \to -\sqrt{\frac{2\pi}{|\gamma_{22}|}}\frac{C(\rho, s_e)}{\gamma_1}\left[\Theta_{h(-S_2)} * \mathcal{I}\right](\gamma_0). \tag{G.95}$$

where

$$\begin{aligned} \Theta_{h(-S_2)} &= \Theta_{h-}, \ \gamma_{22} > 0, \\ &= \Theta_{h+}, \ \gamma_{22} < 0. \end{aligned} \tag{G.96}$$

An end-inflection

If the event at s_e is an inflection point, then the resulting event combines an inflection filter with the termination event:

$$\mathcal{O}_e(\rho) \simeq C(\rho, s_e) \int\int_{N_e} dr_1 dr_2 \bar{\mathcal{I}}\left(\gamma_0 + \gamma_1 r_1 + \gamma_2 r_2 + \gamma_{222}\frac{r_2^3}{6}\right). \tag{G.97}$$

which becomes

$$\mathcal{O}_e(\rho) \simeq -\frac{1}{\gamma_1}\frac{C(\rho, s_e)\gamma_d}{\gamma_2}\left[\Theta * W_0(\gamma_d) * f * \mathcal{I}\right](\gamma_0). \tag{G.98}$$

with

$$\gamma_d = \frac{2}{3}\sqrt{\frac{2\gamma_2^3}{\gamma_{222}}}. \tag{G.99}$$

Appendix H

Wave-based, ray-based, and reflector-based coordinates

H.1 Wave-based coordinates for constant-velocity wave propagation

Wavenumber and frequency

In a 3-D constant-velocity medium, a plane wave is characterized by its three-component wavenumber vector $\boldsymbol{k} = \left(k_x, k_y, k_z\right)$. Angular frequency ω of the wave is determined from the magnitude of the wavenumber vector:

$$|\omega| = c|\boldsymbol{k}|, \tag{H.1}$$

where c is the medium velocity.

The sign of frequency $\mathrm{S}(\omega)$ depends upon $\boldsymbol{k}$ and the direction the wave is traveling. For positive frequencies, $\boldsymbol{k}$ points in the direction of travel, but for negative frequencies, $\boldsymbol{k}$ points in the opposite direction.

If we consider ω to be an independent variable, then the plane wave is characterized by ω and the x- and y-components of $\boldsymbol{k}$. For the wave to propagate, the x- and y-components of $\boldsymbol{k}$ are constrained to obey

$$\sqrt{{k_x}^2 + {k_y}^2} < \frac{|\omega|}{c}, \tag{H.2}$$

and the z-component of $\boldsymbol{k}$ is given to be

$$k_z = \pm\frac{\omega}{c}\sqrt{1 - \left({k_x}^2 + {k_y}^2\right) c^2/\omega^2}, \tag{H.3}$$

with the plus (minus) sign pertaining to downward (upward) traveling waves. Defining the 2-vector $\tilde{\boldsymbol{k}}$ as the horizontal components of $\boldsymbol{k}$,

$$\tilde{\boldsymbol{k}} = \left(k_x, k_y\right), \tag{H.4}$$

we can characterize the wave with the three variables $(\omega, \tilde{\boldsymbol{k}})$.

Slowness vector and frequency

Instead of wavenumber, the wave can alternatively be described in terms of a slowness vector $\boldsymbol{p}$, defined as

$$\boldsymbol{p} = \boldsymbol{k}/\omega. \tag{H.5}$$

No matter the sign of ω, $\boldsymbol{p}$ always points in the direction of travel. The slowness vector has magnitude $1/c$:

$$p = \sqrt{{p_x}^2 + {p_y}^2 + {p_z}^2} = 1/c, \tag{H.6}$$

and z-component

$$p_z = \pm\frac{1}{c}\sqrt{1 - \left({k_x}^2 + {k_y}^2\right)c^2}. \tag{H.7}$$

Defining

$$\tilde{\boldsymbol{p}} = \left(p_x, p_y\right), \tag{H.8}$$

an alternative characterization of a plane wave is $(\omega, \tilde{\boldsymbol{p}})$.

Spherical coordinates and frequency

A third possible parameterization consists of frequency and the spherical coordinates of the wave direction. We can write

$$\boldsymbol{k} = \frac{\omega}{c}\begin{pmatrix} -\sin\phi\cos\theta \\ -\sin\phi\sin\theta \\ \pm\cos\phi \end{pmatrix}, \tag{H.9}$$

with the plus (minus) sign in the z-component for waves traveling downward (upward). The negative sign in the x- and y-components renders ϕ positive for travel in the negative x- and y-directions, which proves a convenience for experimental configurations where sources lead receivers. With these conventions, ϕ and θ can be constrained between $-\pi/2$ and $\pi/2$, and we can solve unambiguously for ϕ and θ:

$$\cos\phi = \left|k_z c/\omega\right|, \tag{H.10}$$

$$\cos\theta = \frac{\left|k_x c/\omega\right|}{\sqrt{1 - (k_z c/\omega)^2}}, \tag{H.11}$$

$$\sin\phi = -\text{Sign}\,(k_x/\omega)\sqrt{1 - (k_z c/\omega)^2}. \tag{H.12}$$

and

$$\sin\theta = \frac{-\text{Sign}\,(k_x/\omega)\,k_y c/\omega}{\sqrt{1 - (k_z c/\omega)^2}}. \tag{H.13}$$

Reflections

A reflection is characterized by both an incident wave and a reflected wave. Both waves will have the same temporal frequency ω, but (for converted waves) may travel at different velocities c_i and c_r. Considering only singly reflected waves, the incident wave necessarily travels downward and the reflected wave upward. An appropriate set of coordinates to describe the reflection is the 5-D $(\omega, \tilde{\boldsymbol{k}}_r, \tilde{\boldsymbol{k}}_i)$, with

$$\left|\tilde{\boldsymbol{k}}_i\right| < \frac{|\omega|}{c_i},$$

$$\left|\tilde{\boldsymbol{k}}_r\right| < \frac{|\omega|}{c_r}, \tag{H.14}$$

and the z-components of the wavenumbers given by

$$k_{iz} = +\frac{\omega}{c_i}\sqrt{1 - \left(k_{ix}{}^2 + k_{iy}{}^2\right) c_i{}^2/\omega^2},$$
$$k_{rz} = -\frac{\omega}{c_r}\sqrt{1 - \left(k_{rx}{}^2 + k_{ry}{}^2\right) c_r{}^2/\omega^2}. \tag{H.15}$$

In a constant-velocity medium, the incident and reflected waves each maintain a constant direction, and we can identify $\tilde{\boldsymbol{k}}_i$ with source wavenumber $\tilde{\boldsymbol{k}}_s$, and $\tilde{\boldsymbol{k}}_r$ with receiver wavenumber $\tilde{\boldsymbol{k}}_g$.

The slowness vectors provide an equivalent set of five parameters $(\omega, \tilde{\boldsymbol{p}}_r, \tilde{\boldsymbol{p}}_i)$, where

$$\left|\tilde{\boldsymbol{p}}_i\right| < \frac{1}{c_i},$$
$$\left|\tilde{\boldsymbol{p}}_r\right| < \frac{1}{c_r}, \tag{H.16}$$

and

$$p_{iz} = +\frac{1}{c_i}\sqrt{1 - \left(p_{ix}{}^2 + p_{iy}{}^2\right) c_i{}^2},$$
$$p_{rz} = -\frac{1}{c_r}\sqrt{1 - \left(p_{rx}{}^2 + p_{ry}{}^2\right) c_r{}^2}. \tag{H.17}$$

We may if we wish express wavenumbers or slownesses in spherical coordinates, in which case the equivalent set is $(\omega, \phi_r, \theta_r, \phi_i, \theta_i)$, where

$$\boldsymbol{k}_r = \frac{\omega}{c_r}\begin{pmatrix} -\sin\phi_r\cos\theta_r \\ -\sin\phi_r\sin\theta_r \\ -\cos\phi_r \end{pmatrix}, \tag{H.18}$$

and

$$\boldsymbol{k}_i = \frac{\omega}{c_i}\begin{pmatrix} -\sin\phi_i\cos\theta_i \\ -\sin\phi_i\sin\theta_i \\ +\cos\phi_i \end{pmatrix}. \tag{H.19}$$

Two-dimensional media

For a 2-D medium, the above relations still apply, with the y-components of the wavenumber or slowness vectors removed. Thus, a reflection is described by three coordinates (ω, k_{rx}, k_{ix}), or equivalently by (ω, p_{rx}, p_{ix}) or by (ω, ϕ_r, ϕ_i).

H.2 Ray-based coordinates for variable-velocity wave propagation

If velocity is variable, solutions to the wave equation are not plane waves. However, if velocity varies slowly, wavefronts may be locally planar, and we can still define local wavenumber or slowness vectors which satisfy, to a good approximation, the above equations. Ray theory provides a means of tying these wavenumbers to specific source, receiver, and reflection points. At the source point, $\tilde{\boldsymbol{k}}_i \to \tilde{\boldsymbol{k}}_s$, and at the receiver point $\tilde{\boldsymbol{k}}_r \to \tilde{\boldsymbol{k}}_g$. Elsewhere, a locally planar wave function traveling at velocity $c(\boldsymbol{x})$ is assumed to have the form

$$\psi(x, \omega) \simeq \mathcal{A}(x, \omega)e^{\pm i\omega\tau(x)}, \tag{H.20}$$

with $\mathcal{A}$ sufficiently slowly varying in $\boldsymbol{x}$ that

$$\nabla\psi(x,\omega) \simeq \pm i\omega\nabla\tau(x)\,\psi(x,\omega), \tag{H.21}$$

and $\tau(\boldsymbol{x})$ satisfying the Eikonal equation

$$|\nabla\tau(x)| = \frac{1}{c(x)}. \tag{H.22}$$

There are two possible directions of travel for the wave, the direction in which τ increases (i.e. the direction of the gradient of τ), or the direction in which τ decreases.

If we compare a local plane wave to an exact plane wave of wavenumber $\boldsymbol{k}$,

$$\psi_{\boldsymbol{k}}(x,\omega) = \mathcal{A}_{\boldsymbol{k}} e^{ik\cdot x} \tag{H.23}$$

such that

$$\nabla\psi_{\boldsymbol{k}}(x,\omega) = i\boldsymbol{k}\psi_{\boldsymbol{k}}(x,\omega), \tag{H.24}$$

we can identify in the locally planar wave a local wavenumber

$$\boldsymbol{k}(x,\omega) = \pm\omega\nabla\tau(x). \tag{H.25}$$

The sign of this relation is positive for a wave traveling in the direction of the gradient of τ. Because $\tau(\boldsymbol{x})$ obeys the Eikonal equation, the magnitude of $\boldsymbol{k}$ is

$$k(x,\omega) = \frac{|\omega|}{c(x)}. \tag{H.26}$$

We can also define a local slowness vector $\boldsymbol{p}(\boldsymbol{x})$ as

$$\boldsymbol{p}(x) \equiv \frac{\boldsymbol{k}(x,\omega)}{\omega} = \pm\nabla\tau(x). \tag{H.27}$$

For a seismic reflection experiment, in which an incident wave ψ_i travels downward in the direction of increasing τ_i and an upward traveling reflected wave ψ_r travels in the direction of decreasing τ_r,

$$\boldsymbol{k}_i(x,\omega) = +\omega\nabla\tau_i(x), \tag{H.28}$$

and

$$\boldsymbol{k}_r(x,\omega) = -\omega\nabla\tau_r(x). \tag{H.29}$$

The z-components of wavenumber are

$$k_{iz}(x,\omega) = +\frac{\omega}{c_i(x)}\sqrt{1-\left(k_{ix}{}^2+k_{iy}{}^2\right)c_i{}^2(x)\,/\omega^2}, \tag{H.30}$$

and

$$k_{rz}(x,\omega) = -\frac{\omega}{c_r(x)}\sqrt{1-\left(k_{rx}{}^2+k_{ry}{}^2\right)c_r{}^2(x)\,/\omega^2}. \tag{H.31}$$

Alternatively, we may write

$$\boldsymbol{p}_i(x) = +\nabla\tau_i(x), \tag{H.32}$$

$$\boldsymbol{p}_r(x) = -\nabla\tau_r(x), \tag{H.33}$$

with

$$p_{iz}(x) = +\frac{1}{c_i(x)}\sqrt{1 - \left(p_{ix}{}^2 + p_{iy}{}^2\right) c_i{}^2(x)}, \tag{H.34}$$

$$p_{rz}(x) = -\frac{1}{c_r(x)}\sqrt{1 - \left(p_{rx}{}^2 + p_{ry}{}^2\right) c_r{}^2(x)}. \tag{H.35}$$

H.3 Reflection centered coordinates

We described above ways to parameterize a reflection in terms of what went in, and what came out: the information is carried by the five parameters ω, k_{ix}, k_{iy}, k_{rx}, and k_{ry}. Another possible parameterization is in terms of reflector orientation. The incident and reflected wavenumbers together define a plane, which we call the *plane of incidence and reflection* (see Figure H.1), or, more succinctly, the *i–r* plane. Linear combinations of the two wavenumber vectors all lie in this plane. In particular, we are interested in the vector $\boldsymbol{k}_m$, defined as the linear combination

$$\boldsymbol{k}_m \equiv \boldsymbol{k}_r - \boldsymbol{k}_i = \omega\left(\boldsymbol{p}_r - \boldsymbol{p}_i\right) \equiv \omega \boldsymbol{p}_m. \tag{H.36}$$

According to Snell's law, at a reflector, the components in the plane of the reflector of the incident and reflected wavenumbers must be equal. Consequently, the difference $\boldsymbol{k}_m$ between the incident and the reflected wavenumbers must be normal to the reflector at the reflection point. Thus, $\boldsymbol{k}_m$ defines a second plane, which we call the *reflector tangent plane* (again see Figure H.1), tangent to the reflecting surface at the reflection point.

By itself, $\boldsymbol{k}_m$ does not carry all the information that parameters ω, k_{ix}, k_{iy}, k_{rx} and k_{ry} do. At least two additional independent parameters are needed. Consider as candidates the opening angle σ between the incident and reflected waves, and the azimuthal angle χ_{ri} made by the plane of incidence and reflection with respect to the x and y axes. These angles are illustrated in Figure H.1.

Opening angle σ is related to the dot product of the incident and reflected wavenumber vectors:

$$\cos\sigma = -\frac{\boldsymbol{k}_i \cdot \boldsymbol{k}_r\, c_i c_r}{\omega^2} = -\boldsymbol{p}_i \cdot \boldsymbol{p}_r c_i c_r. \tag{H.37}$$

The sign in this relation is negative because $\boldsymbol{p}_i$ points downward, while $\boldsymbol{p}_r$ points upward. Since the cosine is a symmetric function, this relation does not determine the sign of σ.

The azimuthal angle χ_{ir} can be determined from a weighted sum of $\boldsymbol{k}_r$ and $\boldsymbol{k}_i$. Define the vector $\boldsymbol{q}_0$ as

$$\boldsymbol{q}_0 \equiv \boldsymbol{p}_i p_{rz} - \boldsymbol{p}_r p_{iz} = \begin{pmatrix} p_{ix}p_{rz} - p_{rx}p_{iz} \\ p_{iy}p_{rz} - p_{ry}p_{iz} \\ 0 \end{pmatrix} = \begin{pmatrix} +\left(\boldsymbol{p}_r \times \boldsymbol{p}_i\right)_y \\ -\left(\boldsymbol{p}_r \times \boldsymbol{p}_i\right)_x \\ 0 \end{pmatrix}. \tag{H.38}$$

As a linear combination of $\boldsymbol{p}_i$ and $\boldsymbol{p}_r$, $\boldsymbol{q}_0$ lies in the *i–r* plane. Since it has no z-component, it also lies in the x–y plane. Accordingly, we can define azimuthal angle χ_{ir} so that the x-component of $\boldsymbol{q}_0$ is proportional to the cosine of χ_{ir}, and the y-component of $\boldsymbol{q}_0$ is proportional to the sine of χ_{ir}. Then

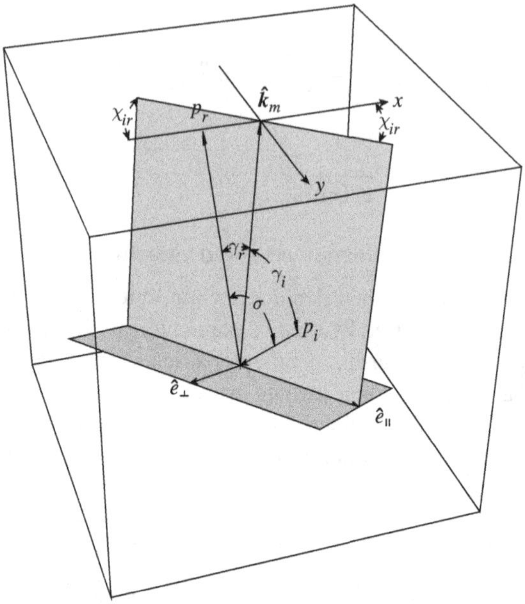

Figure H.1 Schematic of a reflection. An incident ray approaches a reflection point in the direction of the slowness vector $\boldsymbol{p}_i$, and a reflected ray leaves in the direction of its slowness vector $\boldsymbol{p}_r$. The two slowness vectors together define the plane of incidence and reflection, shown as the upper plane in this figure. The difference between the two slowness vectors, multiplied by frequency, form the reflection wavenumber $\boldsymbol{k}_m$, whose unit vector $\hat{\boldsymbol{k}}_m$ is shown in this figure. The wavenumber $\boldsymbol{k}_m$ lies in *the plane of incidence and reflection* or *i–r plane*, and is normal to the *reflector tangent plane*, shown as the lower plane in this figure, and so named because, if a specular reflector lies at the reflection point, the plane is tangent to it at this point. The angle between the two slowness vectors is the opening angle σ. The angle between $\boldsymbol{p}_i$ and $\boldsymbol{k}_m$ is the angle of incidence γ_i, and the angle between $\boldsymbol{k}_m$ and $\boldsymbol{p}_r$ is γ_r, the angle of reflection. The sum of γ_i and γ_r equals σ. Any horizontal line in the plane of incidence and reflection makes the angle χ_{ir} with respect to the x-axis. We define χ as the reflection point azimuthal angle. For straight rays, χ will be the same as the source–receiver azimuthal angle χ_{gs} on the measurement surface, but for curved rays, χ and χ_{gs} will be different. Two orthogonal direction vectors $\hat{\boldsymbol{e}}_\perp$ and $\hat{\boldsymbol{e}}_\parallel$ are shown in the reflector tangent plane. $\hat{\boldsymbol{e}}_\perp$ is normal to the *i–r* plane, and, for shear waves, lies in the SH direction. $\hat{\boldsymbol{e}}_\parallel$ lies in the *i–r* plane in the direction normal to $\boldsymbol{k}_m$ and $\hat{\boldsymbol{e}}_\perp$. Since $\hat{\boldsymbol{e}}_\perp$ is normal to horizontal lines in the *i–r* plane, its horizontal components also measure azimuthal angle as per equation (H.61).

$$\tan \chi_{ir} \equiv \frac{q_{0y}}{q_{0x}} = \frac{p_{iy}p_{rz} - p_{ry}p_{iz}}{p_{ix}p_{rz} - p_{rx}p_{iz}} = -\frac{\left(\boldsymbol{p}_r \times \boldsymbol{p}_i\right)_x}{\left(\boldsymbol{p}_r \times \boldsymbol{p}_i\right)_y}. \tag{H.39}$$

As was the case for opening angle, this relation only determines χ within a 180° range. To extend these relations to all angles, look at the vector

$$\boldsymbol{q}_\perp \equiv \boldsymbol{p}_r \times \boldsymbol{p}_i = \frac{1}{\omega^2}\begin{pmatrix} (\boldsymbol{k}_r \times \boldsymbol{k}_i)_x \\ (\boldsymbol{k}_r \times \boldsymbol{k}_i)_y \\ (\boldsymbol{k}_r \times \boldsymbol{k}_i)_z \end{pmatrix} = \begin{pmatrix} -q_{0y} \\ +q_{0x} \\ q_{\perp z} \end{pmatrix}. \tag{H.40}$$

This vector is normal to the *i–r* plane, hence lies in the reflector tangent plane, and in fact, for shear waves, lies in the SH direction. The magnitude $q_\perp$of $\boldsymbol{q}_\perp$ is proportional to the sine of opening angle σ:

$$q_\perp{}^2 = q_0{}^2 + q_{\perp z}{}^2 = \frac{\sin^2\sigma}{c_r{}^2 c_i{}^2}. \tag{H.41}$$

Define an angle φ by the relations

$$\boldsymbol{q}_\perp = \frac{\sin\sigma}{c_r c_i}\begin{pmatrix} -\cos\varphi\sin\chi_{ir} \\ +\cos\varphi\cos\chi_{ir} \\ \sin\varphi \end{pmatrix}, \tag{H.42}$$

with $\cos\varphi$ constrained to be positive. Since $\boldsymbol{q}_\perp$ is normal to the *i–r* plane, it is normal to $\boldsymbol{k}_m$:

$$0 = \boldsymbol{q}_\perp \cdot \boldsymbol{k}_m = \frac{\sin\sigma}{c_r c_i}\left(\cos\varphi\left(k_{my}\cos\chi_{ir} - k_{mx}\sin\chi_{ir}\right) + k_{mz}\sin\varphi\right), \tag{H.43}$$

so that φ is determined by $\boldsymbol{k}_m$ and χ_{ir}:

$$\tan\varphi = (k_{mx}/k_{mz})\sin\chi_{ir} - \left(k_{my}/k_{mz}\right)\cos\chi_{ir}, \tag{H.44}$$

$$\cos\varphi = \frac{1}{\sqrt{1 + ((k_{my}/k_{mz})\cos\chi_{ir} - (k_{mx}/k_{mz})\sin\chi_{ir})^2}}, \tag{H.45}$$

and

$$\sin\varphi = \frac{(k_{mx}/k_{mz})\sin\chi_{ir} - \left(k_{my}/k_{mz}\right)\cos\chi_{ir}}{\sqrt{1 + ((k_{my}/k_{mz})\cos\chi_{ir} - (k_{mx}/k_{mz})\sin\chi_{ir})^2}}. \tag{H.46}$$

We can write $\boldsymbol{q}_0$ in terms of σ, χ_{ir}, and φ:

$$\boldsymbol{q}_0 = \frac{\sin\sigma\cos\varphi}{c_r c_i}\begin{pmatrix} \cos\chi_{ir} \\ \sin\chi_{ir} \\ 0 \end{pmatrix}. \tag{H.47}$$

From this equation for $\boldsymbol{q}_0$ it is clear that the ambiguities in σ and χ_{ir} are related. If q_{0x} is negative, then either $\sin\sigma$ or $\cos\chi_{ir}$ must be negative. If both can become negative, then we are left with an unresolvable ambiguity. We choose to define the sign of σ to be the sign of q_{0x} (or equivalently, the sign of $q_{\perp y}$), which determines σ between $-\pi$ and π. The cosine of χ_{ir} is then always positive, corresponding to an angle between $-\pi/2$ and $\pi/2$. Within this range, χ_{ir} is fully defined by its tangent (H.39). To completely specify σ, we define its cosine by (H.37), and its sine by

$$\sin\sigma = c_i c_r|\boldsymbol{p}_r \times \boldsymbol{p}_i|\,\mathrm{Sign}\left(\left(\boldsymbol{p}_r \times \boldsymbol{p}_i\right)_y\right). \tag{H.48}$$

To establish $(\boldsymbol{k}_m, \sigma, \chi_{ir})$ as an alternative to the coordinates $(\omega, \tilde{\boldsymbol{k}}_r, \tilde{\boldsymbol{k}}_i)$ or $(\omega, \tilde{\boldsymbol{p}}_r, \tilde{\boldsymbol{p}}_i)$, we must confirm that there is a one-to-one correspondence between the two sets of parameters. We have seen above that $(\boldsymbol{k}_m, \sigma, \chi_{ir})$ is fully determined by $(\omega, \tilde{\boldsymbol{p}}_r, \tilde{\boldsymbol{p}}_i)$, with σ an angle between $-\pi$ and π, and χ_{ir} an angle between $-\pi/2$ and $\pi/2$. We must still establish that $(\omega, \tilde{\boldsymbol{p}}_r, \tilde{\boldsymbol{p}}_i)$ are fully determined by $(\boldsymbol{k}_m, \sigma, \chi_{ir})$.

Frequency as a function of $\boldsymbol{k}_m$ and σ

The magnitude of $\boldsymbol{k}_m$ is

$$k_m = |\boldsymbol{k}_r - \boldsymbol{k}_i| = \sqrt{k_i{}^2 + k_r{}^2 - 2\boldsymbol{k}_i \cdot \boldsymbol{k}_r} = |\omega| s_e(\sigma), \tag{H.49}$$

where $s_e(\sigma)$ is the effective slowness

$$s_e(\sigma) \equiv \sqrt{\frac{1}{c_i{}^2} + \frac{1}{c_r{}^2} + 2\frac{\cos\sigma}{c_i c_r}}. \tag{H.50}$$

In lieu of slowness, we can define an angle-weighted average velocity $c_e(\sigma)$ as

$$c_e(\sigma) \equiv \frac{1}{2}\sqrt{c_i{}^2 + c_r{}^2 + 2c_i c_r \cos\sigma} = \frac{c_i c_r}{2} s_e(\sigma). \tag{H.51}$$

Frequency can be given in terms of slowness or average velocity:

$$\omega = -\text{Sign}\,(k_{mz})\,\frac{k_m}{s_e(\sigma)} = -\text{Sign}\,(k_{mz})\,\frac{k_m c_i c_r}{2c_e(\sigma)}. \tag{H.52}$$

Frequency depends upon wavenumber $\boldsymbol{k}_m$ and upon opening angle σ, but not on azimuthal angle χ_{ir}.

Incident and reflected angles as a function of opening angle

Define $\hat{\boldsymbol{k}}_i$ and $\hat{\boldsymbol{k}}_r$ as unit vectors pointing in the direction of $\boldsymbol{k}_i$ and $\boldsymbol{k}_r$:

$$\hat{\boldsymbol{k}}_r = \frac{\boldsymbol{k}_r c_r}{|\omega|}, \tag{H.53}$$

$$\hat{\boldsymbol{k}}_i = \frac{\boldsymbol{k}_i c_i}{|\omega|.} \tag{H.54}$$

The unit vector $\hat{\boldsymbol{k}}_m$ pointing in the direction of $\boldsymbol{k}_m$ is thus

$$\hat{\boldsymbol{k}}_m = (\boldsymbol{k}_r - \boldsymbol{k}_i)\,\frac{c_i c_r}{2|\omega| c_e(\sigma)} = \frac{(\boldsymbol{k}_r - \boldsymbol{k}_i)}{|\omega| s_e(\sigma)} = \frac{\hat{\boldsymbol{k}}_r c_i - \hat{\boldsymbol{k}}_i c_r}{2c_e\,(\sigma)} = \frac{\hat{\boldsymbol{k}}_r / c_r - \hat{\boldsymbol{k}}_i / c_i}{s_e(\sigma)}. \tag{H.55}$$

The angles of incidence and reflection γ_i and γ_r can be defined as the angles between $\boldsymbol{k}_m$ and the incident and reflected ray directions $\hat{\boldsymbol{k}}_i$ and $\hat{\boldsymbol{k}}_r$:

$$\begin{aligned}
\cos\gamma_i &= -\hat{\boldsymbol{k}}_m \cdot \hat{\boldsymbol{k}}_i \\
&= \frac{c_r - c_i\, \hat{\boldsymbol{k}}_r \cdot \hat{\boldsymbol{k}}_i}{2c_e(\sigma)} \\
&= \frac{c_r + c_i \cos\sigma}{2c_e\,(\sigma)} \\
&= \frac{c_r + c_i \cos\sigma}{c_i c_r s_e(\sigma)},
\end{aligned} \tag{H.56}$$

$$\begin{aligned}
\sin\gamma_i &= \left|\hat{\boldsymbol{k}}_m \times \hat{\boldsymbol{k}}_i\right| \operatorname{Sign}(\hat{\boldsymbol{k}}_m \times \hat{\boldsymbol{k}}_i)_y) \\
&= \frac{c_i |\hat{\boldsymbol{k}}_r \times \hat{\boldsymbol{k}}_i|}{2c_e(\sigma)} \operatorname{Sign}\left((\hat{\boldsymbol{k}}_r \times \hat{\boldsymbol{k}}_i)_y\right) \\
&= \frac{c_i \sin\sigma}{2c_e(\sigma)},
\end{aligned} \tag{H.57}$$

$$\begin{aligned}
\cos\gamma_r &= +\hat{\boldsymbol{k}}_m \cdot \hat{\boldsymbol{k}}_r \\
&= -\frac{c_r \hat{\boldsymbol{k}}_r \cdot \hat{\boldsymbol{k}}_i - c_i}{2c_e(\sigma)} \\
&= \frac{c_i + c_r \cos\sigma}{2c_e(\sigma)} \\
&= \frac{c_i + c_r \cos\sigma}{c_i c_r s_e(\sigma)},
\end{aligned} \tag{H.58}$$

$$\begin{aligned}
\sin\gamma_r &= \left|\hat{\boldsymbol{k}}_m \times \hat{\boldsymbol{k}}_r\right| \operatorname{Sign}(\hat{\boldsymbol{k}}_m \times \hat{\boldsymbol{k}}_r)_y) \\
&= \frac{c_r |\hat{\boldsymbol{k}}_r \times \hat{\boldsymbol{k}}_i|}{2c_e(\sigma)} \operatorname{Sign}(\hat{\boldsymbol{k}}_r \times \hat{\boldsymbol{k}}_i)_y) \\
&= \frac{c_r \sin\sigma}{2c_e(\sigma)}.
\end{aligned} \tag{H.59}$$

The incident and reflected angles γ_i and γ_r have the same sign as σ, and obey Snell's law $\sin\gamma_i/c_i = \sin\gamma_r/c_r = \sin\sigma/2c_e(\sigma)$.

$\hat{\boldsymbol{e}}_\perp$ as a function of k and χ_{ir}

A unit vector proportional to $\boldsymbol{q}_\perp$ is

$$\hat{\boldsymbol{e}}_\perp \equiv \boldsymbol{q}_\perp \frac{c_r c_i}{\sin\sigma} = \frac{\hat{\boldsymbol{k}}_r \times \hat{\boldsymbol{k}}_i}{\sin\sigma} = \begin{pmatrix} -\cos\varphi \sin\chi_{ir} \\ +\cos\varphi \cos\chi_{ir} \\ \sin\varphi \end{pmatrix}. \tag{H.60}$$

Since this vector is normal to both the incident and reflected wavenumbers, it is normal to their linear combinations, and in particular, normal to $\boldsymbol{k}_m$. That places it in the reflector tangent plane. Since it is normal to both the incident and reflected rays, it points in the direction of SH wave motion.

Invoking the relations above for φ as a function of $\boldsymbol{k}_m$ and χ_{ir},

$$\hat{\boldsymbol{e}}_\perp \equiv \frac{\begin{pmatrix} -\sin\chi_{ir} \\ +\cos\chi_{ir} \\ (k_{mx}/k_{mz})\sin\chi_{ir} - (k_{my}/k_{mz})\cos\chi_{ir} \end{pmatrix}}{\sqrt{1 + ((k_{my}/k_{mz})\cos\chi_{ir} - (k_{mx}/k_{mz})\sin\chi_{ir})^2}}. \tag{H.61}$$

The SH direction vector $\hat{\boldsymbol{e}}_\perp$ is a function of $\boldsymbol{k}_m$ and χ_{ir} only.

$\hat{\boldsymbol{e}}_\parallel$ as a function of $\boldsymbol{k}_m$ and χ_{ir}

We can define a third direction vector normal to both $\hat{\boldsymbol{k}}_m$ and $\hat{\boldsymbol{e}}_\perp$:

$$\hat{\boldsymbol{e}}_\parallel \equiv \hat{\boldsymbol{k}}_m \times \hat{\boldsymbol{e}}_\perp \tag{H.62}$$

$$= \frac{-k_{mz}/|\omega|}{s_e(\sigma)\sqrt{1 + \left(\frac{k_{my}}{k_{mz}}\cos\chi_{ir} - \frac{k_{mx}}{k_{mz}}\sin\chi_{ir}\right)^2}} \begin{pmatrix} \frac{(k_{my}{}^2 + k_{mz}{}^2)}{k_{mz}{}^2}\cos\chi_{ir} - \frac{k_{mx}k_{my}}{k_{mz}{}^2}\sin\chi_{ir} \\ \frac{(k_{mx}{}^2 + km_z{}^2)}{k_{mz}{}^2}\sin\chi_{ir} - \frac{k_{mx}k_{my}}{k_{mz}{}^2}\cos\chi_{ir} \\ \left(\frac{k_{mx}}{k_{mz}}\cos\chi_{ir} + \frac{k_{my}}{k_{mz}}\sin\chi_{ir}\right) \end{pmatrix}.$$

This vector lies in both the reflector tangent plane and the *i–r* plane. As for any vector in the *i–r* plane, it can be expressed as a linear combination of $\hat{\boldsymbol{k}}_i$ and $\hat{\boldsymbol{k}}_r$:

$$\begin{aligned}
\hat{\boldsymbol{e}}_\parallel &= \frac{\hat{\boldsymbol{k}}_r c_i - \hat{\boldsymbol{k}}_i c_r}{2c_e} \times \frac{\hat{\boldsymbol{k}}_r \times \hat{\boldsymbol{k}}_i}{\sin\sigma} \\
&= \frac{c_i\hat{\boldsymbol{k}}_r \times \left(\hat{\boldsymbol{k}}_r \times \hat{\boldsymbol{k}}_i\right) - c_r\hat{\boldsymbol{k}}_i \times \left(\hat{\boldsymbol{k}}_r \times \hat{\boldsymbol{k}}_i\right)}{2c_e\sin\sigma} \\
&= \frac{c_i\hat{\boldsymbol{k}}_r\left(\hat{\boldsymbol{k}}_r\cdot\hat{\boldsymbol{k}}_i\right) - c_i\hat{\boldsymbol{k}}_i\left(\hat{\boldsymbol{k}}_r\cdot\hat{\boldsymbol{k}}_r\right) - c_r\hat{\boldsymbol{k}}_r\left(\hat{\boldsymbol{k}}_i\cdot\hat{\boldsymbol{k}}_i\right) + c_r\hat{\boldsymbol{k}}_i\left(\hat{\boldsymbol{k}}_r\cdot\hat{\boldsymbol{k}}_i\right)}{2c_e\sin\sigma} \\
&= \frac{-c_i\hat{\boldsymbol{k}}_r\cos\sigma - c_i\hat{\boldsymbol{k}}_i - c_r\hat{\boldsymbol{k}}_r - c_r\hat{\boldsymbol{k}}_i\cos\sigma}{2c_e\sin\sigma} \\
&= -\frac{\hat{\boldsymbol{k}}_r(c_r + c_i\cos\sigma) + \hat{\boldsymbol{k}}_i(c_i + c_r\cos\sigma)}{2c_e\sin\sigma} \\
&= -\frac{\hat{\boldsymbol{k}}_r\cos\gamma_i + \hat{\boldsymbol{k}}_i\cos\gamma_r}{\sin\sigma}.
\end{aligned} \tag{H.63}$$

$\hat{\boldsymbol{k}}_r$ and $\hat{\boldsymbol{k}}_i$ as a function of $\boldsymbol{k}_m$, σ, and χ_{ir}

Since the two vectors $\hat{\boldsymbol{k}}_m$ and $\hat{\boldsymbol{e}}_\parallel$ are both linear combinations of the incident and reflected wavenumbers, we can solve for $\hat{\boldsymbol{k}}_i$ and $\hat{\boldsymbol{k}}_r$. Solving first for $\hat{\boldsymbol{k}}_i$,

$$\hat{\boldsymbol{k}}_i = -\frac{\hat{\boldsymbol{e}}_\parallel c_i\sin\sigma + 2c_e(\sigma)\hat{\boldsymbol{k}}_m\cos\gamma_i}{c_i\cos\gamma_r + c_r\cos\gamma_i}. \tag{H.64}$$

Now, from (H.57) and (H.58), $c_i \sin\sigma = 2\, c_e(\sigma)\sin\gamma_i$, and $c_i \cos\gamma_r + c_r \cos\gamma_i = 2\, c_e(\sigma)$. With these substitutions,

$$\hat{\boldsymbol{k}}_i = -\hat{\boldsymbol{k}}_m \cos\gamma_i - \hat{\boldsymbol{e}}_\parallel \sin\gamma_i. \tag{H.65}$$

Similarly, solving for $\hat{\boldsymbol{k}}_r$,

$$\hat{\boldsymbol{k}}_r = \hat{\boldsymbol{k}}_m \cos\gamma_r - \hat{\boldsymbol{e}}_\parallel \sin\gamma_r. \tag{H.66}$$

Relations for vectors $\tilde{\boldsymbol{k}}_i$, $\tilde{\boldsymbol{k}}_r$, $\tilde{\boldsymbol{p}}_i$, and $\tilde{\boldsymbol{p}}_r$ are found by applying the appropriate scale factors to equations (H.65) and (H.66). The i–r plane, characterized by the unit vectors $\hat{\boldsymbol{k}}_m$ and $\hat{\boldsymbol{e}}_\parallel$, is determined by $\boldsymbol{k}_m$ and χ_{ir}. The positions of the ray or wavenumber vectors within that plane are determined by σ. The three parameters $\boldsymbol{k}_m$, σ, and χ_{ir} suffice to determine ω, $\tilde{\boldsymbol{p}}_i$, and $\tilde{\boldsymbol{p}}_r$.

Spherical $\boldsymbol{k}_m$ coordinates

If we write $\boldsymbol{k}_m$ in spherical coordinates,

$$\boldsymbol{k}_m = \omega p \begin{pmatrix} \sin\alpha\cos\beta \\ \sin\alpha\sin\beta \\ -\cos\alpha \end{pmatrix}, \alpha\epsilon\left(-\frac{\pi}{2}, \frac{\pi}{2}\right), \beta\epsilon(-\pi, \pi), \tag{H.67}$$

then α represents the dip of the reflector tangent plane and β the dip direction relative to the x-axis.

The angle φ (equations (H.44), (H.45), and (H.46)) can be expressed in terms of α and β as

$$\tan\varphi = \tan\alpha \sin(\beta - \chi_{ir}), \tag{H.68}$$

$$\cos\varphi = \frac{1}{\sqrt{1 + (\tan\alpha\sin(\beta - \chi_{ir}))^2}}, \tag{H.69}$$

$$\sin\varphi = \frac{\tan\alpha\sin(\beta - \chi_{ir})}{\sqrt{1 + (\tan\alpha\sin(\beta - \chi_{ir}))^2}}. \tag{H.70}$$

H.4 Two-dimensional media

In two dimensions, things simplify slightly: $\boldsymbol{k}_i$ and $\boldsymbol{k}_r$ have no y-component, so neither does $\boldsymbol{k}_m$ (equation (H.36)). The azimuthal line vector (H.38) has only an x-component, and azimuthal angle, defined in (H.39), is zero.

The cross product $\boldsymbol{k}_r \times \boldsymbol{k}_i$ has only a y-component, and the relation (H.48) for the sine of opening angle becomes

$$\sin\sigma = \frac{(k_{rz}k_{ix} - k_{rx}k_{iz})\, c_i c_r}{\omega^2}. \tag{H.71}$$

If we spherically decompose k_r and k_i into angles ϕ_r, θ_r, ϕ_i, and θ_i as per equations (H.18) and (H.19), then, since in 2-D, $\theta_r = \theta_i = 0$,

$$\boldsymbol{k}_r = \frac{\omega}{c_r}\begin{pmatrix} -\sin\phi_r \\ 0 \\ -\cos\phi_r \end{pmatrix}, \tag{H.72}$$

and

$$\boldsymbol{k}_i = \frac{\omega}{c_i}\begin{pmatrix} -\sin\phi_i \\ 0 \\ +\cos\phi_i \end{pmatrix}, \tag{H.73}$$

and opening angle becomes

$$\sigma = \phi_r + \phi_i. \tag{H.74}$$

The relation (H.52) between k, σ, and ω is unchanged, as are the relations (H.56) to (H.59) between opening angle and incident and reflected angles. The unit vector $\hat{\boldsymbol{e}}_\perp$ (equation (H.60)) is just the $\hat{\mathbf{y}}$ unit vector in the $+y$ direction. The unit vector $\hat{\boldsymbol{e}}_\parallel$ (equation (H.62)) reduces to

$$\hat{\boldsymbol{e}}_\parallel = \begin{pmatrix} -\hat{k}_{mz} \\ 0 \\ \hat{k}_{mx} \end{pmatrix}. \tag{H.75}$$

The relations (H.65) and (H.66) for incident and reflected wavenumber as a function of $\boldsymbol{k}_m$ and σ still hold. Thus, in two dimensions, there is a one-to-one relation between the variable sets (ω, p_{rx}, p_{ix}) and $(\boldsymbol{k}_m, \sigma)$.

H.5 Wave-type preserving reflections

For reflections that preserve wave type, $c_r = c_i = c$, and some of the above relations simplify. Incident and reflected angles become $\gamma_r = \gamma_i = \gamma = \sigma/2$. The effective velocity $c_e(\sigma)$ becomes

$$c_e(\sigma) \rightarrow \frac{c}{2}\sqrt{2(1+\cos\sigma)} = c\cos\gamma. \tag{H.76}$$

Equation (H.63) for the unit vector $\hat{\boldsymbol{e}}_\parallel$ in terms of incident and reflected wavenumbers reduces to the simple sum

$$\hat{\boldsymbol{e}}_\parallel \rightarrow -\frac{\hat{\boldsymbol{k}}_r + \hat{\boldsymbol{k}}_i}{2\sin\gamma}, \tag{H.77}$$

and expressions (8.65) and (8.66) simplify slightly to

$$\hat{\boldsymbol{k}}_i \rightarrow -\hat{\boldsymbol{k}}_m \cos\gamma - \hat{\boldsymbol{e}}_\parallel \sin\gamma \tag{H.78}$$

and

$$\hat{\boldsymbol{k}}_r \rightarrow \hat{\boldsymbol{k}}_m \cos\gamma - \hat{\boldsymbol{e}}_\parallel \sin\gamma. \tag{H.79}$$

References

Section numbers are cited in parentheses.

Aki, K. and Richards, P. (1980). *Quantitative Seismology* Volume 1: *Theory and Methods.* San Francisco: W.H. Freeman. (S 5.2.3, 8.4.1, 8.4.2, 8.4.3, 8.4.4, 8.7)

Alkhalifah, T. and Bagaini, C. (2006). Straight-rays redatuming: A fast and robust alternative to wave-equation-based redatuming. *Geophysics*, **71**, 37–46. (S 12.3)

Aveni, G. and Biondi, B. (2010). Target-oriented joint least-squares migration/inversion of time-lapse seismic data sets. *Geophysics*, **75**, 61–73. (S 14.1)

Baysal, E., Kosloff, D. and Sherwood, J. (1983). Reverse time migration. *Geophysics*, **48**, 1514–24. (S 2.4)

Berkhout, A. (1981). Wave field extrapolation techniques in seismic migration, a tutorial. *Geophysics*, **46**, 1638–56. (S 2.3)

Berkhout, A. (1982). *Seismic Migration: Imaging of Acoustic Energy by Wave Field Extrapolation, Part A: Theoretical Aspects.* Amsterdam: Elsevier. (S 1.5)

Berkhout, A. (1984). *Seismic Resolution, a Quantitative Analysis of Resolving Power of Acoustical Echo Techniques.* Amsterdam: Geophysical Press. (S 4.2)

Beydoun, W. and Keho, T. (1987). The paraxial ray method. *Geophysics* **52**, 1639–53. (S 6.2.3)

Beylkin, G. (1985). Imaging of discontinuities in the inverse scattering problem by inversion of a causal generalized radon transform. *J. Math. Phys.*, **26**, 99–108. (S 2.7.1, 10.4.4)

Bisset, D. and Durrani, T. (1990). Radon transform migration below plane sloping layers. *Geophysics*, **55**, 277–83. (S 2.7.1, 10.4.4)

Black, J., Schleicher, K. and Zhang, L. (1993). True-amplitude imaging and dip moveout. *Geophysics*, **58**, 47–66. (S 1.10, 4.3. 10.2)

Bleistein, N. (1984). *Mathematical Methods for Wave Phenomena.* London: Academic Press. (S 2.6.1)

Bleistein, N. (1987). On the imaging of reflectors in the earth. *Geophysics*, **52**, 931–42. (S 8.2)

Bleistein, N. and Handelsman, R. (1986). *Asymptotic Expansions of Integrals.* Mineola: Dover. (S F)

Bleistein, N., Cohen, J. and Hagin, F. (1985). Two and one-half dimensional Born inversion with an arbitrary reference. *Geophysics*, **52**, 26–36. (S 5.4.2)

Bleistein, N., Cohen, J. and Jaramillo, H. (1999). True-amplitude transformation to zero offset of data from curved reflectors. *Geophysics*, **64**, 112–29. (S 10.2)

Bleistein, N., Cohen, J. and Stockwell, J. (2001). *Mathematics of Multidimensional Seismic Imaging, Migration, and Inversion.* New York: Springer Verlag. (S 2.2.1, 11.2, Appendix E, F)

Born, M. and Wolf, E. (1980). *Principles of Optics: Electromagnetic Theory of Propagation, Interference, and Diffraction of Light*, 6th Edn. Oxford: Pergamon Press. (S 2.2.2, 2.2.6)

Cerveny, V. (2001). *Seismic Ray Theory.* Cambridge: Cambridge University Press. (S 6.2.3)

Chapman, C. (1978). A new method for computing synthetic seismograms. *Geophys. J. R. Astr. Soc.*, **54**, 481–518. (S 2.6.1)

Chapman, C. and Coates, R. (1994). Generalized Born scattering in anisotropic media. *Wave Motion*, **19**, 309–41. (S 5.2.3)

Claerbout, J. (1971). Toward a unified theory of reflector mapping. *Geophysics*, **36**, 467–81. (S 1.5, 2.3)

Claerbout, J. (1976). *Fundamentals of Geophysical Data Processing: With Applications to Petroleum Prospecting.* New York: McGraw-Hill. (S 1.5, 2.3, 2.5, 3.1, 10.2, Appendix E)

Clayton, R. and Stolt, R. (1981). A Born–WKBJ inversion method for acoustic reflection data. *Geophysics*, **46**, 1559–67. (S 2.6.1, 3.2, 5.4.2)

Cohen, J. and Bleistein, N. (1977). An inverse method for determining small variations in propagation speed. *SIAM J. Appl. Math.*, **32**, 784–99. (S 5.4.2)

Cohen, J., Hagin, F. and Bleistein, N. (1986). Three-dimensional Born inversion with an arbitrary reference. *Geophysics*, **51**, 1552–8. (S 5.4.2)

de Hoop, M. and Bleistein, N. (1997). Generalized Radon transform inversions for reflectivity in anisotropic elastic media. *Inverse Problems*, **13**, 669–90. (S 5.2.3)

Dix, C. (1955). Seismic velocities from surface measurements. *Geophysics*, **20**, 68–86. (S 2.8.1)

Fomel, S. (2003). Velocity continuation and the anatomy of residual prestack time migration. *Geophysics*, **68**, 1650–61. (S 1.7, 10.3, 12.1)

Gazdag, J. (1978). Wave equation migration with the phase-shift method. *Geophysics*, **43**, 1342–51. (S 2.6.1)

Gazdag, J. and Sguazzero, P. (1984). Migration of seismic data by phase shift plus interpolation. *Geophysics*, **49**, 124–31. (S 2.6.1)

Jackson, J. (1962). *Classical Electrodynamics.* New York: McGraw-Hill. (S 2.2.2, 2.3.3)

Keller, J. (1953). The geometrical theory of diffraction. *Proc. Symp. Microwave Opt., Eaton Electronics Research Laboratory.* McGill University:, Montreal. (S 2.2.6, Appendix G)

Keller, J. (1978). Rays, waves, and asymptotics. *Bull. Am. Math. Soc.*, **84**, 727–50. (S 2.2.6, Appendix G)

Lambar, G., Operto, S., Podvin, P. and Thierry, P. (2003). 3D ray + Born migration/inversion. Part 1: Theory. *Geophysics*, **68**, 1348–56. (S. 11.2, 14.1)

Levin, S. (1980). A frequency-dip formulation of wave-theoretic migration in stratified media. In Wang, K. Y., ed. *Acoustical Imaging: Visualisation and Characterisation* Volume **9**, 681–97. New York: Plenum. (S 10.4.4)

Liu, Z. and Bleistein, N. (1995). Migration velocity analysis: Theory and an iterative algorithm. *Geophysics*, **60**, 142–53. (S 1.7)

Marfurt, K. (1978). *Elastic Wave Equation Migration-Inversion.* Unpublished Ph.D thesis, Columbia University, New York. (S 5.2.3)

Newton, R. (1966). *Scattering Theory of Waves and Particles.* New York: McGraw-Hill. (S 5.4.2, Appendix F)

Operto, M., Xu, S. and Lambar, G. (2000). Can we quantitatively image complex structures with rays? *Geophysics*, **65**, 1223–38. (S 14.1, 14.4.3)

Ottolini, R. and Claerbout, J. (1984). The migration of common midpoint slant stacks. *Geophysics*, **49**, 237–49. (S 2.7.1, 10.4.4)

Ramirez, A. and Weglein, A. (2009). Green's theorem as a comprehensive framework for data reconstruction, regularization, wavefield separation, seismic interferometry, and wavelet estimation: A tutorial. *Geophysics*, **74**, 35–62. (S 12.3)

Raz, S. (1981). Three-dimensional velocity profile inversion from finite-offset scattering data. *Geophysics*, **46**, 837–42. (S 5.4.2)

Ronen, S. and Liner, C. (2000). Least-squares DMO and migration. *Geophysics*, **65**, 1364–71. (S 14.1)

Rothman, D., Levin, A. and Rocca, F. (1985). Residual migration: Applications and limitations. *Geophysics*, **50**, 110–26. (S 10.3, 12.1)

Sava, P. (2003). Prestack residual migration in the frequency domain. *Geophysics*, **68**, 634–40. (S 10.2, 10.3, 12.1)

Sava, P., Biondi, B. and Elgen, J. (2005). Wave equation migration velocity analysis by focussing diffractions and reflections. *Geophysics*, **70**, 19–27. (S 1.7)

Schleicher, J., Tygel, M. and Hubral, P. (1993). 3-D true-amplitude finite-offset migration. *Geophysics*, **58**, 1112–26. (S 11.2)

Schleicher, J., Tygel, M. and Hubral, P. (2007). *Seismic True-Amplitude Imaging.* Tulsa: Society of Exploration Geophysicists. (S 1.10, 2.2.1, 5.2.3, 6.2.3, 10.2, 11.2, 12.1, 12.3)

Schneider, W. (1978). Integral formulation for migration in two and three dimensions. *Geophysics*, **43**, 49–76. (S 2.2.1, 11.2)

Schultz, P. and Sherwood, J. (1980). Depth migration before stack. *Geophysics*, **45**, 376–93. (S 1.6)

Shuey, R. (1985). A simplification of the Zoeppritz equations. *Geophysics*, **50**, 609–14. (S 10.1.1)

Stolt, R. (1978). Migration by Fourier transform. *Geophysics*, **43**, 23–48. (S 2.6.2, 2.8.2)

Stolt, R. (2002). Seismic data mapping and reconstruction. *Geophysics*, **67**, 890–908. (S 12.3)

Stolt, R. and Benson, A. (1986). *Seismic Migration: Theory and Practice.* Amsterdam: Geophysical Press. (S 1.5, 1.6, 8.2, 10.2, 10.3)

Stolt, R. and Weglein, A. (1985). Migration and inversion of seismic data. *Geophysics*, **50**, 2458–72. (S 5.4.2, 8.2)

Tang, Y. (2009). Target-oriented wave-equation least-squares migration/inversion with phase-encoded Hessian. *Geophysics*, **74**, 95–107. (S 14.1)

Taylor, J. (1972). *Scattering Theory: The Quantum Theory of Nonrelativistic Collisions.* New York: John Wiley and Sons. (S 5.4.2)

Tygel, M., Schleicher, J. and Hubral, P. (1994). Pulse distortion in depth migration. *Geophysics*, **59**, 1561–69. (S 4.2.2)

Tygel, M., Schleicher, J. and Hubral, P. (1998). 2.5-D true-amplitude Kirchhoff migration to zero offset in laterally inhomogeneous media. *Geophysics*, **63**, 557–73. (S 10.2)

Weglein, A., Stolt, R. and Mayhan, J. (2011a) Reverse time migration and Green's theorem. Part I: The evolution of concepts and setting the stage for the new RTM method. *J. Seism. Explor.*, **20** (in press). (S 2.4)

Weglein, A., Stolt, R. and Mayhan, J. (2011b) Reverse time migration and Green's theorem. Part II: A new and consistent theory that progresses and corrects current RTM concepts and methods. *J. Seism. Explor.*, **20**, (in press). (S 2.4)

Whitmore, D. (1983). Iterative depth migration by backward time propagation. *53rd Annual International Meeting, SEG Expanded Abstracts*, 382–85. (S 2.4)

Yilmaz, O. (2001). *Seismic Data Analysis: Processing, Inversion, and Interpretation of Seismic Data.* Tulsa: SEG Investigations in Geophysics. (S 1.7)

Zhang, Y. and Zhang, G. (2009). One-step extrapolation method for reverse time migration. *Geophysics*, **74**, A29–A33. (S 2.4)

Index

Printed in the United States
By Bookmasters